T0179209

A Water Quality Assessment of the Former Soviet Union

Also available from E & FN Spon

Water Quality Assessments, 2nd edition
A guide to the use of biota, sediments and water in environmental monitoring
Edited by D. Chapman

Water Quality Monitoring
A practical guide to the design and implementation of freshwater quality studies and monitoring programmes
Edited by J. Bartram and R. Ballance

Water Pollution Control
A guide to the use of water quality management principles
R. Helmer and I. Hespanhol

Toxic Cyanobacteria in Water
A guide to public health consequences and their managment in water resources and their supplies
Edited by I. Khorus and J. Bartram

Water Policy
Allocation and management in practice
Edited by P. Howsam and R.C. Carter

Water And The Environment
Innovative issues in irrigation and drainage
Edited by L.S. Pereira and J. Gowing

Water Resources
Health, environment and development
Edited by B.H. Kay

The Coliform Index and Waterborne Disease
Problems of microbial drinking water assessment
C. Gleeson and N. Gray

International River Water Quality
Pollution and restoration
Edited by G. Best, E. Niemirycz and T. Bogacka

Ecological Effects Of Waste Water, 2nd edition
Applied limnology and pollutant effects
E.B. Welch

Water: Economics, Management and Demand
Edited by M. Kay, L.E.D. Smith and T. Franks

To order or obtain further information on any of the above or receive a full catalogue please contact:
The Marketing Department, E & FN Spon, 11 New Fetter Lane, London EC4P 4EE.
Tel: 0171 842 2400; Fax 0171 842 2303

A Water Quality Assessment of the Former Soviet Union

Edited by
**Vitaly Kimstach, Michel Meybeck
and Ellysar Baroudy**

Taylor & Francis Group

LONDON AND NEW YORK

First published 1998
by Routledge

2 Park Square, Milton Park, Abingdon, Oxfordshire OX14 4RN
52 Vanderbilt Avenue, New York, NY 10017

Routledge is an imprint of the Taylor & Francis Group, an informa business

First issued in paperback 2019

Publisher's Note
This book has been prepared from camera-ready copy provided by the editors.

British Library Cataloguing in Publication Data
A catalogue record for this book is available
from the British Library

Library of Congress Cataloging in Publication Data
A catalogue record for this book has been requested

ISBN 978-0-419-23920-8 (hbk)
ISBN 978-0-367-86593-1 (pbk)

TABLE OF CONTENTS

FOREWORD

The Union of the Soviet Socialist Republics (USSR), which existed until 1991, was the world's largest country and occupied almost one sixth of the inhabited land territory of the earth. By the time of its dissolution, it consisted of 15 formally equal Union Republics with their own national, cultural and religious traditions and levels of economic development.

The territory of the former USSR is exclusively rich with natural resources. All types of minerals are represented in its reserves of raw materials. In 1990, 570 million tonnes of crude oil, 703 million tonnes of coal and 236 million tonnes of iron ore were extracted in the USSR. The territory has a quarter of the world's forest reserves and the world's largest freshwater stock, being second to Brazil for renewable freshwater resources. Long-term freshwater resources in the USSR are estimated to be 41,400 km^3, including 27,000 km^3 in lakes, 11,400 km^3 in glaciers and 3,000 km^3 in bogs. River runoff of the former USSR comprises approximately 10 per cent of global river runoff.

By the late 1920s, the political, social and economic systems in the country had been transformed. A directive system of State management based on nationalised property as the means for production was adopted. The market economy had been forced out. Numerous long-term factors created extremely unfavourable environmental conditions: the focus to accelerated industrialisation of the economy and the intensive construction of thousands of industrial enterprises in the 1930s; the relocation of many plants from the western part of the country to the east and construction there of new military industrial enterprises during World War II; the restoration of the economy destroyed by the War in the western part with no external economic aid and an exhausting arms race during the Cold War. As a consequence, there was an increase in negative environmental impacts on human health. As a rule, plans for economic development did not take into consideration the physico-geographic, demographic, social, cultural and environmental characteristics of the regions.

In the 1970–80s, the rate of economic development in the Soviet Union slowed down significantly. Attempts to support it without changing fundamental structures led to an even more intensive use of natural resources and a sharp increase in the anthropogenic impact on the environment. These processes, being magnified by physico-geographic, social and economic characteristics of the country, caused large-scale environmental consequences. Exhausting some types of natural resources, decreasing biodiversity and agricultural and forestry productivity, pollution of water, atmosphere and soils and other environmental consequences due to the pursued economic policy resulted in an increase in morbidity of the population.

During this period, Party leaders and the government began paying attention to the issues of environmental protection and the optimal use of natural resources. A number of decrees highlighted the necessity to increase responsibility of both central executive bodies and industrial enterprises for rational use of natural resources, pollution of the environment and protection of flora and fauna.

In the USSR, as in other countries, the decisions of the Stockholm Conference (1972) were taken as a background for environmental protection measures. The USSR Supreme Council (Parliament) adopted the Decree "On the measures on further improvement of environmental protection and rational use of natural resources" (September 1972) based on the implementation of the Conference resolutions. However this decree, as well as other decrees and resolutions at that time in the field of environmental protection did not foresee the fundamental changes in economic policy from an environmental point of view, and provided no mechanism for implementation of these measures.

Under such conditions of extensive economic development, freshwaters, particularly in the European part of the country and in Central Asia, became the most exploited natural resource. The discrepancy between industrial needs and resource potential intensified, exhausting the use of natural resources and having a strong impact on the state of water resources. As a result, some regions, which had enough water resources in their natural state, began to suffer from a water deficit due to the depletion and pollution of these resources. The state of water resources in these regions became a critical factor that limited further economic development.

All global water quality issues, which appeared in the 20th century due to urbanisation and industrial and agricultural development, were also issues in the former USSR. However, it must be considered that the former USSR underwent industrial development in a few decades compared with Western European countries that had taken more than a century to achieve a similar level of development. This factor, in combination with the stormy economic development of the USSR, has determined the intensive growth of environmental problems, especially in regions with a high population density. By the time of the dissolution of the USSR, the anthropogenic impact on many water bodies had reached such levels that the degradation of water resources and ecosystems became irreversible or, at least, needed investments that significantly exceeded the economic capabilities of the State.

Effective environmental protection measures and sustainable development of water resource management can be implemented only where reliable information on the state of these resources, including water quality, is available. However, according to USSR regulations, some types of data on pollution of water bodies were classified. In addition, a large amount of non-classified information, including scientific results, were published during the Soviet period in limited editions by different departments or institutes, which were not available to the international scientific community. All this, in combination with language difficulties, limited both the access of foreign scientists and experts to the Soviet publications and the publication of Soviet scientists in international journals. Thus the issues of freshwater quality in the former USSR were scarcely known abroad, and monitoring data were practically excluded from global water quality assessments made by United Nations' specialised agencies.

After the dissolution of the former USSR, practically all the independent states, which were formed from the former Soviet Socialist Republics, suffered a sharp decrease in economic performance, particularly industrial production. To a certain extent, this decreased the anthropogenic impact on water bodies but did not cause a significant improvement in the environmental situation. At present, the complex water quality problems, which existed in the regions of the former USSR remain and, in some cases, are even growing.

The situation is complicated by the fact that many river basins of the former USSR have become international basins following the dissolution of the Union. Therefore, solution of water protection issues and problems of water resource management in these basins need to be co-ordinated at an international level, including the signing of international treaties, the establishment of joint systems of water management and monitoring of water resources and other steps based on sustainable use of water resources by the countries situated in these basins. For countries with transitional economies, particularly in basins with poor water availability, this task is extremely difficult. The solution in these countries can be realised with active technical and economic assistance of the international community provided these countries adopt the appropriate legislation. The United Nations and its specialised agencies, international financial organisations, and economically developed countries, being aware of the international nature of the problem and its importance for the populations of the republics of the former USSR and for the protection of the total environment, have started to implement some actions to assist the governments of new independent states in solving environmental problems.

In support of this work, information on the state of the water resources of the former USSR, as well as the retrospective analysis of trends in the development of water quality issues, became urgently needed. Taking into account the above mentioned limitations of the international community to access this type of information in previous years, a special publication relating to this problem became necessary. Favourable conditions for its preparation appeared after the Russian

Federation joined the freshwater programme of the Global Environment Monitoring System (GEMS/Water). On the initiative, and with the support, of the World Health Organization and the United Nations Environment Programme, this special project on the regional assessment of the Former USSR was launched in 1993. Leading experts from the Commonwealth of Independent States (CIS) took part in the project's implementation.

The present assessment covers issues of both natural water quality and major pollution of water bodies such as salinisation, acidification, contamination with heavy metals and organic compounds, microbiological pollution, and groundwater pollution with nutrients, particularly nitrates. Specific chapters are dedicated to the description of the water quality monitoring system in the former USSR and to the application of ecological monitoring for the assessment of natural water quality. In addition to general water quality issues, the assessment also considers regional problems related to the pollution of some of the most important water bodies in the former USSR.

I believe that the present assessment should be interesting and useful for scientists and specialists of the newly independent states, and for the acquaintance of the international scientific community with water quality issues of the former USSR. The findings could also be used by experts seeking solutions for water quality problems both in the countries of the region and in other parts of the world.

Professor Victor I. Danilov-Danilyan
Chairman of the State Committee of the Russian Federation
for Environmental Protection
Moscow, January 14, 1998

ACKNOWLEDGEMENTS

The co-sponsoring organisations are grateful for the contributions made by all those who have participated in the production of this book. Their enormous effort made at all levels has resulted in this unique assessment. Because it is difficult to identify precisely the contributions of different authors, the names of principal authors and contributors are listed below in alphabetical order:

V. Abakumov, Institute of Global Climate and Ecology, Moscow (Chapters 10, 11 and 13)

I.K. Akuz, North Caucasus Branch of Geological Service, Rostov-on-Don (Chapters 4 and 14)

E.N. Bakaeva, Hydrochemical Institute, Rostov-on-Don (Chapter 14)

V.F. Baron, Russian Institute of Hydrogeology and Engineering Geology, Moscow (Chapter 5)

L.V. Boeva, Hydrochemical Institute, Rostov-on-Don (Chapter 9)

V.A. Bryzgalo, Hydrochemical Institute, Rostov-on-Don (Chapter 14)

A.A. Bylinkina, Institute of Biology of Inland Waters, Borok (Chapter 16)

G.M. Chernogaeva, Institute of Global Climate and Ecology, Moscow (Chapters 3, 11 and 20)

A.I. Denisova, Ukrainian Institute of Aquatic Environmental Problems, Kiev (Chapter 17)

V.G. Drabkova, Institute of Lake Research, St Petersburg (Chapters 2, 6 and 19)

V.M. Emetz, Voronezh State Biosphere Reserve, Grafskaya (Chapter 8)

Y.A. Fedorov, Hydrochemical Institute, Rostov-on-Don (Chapter 15)

V.Y. Georgievsky, State Hydrological Institute, St Petersburg (Chapters 1 and 3)

V.V. Gordeev, Institute of Oceanology, Moscow (Chapter 12)

V.A. Kimstach, Arctic Monitoring and Assessment Programme, Oslo (Chapters 9 and 11)

L. Korotova, Hydrochemical Institute, Rostov-on-Don (Chapter 12)

O.M. Kozhova, Irkutsk University, Irkutsk (Chapter 18)

V.I. Kozlovskaya, Institute of Biology of Inland Waters, Borok (Chapter 16)

R.A. Kulmatov, Institute of Nuclear Physics, Tashkent (Chapter 15)

A.P. Lvov, State Committee for Environmental Protection of the Russian Federation, Moscow (Chapter 3)

A.A. Matveev, Hydrochemical Institute, Rostov-on-Don (Chapter 18)

N. Matveeva, Hydrochemical Institute, Rostov-on-Don (Chapters 12 and 18)

L.A. Ostrovsky, Russian Institute of Hydrogeology and Engineering Geology, Moscow (Chapter 5)

N.N. Osipov, Institute of Biology of Inland Waters, Borok (Chapter 16)
V.I. Peleshenko, Kiev State University, Kiev (Chapter 17)
V.S. Petrosyan, Moscow State University, Moscow (Chapter 9)
M.P. Polkanov, Ministry of Natural Resources of the Russian Federa-
tion, Moscow (Chapters 1, 2, 4 and 7)
F.E. Rubinova, Central Asian Hydrometeorological Research Institute,
Tashkent (Chapter 15)
V.A. Rumyantsev, Institute of Lake Research, St Petersburg (Chapter 19)
A.D. Semenov, Azov Fisheries Research Institute, Rostov-on-Don
(Chapter 14)
T.A. Sergeeva, Moscow State University, Moscow (Chapter 9)
I.A. Shiklomanov, State Hydrological Institute, St Petersburg (Chapter 1)
V. Shlychkova, Hydrochemical Institute, Rostov-on-Don (Chapter 12)
Y. Talayeva, Institute of Communal Hygiene, Moscow (Chapter 10)
V.V. Tsirkunov, Moscow Office of the World Bank, Moscow (Chapters 2,
4, 5, 12 and 14)
Y.Y. Vinnikov, Hydrochemical Institute, Rostov-on-Don (Chapter 9)
V.P. Zakutin, SPA 'GIREK', Moscow (Chapter 7)
A.A. Zenin, Hydrochemical Institute, Rostov-on-Don (Chapter 14)
A.V. Zhulidov, Centre for Preparation and Implementation of Interna-
tional Projects on Technical Assistance, North Caucasus Branch,
Rostov-on-Don (Chapters 8 and 14)

In addition to specific authors, the contribution of staff from the former USSR national freshwater monitoring network must also be acknowledged. For many years, these personnel have been responsible for collecting and analysing thousands of samples from all over the vast country. Scientists and technicians of the Hydrochemical Institute in Rostov-on-Don, which was the lead institution responsible for the opera-tion of the national freshwater monitoring system, provided many of the data for use in this assessment, some of which has not been published before. We are also grateful to Professor A.M. Nikanorov, Director of the Hydrochemical Institute, who kindly facilitated the use of the archives and databases of the Institute and who was personally involved in the organisation of case studies.

Every effort has been made to ensure that appropriate acknowledge-ment has been made for data and information used and that appropriate permissions have been obtained; we apologise for any unintentional omissions in this respect. Information not specifically attributed to a published source has been reproduced from original data with the permission of contributing authors, their institutes and the national monitoring network.

The publication of this unique assessment is the result of the dedica-tion of the editors, V.A. Kimstach (Arctic Monitoring and Assessment Programme, Oslo), M. Meybeck (Université de Pierre et Marie Curie,

Paris) and E. Baroudy (Monitoring and Assessment Research Centre (MARC), London), to whom we are very grateful.

The role of the Centre for International Projects of the State Committee of the Russian Federation for Environmental Protection is also gratefully acknowledged for providing organisational support for the project including two expert meetings, communication between editors and lead authors, and the primary translation of the chapters into English.

Additional support for this assessment, including the organisation of meetings, was provided by R. Helmer, J. Kenny and U. Enderlein of the World Health Organization, Geneva. The assessment was undertaken within the framework of the Global Environment Monitoring System (GEMS/Water).

Advice and comments relating to various chapters were gratefully received from W.P. Williams (MARC) and R.J. Miles (King's College London (KCL)). Production of illustrations was by D. Cartwright (MARC) with the assistance of R. Beaumont (KCL) and her team who were responsible for the initial stages of map production. Table layout was kindly undertaken by A. Willcocks and L. Willcocks. Administrative support was provided by F. Preston (MARC) and D. Wilkin (MARC) and production assistance by L. Chapman and B. Lescovec.

Production and invaluable advice were provided by Deborah Chapman, to whom we are gratefully indebted. Dr Chapman is the series editor for the UNEP/WHO co-sponsored guidebooks dealing with various aspects of water quality management.

Chapter 1[*]

NATURAL WATER RESOURCES

The territory of the former USSR covers 22.4 million km^2, equivalent to almost one sixth of the world's habitable land. Vast areas, complex geographic relief and substantial climatic differences determine the great diversity of nature in the former USSR (Figure 1.1).

1.1   Landscape

The western part of the former Soviet Union generally is characterised by vast lowlands and plains. The eastern part, from the River Yenisey to the Pacific coast is dominated by mountains and highlands. Mostly uninterrupted mountains form the southern border of the former USSR.

The European territory of the former USSR consists mainly of the Russian Plain. Its surface is heterogeneous with hills up to 300–400 m alternating with the Dnieper, Black Sea and Caspian lowlands. The northern part of the plain, which was glaciated during the Quaternary period, is dominated by hilly watersheds with numerous depressions often occupied by lakes and wetlands. In the south, watersheds form elevated areas (highlands) divided by a network of erosion valleys and gullies. To the north-west of the Russian Plain are the highlands of Karelia and the mountains of the Kola Peninsula. In the south, the Russian Plain is bordered by the mountain systems of the Eastern Carpathians, the Crimea and the Caucasus. In the east, the Ural mountains form the boundary between Europe and Asia.

In the Asian territory of the former USSR, between the Urals and the River Yenisey, there is the West Siberian Plain with a monotonous relief. Between the Rivers Yenisey and Lena, there is one of the world's largest plateaux, the Central Siberian Plateau, with high-plain watersheds. In the north between the plateau and the Byrranga mountains, there is the hilly North-Siberian Lowland. The south and east of the Asian territory are dominated by mountainous terrain.

---

[*]   *This chapter was prepared by I.A. Shiklomanov, V.Y. Georgievsky and M.P. Polkanov*

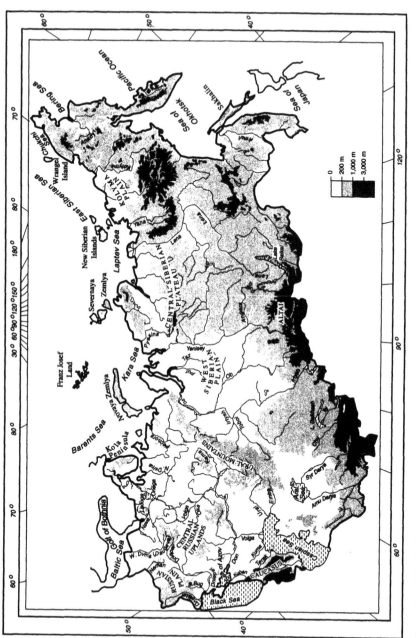

Figure 1.1 Physical map of the former USSR

1.2 Geological structure

The former Soviet Union can be divided into six geological regions: the Russian and Siberian platforms, the Ural-Altai, East-Asian, Alpine and Pacific regions.

The Russian and Siberian platforms are formed of sedimentary rocks of Precambrian age which underwent metamorphism. These are predominantly gneisses and metamorphic slates, with plutonic intrusions mainly of granite. The fold has weakly fractured sedimentary rocks of Palaeozoic, Mesozoic and Cenozoic eras.

The Ural-Altai region is situated between the Russian and Siberian platforms. The Palaeozoic orogenesis forms complex folded structures — anticlinal elevations and synclinal depressions of various orientations and forms. The most ancient formations are the Sayans, the eastern part of Tuva, Kuznetsk Alatau and the eastern part of Altai. Younger rocks (Upper Mesozoic or Hercynian) are found in the regions of the Urals, Kazakhstan, Tian-Shan and Taimyr.

Southern Siberian, Altai and Kazakhstan folded structures descend beneath the cover of Mesozoic and Cenozoic rocks of Western Siberia and the Tian-Shan structures of Kyzylkum and Aral lowland.

The Pacific region is divided into the Verkhoyansk-Chukchi and Amur-Primorye sub-region by the Sea of Okhotsk. The Verkhoyansk-Chukchi sub-region includes the north-eastern edge of the Asian continent. The Amur-Primorye sub-region covers the Eastern Trans-Baikal area, lower reaches of the River Amur and the Primorye. The Verkhoyansk-Chukchi sub-region is a vast folded arc with anticlinal elevations in the west. The Amur-Primorye sub-region consists of two branches, the Eastern Trans-Baikal and the second branch which stretches from lower reaches of the River Amur through the coastal Sikhote-Alin range to the Sea of Japan.

Synclinal structures in this region are filled with sediments, in some places with rocks of volcanic origin of the Mesozoic and Cenozoic era. In some regions, coal-bearing layers occur belonging to Jurassic and Cretaceous continental sediments. On the coast, Palaeocene and Quaternary volcanic rock occurs.

The mountain range of the Pacific Coast includes the Kamchatka Peninsula, Sakhalin Island and the Kuril Islands. In the northern continental part is the Koryak range. The Pacific coastal region is in an early stage of geosynclinal development. It is formed mainly by young Meso-Cenozoic rocks and is characterised by active tectonic movements and volcanic activity.

The Alpine orogenesis region of the former USSR encompasses a vast area from the Carpathians in the western part of the country to the Pamir mountains in Central Asia. The core of the Eastern Carpathians is formed by metamorphic slate of Precambrian and Palaeozoic age. To the east, there is the anticlinorium of the Crimea in the Black Sea and the major anticlinorium of the Caucasus mountains. The core of the Caucasus mountains is formed by Precambrian and Palaeozoic granite, gneisses and other metamorphic rocks. At the margins Jurassic, Cretaceous, Palaeocene and Neogene sediments developed and crumbled into folds.

East of the Caucasus mountains there is the elevation of the Great Balkhans, formed mainly by Jurassic, Cretaceous and Palaeocene sandy-argillaceous rocks and limestone. Northern Pamir has a complex structure composed mainly of Palaeozoic rocks. Between the Pamir and Tian-Shan mountain ranges there is the Tadzyk Depression, formed by folded deposits of Jurassic, Cretaceous, Palaeocene and Neogene sediments. Unlike the Pacific region, the Alpine region is believed to be in its final orogenic phase.

Practically all types of Quaternary facies and lithologies are found in the former USSR.

1.3   Climate

The vast territory of the former USSR and its complex orographic structure determine the great variety of climatic conditions. The extreme north of the country and the majority of islands of the Arctic Ocean belong to the Arctic climatic zone. The north of the European territory and Western Siberia up to the Arctic circle and Central and Eastern Siberia belong to the sub-Arctic zone. The remainder of the European territory and Siberia, as well as Central Asia above 40 °N belong to temperate zones. Parts of the Trans-Caucasus and Southern Central Asia belong to the subtropical zone. Almost everywhere the climate can be defined primarily as continental and there is a marked seasonal division of the year into cold and warm periods. Continentality of the climate strengthens from west to east and is especially marked in Central Asia and Eastern Siberia. Climates of extreme Western regions, the Black Sea coast of the Caucasus and the southern coast of Crimea are more mild. Southern regions of the Far East are under the influence of the Pacific Ocean and have a monsoon climate.

Atmospheric circulation over the territory of the former USSR shows strong seasonal variations. In winter the Asian anticyclone is formed over the interior of the Asian continent. During this season, the northern half of the European territory and Western Siberia are under the influence of repeated Atlantic cyclones, which raise significantly both the air temperature and the amount of precipitation. The Far East, in winter, is influenced by an anticyclone regime with cold and dry west and north-west winds blowing in from the Asian anticyclone. The intensive cooling of the land coupled with the cold continental air in winter leads to relatively low air temperatures and small amounts of precipitation. In summer, there is a strong warming of Asia and the formation of relatively low atmospheric pressure. The southern half of the European

Table 1.1   Number of rivers in different length categories

River length (km)	European territory	Asian territory	Total
< 10	630,151	2,182,436	2,812,587
11–25	27,059	86,915	113,974
26–50	6,000	18,110	24,110
51–100	2,414	6,209	8,623
101–200	808	2,049	2,857
201–300	170	460	630
301–500	87	270	357
501–1,000	52	145	197
> 1,000	18	45	63

Source: USSR Water Resources and their Use, 1987

territory is under the influence of an Azores anticyclone ridge and the Far East is dominated by summer monsoons. A result of severe continental climate in Siberia and the northern Far East is the deeply frozen ground in winter and the wide occurrence of permafrost.

The vegetation of the country belongs entirely to the Holarctic biome with clearly defined vegetation zones from north to south including arctic deserts, tundra, forest tundra, forest, forest steppe, steppe, semi-desert and desert (Figure 1.2).

1.4   Main characteristics of water bodies

1.4.1   Rivers

More than 2,900,000 rivers, streams and temporary running waters have been mapped over the territory of the former USSR (Table 1.1). However, the main part of the flow (80 per cent) is formed by the 36 largest rivers. The density of the river network changes significantly from zero in deserts of Central Asia to 2 km km^{-2} in the Caucasus and Carpathians mountains. The average river network density is 0.45 km km^{-2} for the entire country (USSR water resources and their use, 1987).

Rivers in the former USSR belong to the catchments of the Pacific, Atlantic and Arctic Oceans. The majority of river water (64 per cent) flows into the basin of the Arctic Ocean. The largest of these rivers are: Northern Dvina, Pechora, Ob, Yenisey, Khatanga, Lena, Yana, Indigirka and Kolyma (Figure 1.1). The Pacific Ocean catchment includes rivers in the east of which the largest is the Amur. The Atlantic Ocean catchment includes rivers of the central and western parts of the European territory. The main rivers of this catchment area are Neva,

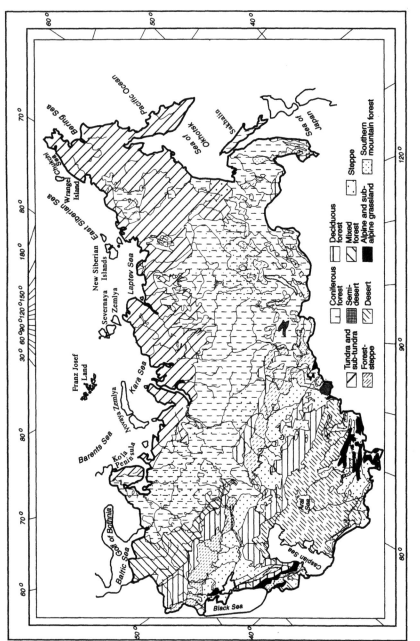

Figure 1.2 Geo-botanical map of the former USSR

Table 1.2   Number of rivers with water catchment areas greater than 50,000 km^2

Water catchment area (10^3 km^2)	European territory	Asian territory	Total
50–100	20	43	63
101–300	10	25	35
301–500	4	7	11
501–1,000	3	3	6
> 1,000	1	6	7

Source: USSR Water Resources and their Use, 1987

Western Dvina (Daugava), Neman, Kuban, Don, Dnieper, Dniester and Danube. Approximately one quarter of the European territory belongs to the Aral-Caspian catchment which is an endorheic region (i.e. internal drainage). About 9 per cent of total river flow is formed here. The catchment of the Caspian Sea includes the Rivers Volga, Ural, Emba, Terek and Kura. The catchment of the Aral Sea includes the Amu Darya and Syr Darya. Lake Balkhash receives the waters of the River Ily. Many rivers of this region enter small lakes or lose their flow in desert areas. These rivers include Tedjen, Murgab, Sarysu and others. The distribution of rivers with different catchment areas is given in Table 1.2. Data regarding the number of rivers of different lengths are given in Table 1.1. The total average volume of water in the principal rivers is estimated at 122 km^3.

1.4.2 Lakes
In the former USSR, there are about 2,854,200 lakes, of which 2,686 are saline. The total surface area of lakes, including the Caspian Sea, is 892,850 km^2 and they occupy 4 per cent of the territory.

The characteristics of the largest 19 lakes in the former USSR, some of which rank amongst the largest lakes in the world, are given in Table 1.3. The lakes are distributed unevenly over the territory: 14 per cent of lakes are in the north-west, 8.6 per cent occur on the West Siberian Plain and approximately 6 per cent are in the Kola and Taimyr Peninsulas.

Most lakes (2,815,287) are small with a surface area < 1 km^2. They correspond to 18 per cent of the total surface area of lakes. The 21 largest lakes (with a surface area > 1,000 km^2) correspond to 62 per cent of the total surface area. The distribution of lakes in both the European territory and the Asian territory is given in Table 1.4.

The total volume of water in lakes is 27,160 km^3. Including the largest endorheic lakes of the world, the Caspian Sea and the Aral Sea, the total volume is 106,380 km^3. In the 21 largest lakes (with surface areas

8 Water Quality Assessment of the Former USSR

Table 1.3 Characteristicsof the largest lakes of the former USSR

	Catchment area (km²)	Water surface area (km²)	Average depth (m)	Maximum depth (m)	Volume (km³)
Caspian Sea[1]	3,100,000	376,300	200.0	1,025.0	78,200
Aral Sea[1]	940,000	66,100	16.1	69.0	1,020
Baikal	540,000	31,500	730.0	1,741.0	23,000
Balkhash[1]	395,000	18,200	6.2	26.0	112
Ladoga	258,000	17,700	51.4	230.0	908
Onega	51,500	9,630	30.6	127.0	295
Issyk-Kul[1]	15,700	6,240	278.0	668.0	1,738
Taimyr	43,920	4,560	2.8	26.0	13.0
Khanka	16,900	4,190	4.4	10.6	18.5
Pskov-Chud	44,200	3,560	7.1	15.3	25.1
Sarykamysh[1]	–	2,850	10.0	39.5	28.5
Alakol	68,700	2,650	21.1	54.0	58.6
Chany (natural conditions)	23,600	2,270	1.7	8.5	4.8
Chany (excluding Yudynsky reach)	23,600	1,294	2.0	8.5	2.6
Arnasai[1]	–	2,000	6.5	22.0	13.0
Zaisan (natural conditions)	–	1,800	3.6	10.0	6.6
Tengyz[1]	94,900	1,590	–	8.0	–
Sevan (natural conditions)	3,580	1,420	41.0	106.0	58.5
Sevan (after lowering lake level 20 m)	3,770	1,230	31.0	86.0	38.0
Beloye (natural conditions)	14,000	1,284	4.1	5.5	5.3
Ilmen	66,400	1,100	2.6	4.3	2.8

[1] Saline and sub-saline lakes   Source: USSR water resources and their use, 1987

> 1,000 km²) water volume is 105,485 km³. Most of this volume (96 per cent) belongs to the Caspian Sea and Lake Baikal. The total volume of freshwater lakes is 25,200 km³, 91 per cent of which (23,000 km³) is found in Lake Baikal. The total volume of saline waters is 1,960 km³ (excluding the Caspian Sea and Aral Sea) of which 89 per cent is concentrated in the largest mountain lake, the Issyk-Kul.

The theoretical residence time of lake water ranges from 326 years (Lake Issyk-Kul) to < 1 year (small shallow lakes); the volume weighted average residence time is 140 years.

Table 1.4   Number of lakes and their surface area in the former USSR

Lake surface area (km²)	European territory	Asian territory	Total number	Total surface area (km²)
< 1	537,051	2,278,236	2,815,287	159,667
1–10	5,445	31,278	36,723	86,641
10–50	509	1,549	2,058	38,623
50–100	59	162	221	15,013
100–1,000	45	100	145	37,964
> 1, 000	6	13	19	554,940

Source: USSR Water Resources and their Use, 1987

1.4.3 Hydrogeological conditions

The formation and distribution of groundwater in the former USSR is described in 45 regional and 6 combined volumes of the monographs 'Hydrogeology of the USSR' (1966–1972, 1976). Analysis of this comprehensive material shows that the distribution, formation and accumulation of groundwater depend on the origin and distribution of rocks of different composition.

The hydrogeological regions reflecting the most common hydrogeological features in the former USSR are shown in Figure 1.3 and Appendix 1.1 (Hydrogeological Regional Map, 1973). The map shows the hydrogeological regions of the East-European (Russian) and Turan platforms as well as the hydrogeological folded regions, such as Carpathian-Crimean-Caucasian region, Kopetdag-Great Balkhan region, Tian-Shan-Djungar-Pamir region, Central Kazakhstan region, Timan-Ural region, Sayan-Ural-Yenisey region, etc. In sedimentary platforms the hydrogeological regions are characterised by developed artesian basins. The folded regions are characterised by the abundance of fractured aquifers. Hydrogeological regions are classified into first order hydrogeological regions based on the large structural elements of platforms and folded regions (e.g. anticlines, synclines, shields, plates and edge depressions), and second order zones determined by borders of aquifers containing complex, relative pathways of underground flow and other hydrodynamic features. Third order classes consider regional distribution and the overall water cycle.

1.5   Water resources

1.5.1 Surface water

The distribution of surface water resources over the former USSR is extremely unequal and in many cases does not meet the basic requirements of the population, industry and agriculture. In the European

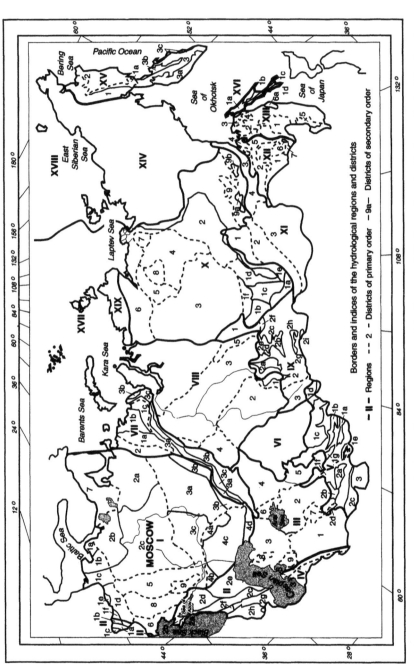

Figure 1.3 Scheme of the hydrogeological regions and districts of the former USSR (Based on the Hydrogeological Regional Map of the USSR, 1973). Districts of tertiary order are not shown. See detailed legend in Appendix 1

Borders and indices of the hydrological regions and districts

– II – Regions   **– – 2 –** Districts of primary order   **– 9a–** Districts of secondary order

territory, where approximately 70 per cent of the population is concentrated, the total river discharge is 1,050 km^3 a^{-1} or 24 per cent of total water resources. Only 2 per cent of water resources occur in the dry southern regions of the European territory, and in Kazakhstan and Central Asia. These areas occupy more than a quarter of the territory and their climate is suitable for the production of many valuable agricultural products. The largest rivers such as Northern Dvina, Pechora, Ob, Yenisey, Khatanga, Lena, Yana, Indigirka, Kolyma and Amur flow in economically less developed and less populated regions.

In the former Soviet Union, specific runoff of lowland rivers decreases from northern to southern regions and increases with elevation in the mountains. Geographic variations in annual runoff range from 2,000–3,000 mm a^{-1} (Caucasus Mountains) to 5 mm a^{-1} or less in the deserts of Central Asia (Figure 1.4). In the north and north-west of the European region the annual average runoff is 200–400 mm a^{-1} whereas in the south and south-east it is 10–20 mm a^{-1}. In West Siberia, runoff varies significantly from 300 mm a^{-1} in the basin of the River Pura to 10 mm a^{-1} in the upper reaches of the River Ishima. In Eastern Siberia, the Primorye region, Yakutia and Kamchatka the general north-south latitudinal gradient of runoff changes into an east-west lateral gradient corresponding to the location of numerous mountain ranges which hold back moisture from the Pacific Ocean. Here, annual runoff varies from 500–800 mm a^{-1} in mountainous regions to 10–20 mm a^{-1} in some regions of Yakutia and Southern Baikal. On Arctic islands, annual runoff is 100–300 mm a^{-1}.

In terms of water supply, the former Soviet Union may be divided into three zones:

- Areas of high water availability (specific discharge is > 6 l s^{-1} km^{-2}) which occupy 48 per cent of the country and within which 80 per cent of river discharge is formed.
- Areas of intermediate water availability (specific discharge is 2–6 l s^{-1} km^{-2}) which occupy 25 per cent of the territory and within which 18 per cent of water resources are formed.
- Areas of insufficient water availability (specific discharge is < 2 l s^{-1} km^{-2}) which occupy 27 per cent of the country and within which 2 per cent of water resources are formed.

The natural water resources of principal river basins are presented in Table 1.5. The total water runoff over the basins drained in the former USSR is 4,771 km^3 a^{-1}. However, not all water resources drain into oceans, some may be held in catchment areas, some may be lost through evaporation or may percolate into deep aquifers which do not drain into the oceans. This total loss is estimated at 484 km^3 a^{-1}, or 10 per cent of water resources (Table 1.6).

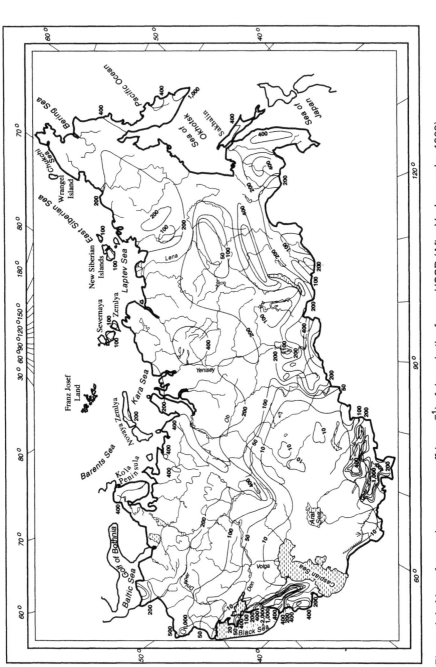

Figure 1.4 Map of annual average runoff (mm a^{-1}) of rivers in the former USSR (After Voskresensky, 1962)

Table 1.5   Long-term river discharges in the former USSR

River	Water catchment area (km²)	Average discharge (km³ a⁻¹)	Coefficient of variation C_v [1]	Coefficient of asymmetry C_s [2]
Pechora	322,000	130.0	0.13	0.27
Northern Dvina	357,000	109.0	0.28	0.56
Vychegda[3]	121,000	33.4	0.17	0.34
Onega	56,900	15.7	0.21	0.63
Neva	281,000	78.5	0.18	0.30
Western Dvina (Daugava)	87,900	20.3	0.25	0.58
Neman	98,200	19.7	0.17	0.45
Dniester	72,100	10.7	0.27	0.64
Prut	28,400	2.8	0.32	0.44
Dnieper	504,500	53.9	0.26	0.80
Don	422,000	28.1	0.38	0.76
Seversky Donets[4]	98,900	5.8	0.46	0.94
Kuban	57,900	13.5	0.17	0.34
Kura[5]	188,000	26.8	0.17	0.34
Terek[5]	37,400	11.0	0.11	0.22
Volga[5]	1,380,000	254.0	0.18	0.30
Oka[5]	245,000	38.5	0.21	0.42
Kama[5]	507,000	117.0	0.22	0.36
Ural[5]	236,000	10.1	0.60	1.20
Ob	2,990,000	404.0	0.17	0.28
Irtysh	1,643,000	88.3	0.20	1.20
Tobol	426,000	26.8	0.41	0.90
Yenisey	2,580,000	630.0	0.08	0.00
Selenga[6]	444,000	27.9	0.26	0.76
Olenek	219,000	34.3	0.24	0.00
Lena	2,490,000	532.0	0.10	0.24
Yana	238,000	30.7	0.20	0.24
Indigirka	360,000	53.6	0.18	0.50
Kolyma	647,000	128.0	0.22	0.35
Anadyr	191,000	64.1	0.19	0.28
Penzhyna	73,500	22.6	0.27	0.31
Kamchatka	55,900	33.1	0.10	0.52
Amur	1,855,000	344.0	0.20	0.05
Syr Darya[5]	–	37.0	0.21	0.26
Naryn[5]	59,900	13.8	0.20	0.69
Amu Darya[5]	–	69.5	0.15	1.26
Ily[5]	129,000	18.1	0.18	0.85

[1] $C_v = \sigma/Q$
[2] C_s = Coefficient of asymmetry usually fitted by a Pearson or Kritsky-Menke distribution
[3]   Tributary of Northern Dvina
[4]   Tributary of River Don

[5]   River basins with endorheic discharge
[6]   Discharges to Lake Baikal

Sources: USSR Water Resources and their Use 1987; Shiklomanov, 1988

Table 1.6   River water resources discharged to regional seas

Sea and endorheic regions	Catchment area (10^3 km^2) Within USSR	Total[1]	Water resources (km^3) Within USSR	Total[1]
White Sea	709.8	717.6	223.2	225.7
Barents Sea	525.7	542.4	207.7	213.4
Baltic Sea	568.2	637.9	142.7	161.2
Sea of Azov	579.3	579.3	42.1	42.1
Black Sea	787.8	1,613.3	136.6	265.3
Kara Sea	6,248.2	6,649.7	1,344.0	1,375.0
Laptev Sea	3,692.9	3,692.9	783.0	783.0
East Siberian Sea	1,295.5	1,295.5	268.0	268.0
Chukchi Sea	101.0	101.0	24.4	24.4
Bering Sea and Pacific Ocean[2]	569.7	569.7	202.0	202.0
Sea of Okhotsk	1,695.4	2,547.4	563.8	660.0
Sea of Japan	124.3	34.3	33.7	35.7
Caspian Sea[3]	2,800.0	2,950.0	317.0	329.0
Kazakhstan and Central Asia[3]	2,577.2	2,893.3	148.9	186.2
Total in the basins	**22,274.9**	**24,942.2**	**4,437.0**	**4,771.0**

[1]   Including foreign territories
[2]   The Pacific Ocean includes the coast of the Kamchatka Peninsula and eastern slopes of the Kuril Islands

[3]   Endorheic regions

Source: USSR water resources and their use, 1987

1.5.2   Groundwater

Groundwater suitable for economic use should be considered as a mineral resource (Hydrogeological Regional Map, 1973; Guidelines, 1986). According to their type of use, groundwaters are usually divided into the following categories: i) fresh water (dissolved solids up to 1 g l^{-1}), sub-saline (from 1 to 10 g l^{-1}) and saline (from 10 to 35 g l^{-1}); ii) mineral waters used for hydrotherapy; iii) thermal waters used for heating and electric power; iv) industrial waters used for the production of valuable components (e.g. iodine, bromine).

Regional estimates of exploitable groundwater resources in the former USSR were made using the methods of Bindeman and Bochever (1964) to work out a general scheme for the use and protection of water resources. The result of this was a 1/5,000,000 map of exploitable resources of fresh and sub-saline groundwaters. In later years this estimate was made more precise following surveys in several artesian basins including Moscow, Sura-Khoper, Volga-Kama, West-Siberia (Table 1.7).

Table 1.7 Distribution of resources of fresh and brackish groundwaters in principal hydrogeological regions[1]

Hydrogeological region	Region[2]	Groundwater resources ($m^3\ s^{-1}$)		
		Natural	Exploitable	
			Total	Renewable
Artesian aquifers in sedimentary platforms				
East-European artesian basins:	I	5,100	3,540	1,670
Baltic-Polish	I.1	700	430	200
Central-Russian	I.2	2,000	1,600	770
Eastern-Russian	I.3	1,100	640	340
Caspian	I.4	110	180	70
Dnieper-Donets	I.5	540	540	200
Pri-Black Sea	I.6	70	50	10
Basins of fractured waters:				
Baltic	I.7	500	35	30
Ukrainian	I.8	60	50	30
Donets	I.9	20	30	20
Turan region	III	430	630	330
West-Siberian region	VIII	1,200	1,290	320
East-Siberian region	X	2,400	400	200
Total in artesian regions		**9,130**	**5,860**	**2,520**
Hydrogeological folded regions				
Carpathian-Crimean-Caucasian	II	1,860	530	380
Kopetdag-Great Balkhan	IV	10	15	5
Tian-Shan-Djungar-Pamir	V	2,560	1,890	1,300
Central-Kazakhstan	VI	260	210	150
Timan-Ural	VII	1,450	385–412	250–290
Sayan-Altai-Yenisey	IX	2,300	80	45
East-Siberian	XI	400	185	80
Zee-Burein and Sikhote-Alin	XII, XIII	1,500–1,600	335	51
Koryak-Kamchatka-Kuril	XV	4,600	200	80
Sakhalin	XVI	560	310	210
Taimyr, Nansena, Eastern, Verkhoyansk-Chukchi	XVIII, XIX, XIV	1,650	–	–
Total (in listed hydrogeological regions)		**16,450**	**4,225**	**2,620**
Total (in the former USSR)		**25,580**	**10,100**	**5,140**

1   Fresh groundwater is defined as < 1 g l^{-1} TDS and brackish groundwater as 1–10 g l^{-1} TDS

2   See Figure 1.3 and Appendix 1 for the location of regions

Source: Hydrogeology of the USSR, 1977

The total exploitable resources of freshwater and slightly sub-saline groundwater are approximately 10,100 m^3 s^{-1}, of which about one half are renewable. Natural groundwater resources vary from 25,000 to 35,000 m^3 s^{-1}, which is approximately 24 per cent of the total river flow. Several southern regions (e.g. the south of the European part of Russia and Central Asia) are characterised by the presence of sub-saline and saline ground waters which are used for economic purposes when there is no fresh water. The total amount of such water is estimated at 2,180 m^3 s^{-1} (Nikitin and Tsyganova, 1973). Actual total groundwater use is about 800 m^3 s^{-1}, of which 350 m^3 s^{-1} is used to supply towns and 450 m^3 s^{-1} is used in agriculture and irrigation.

1.5.3 Thermal, mineral and industrial waters

There are 97 types of mineral water used in about 300 resorts, sanatoriums and regional hydrotherapy resorts and 120–130 water-bottling plants in the former USSR. Total resources of mineral waters are estimated at 195,000–200,000 m^3 d^{-1}. Present consumption is 30,000–40,000 m^3 d^{-1} with future use predicted at 90,000–100,000 m^3 d^{-1}.

Thermal waters are defined as groundwaters with a temperature > 20 °C. Such waters are widespread and usually have a high mineral content up to 30–450 g l^{-1} especially at a depth of 1–4 km. The total exploitable thermal water resources in sedimentary aquifers are estimated at 234 m^3 s^{-1} and for fractured waters 7 m^3 s^{-1}. Of the latter, two thirds are found in the Kamchatka folded region where the Pauzhetsk geothermal power plant is located. In other regions (e.g. Caucasus, Central Asia, Siberia), thermal waters are used for domestic heating and in agriculture, specifically in greenhouses.

Groundwater which contains high concentrations of boron, bromine, iodine, lithium, strontium, caesium, rubidium, etc., may be used for chemical extraction or for hydrotherapy purposes. Iodo-bromide waters are popular and widely used. They are mainly found in platform regions, but are less common in geologically folded regions. The mineral content of such waters is 10–320 g l^{-1} with concentrations of iodine and bromide at 1–26 mg l^{-1} and 10–2,500 mg l^{-1} respectively. The impact of these industrial waters on the quality and pollution of surface waters and shallow aquifers is almost negligible.

1.6 Water balance

The average long-term annual water balance in different regions of the former USSR is given in Table 1.8 (USSR Water Resources and their Use, 1987). The average annual amount of precipitation for the whole territory is 503 mm a^{-1}, or 11,200 km^3. The runoff value is 197 mm a^{-1}, or 4,387 km^3, and annual surface evaporation is 306 mm a^{-1}, or

Table 1.8 Water balance (total and for various regions) of the former USSR

Region	Area (10^3 km^2)	Precipitation (km^3 a^{-1})	Runoff (km^3 a^{-1})	Evaporation (km^3 a^{-1})	Precipitation (mm a^{-1})	Runoff (mm a^{-1})	Evaporation (mm a^{-1})
Baltic States and Byelorussia	396.7	295.8	83.6	212	746	210	536
North European USSR	1,926.1	1,278.4	601.1	677	664	312	352
Central European USSR	652.9	463.5	111.1	352	710	170	540
South-west USSR	634.7	393.0	50.2	324	620	79	540
Northern Caucasus	355.1	221.2	43.7	177	623	123	500
Trans-Caucasus	186.1	148.9	67.8	81	800	364	436
Ural	680.4	350.9	102.0	248	516	150	366
Volga	680.1	355.9	68.3	287	523	100	423
Kazakhstan	2,717.3	800.0	53.5	746	294	20	274
Central Asia	1,277.1	297.9	133.3	184	233	89	144
Western Siberia	2,427.2	1,332.4	482.6	849	549	199	350
Eastern Siberia	4,122.8	2,020.2	1,070.8	949	490	260	230
Far East	6,215.9	3,242.8	1,538.6	1,704	522	248	274
Total	**22,275**	**11,200**	**4,387**	**6,813**	**503**	**197**	**306**

Source: USSR Water Resources and their Use, 1987

6,813 km^3. The influence of vegetation zones on water runoff is presented in Chapter 2 (Table 2.1).

1.7   Characteristics of river water regimes

The following phases of water regime are characteristic for most lowland rivers of the former USSR: spring snow-melt flood, summer low water, summer and autumn rain flood and winter low water.

Different types of river water regime are defined principally by the origins of their flow, dominant source of runoff and seasonal variability (Figure 1.5). There are three main groups:

1.   Snow-melt rivers with spring flood formed by melting snow at altitudes below the level of permanent snow fields.
2.   Snow and ice melt rivers with spring-summer and summer floods formed mainly by the melting of seasonal snow and of permanent snow and glaciers.
3.   Rain-fed rivers with flood regimes determined by rain precipitation and where snow-melt water is insignificant or absent.

Each of these is, in turn, divided into subtypes which are defined by the seasonal pattern of river flow (i.e. periods of high and low flows).

1.7.1   Snow-melt dominance

The majority of rivers belong to the snow-melt group with spring flood. Most of the total annual flow originates from the snow melt. In forest areas, this share is 50–70 per cent. In many rivers of dry steppe and semi-steppe regions of the North Caspian Lowland and Kazakhstan, nearly 100 per cent of the annual flow is during flood period. In rivers of the Baltic States and especially of the Far East, spring flow is relatively less important. Considering the characteristics of spring-flood rivers, this group is further subdivided into several subtypes: Kazakhstan, East-European, West-Siberian and East-Siberian (Figure 1.5).

Rivers of the Kazakhstan subtype are characterised by an extremely high spring flood because snow is the principal and almost only source of river water. During the rest of the year the water is very low and many rivers dry up completely. These rivers are usually found in the dry semi-desert and steppe regions of Kazakhstan, the region east of the Lower Volga and north of the Aral-Caspian lowland.

The East European subtype is characterised by high spring floods from snow melt, low waters in summer and winter and high waters in autumn due to rainfall. The most obvious example is the River Volga. On rivers situated in more southern regions of the East-European plain, the autumn flood is moderate (e.g. the River Don) or completely absent.

The West Siberian subtype is characterised by lower but prolonged spring flood, higher summer-autumn flow and low winter flow. The smoothed flood peaks are due to both the plain relief and to the presence

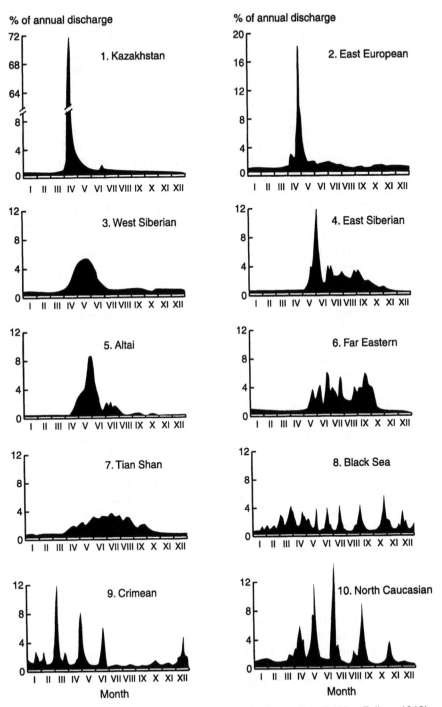

Figure 1.5 Types of water regimes of rivers in the former USSR (After Zaikov, 1946)

of numerous wetlands in the West Siberian Plain. This subtype is especially characteristic of rivers in forest areas of West Siberia. Rivers of the East Siberian subtype are characterised by high spring flood, summer-autumn high runoff and extremely low flow in winter until complete freezing and cessation of the river base-flow, even on large rivers. This is explained by the very low groundwater input during this period due to the permafrost.

Rivers with spring-summer and summer floods are found in high mountain regions, the Caucasus, Central Asia, Altai and Sayan. Amongst this group of rivers, there are two main subtypes: Altai (mostly snow-melt) and Tian-Shan (glacier-melt).

The Altai subtype is characteristic of mountain regions where the spring-summer flood is determined mainly by the melting of seasonal snow. It has a relatively low and prolonged flood, and increased summer-autumn flow, caused by the combination of mountain snow melt and rainfall. Due to the marked differences in altitude the mountain snow melt, even within small catchments, does not occur simultaneously over the whole basin and this prolongs the flood and causes relatively low amplitudes in water fluctuations. When snow melt is intensive there are high peaks preceding the main wave of the flood. Autumn rainfall also increases flow but the subsequent rise in water level is usually characteristic of the tail-end of the main flood wave. Besides Altai, rivers with this type of flow-regime are found in the Trans-Caucasus region, in Central Asia and on Sakhalin Island.

1.7.2   Glacier-melt dominance

The Tian-Shan subtype is characterised by summer flood formed by the melting of snow and glaciers in high mountains. The flow regime is closely related to temperature, with maximum flows occurring in periods with the highest temperatures. This influence is noticeable both over seasonal cycles and over daily cycles. There is a slight time-delay between maximum temperatures and maximum flow. The Tian-Shan subtype is found in mountain systems of Central Asia: Tian-Shan and Pamir, and for rivers of the high mountain regions of Caucasus and Kamchatka.

1.7.3   Rain-fed regimes

The rain-fed regimes are noted in the Far East, parts of Trans-Caucasus, Crimea and the Baltic States. These rivers can be divided into the following subtypes: Far Eastern, Black Sea, Crimean and North Caucasian. The Far Eastern type is characterised by low and very prolonged multi-peak floods during warm periods of the year and very low flow during the other periods. The main source of water for rivers of this subtype is rain. In winter, many rivers freeze completely and their river base-flow stops. Rivers of this subtype are found in the Far East, Eastern

Sayan, Trans-Baikal, the Vitim-Olekma mountains and the Yana-Indigirka region.

The Black Sea subtype is found in the warm and moist climate conditions of the Eastern Trans-Caucasus (Azerbaijan) and is characterised by floods that may occur over the whole year, caused by heavy precipitation from moist sea winds.

The Crimean subtype has winter and spring floods; the low water stage occurs during summer and autumn and many rivers dry up.

The North Caucasian subtype has stable low water during cold periods of the year and in summer is characterised by frequent floods. This subtype includes mainly rivers in the eastern part of the northern slopes of the Great Caucasus mountains.

1.8   Suspended matter

River suspended matter results from erosion in river catchments and river beds. Under the influence of water, the upper and most fertile layer of soil is washed off, its structure deteriorates and valuable nutrients are flushed out. The results of erosion cause an increase in the turbidity of river water (Karaushev, 1977; USSR Water Resources and their Use, 1987). Schematic variations in the discharge of suspended matter are shown in Figure 1.6. In most areas, sediment yield is quite low on a global scale, not greater than 20 t km^{-2} a^{-1}. The lowest value is 5 t km^{-2} a^{-1} and is characteristic of rivers in the western half of the European territory within forest areas, in rivers of North Caucasus, West Siberia, Kazakhstan, the Central Siberian Plateau, and areas of the northern permafrost margin.

The lowest average sediment yield (5–20 t km^{-2} a^{-1}) is characteristic of the eastern part of the European territory as for the Western Dvina due to a gentle relief, high humidity, numerous wetlands and lakes in the catchment and, in the Urals, mountains of South Siberia, the northeast and Far East and also for the desert regions of Central Asia. Within the elevated steppe zone of the European territory and West Siberia, in mountain areas of the Carpathians, Crimea, Primorye, Kamchatka and Sakhalin, sediment yield varies between 20–100 t km^{-2} a^{-1}.

Highest sediment yields attaining or exceeding 2,500 t km^{-2} a^{-1} are found in the alpine orogenesis mountains of Eastern Caucasus and Central Asia as in the Samur river (1,100 t km^{-2} a^{-1}) in the Eastern part of the Caucasus. Active erosion processes are facilitated by the occurrence of easily erodible rocks (shale and sandstone), steep slopes and lack of vegetation.

The grain size composition of particulate load in rivers, both as suspended load and bed load, depends on the processes of surface erosion in the catchment area and on river flow itself. These processes reflect physico-geographical, particularly soil and geomorphological, features of

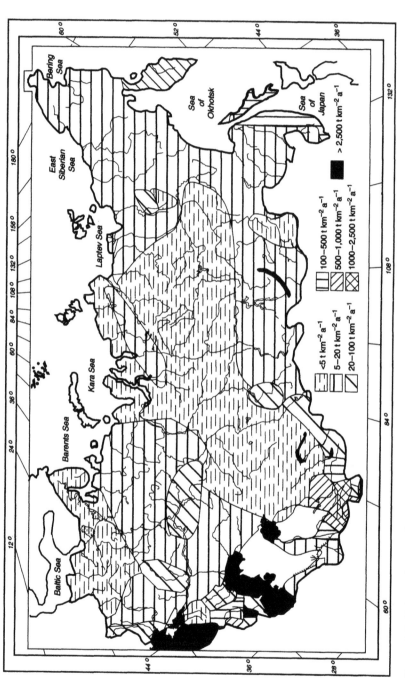

Figure 1.6 Schematic map of sediment yield in rivers of the former USSR (After USSR Water Resources and their Use, 1987)

the territory. Changes in composition of suspended matter and sediment deposition are caused by seasonal variations in the river water regime.

The suspended load of rivers is usually dominated by fine fractions with particle diameters < 0.05–0.10 mm. The sediment composition of the surface layer of the river bed is usually formed by particles with a diameter range between 0.001 and 20 mm for lowland rivers and between 1 and 200 mm for mountain rivers.

The overall sediment load transported to regional seas as well as the general level of total suspended solids in rivers of the former USSR are presented in Chapter 12.

1.9   References

(All references are in Russian unless otherwise stated).

Bindeman, N.N. and Bochever, F.M. 1964 Regional assessment of exploitable resources of ground waters. *Soviet Geology*, **1**, 65–78.

Guidelines on application of the classification of exploitable resources for drinking and industrial waters sources. A set of guidelines on geology and economic assessment of minerals deposits 1986. Volume 3, USSR Ministry of Geology, Moscow, 100–117.

Hydrogeological Regional Map of the USSR 1973 VSEGINGEO, Moscow.

Hydrogeology of the USSR 1966-1972 45 Volumes, Nedra, Moscow.

Hydrogeology of the USSR 1976 The basic trends of groundwater distribution on the territory of the USSR. Summary Volume 1, Nedra, Moscow, 656 pp.

Hydrogeology of the USSR 1977 Summary Volume 3, Nedra, Moscow, 280 pp.

Karaushev, A.V. [Ed.] 1977 *Suspended Matter - Its Study and Geographic Distribution*. Gidrometeoizdat, Leningrad, 240 pp.

Nikitin, M.R. and Tsyganova, K.N. 1973 Regional assessment of operational resources of brackish and saline ground waters. *Water Resources*, **2**, 134–149.

Shiklomanov, I.A. 1988 *Inland Water Resources Research: Results, Problems, Perspective*. Gidrometeoizdat, Leningrad, 170 pp.

USSR Water Resources and Their Use 1987 Gidrometeoizdat, Leningrad, 301 pp.

Voskresensky, K.P. 1962 *Normal Annual Runoff of the USSR Rivers and its Variability*. Gidrometeoizdat, Leningrad, 546 pp.

Zaikov B.D. 1946 *Average Runoff and its Annual Distribution in the USSR*. Gidrometeoizdat, Leningrad, 240 pp.

Appendix 1

Detailed legend of hydrogeological regions in Figure 1.3

Hydrogeological regions of platforms

I. *East-European (Russian)*: 1-Baltic-Polish, 2-Central Russian, 3-Eastern Russian, 4-Caspian, 5-Dnieper-Donets, 6-Black Sea, 7-Baltic, 8-Ukrainian, 9-Donets; 1a-Estonian, 1b-Latvian, 1c-Polish-Lithuanian, 1d-Volyn-Podol, 1e-Brest, 1f-Lvov, 2a-North-Dvina, 2b-Leningrad, 2c-Moscow, 3a-Volga-Kama, 3b-Western Ural, 3c-Sura-Khoper, 4a-Southern Syrt, 4b-Yergeni, 4c-Northern Caspian, 4d-Emba. III. *Turan*: 1-Amu Darya, 2-Syr Darya, 3-Ustyurt, 4-Turgai, 5-Chu-Sarysu, 6-Northern Aral, 7-Central Kyzylkum, 8-Mangyshlak, 9-Tuarkyr. VIII. *West-Siberian*: 1-Upper Ob, 2-Irtysh, 3-Central Ob, 4-Tobol, 5-Central Yenisey. X. *East-Siberian*: 1-Angara-Lena, 2-Yakutia, 3-Tungus, 4-Olenek, 5-Kotui, 6-Khatanga, 7-Lower Olenek, 8-Anabar, 9-Aldan; 1a-Irkut, 1b-Kan, 1c-Angara, 1d-Upper Lena, 1e-Kirenga, 1f-Mur, 9a-Chulma, 9b-Tokkin

Hydrogeological folded regions

II. *Carpathian-Crimean-Caucasian*: 1-Carpathian, 2-Crimean-Caucasian; 1a-Carpathian, 1b-Eastern Carpathian, 1c-Western Carpathian, 2a-Crimea Mountain, 2b-Western Crimea, 2c-Great Caucasus, 2d-Azov-Kuban, 2e-East-Northern Caucasian, 2f-East Black Sea, 2g-Kurin, 2h-Small Caucasus. IV. *Kopetdag-Great Balkhan*: 1-Kopetdag, 2-Great Balkhan, 3-South Caspian, 4-Trans-Kopetdag, 5-Trans-Great Balkan. V. *Tian-Shan-Djungar-Pamir*: 1-East Tian-Shan and Djungar, 2-West Tian-Shan, 3-Pamir, Alatau; 1a-Issyk-Kul, 1b-Ily, 1c-Balkhash-Alakol, 1d-Zaisan, 1e-Naryn, 1f-Talas, 1g-Chu, 2a-Fergana, 2b-Trans Tashkent, 2c-South-Tajikistan, 2d-Zeravshan, West-Tian-Shan. VI. *Central-Kazakhstan*. VII. *Timan-Ural*: 1-Pechora system, 2-Timan system, 3-Ural system: 1a-Izhersko-Pechora, 1b-Bolshezemel, 1c-Trans-Ural, 3a-West-Ural, 3b-East slope of the Urals, 3c-Central Ural. IX. *Sayan-Altai-Yenisey*: 1-Yenisey, 2-Sayan-Altai, 3-Zharminsko-Rudnoaltai, 2a-Kuznetsk, 2b-South-Minusin, 2c-Sydo-Erbin, 2d-Chebakovo-Balakhtin, 2e-Nazarov, 2f-Rybinsk, 2g-Khemchinsk, 2h-Ulugkhem, 2i-Ubsunur-Tekhsem. XI. *East-Siberian*: 1-Vitimo-Patom, 2-Trans-Baikal, 3-West-Baikal, 4-Stanovoy ridge. XII. *Zee-Burein*: 1-Amur-Zeisk, 2-Upper Zeisk, 3-Uda, 4-Torom, 5-Burein, 6-Kimkan, 7-South-Khingan. XIII. *Sikhote-Alin*: 1-Central Amur, 2-Tuguro-Nimelen, 3-Chlya-Orel, 4-Udyl-Kizin, 5-Prikhankai, 6a-Sovetsk-Gavan. XIV. *Upper Chukotsk*. XV. *Koryak-Kamchatka-Kuril*: 1-Penzheno-Anadyr, 2-Koryak, 3-Kuril-Kamchatka; 1a-Parapol, 3a-West-Kamchatka, 3b-Central Kamchatka, 3c-East-Kamchatka. XVI. *Sakhalin*: 1a-North Sakhalin, 1b-Poronay, 1c-Susunai, 1d-Tatar. XVII. *Central Arctic*. XVIII. *Eastern*. XIX. *Taimyr*

Chapter 2[*]

NATURAL COMPOSITION OF SURFACE WATER AND GROUNDWATERS

The previous chapter showed that the former USSR is characterised by a variety of physico-geographic and landscape conditions from arctic tundra in the north and north-east to deserts in the south and subtropics in the south-west. Specific features of the territory which influence the chemical composition of natural waters include remoteness of most regions from the sea, continental climate, sharp seasonal changes in temperature, relatively low runoff in most river basins and widespread permafrost.

The subsequent chapters will show that the present composition of natural waters have been greatly influenced by anthropogenic activities. In this chapter natural water composition is described based on data published prior to 1960, when anthropogenic influence was still limited or non-existent.

2.1   Rivers

There are many publications devoted to the study of chemical composition of rivers in the former USSR. The classic works of Soviet hydrochemistry include those of Almazov (1952), Skopintsev (1941), Alekin (1948, 1950a,b), Lazarev (1957), Bochkarev (1959), Zenin (1965), Voronkov (1970), Veselovsky et al. (1975), Konovalov et al. (1977) and Tarasov and Beschetnova (1987) amongst others.

According to Alekin (1950a,b, 1970) the principal features which determine the chemical composition of river water are:

- Water flow, substrate and catchment hydrogeological properties;
- Formation of water in surface layers;
- Dependence of water regime on climatic and weather conditions;
- The interaction of water and atmosphere;
- The influence of biotic factors on water.

River waters are therefore characterised by low mineralisation, rapid variability of composition under the influence of hydrometeorological

---

[*]   *This chapter was prepared by V.V. Tsirkunov, M.P. Polkanov and V.G. Drabkova*

conditions and the permanent presence of gases of atmospheric origin in the water, particularly dissolved oxygen.

2.1.1 Temperature, dissolved gases and pH

Water temperature is of significant importance for chemical reactions, self-purification of water bodies and for the development of biota. Extremely low temperatures in the north and east of the country should be considered as a specific feature of the thermal regime of rivers in the former USSR. The annual mean temperature of many rivers of East Siberia does not exceed 2–3 °C (USSR Surface Water Resources, 1972a), which results in the complete freezing of most small and medium-sized rivers in these regions from November to April. Transfer of large amounts of heat from south to north by the largest rivers (e.g. Yenisey, Lena, Ob) draining to the Arctic Ocean also occurs.

Besides temperature, the photosynthetic activity of micro-organisms, the duration of ice cover and the origins of river waters influence the concentration of dissolved gases (mainly oxygen and carbonic acid) in river waters. In winter, the concentration of oxygen in river water is generally low. Water saturation with oxygen decreases as a result of ice cover and the increase in groundwater recharge which is characterised by low oxygen content and very low photosynthetic activity. The very low oxygen concentrations found in winter may cause fish kills, as occurred under ice in the River Volga in 1939–40 following exceptionally low water levels in the autumn of 1939 and the subsequent input of ground-waters with low oxygen concentrations (Skopintsev, 1941). Low oxygen concentrations in winter were also found in the River Ob (downstream of the River Ket), into which flow tributaries draining wetland areas (Alekin, 1970). Fish kills occurred as a result of this sharp decrease in oxygen to < 5 per cent saturation. A similar phenomenon was also noted occasionally in the Northern Dvina, Pechora and other northern rivers (USSR Surface Water Resources, 1972b). Once ice cover melts, fast aeration of river waters occurs leading to a rapid increase in oxygen concentration (Figure 2.1). Oxygen saturation of river waters promotes increased photosynthesis in the spring which may lead, in turn, to super-saturation ($\geq$ 150 per cent) of water with oxygen in the summer.

The CO_2 regime in water is the inverse to that of oxygen. In winter, oxidation processes in the water, coupled with an increase in input of groundwater with a high CO_2 content, leads to the accumulation of CO_2 under ice, with concentrations reaching $\geq$ 50 mg l^{-1}. Following ice melt, the concentration of CO_2 decreases rapidly due to its dissipation to the atmosphere and its consumption during intensified photosynthesis (Figure 2.1). Thus, the concentration of CO_2 in rivers in summer is usually 1–5 mg l^{-1} where there is intensive growth in aquatic vegetation, and CO_2 may even disappear completely.

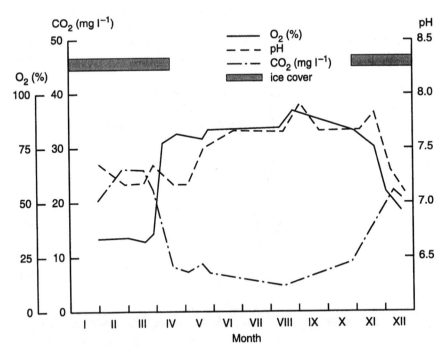

Figure 2.1 Changes in O_2, pH and CO_2 in the River Vyatka (Volga basin) during 1953 (After Alekin, 1970)

Concentrations of hydrogen ions vary in river waters of the former USSR, with pH generally in the range of 6.5–8.5 (Alekin, 1970). Waters with lower pH are typical of northern rivers, whereas southern rivers have higher pH. In general, the pH regime is similar to that of CO_2. In winter, pH values for most major rivers are 6.8–7.4, whereas in summer pH values are 7.4–8.3. Rivers fed by waters draining wetlands are characterised by a much lower pH, less than 6.0, because the pH of wetland waters may be as low as 3.1–3.5 (Kalyuzhny and Levandovskaya, 1976). At the other end of the pH scale, rivers flowing in regions where carbonate rock is dominant have pH values exceeding 8.0–8.5.

2.1.2 Formation of ionic composition in small catchments (local runoff)

According to Voronkov (1970), river networks in small catchments (i.e. < 12,000 km^2) contain waters which reflect the physico-geographic features of that area. In larger, more heterogeneous catchments, river networks combine waters with very different chemical composition reflecting various natural conditions.

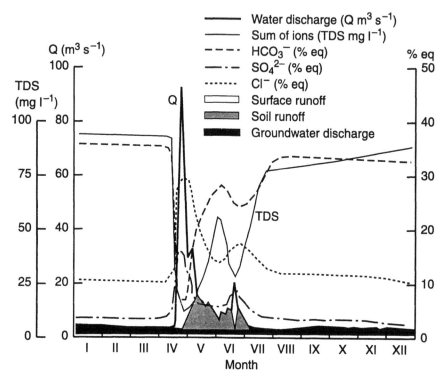

Figure 2.2 Hydrochemical regime of the River Pyalitsa, 1960 (Pyalitsa village, drainage area = 946 km^2) (After Skakalsky, 1983)

Studies of the water chemistry of small river basins, under different conditions (lithology, climate, water discharge regimes), allowed the determination of the original, principal components of river water, namely surface runoff, soil runoff and groundwater discharge (Voronkov, 1963). The periodic alterations in the principal types of water feeding into a river network together with seasonal changes, determine the hydrochemical regime of a river. Surface runoff prevails in rivers of small water catchments during the spring flood peak (Figure 2.2). Waters of soil runoff and groundwater origin prevail during the receding stages of a flood. In summer and winter low-flow, river discharge is due predominantly to drainage of ground aquifers.

2.1.3 Ionic composition of water from different vegetation zones
An assessment based on long-term monitoring of small catchments in the European part of the former USSR, showed the typical characteristics of local water runoff for different vegetation zones (Skakalsky, 1983; see also section 1.6). The study demonstrated that for all zones local runoff is predominantly formed by surface runoff, which comprises 40 to

Table 2.1   Proportions of surface, soil and groundwater contributing to local runoff in different vegetation zones

Vegetation zone	Surface		Soil		Ground		Average annual discharge (mm a^{-1})
	Runoff (mm a^{-1})	Proportion of annual discharge (%)	Runoff (mm a^{-1})	Proportion of annual discharge (%)	Runoff (mm a^{-1})	Proportion of annual discharge (%)	
Tundra	150–250	50–65	90–110	20–30	80–150	15–25	340–450
Forest							
Taiga	125–200	50–65	75–125	20–35	50–100	15–30	250–400
Mixed forest	50–150	40–60	25–75	20–30	50–100	20–35	150–300
Forest steppe	40–80	50–65	10–30	15–20	10–50	15–35	70–150
Steppe	10–50	65–90	5–10	10–20	2–10	5–20	10–70

Source: Skakalsky, 1983

90 per cent of the total annual discharge of small rivers (Table 2.1). This surface runoff is predominantly formed by spring snow melt. In western regions, in mixed forest, forest-steppe and steppe zones, a large amount of surface runoff is formed at other times of the year during heavy rains or winter thaws. The predominance of surface runoff in the annual river flow is greatest in steppe catchments (> 70 per cent) and least in forest zones (40–50 per cent).

Soil runoff provides the highest input to total river flow in tundra and taiga catchments (20–30 per cent of annual river discharge). This results both from the high natural humidity of soil and rocks which reduces infiltration loss, and from the high variability of filtration features of soil in those catchments. In mixed forest and forest-steppe zones, the input of soil runoff to annual river discharge decreases and is often less than the input of groundwater discharge. In comparison with other zones, groundwaters form the largest proportion of river water in forest zones, especially in mixed forests where they reach 30–40 per cent of total river discharge.

According to Voronkov (1970) and Skakalsky (1983), attributes of chemical composition of river waters formed by local runoff are determined mainly by the relative proportions of waters of different origin throughout the year and their interactions. Some general hydrochemical characteristics of waters of local runoff in different vegetation zones in the European part of the former USSR are given in Table 2.2.

Formation of local runoff and the proportions of its components are influenced mainly by climatic conditions. Hence spatial changes in hydrochemical characteristics of local runoff occur along latitudinal zones. In particular, the mineral content of waters of all origins regularly increases from north to south, which is inversely related to the volume of water runoff. According to Skakalsky (1983), the mineralisation of

Table 2.2   General hydrochemical characteristics of local runoff waters in the European territory of the former USSR

Vegetation zone	Origin of waters	Mineralisation (mg l^{-1})	Dominant anions and cations (%)		COD (mg O$_2$ l^{-1})	Total hardness (mmol l^{-1})
Tundra	Surface	10–30	HCO$_3^-$	25–28	20–30	0.2–0.5
			Ca^{2+}	< 25		
			Mg^{2+}	< 25		
			Na$^+$	< 25		
	Soil	30–60	HCO$_3^-$	25–36	15–20	0.2–0.8
			Ca^{2+}	< 25		
			Mg^{2+}	< 25		
	Ground	40–200	HCO$_3^-$	25–36	8–10	0.5–2.5
			Ca^{2+}	25–28		
			Na$^+$	25–28		
Forest	Surface	30–100	HCO$_3^-$	28–44	30–40	0.5–1.0
			Ca^{2+}	28–44		
	Soil	50–200	HCO$_3^-$	28–50	40–60	0.5–2.0
			Ca^{2+}	28–44		
	Ground[1]	100–500	HCO$_3^-$	36–50	10–40	2.0–4.0
			Ca^{2+}	28–36		
			Mg^{2+}	< 25		
Forest steppe	Surface	100–200	HCO$_3^-$	28–44	20–25	1.5–2.5
			Ca^{2+}	28–44		
	Soil	250–450	HCO$_3^-$	28–44	25–30	3.0–5.0
			Ca^{2+}	28–44		
	Ground[1]	350–1,500	HCO$_3^-$	28–50	10–20	4.0–10.0
			Ca^{2+}	25–36		
Steppe	Surface	200–600	HCO$_3^-$	25–28	20–30	2.0–6.0
			SO$_4^{2-}$	< 25		
			Cl$^-$	< 25		
	Soil	300–1,000	SO$_4^{2-}$	25–28	30–40	3.0–10.0
			Cl$^-$	25–28		
			Ca^{2+}	< 25		
			Mg^{2+}	< 25		
	Ground[1]	1,000–4,000	SO$_4^{2-}$	25–36	20–40	6.0–20.0
			Cl$^-$	25–36		
			Na$^+$	25–28		

[1]   Winter low water   Source: Skakalsky, 1983
COD   Chemical oxygen demand

surface waters increases successively from low values of 10–30 mg l^{-1}, typical of tundra and forest-tundra catchments to 200–600 mg l^{-1} for steppe catchments (Figure 2.3). For soil waters there is a successive change in mineralisation from 30–60 mg l^{-1} in tundra areas to 300–1,000 mg l^{-1} in steppe areas.

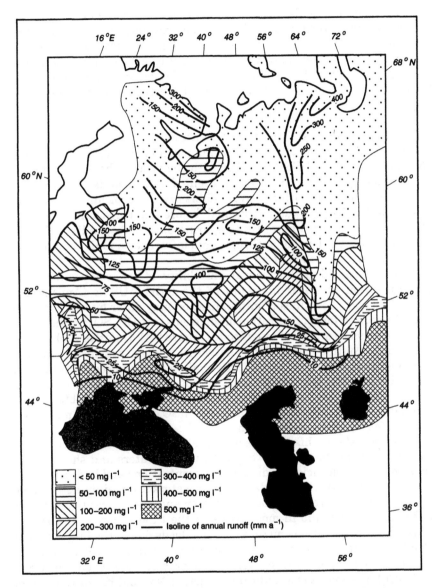

Figure 2.3 Annual runoff and changes in mineralisation of river waters of surface origin in the European territory of the former Soviet Union (After Skakalsky, 1983)

Changes in mineral content with latitude also occur in waters of groundwater origin. Mineral content ranges between 40–200 mg l^{-1} in northern catchments to 1,000–4,000 mg l^{-1} in the southern steppe catchments (Table 2.2). This is related to a decrease from north to south in the level of flushing of soils and rocks in the unsaturated zone of ground-water through precipitation. Some other parameters of water, e.g. hardness, follow a similar pattern (Table 2.2). However, a decrease in organic matter content in waters of different origin usually occurs from south to north. Thus, the content of dissolved organic matter, as recorded by the chemical oxygen demand (COD), decreases with a decrease in precipitation.

2.1.4 Ionic composition of river waters

Long-term data on the average ionic composition of water in the largest rivers of the former USSR together with the typical chemical composition of some 'pristine' rivers are presented in Appendix 2 Table I. It should be noted that for large rivers liable to anthropogenic salinisation (see Chapter 5), only data obtained before intensive anthropogenic impact are considered here. By comparing data in Appendix 2 Table I and Table 5.1, the level of anthropogenic influence on ionic composition of large rivers can be assessed.

There is considerable natural variability of ion composition and mineralisation of water in rivers of the former USSR (Appendix 2 Table I). Thus, the values of mineralisation may differ by two orders of magnitude, and for concentration of some ions (e.g. Cl^-, SO_4^{2-}, $Na^+ + K^+$) by three or more orders of magnitude (Figure 2.4). Unfortunately, since the State routine monitoring network seldom provided separate chemical analyses of Na^+ and K^+, the available data series are relatively limited (since the late 1970s) and are not reliable for these elements. Hence, the data on total concentration of $Na^+ + K^+$ were obtained by evaluation of the difference in anions ($Cl^- + SO_4^{2-} + HCO_3^-$) and cations ($Ca^{2+} + Mg^{2+}$) in equivalent form.

To some extent, this drawback was eliminated by Koreneva (1977) who studied the Na^+ and K^+ content in river waters and showed that these ions have sufficiently different concentrations and hydrochemical regimes. Concentrations of Na^+ in river waters are mainly determined by the origin and recharge of river water and are inversely related to water discharge while in most rivers in the tundra, forest and steppe zones, K^+ concentrations increase during floods due to the large content of K^+ in plants and to the easy leaching of K^+ from plant debris into water. The Na^+ to K^+ ratio in rivers varied greatly and ranged from 0.8 to 75 (eq/eq). Estimates of specific discharges of Na^+ and K^+ for the former USSR were 1.49 and 0.22 t km^{-2} a^{-1} respectively (Koreneva, 1977).

Alekin (1948, 1950b) found that waters of most rivers of the former USSR were dominated by bicarbonate ions. According to the cation composition this water belongs to the calcium group. Bicarbonate waters of magnesium and sodium groups are extremely rare. In the north of the European territory and most of the Asian territory of the former USSR, waters of low (< 200 mg l^{-1}) and very low (< 100 mg l^{-1}) mineralisation

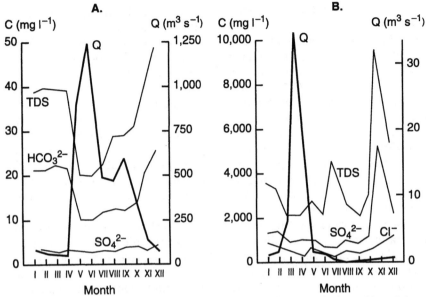

Figure 2.4 Examples of variations in mineralisation and chemical composition of river waters of the USSR. **A.** Ultra-fresh waters (Taui river at Talon); **B.** Highly mineralised waters (Kalaus river at Svetlograd)

prevail (Appendix 2 Table I). Their distribution coincides with widespread tundra soils which are poor in soluble salts but rich in organic matter. Soil organic matter and atmospheric precipitation play an important role in the formation of the chemical composition of water in these regions.

Waters of medium mineralisation (200–500 mg l^{-1}) are less widespread. They occupy a central band of the European territory, forests and forest-steppe zones. River basins with waters of higher mineral content (500–1,000 mg l^{-1}) occupy an even smaller area and are found mainly in the south-east of the European part of the country in the forest-steppe zone. These waters are often associated with chernozem soil frequently found on carbonate rocks.

Rivers with waters of high sulphate content are comparatively rare and mainly found in steppe zones and occasionally in semi-deserts. They also include some northern rivers, for example in basins of the Onega and Chusovaya and in parts of the Volga, where they are related to the occurrence of pyrite-containing rocks, gypsum, etc. The principal cations of these rivers are Ca^{2+} and to a lesser extent Na^+. The mineral content of these rivers is relatively high and in some cases the total ion concentration exceeds 4,000–5,000 mg l^{-1} (e.g. the Kuma and Kalaus rivers) (Figure 2.4B). It is of interest to note that if Gibbs' (1970) diagram is considered, the chemical composition of the Rivers Taui and Kalaus

(Figure 2.4) should be influenced by atmospheric precipitation and evaporation and have a chloride–sodium composition. However, the River Taui has a bicarbonate–calcium composition and the River Kalaus a sulphate-sodium composition probably reflecting the greater influence of soils and rocks than was predicted by Gibbs (1970).

Rivers with high Cl⁻ content are almost as rare as rivers with prevailing sulphate. They are found from the Lower Volga basin to the upper basin of the Ob where steppes, semi-deserts and deserts occur. The dominant cations in waters of this type are usually Na⁺. During summer and winter low-flow, these waters are characterised by high mineralisation (usually > 1,000 mg l⁻¹ and less frequently 500–1,000 mg l⁻¹). During periods of spring flood, concentrations of major ions are significantly lower and are close to those shown in Figure 2.3 for surface runoff in small catchments.

The distribution of total dissolved salts (TDS), which is an indicator of mineralisation of rivers, follows a certain geographic pattern. There is a general trend of increased mineralisation in most rivers of the European part of the former USSR from north to south and from west to east. In the Asian territory, there is an increase from north to south with maximum mineralisation in rivers of North Kazakhstan, South West Siberia and some parts of the Southern Urals. High moisture in the north, predominant tundra, wetland and forest soils and permafrost lead to lower mineralisation. In the south, precipitation decreases, climate aridity increases and podsolic soils are substituted by chernozem chestnut soils and sometimes saline soils which enrich the water with minerals.

Trends in chemical composition of river water are affected by differences in rock composition and their distribution. This is evident for basins where evaporite deposits are found. Such rivers are situated in the European territory in the basins of Northern Dvina and Onega (Lodma and Kuloi rivers) (Figure 2.5A), in the basin of the Middle Volga (Ilet, Kazanka, Alatyr rivers), in the Kama basin (Sylva river), where Permian sediments containing evaporites including NaCl are found. The influence of rock-types is most evident in the basin of the Lena where evaporite rock of the Cambrian East-Siberian platform lie close to the surface. The discharge of groundwaters into the basins of Biryuk, Olekma, Kirenga, Namana, Peledui, Chara, Kampandaee and other rivers leads to the formation of highly mineralised waters, with high chloride and sulphate concentrations during low-flow periods (USSR Surface Water Resources, 1972a,b).

Extremely high natural concentrations of salts are found in very few rivers (e.g. River Kampandaee) (Figure 2.5B), mainly among right bank tributaries of the middle reaches of the River Vilyui. The high salt content is due to the dissolution of fractured Kampandaee salt domes. In winter low-flow, waters of such rivers have a particularly high chloride-sodium brine content with concentrations of salts sometimes exceeding 100 g l⁻¹.

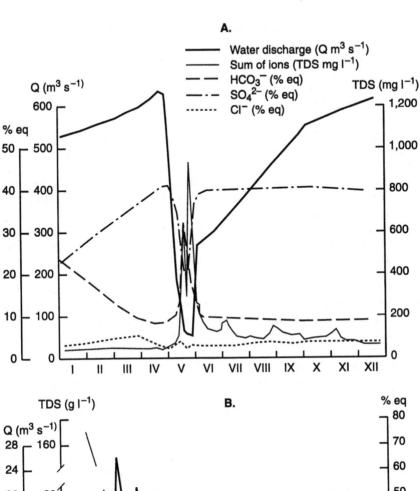

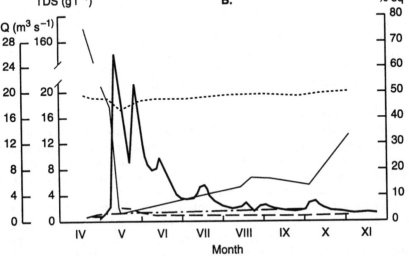

Figure 2.5 Seasonal variations in mineralisation and ion composition of water.
A. River Kuloi at Kuloi in 1961 (USSR Surface Water Resources, 1972b); **B.** River Kampandaee at Kempendyai in 1961 (After Ivanik and Veselovsky, 1974)

One of the features of the permafrost zone is massive frazil ice formation which influences river discharge throughout the year. In winter, frazil ice promotes the accumulation of groundwater and significantly reduces river discharge but in spring and summer accumulated waters increase river discharge. The influence of these waters is especially visible if groundwaters have, as shown above, high mineralisation and a chemical composition which is different from those for surface waters.

Geographical trends in chemical composition of small and medium rivers do not always apply to large rivers (i.e. > 100,000 km^2). These rivers are complex systems integrating basins of many smaller rivers characterised by varying chemical composition. When rivers are very long, the chemical composition of their waters, especially when they flow through different geographic zones, often does not correspond to the local chemical composition of water formed in that area. This applies to the lower reaches of almost all large rivers flowing in the south of the European part of the former USSR (e.g. Ural, Volga, Dnieper, Don).

The chemical composition of waters of the large rivers have been studied both in fixed cross-sections and along the whole river course. For large rivers of the European part flowing south, and for rivers of Central Asia (e.g. Amu Darya, Syr Darya, Ily, Chu), concentrations of major ions increase as the aridity of the climate increases. This trend was not observed for rivers of northern and eastern parts of the country. However, a decrease in the concentration of nutrients and organic matter occurs in river flow on the southern slope of the Russian platform.

When cross-sections were studied at fixed stations, the chemical composition of large rivers was generally found to be homogeneous. The exceptions to this were noted in river stretches downstream of large confluences with tributaries having different chemical composition. This is typical of the River Volga, downstream of its confluences with the Oka and Kama rivers, where more dilute Volga waters are mixed with more mineralised waters of the larger tributaries (Zenin, 1965).

2.1.5 Hydrochemical regime

Of surface waters, rivers have the greatest temporal variability in chemical composition. Seasonal variability is mostly caused by the various proportions of water sources (surface and ground, different tributaries). In addition to seasonal variations, other fluctuations of lower duration and amplitude, including daily fluctuations can be noted.

Daily variations are well documented, especially for high mountain rivers during warmer seasons where daytime snow and ice-melt waters are replaced by groundwaters at night (e.g. Veselovsky et al., 1975; Fadeev et al., 1976; Konovalov et al., 1977). Studies in the Caucasus showed that the amplitude of fluctuations within 24 hours could reach 50–100 per cent for the Byelaya river (Korobeinikova et al., 1977) and the Baksan river (Fadeev et al., 1989) and 200–250 per cent for the

Garabashi river (Konovalov *et al.*, 1977). These variations may be considered as extreme in rivers with natural hydrochemical regimes.

In the 1940s, an attempt was made to correlate the different types of hydrochemical regime of rivers with their water regime (Alekin, 1948, 1950a). A classification of river hydrochemical regimes was proposed based on two features: the type of seasonal TDS fluctuations and the prevailing anions in the water. The time and recurrence of maximum and minimum values of mineralisation are closely related to the water regime of a river. The amplitude of TDS fluctuations during a year depends on both the water regime and the composition of groundwaters feeding the river during periods of ice-cover and in summer low-flow. Based on the character of seasonal changes of river water mineralisation in the former USSR, Alekin (1970) determined six types of hydro-chemical regime named after territories for which each is characteristic, illustrated either by long-term monthly averages (Figure 2.6) or by variations within a specific year (Figure 2.7): EE, Eastern European type of hydrochemical regime (Figure 2.6A), S, Siberian (Figure 2.6B), F, Far Eastern (Figure 2.6C), K, Kazakhstan (Figure 2.7A), B, Black Sea (Figure 2.7B) and T, Tian-Shan (Figure 2.7C).

During a year, prevailing ions are determined mainly by atmospheric and climatic conditions, soil, geological and landscape conditions, etc. There are three possible classes determined by prevailing anions: bicar-bonate (referred to as class C), sulphate (class S) and chloride (class Cl). According to Alekin (1970), seven combinations of the different classes of river water composition are possible during the year: C; C,S; C,Cl; C,S,Cl; S; S,Cl and Cl depending on the dominance of each anion expressed in equivalents per litre (eq l^{-1}). By combining six types of seasonal mineral fluctuations with the seven types of possible classes of water composition, there are a possible 42 types of hydrochemical regime for rivers.

In rivers of the former USSR, only nine types of hydrochemical regime are most common: EE-C; EE-S,C (Figure 2.5A); K-C, Cl; K-C, S, Cl; S-C; F-C; B-C; T-C and T-C,S (Alekin, 1970). Most of these types of hydro-chemical regime are further divided into subtypes. In particular, for Eastern European bicarbonate types there are the Kola, Karelian, Valdai and Volga sub-types of hydrochemical regime.

Following Alekin's studies (Alekin, 1970), some types and sub-types of river hydrochemical regimes were modified. For the S-C type, in addition to the East-Siberian subtype defined, a Central Yakutian subtype and three other sub-types — Middle Lena C,Cl and C,S sub-types and Middle Vilyui Cl subtype (Figure 2.5B) — were defined (Ivanik and Veselovsky, 1974). Tsirkunov (1985) proposed a Turkmenian type of hydrochemical regime, found in particular on the Murgab river of Central Asia. It is characterised by a sharp increase in mineral content in spring on the rise and peak of the flood, which is probably related to a flushing of salts

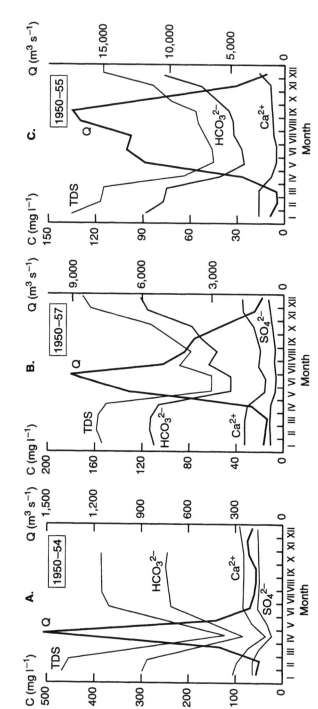

Figure 2.6 Long-term average of seasonal variations of major ion concentrations and water discharge. **A.** East European river type (River Oka at Kaluga); **B.** Siberian river type (River Yenisey at Yeniseisk); **C.** Far Eastern river type (River Amur at Khabarovsk)

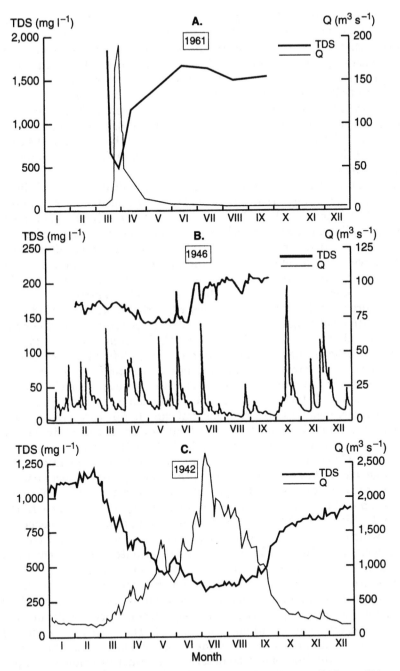

Figure 2.7 Annual variations in mineralisation (TDS) and water discharge (Q).
A. Kazakhstan river type (River Nura at Sergiopolskoye); **B**. Black Sea river type (River
Sochi at Plastunka); **C.** Tian-Shan river type (River Vakhsh at Tutkaul) (After Alekin, 1970)

which had accumulated during the dry period. Such a phenomenon is also described for the rivers of the Fergana valley (Burkaltseva, 1964). Rivers of this type are characterised by an increase in mineralisation of water, with sulphates dominating at the rise and peak of the flood, while at other times bicarbonates dominate, though to a lesser extent.

Alekin (1950a) produced distribution maps of the different types of hydrochemical regime for the former USSR, as well as hydrochemical maps of rivers. These were later improved both for the whole country (Voronkov, 1970; Emelyanova and Danilova, 1979; Hydrochemical Atlas, 1990) and for separate regions (USSR Surface Water Resources, 1972a,b, 1973).

The relationship between water and hydrochemical regimes of rivers was studied by Blinov (1946), Alekin (1950b), Almazov (1952), Fadeev et al. (1976), amongst others. For rivers in the former USSR, as for most rivers globally, a negative correlation between concentrations of major ions and water discharge is usually noted. Fadeev et al. (1989) showed that the overall ionic concentrations and water discharge is greatly influenced by several factors including the size of the catchment area, the presence of karst, permafrost and the annual water regime.

Nikanorov and Tsirkunov (1991) described the relationships between concentrations of major ions and water discharge for rivers of the principal hydrochemical types. The inverse relationships between HCO_3^-, SO_4^{2-} and other minerals and discharge is well defined and relatively constant for the rivers of EE, S and F types (correlation coefficients were generally > 0.6). Other ions were characterised by weaker correlations. In rivers of B and T types, the negative relationships of most ions with discharge were usually statistically insignificant. In rare cases, these rivers and rivers of the Turkmenian type had a tendency for an increase in concentrations of some ions (e.g. Mg^{2+}, $Na^+ + K^+$, SO_4^{2-}) with an increase in water discharge. Generally, the relationship between concentration and discharge was best described by exponential and hyperbolic functions.

The study of long-term changes in chemical composition of rivers makes it possible to estimate the level of anthropogenic influence. In the former USSR, over 30 'background' monitoring points were chosen on rivers which, in terms of major ions, were not particularly liable to anthropogenic influence. These points are situated on the rivers of Siberia and mountain rivers of the Caucasus and Central Asia.

Nikanorov and Tsirkunov (1991) found that concentrations of most major ions at 'background' monitoring stations were characterised by small, long-term changes with relatively strong fluctuations around long-term mean concentrations (Figure 2.8). The long-term series of concentrations of major ions and mineralisation enables the upper and lower limits of fluctuations to be determined: mineralisation 15–25 per cent, Ca^{2+} and HCO_3^- 15–35 per cent, Mg^{2+} and SO_4^{2-} 30–50 per cent, $Na^+ + K^+$ and Cl^- 40–70 per cent (Tsirkunov, 1985).

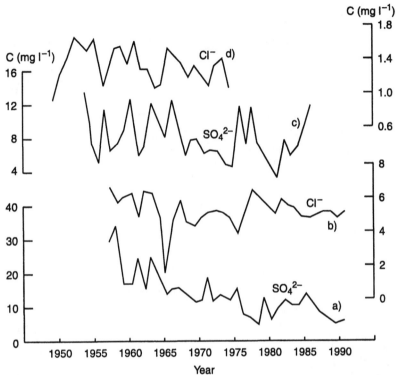

Figure 2.8 Long-term variations of natural background concentrations of two major ions. a) and b) Kamchatka river at Klyuchi; c) Aldan river at Tommot; d) Vitim river at Bodaibo

2.1.6 Nutrients and organic matter

There are less data concerning the content of nutrients and organic matter in rivers of the former USSR before the start of intensive anthropogenic impact than there are for major ions. Concentrations of nutrients for some rivers since the 1980s are presented in Appendix 2 Table II. For many rivers, especially in basins of the Atlantic Ocean, Caspian and Aral Seas and in exorheic basins, concentrations of nitrates, ammonium and phosphates exceed natural concentrations. This is highlighted by the comparison of nitrate, ammonium and phosphate values in rivers of the Pacific and Arctic Ocean basins.

Concentrations of silica and iron do not undergo such obvious changes under anthropogenic impact. However, rivers subject to significant anthropogenic impact have higher maximum concentrations of iron (e.g. Don, Terek, Volga, Murgab rivers) (Appendix 2 Table II).

Nitrate concentrations in unpolluted river waters vary generally in the range 0–1.0 mg l^{-1} N (Figure 2.9A). For rivers, the two important sources of nitrate compounds are atmospheric precipitation and nitrogen

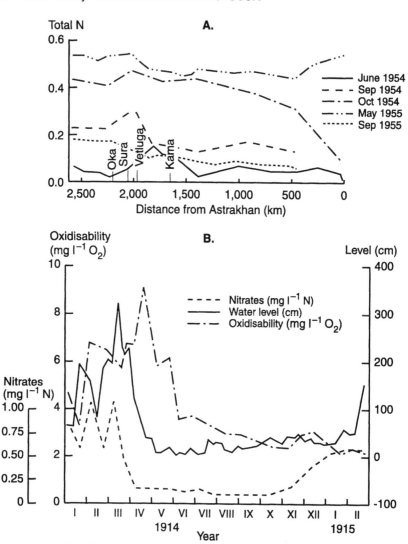

Figure 2.9 A. Variation in total nitrogen (nitrates, nitrites, ammonium) along the course of the River Volga (After Zenin, 1965); **B.** Variation in nitrates and susceptibility to oxidation depending on the water level in the River Moskva (Rublyevo village) (After Alekin, 1970)

recycled through the decay of vegetation debris and animal excretion. The nitrate regime is characterised by minimum concentrations during periods of vegetation growth and photosynthesis (0.01 mg l^{-1} N). In autumn, nitrate concentrations in water begin to increase and reach their maximum in winter when the minimal consumption coincides with the decomposition of organic matter and the transformation of nitrogen from organic form into mineral form. In spring, when temperature and light levels increase, plant growth is accompanied by an increase in uptake of nitrates and, consequently, a decrease in nitrate concentration in water (Figure 2.9B).

Concentration of nitrite ions in river water is much lower than those of nitrates (c. 0.01 mg l^{-1} N). In polluted waters, nitrite concentrations increase to 0.1 mg l^{-1} N. Nitrites appear during the natural cycle of organic matter decay at the stage of nitrification, usually at the end of summer and autumn. Ammonium concentrations in river waters vary in the range of 0.01–0.1 mg l^{-1} (Appendix 2 Table II).

The concentration of inorganic phosphorus in river waters does not generally exceed 0.1 mg l^{-1}, and is frequently 0.001–0.01 mg l^{-1}. The phosphorus regime is similar to that of nitrates, when minimum concentrations are also found during periods of vegetation growth.

Iron concentrations in river waters of the former USSR can vary but are generally in the range 0.1–0.5 mg l^{-1}. In Northern rivers, iron concentrations are much higher due to the formation of organic complexes with humus substances (Alekin, 1970). When the concentration of humus is high, iron concentrations reach several milligrams per litre. Iron concentrations usually increase in spring due to the inflow of surface runoff enriched with humus. For some rivers, an increase in iron concentrations occurs in summer due to iron-rich groundwater inputs.

Silica concentration in river waters is generally 1–5 mg l^{-1}. In winter, an increase in silica concentration usually occurs due to increased groundwater inputs. However, in some rivers in arid areas (e.g. Syr Darya, Amu Darya, Sarysu, Terek, Nura, Murgab) and in some mountain rivers (e.g. Chamlyk, Akhuryan, Kamchatka) average concentrations exceed 3–4 mg l^{-1}, and maximum concentrations may exceed 10 mg l^{-1} (Appendix 2 Table II).

Organic matter in river waters is usually present as substances of humus origin flushed from soils and wetlands and as products of decay of organic substances mainly of plant origin. Data on concentrations of organic matter in river waters are obtained using indicators such as colour index and the susceptibility of organic matter to oxidation by permanganate and dichromate reflecting chemical oxygen demand.

Smirnov and Tarasov (1977, 1982) produced maps of river long-term seasonal and annual mean values of colour index and the susceptibility of organic matter to oxidation by permanganate and dichromate. These showed that, generally, organic matter distribution is related to physico-geographic features of the environment. The formation and distribution of organic matter depend on vegetation, morphology and drainage pattern. For example, in the tundra, forest tundra and open tundra forests the mean annual COD values of river water ranges from 10–20 mg l^{-1} O_2 to 20–30 mg l^{-1} O_2 respectively, and in the Central Siberian upland to 30–40 mg l^{-1} O_2. The mean annual COD values for river waters of Northern and Southern taiga and mixed forests vary between 40–60 mg l^{-1} O_2. South of the taiga landscapes, in the zone of deciduous forests, forest steppes and steppe, COD values are somewhat reduced (mean annual values 20–30 mg l^{-1} O_2 with extremes to 60 mg l^{-1} O_2). During the flood period, higher COD values are obtained whereas in winter low-flow water the COD is lower. In semi-desert and desert, where river runoff occurs during the year, slightly higher COD values prevail. Sub-tropic areas are situated between mountain ranges and their nature is, to a great extent, determined by the mountain landscape. Low levels of organic matter are recorded in these rivers. In humid subtropics (e.g. West Trans-Caucasus) low values (5–10 mg l^{-1} O_2) dominate during the year whereas in dry subtropics (e.g. East Trans-Caucasus) mean COD values are 10–20 mg l^{-1} O_2. There are few data on COD values for mountain rivers in high-altitude tundra-arctic zones.

Maps of the long-term mean annual values of colour index and the susceptibility of organic matter to oxidation by permanganate of river waters showed that latitudinal zones of tundra, forest steppe, semi-desert and desert correspond to hydrochemical zones of medium colour index (25–50°) and susceptibility to oxidation (5–10 mg l^{-1} O_2). Latitudinal zones of northern, central and large parts of southern taiga correspond to hydrochemical zones of higher colour index (75–100°) and susceptibility to oxidation (10–20 mg l^{-1} O_2).

Trends in the distribution of components of organic matter (concentrations of organic carbon (C_{org}), humic acids, fulvic acids and other indicators of organic substances) in river waters occur in different vegetation zones (Table 2.3). Organic matter content in waters is high in humid zones and significantly lower in arid zones. In northern latitudes, natural water properties are determined mainly by organic substances whereas in southern arid regions mineral components are of prime importance. The organic matter regime is closely related to water discharge. In an annual cycle, values of colour index, the ratio of susceptibility to oxidation by permanganate to COD values, the susceptibility to oxidation by permanganate to C_{org} values, humic acids to fulvic acids and the ratio of organic matter to mineral content reach maximum values during floods and high waters and are minimal at low waters, particularly in winter.

2.1.7   Trace elements

Compared with nutrients and organic data available (section 2.1.6), data on concentrations of trace elements in river waters of the former Soviet Union are less reliable. The first reliable results were obtained using ultra-clean methods within the framework of the Joint Russian-French-Dutch Scientific Program on Arctic and Siberian Aquatic Systems (SPASIBA). The main goal of this programme was to provide biogeochemical data from the mouth of the largest Arctic rivers, Lena, Yenisey and Ob. Considering the variability of landscape and physico-

Table 2.3 Average long-term values of some indicators of organic matter in river water of the former USSR

Latitudinal zone	Colour Index (°)	OP (mg l^{-1} O_2)	OP/COD (%)	C_{org} (mg l^{-1})	OP/C_{org}	OM/TDS (%)	Humic acids (mg l^{-1} C)	Fulvic acids (mg l^{-1} C)
Tundra	38–67	8.3–13.7	40–49	4.7–7.0	0.86–1.13	20–80	0.12–0.35	1.01–2.32
Forest tundra and thin tundra forest	21–65	7.0–14.0	40–47		1.08–1.37	8–70	0.08–0.43	1.69–3.33
Northern taiga	45–86	8.9–16.0	45–50	5.9–9.8	0.96–1.20	10–56	0.11–0.93	1.38–5.43
Central taiga	39–107	10.4–22.0	45–52	7.6–12.7	0.96–1.13	7–56	0.05–1.05	2.58–7.01
Southern taiga and mixed forest	41–86	9.0–18.3	41–45	8.2–12.2	0.85–1.02	5–19	0.07–0.66	2.10–4.93
Deciduous forest and forest steppe	22–49	6.6–10.1	37–41	3.7–7.8	0.75–0.89	2–7	0.07–0.35[2]	0.98–2.98[2]
Steppe	24–38	7.6–9.5	34–38	4.0–5.0	0.82[1]–0.86[2]	1.8–4.2	0.02–0.10	0.58–1.04
Semi-desert	22–36	7.0–9.7	27–37	5.0[1]–6.6[2]	0.66[1]–0.78[2]	1.1–5.2	0.004–0.01	0.40–0.53[2]
Desert	12–17	5.0–7.0	28[1]–33	2.4[1]–3.4[2]		1.0–3.7	0.003–0.01	0.31–0.60

OP Susceptibility of organic matter to oxidation by permanganate
OM Organic matter
TDS Total dissolved solids
[1] Data for floods
[2] Data for low water (winter, summer)

Source: Nikanorov, 1989

geographical conditions of the country, the three rivers could not provide a comprehensive picture of background concentrations of micro-elements in river waters. However, Arctic rivers are least affected by anthropogenic impact compared with rivers in other regions of the country. Therefore the data in Table 2.4 can be considered as close to background metal concentrations in the former USSR. In waters of the three Arctic rivers, concentrations of heavy metals such as Fe, Zn, Cu, Ni, Cd are equal to, or less than, the average for world rivers. Concentrations of Pb, Hg, As are considerably lower than mean global values (Table 2.4).

2.2   Lakes

The chemical composition of lakes shows greater variation than rivers due to the specific features of lakes. An increase in water residence time in a lake strengthens the influence of local climatic conditions (e.g. precipitation/evaporation) on chemical composition, primarily the mineralisation of lake water compared with rivers (Alekin, 1970). For endorheic lakes, the ratio of water inflow to a lake from the catchment area to evaporation rate is important.

For a comprehensive study of the variability of lake hydrochemistry, the relationship between chemical processes occurring in natural waters and conditions of the environment which may change the direction and intensity of these processes need to be analysed. This is especially important because of the great variety of physico-geographical features of the former USSR. This relationship is the basis for lake classification by hydrochemical parameters which were defined by Stchukarev (1924), Alekin (1946) and Baranov (1961). According to the classification, five geographic zones may be recognised:

1. A tundra zone usually dominated by siliceous and bicarbonate-siliceous lake waters;
2. A forest zone with dominant bicarbonate–calcium lake waters;
3. A steppe zone with dominant sulphate or bicarbonate–sodium lake waters;
4. Desert and semi-desert zones with dominant chloride–sodium lake waters;
5. Tropic and sub-tropic zones with dominant bicarbonate–siliceous lake waters.

Although each geographic zone corresponds to specific dominant hydrochemical features of lake waters, these features also depend on other factors and can result in a wide range of ion concentrations. According to Kuznetsov (1970), the range of total concentration of ions in lake waters of the tundra zone was 47–62 mg l^{-1}, in the forest zone was 30–1,770 mg l^{-1}, in the steppe zone was 1,230–12,745 mg l^{-1} and in desert and semi-desert zones was 192,000–265,000 mg l^{-1}.

Table 2.4 Concentrations of selected dissolved micro-elements in the delta areas of the Lena, Yenisey and Ob rivers compared with the world's largest rivers

River	Cu (nM)	Pb (nM)	Zn (nM)	Ni (nM)	Cd (nM)	As (nM)	Hg (nM)	Fe (nM)	Source(s)
Lena	9.4	0.08	5.4	5.1	0.03–0.07	2.0	3.0–5.4	410	1,2
Yenisey	21.5–29.5	0.025–0.030	7.8–32.0	8.8–9.4	0.011–0.016	–	0.8–2.1	251–317	3,4,5
Ob	29.1–38.0	0.055–0.083	<1.5–12.3	21.0–23.7	0.005–0.008	–	2.4–3.2	430–654	3,4,5
Amazon	23.8	–	0.3–3.7	5.1	0.09	–	–	117–1,270	6,7
Huang He	9.4–14.2	0.18	–	–	0.04	27.0	–	–	8
Mississippi	22.7	–	2.9	22.7	0.12	–	–	30	9
Orinoco	18.9	–	2.0	3.4	0.14	–	–	–	7
Nile	14.8	0.16	–	–	0.09	–	–	–	10
Mean global value	23.6	0.15	9.2	8.5	0.09	22.7	350	716	11

Sources:
1 Martin et al., 1993
2 Cossa and Coquery, 1993
3 Dai and Martin, 1995
4 Cossa and Coquery, 1994
5 Kravtsov et al., 1994
6 Boyle et al., 1982
7 Edmond et al., 1985
8 Huange et al., 1988
9 Shiller and Boyle, 1987
10 Dorten et al., 1991
11 Martin and Windom, 1991

The majority of lakes in the former USSR are found in northern and north-western regions. They are characterised by low trophic status and low mineralisation. Lakes with particularly low mineral content occur in the area of the Scandinavian shield, created by Precambrian folds, and on tundra. Soil in these regions (peat-marsh, podsol-marsh, gley-podsol, podsol) are characterised by low base-saturation, with a soil solution of pH 4.0–4.5. The mineral content of these lakes which belong to the bicarbonate type, calcium group, is between 15–100 mg l^{-1} and increases from north to south. The concentration of SiO_2 is generally similar to that of SO_4^{2-} although in the south there is an increase in sulphate concentration and the SiO_2 to SO_4^{2-} ratio is < 1.

Humic lakes are characteristic of northern territories. The inflow of humic waters from wetlands lowers the transparency and pH of lake water and increases susceptibility to oxidation and iron concentrations. Phosphorus availability in the water limits the intensity of phytoplankton development. If the lake water contains iron, phosphorus may precipitate with it, and with the humic complexes, and may be deposited. In lakes with little humic substances, the interactions of phosphorus and iron do not occur.

In forest zones, the mineral content of lake waters varies from 100–500 mg l^{-1} and the waters are mainly bicarbonate with a variable Ca^{2+} to Mg^{2+} ratio, especially during periods of intensive algal growth. The proportion of organic matter in water in relation to total mineral content is lower than that of northern regions.

Where chernozem soils and occasionally chestnut soils occur, lakes are characterised by an increase in mineralisation of lake water to 1,900–4,000 mg l^{-1}. The increase in mineralisation is accompanied by an increase in specific ion concentrations as follows: $HCO_3^- \rightarrow SO_4^{2-} \rightarrow Cl^-$ and $Ca^{2+} \rightarrow Mg^{2+} \rightarrow Na^+$. The higher concentration of nitrogen and phosphorus in water bodies of the chernozem zone is due to the high content of phosphorus and nitrogen in this soil type. In these southern lakes, the role of allochthonous organic matter decreases compared with lakes in more northern latitudes. The ratio allochthonous to autochthonous organic matter in the different geographic zones of the former USSR is 11 in forest tundra, 0.9–1.6 in northern taiga, < 0.2 in southern taiga and < 0.1 in forest steppe (Drabkova and Sorokina, 1979).

In steppe regions with chernozem-carbonate and chestnut soils (e.g. Azov-Manych and Caspian lowlands), and where saline soils frequently occur, there is a significant increase in mineralisation of lake waters leading to the formation of natural brine lakes. Lake waters are usually in the chloride-sodium class.

Although the geographical location of a lake influences its hydrochemical regime, other factors must also be considered. The chemical composition of lakes formed in large river deltas and flood plains is influenced primarily by the chemical composition of the river water forming the lakes whereas for mountain lakes, the chemical composition may also be influenced by altitude. The influence of geographic zones and other factors on the hydrochemical regime of major lakes can be seen in Table 2.5. The data on ion concentrations are those of natural conditions before any significant anthropogenic impact.

The regime of nutrient elements and autochthonous organic matter in lakes depends on the intensity of biochemical processes at all levels of the trophic chain. Quantitative indicators of this group are influenced by factors such as lake morphometry, hydrodynamic processes, temperature and stratification. These parameters are important as they define the trophic status of a water body and the general level of their productivity (see Chapter 6).

2.3 Groundwater

The chemical composition of natural groundwaters depends on numerous factors including conditions of recharge and discharge, geology, depth of aquifers, landscape and climatic conditions. In spite of the variety of chemical composition and mineralisation, groundwaters are characterised by vertical trends at a given location and by spatial variations between regions.

2.3.1 Spatial variations in chemical composition

In terms of groundwater chemistry, two zones have been defined: leaching and continental salinisation. The first zone is typical of regions with excess moisture and of regions with less humidity but good natural drainage; the second is characteristic of dry steppes, semi-deserts and deserts. In the first zone, mineralisation of groundwater ranges from 0.1 to 10 g l^{-1}. In the second zone, the mineral content of groundwaters significantly varies from fresh to saline, depending on the lithology of rocks and geomorphology of the area.

The leaching zone has a southern limit where there is an appropriate balance between evaporation and precipitation and near which groundwaters have a relatively high mineral content with concentrations of $HCO_3^- + CO_3^{2-} > 300$ mg l^{-1} and relatively high Eh-values (up to +400 mV). In northern humid regions, waters have a low mineral content (< 200 mg l^{-1}) but have high concentrations of fulvic and amino acids (up to 100 mg l^{-1}) and a high content of dissolved CO_2. The most important features of groundwaters in this zone are: i) high absolute and relative concentrations of aggressive CO_2, which characterise the high reactivity of these waters with rocks; ii) strong under-saturation with calcium carbonate and sulphate, which determines their high ability to dissolve carbonate rocks; iii) intensive formation of complex compounds with humic organic substances, influencing the migration of iron in water and its accumulation in groundwaters (Krainov and Shvets, 1987).

Table 2.5 Chemical composition of large lakes in different geographic zones of the former USSR

Geographic zone	Lake	Latitude (°N)	Altitude (m)	Area (10³ km²)	Sum of ions (mg l⁻¹)	Ca²⁺ (mg l⁻¹)	Mg²⁺ (mg l⁻¹)	Na⁺ + K⁺ (mg l⁻¹)	HCO₃⁻ (mg l⁻¹)	SO₄²⁻ (mg l⁻¹)	Cl⁻ (mg l⁻¹)	Si (mg l⁻¹)	Source
Tundra	Taimyr	73	6	4.56	–	6.6	2.2	–	10.2	5.4	–	0.7	1
Moderate forest zone	Ladoga	61	4	17.7	58.7	8.0	3.2	4.6	30.6	7.9	4.4	0.9	2
	Onega	61	33	9.63	34.5	4.8	2.0	1.5	20.5	4.2	1.5	0.7	3
	Ilmen	58	19	1.20	128.0	27.4	1.7	10.3	64.3	7.8	16.8	1.4	4
	Pskov-Chud	58	30	3.55	163.0	23.9	5.2	11.5	112.0	4.0	5.2	–	5
	Baikal	53	545	51.5	96.4	15.2	3.1	5.8	66.5	5.2	0.6	1.1	6
	Teletsk	52	434	0.22	68.4	12.4	2.1	1.7	48.6	2.8	0.8	2.3	5
Forest/steppe	Chany	55	105	2.5	6,340	20.0	480	1,770	638	406	3,030	4.6	7
Steppe	Balkhash	46	340	18.2	2,843	26.0	164	694	444	893	574	3.8	8
	Aral Sea	45	100	64.3	10,054	479	538	2,249	158	3,169	3,461	–	8
	Caspian Sea	41	–28	374	12,744	344	730	3,244	111	2,996	5,321	–	8
Mountain/steppe	Issyk-Kul	51	1,610	6.2	5,823	114	294	1,475	240	2,115	1,535	–	8
	Sevan[1]	40	1,914	1.42	682	33.9	55.9	98.7	55.9	16.9	62.9	–	5

[1] Before lowering of water level

Sources:

1 Baranov, 1961

2 Raspletina et al., 1967

3 Forsh-Menshutkina, 1973

4 Smirnova, 1974

5 Alekin, 1970

6 Votintsev and Glazunov, 1963

7 Zhernovskaya, 1982

8 Kuznetsov, 1970

The zone of continental salinisation may be characterised by evaporation rates exceeding precipitation rates which may be followed by concentration of groundwaters. This process is rather uneven and depends on the ratio of horizontal outflow to evaporation. The important geochemical features in this zone are: i) minimum concentration of dissolved and aggressive CO_2 with neutral or weak alkaline pH; ii) low concentrations of organic substances; iii) a shift in equilibrium of the system 'organic acids'-OH^- to OH^- (Krainov and Shvets, 1987).

2.3.2   Vertical variations in chemical composition

The chemical composition of groundwater in a vertical profile is the result of complex interactions of groundwater flow which depend on the horizontal and vertical filtration rates of mountain rocks. Mineralisation and the different types of groundwater reflect, in part, the filtration properties of water-containing rocks.

In different hydrogeological structures, vertical hydrochemical trends vary. Within sedimentary platforms, the trend in geochemical types of groundwater is $HCO_3^--Ca^{2+} \rightarrow SO_4^{2-}-Ca^{2+}-Na^+ \rightarrow SO_4^{2-}-Cl^--Na^+ \rightarrow Cl^--Na^+$. In mountainous areas, the trend is $HCO_3^--Ca^{2+} \rightarrow HCO_3^--Ca^{2+}-Na^+ \rightarrow HCO_3^--Na^+ \rightarrow HCO_3^--Cl^--Na^+ \rightarrow Cl^--HCO_3^--Na^+$.

In arid zones, hydrogeological gradations may be reversed as a result of continental salinisation. In shallow aquifers, sub-saline and saline waters occur whereas in deeper aquifers freshwater occurs. In mountains with folded geological strata, vertical hydrogeochemical gradations vary depending on the age and composition of water-containing rocks. All transformations in chemical composition of groundwaters during the formation of geochemical areas directly influence the concentration of chemical elements (Krainov and Shvets, 1987). As a result, each hydrogeological structure has a specific chemical composition of groundwaters. This in turn limits the use of groundwater.

2.3.3   Regional hydrochemistry

The hydrogeochemical characteristics of groundwaters of the main hydrogeological areas (artesian and folded regions) of the former USSR are presented below (Kirukhin and Tolstikhin, 1987). The main characteristics of major aquifers in the European part of the former USSR are presented in Table 2.6.

East-European artesian region

The East-European artesian region is characterised by three types of groundwater: fresh (TDS < 1 g l^{-1}), brackish (TDS 1–35 g l^{-1}) and saline (TDS 35–350 g l^{-1}). The greatest thickness of the freshwater zone (up to 500 m) is found west of the Dnieper-Pripyat artesian basin. In other basins it is 200–300 m. The least thickness is found in the Caspian artesian basin where it is not more than 50 m. In some areas of this basin

Table 2.6   Components of chemical composition of groundwaters which limit their use in various regions of the European part of the former USSR

Artesian basin	Aquifers and complex aquifers	Components of groundwater
Baltic-Polish[1]	Quaternary	Organic matter, Fe, Mn, NO_3^-, NO_2^-, NH_4^+, SO_4^{2-}, Cl^-, H_2S, F^-, hardness
	Paleogenic	Organic matter, Fe, Mn
	Upper Cretaceous	Cl^-, SO_4^{2-}, mineralisation
	Jurassic	Cl^-, F^-, mineralisation
	Permian	SO_4^{2-}, H_2S, Sr, Fe, F^-, hardness
	Carboniferous	Cl^-, SO_4^{2-}, H_2S, Sr, mineralisation, hardness
	Bureg-Sargaev (Devonian)	SO_4^{2-}, H_2S, Sr, Fe, mineralisation, hardness
	Shventoy-Tartu (Devonian)	SO_4^{2-}, Cl^-, Sr, Fe
	Piarnu (Devonian)	Cl^-, Fe, mineralisation, hardness
	Silurian	Cl^-, SO_4^{2-}, Sr, F^-, Fe, Zn, NO_3^-, NO_2^-, mineralisation, hardness
	Ordovician	Cl^-, SO_4^{2-}, Sr, F^-, Fe, Zn, mineralisation
	Cambro-Ordovician	Cl^-, SO_4^{2-}, Fe, F^-, mineralisation
Central Russian[2]	Quaternary	Organic matter, NO_2^-, NO_3^-, NH_4^+, Fe, Mn, mineralisation, hardness
	Upper Cretaceous	Organic matter
	Lower Cretaceous	Organic matter, Cl^-, SO_4^{2-}, hardness
	Jurassic	SO_4^{2-}, H_2S, Fe, NO_2^-, NO_3^-, NH_4^+, hardness
	Lower Triassic	SO_4^{2-}, Cl^-, Sr, F^-, mineralisation, hardness
	Upper Permian	SO_4^{2-}, Cl^-, Sr, F^-, mineralisation, hardness
	Lower Permian	SO_4^{2-}, Cl^-, Sr, F^-, mineralisation, hardness
	Upper Carboniferous-Upper Gzhel	SO_4^{2-}, Cl^-, Fe, F^-, mineralisation, hardness
	Lower Gzhel	SO_4^{2-}, Cl^-, Fe, F^-, mineralisation, hardness

[1] Estonia, Latvia, Lithuania, Byelorussia, Western Ukraine

[2] Moscow, Tver, Yaroslavl, Vladimir, Ryazan, Tula, Kaluga, Smolensk region

freshwaters are completely absent. In the freshwater of some artesian basins (e.g. Moscow, Sura-Khoper) increased fluorine concentrations (up to 3–5 mg l^{-1}) and increased iron and manganese concentrations (up to 5–10 and 0.2–1 mg l^{-1} respectively) are observed. Where gypsum layers and groundwaters rich in organic matter occur close to the surface, waters with high sulphate concentration occur which may be of therapeutic importance (e.g. Kemeri spa in Latvia). In this artesian region, in deep regions of saliferous formations, brines containing bromide (up to 3 g l^{-1}) and iodide (up to 90 mg l^{-1}) occur. The gaseous composition of groundwater is dominated by nitrogen. In regions of gas and oil deposits, gas saturation levels may reach 1–4 l l^{-1} compared with a background level of 0.3–0.5 l l^{-1}. In these regions, methane dominates the gas

Table 2.6   Continued

Artesian basin	Aquifers and complex aquifers	Components of groundwater
Central Russian continued[2]	Middle Carboniferous Myachkovo-Podolsk	SO_4^{2-}, Cl^-, Fe, F^-, mineralisation, hardness
	Kashirsk (Carboniferous)	SO_4^{2-}, Cl^-, Fe, F^-, mineralisation, hardness
	Lower Carboniferous Protvin	SO_4^{2-}, Cl^-, Fe, Sr, F^-, mineralisation, hardness
	Serpukhov Oka (Carboniferous)	SO_4^{2-}, Cl^-, Fe, Sr, F^-, H_2S, mineralisation, hardness
	Yasnopolyansk (Carboniferous)	SO_4^{2-}, Cl^-, Fe, Sr, F^-, H_2S, mineralisation, hardness
	Upinsk (Carboniferous)	SO_4^{2-}, Cl^-, Fe, Sr, F^-, H_2S, mineralisation, hardness
	Devonian	SO_4^{2-}, Cl^-, mineralisation, hardness
Dnieper-Donetsk[3]	Quaternary	SO_4^{2-}, Cl^-, NO_3^-, mineralisation, hardness
	Poltava (Devonian)	SO_4^{2-}, mineralisation
	Kharkov (Devonian)	Mineralisation, hardness
	Buchagsko-Kanevsk (Devonian)	Cl^-, SO_4^{2-}, F^-, H_2S, mineralisation
	Upper Cretaceous	Cl^-, SO_4^{2-}, mineralisation, hardness
	Lower Cretaceous	Cl^-, SO_4^{2-}, Fe, mineralisation
	Jurassic	SO_4^{2-}, Cl^-, mineralisation
Moldovian[4]	Quaternary	SO_4^{2-}, NO_2^-, NO_3^-, mineralisation, hardness
	Middle Sarmatian (tertiary)	F^-, mineralisation
	Lower Sarmatian (tertiary)	H_2S, F^-, Sr, mineralisation
	Upper Cretaceous	F^-, H_2S, mineralisation
	Silurian	F^-, mineralisation
	Pre-Silurian	Cl^-, F^-, mineralisation

[3]   Ukraine: Kiev, Kharkov, Lugansk, Poltava, Chernigov, Sumsk region

[4]   Moldova

composition with an admixture of heavy hydrocarbons, hydrogen sulphide and carbon dioxide. The temperature of these groundwaters ranges from 0 °C to 10 °C with a geothermal gradient of 30–40 m per °C.

East-Siberian artesian region
The East-Siberian artesian region is influenced by: extreme continental climate with low precipitation and negative mean annual air temperatures; the presence of large rivers (e.g. the Lena, Yenisey, Angara); widespread permafrost with a thickness up to 1,000 m which cover more than 90 per cent of the territory; great capacities of carbonate and saliferous rocks; and widespread tectonic activity.

The zone of freshwaters in this region has a thickness up to 400 m in the southern part and up to 4 km in the region of the Vilyui syncline. Vertical hydrochemical gradation is related to increasing freshwater due to degradation in the permafrost zone. Groundwaters in such areas are ultra-fresh with mineralisation < 100 mg l^{-1}. Groundwaters of alluvial aquifers in river valleys have increased concentrations of iron and manganese. The gaseous composition of groundwater in upper layers are dominated by an oxygen-nitrogen composition, in mid-layers by nitrogen whereas in deep layers methane dominates. In this artesian region, waters with a negative temperature but in a liquid state are widespread and usually have a $Cl^--Ca^{2+}-Mg^{2+}$ composition. The greatest thickness of these zones (up to 1,472 m) is found in the Olenek artesian basin.

West-Siberian artesian region
The West-Siberian artesian region is characterised by a large area (c. 3 million km^2) with a variety of landscapes from tundra to steppe. The total thickness of the sediment layer reaches 7 km. Freshwater occurs mainly in Oligocene-Quaternary formations and in the south in Cretaceous formations. The maximum thickness of this zone is 1.8 km. The anionic composition of the water is mainly bicarbonate with different combinations of cations. In southern arid zones, there is a change in composition of groundwaters: $HCO_3^--Ca^{2+} \rightarrow HCO_3^--Na^+ \rightarrow SO_4^{2-}-Cl^--Na^+ \rightarrow Cl^--Na^+$, and continental salinisation leads to the formation of groundwater of uneven composition ($SO_4^{2-}-Cl^--Na^+$) and mineralisation (5–20 g l^{-1}). In deep aquifers, high concentrations of bromide, iodide and boron often occur. Ammonium, potassium and organic acids are also present. There are also increased concentrations of iron (up to 150 mg l^{-1}) and silica (up to 400 mg l^{-1}). Gas composition of groundwater varies with depth, with upper layers containing oxygen and nitrogen, lower layers containing nitrogen and the deepest layers containing methane which, in some areas, nearly reach the surface.

In the latitudes north of 56–60 °N, a permafrost zone with a thickness of 250–500 m occurs. To the south, there is a zone with two frozen layers, one up to 80 m and the other at 150–300 m with a liquid zone in between. This zone of water (0–20 °C) has a thickness of 250–400 m to 1,000 m in the south-east of the region. There is no permafrost in valleys of large rivers.

Aral artesian region
The Aral artesian region has an area of about 1.8×10^6 km^2 and is characterised by plains, with steppe, semi-desert and desert landscape, extreme continental climate with low precipitation (100–300 mm a^{-1}) and high evaporation rates. Most of the zone belongs to the internal drainage basin of the Aral Sea. The upper aquifers are exposed to salinisation. This is most clearly seen in the upper 20 m of the profile and can

extend up to 100 m. Waters of sodium chloride composition with mineralisation of 10–35 g l^{-1} dominate.

In areas of Neogene-Quaternary sediments, freshwaters with mineralisation of 1–1.5 g l^{-1} of sulphate bicarbonate and calcium composition are distributed. In many artesian basins of this region freshwater aquifers lie below more mineralised aquifers with varied composition and mineralisation. Thus, freshwaters are found 2.2 km deep in the Syr Darya artesian basin and 0.7–1.0 km deep in the Turgai artesian basin.

Deeper aquifers are saline waters of SO_4^{2-}–Cl^- and Cl^-–HCO_3^- composition and brines containing bromide (up to 2 g l^{-1}), potassium (up to 60 g l^{-1}), iodide (up to 60 mg l^{-1}), etc. Brines also contain metals such as zinc, strontium, lithium, barium, rubidium.

Vertical gradation of gases is also observed. Oxygen- and nitrogen-rich waters (with gas saturation of 0.02 1 l^{-1}) have a thickness of 50–100 m and less frequently to 300 m. Nitrogen-rich waters with gas saturation of 0.03–0.04 1 l^{-1} have a thickness of 1 km on the borders of the artesian basins. In southern areas, waters with gases of nitrogen–methane composition are characteristic.

Caspian-Black Sea artesian region
The Caspian-Black Sea artesian region has specific features which include the mountainous area in the south (the Crimea, Caucasus), the presence of deep mountain folds along the alpine orogenesis mountains (Terek-Caspian, Azov-Kuban, Pre-Dobzodzha, etc.), the arid climate, the presence of karst and intensive hydro-volcanic and neotectonic activity.

The greatest thickness of the freshwater zone (up to 1.9 km) is found in the Caucasus and Crimea. For the rest of the territory, it is not more than 200 m. The composition of the water is HCO_3^-–Na^+–Ca^{2+}. In steppe regions the water is enriched with SO_4^{2-} and Cl^-. In areas of continental salinisation (e.g. Sivash plain, Kerch Peninsula and endorheic depressions) there is no freshwater. Saline waters with a thickness of several kilometres are composed of HCO_3^-–Cl^-–Na^+ with increased concentrations of iodide, bromide, boron, organic substances, ammonium, silica and occasionally fluorine (Moldova). The gas composition of waters is greatly influenced by oil and gas-containing rocks in pre-mountain areas of North Caucasus, the Crimea, Northern Black Sea and Caspian Sea areas. Carbon dioxide entering along tectonic breaks forms deposits of mineral carbon dioxide ferrous waters (e.g. Narzan, Yessentuki).

Folded regions
Folded regions may be grouped according to their tectonic age as follows: the oldest (Dorifey) — Baltic, Ukranian, Anabar, Aldan; ancient (Baikal, Caledonian, Hercynian orogeneses) — Yenisey, Kazakhstan, Ural, Novaya Zemlya, Taimyr, Donets; ancient-rejuvenated — Pre- and Trans-Baikal, Sayans, Altai, Tian-Shan, Alai, Amur-Okhotsk; young

(Cimmerian and Alpine orogeneses) — Carpathian, Crimea, Caucasus, Kopetdag, Pamir, Sikhote-Alin, Koryak, Kamchatka, Kuril (Kirukhin and Tolstikhin, 1987).

Specific features of these regions which influence composition and mineralisation of groundwaters include high altitude and rugged physical relief, relatively good water movement and drainage, rock, soil and water interactions, the presence of metamorphic and igneous rocks and numerous tectonic breaks and fissures. Upper aquifers are mainly fresh. The level of mineralisation and composition is determined by geography and altitude. Waters are mainly of the $HCO_3^--Ca^{2+}$ type and mineralisation varies between 0.02 and 1 g l^{-1}. In regions of gypsiferous or saliferous formations, groundwater is enriched by sulphates and chlorides. In arid areas, mineralisation of groundwater increases. In regions of volcanic activity, groundwater quality may be contaminated with volcanic products and gas-hydrothermal processes. The gas component (mainly carbon dioxide, nitrogen, helium, radon) in groundwater is related to gas composition and the gases entering the water through tectonic breaks.

The natural composition of groundwaters may greatly limit their economic use. A summary of these limitations for the European part of the former USSR is presented in Table 2.6

2.4   References

(All references are in Russian unless otherwise stated).

Alekin, O.A. 1946 On the climatic classification of natural waters. Proceedings of NIUGUGMS, **IV**(32), 20–26.

Alekin, O.A. 1948 Hydrochemical classification of USSR rivers. Proceedings of the State Hydrological Institute, 4(58), 32–69.

Alekin, O.A. 1950a Hydrochemical types of USSR rivers. Proceedings of the State Hydrological Institute, **25**(79), 5–21.

Alekin, O.A. 1950b Research on the quantitative relations between mineralisation, ionic content and water regime of USSR rivers. Proceedings of the State Hydrological Institute, **25**(79), 30–43.

Alekin, O.A. 1970 *Fundamentals of Hydrochemistry*. Gidrometeoizdat, Leningrad, 444 pp.

Almazov, A.M. 1952 On relationships between hydrochemical and hydrological river regimes. Academy of Sciences of Ukraine Reports, **3**, 208–212.

Baranov, I.V. 1961 *Limnological Types of USSR Lakes*. Gidrometeoizdat, Leningrad, 276 pp.

Blinov, L.K. 1946 The dependence of river water mineral content on hydrological factors. *Meteorology and Hydrology*, **6**, 43–50.

Bochkarev, P.F. 1959 *Hydrochemistry of Eastern Siberian Rivers*. Book Publishing House, Irkutsk, 155 pp.

Boyle, E.A., Huested, S.S. and Grant, B. 1982 The chemical mass-balance of the Amazon plume. II. Copper, nickel and cadmium. *Deep Sea Research*, **29**(11A), 1355–1364. (In English).

Burkaltseva, M.A. 1964 Methods of ionic river discharge calculation (example: Fergana valley rivers). *Moscow State University Newsletters Geographical Series*, **1**, 121–128.

Cossa, D. and Coquery, M. 1993 Mercury in the Lena delta and Laptev Sea. The Arctic Estuaries and Adjacent Seas: Bio-geochemical Processes and Interaction with Global Change. Third International Symposium, Svetlogorsk, Russia 19–25 April 1993. Abstracts Kaliningrad, p. 11–12. (In English).

Cossa, D. and Coquery, M. 1994 Personal communication.

Dai, M.-H. and Martin, J.-M. 1995 First data on the trace metal level and behaviour in two Arctic river/estuarine systems (Ob and Yenisey) and the adjacent Kara Sea (Russia). *Earth Planet. Sci. Lett.*, **131**, 127–141. (In English).

Dorten, W., Elbaz-Poulichet, F., Mart, L. and Martin, J.-M. 1991 Reassessment of the river input of trace metals into the Mediterranean Sea. *Ambio*, **20**(1), 2–6. (In English).

Drabkova, V.G. and Sorokina, I.N. 1979 *Lakes and their Catchment Area - the Unified Natural System*. Nauka, Leningrad, 194 pp.

Edmond, Y.M., Spivack, A., Grant, B.C., Hum, H., Chen, Z.X., Chen, S. and Zong, X.S. 1985 Chemical dynamics of the Changjiang estuary. *Continental Shelf Research*, **4**(1/2), 17–36. (In English).

Emelyanova, V.P. and Danilova, G.N. 1979 Hydrochemical maps of USSR rivers. *Hydrochemical Materials*, **75**, 3–10.

Fadeev, V.V., Tarasov, M.N. and Pavelko, V.L. 1976 The relationship between hydrochemical and water regimes of lowland and mountain rivers of the USSR. In: Proceedings of the IV Hydrological Congress. Volume 9. Water Quality and the Scientific Basis for its Protection. Gidrometeoizdat, Leningrad, 198–212.

Fadeev, V.V., Tarasov, M.N. and Pavelko, V.L. 1989 *Dependence of River Water Mineralisation and Ionic Content on Water Regime*. Gidrometeoizdat, Leningrad, 176 pp.

Forsh-Menshutkina, T.B. 1973 Hydrochemistry of Lake Onega. In: *Hydrochemistry of Lake Onega and its Tributaries*. Nauka, Leningrad, 130–236.

Gibbs, R. 1970 Mechanisms controlling world water chemistry. *Science*, **170**, 1088–1090. (In English).

Huange, W.W., Martin, J.-M., Seyler, P., Zhang, J. and Zhong, X.M. 1988 Distribution and behaviour of arsenic in the Huang He (Yellow River) Estuary and Bohai Sea. *Marine Chemistry*, **25**, 75–91. (In English).

Hydrochemical Atlas of the USSR 1990 Hydrochemical Institute, Central Board of Geodesy and Cartography, Moscow, 111 pp.

Ivanik, V.M. and Veselovsky, N.V. 1974 Hydrochemical types of rivers in the Lena-Indigirka region. *Hydrochemical Materials*, **61**, 3–13.

Kalyuzhny, I.L. and Levandovskaya, L.Y. 1976 The features of formation of the hydrochemical regime, chemical content and quality of waters of upper bogs of some wetland areas of the USSR. In: Proceedings of the IV Hydrological Congress. Volume 9. Water Quality and the Scientific Basis for its Protection. Gidrometeoizdat, Leningrad, 292–304.

Kirukhin, V.A. and Tolstikhin, N.I. 1987 *Regional Hydrogeology*. Nedra, Moscow, 382 pp.

Konovalov, G.S., Manikhin, V.I. and Melnikova, L.N. 1977 Changes in hydrochemical composition of high-mountain rivers over 24 hours (example: Garabashi river). *Hydrochemical Materials*, **64**, 29–45.

Koreneva, V.I. 1977 Sodium and potassium in river waters of the Soviet Union. Candidate of Science (Geol.) Thesis. Hydrochemical Institute, Novocherkask, 216 pp.

Korobeinikova, N.O., Manikhin, V.I. and Konovalov, G.S. 1977 The ratio of main recharge sources of the Belaya river on the basis of hydrological observations. In: Papers on Hydrology of the State Hydrological Institute, No. 12, 132–136.

Krainov, S.R. and Shvets, V.M. 1987 *Geochemistry of Ground Waters for Industrial and Domestic Use*. Nedra, Moscow, 237 pp.

Kravtsov, V.A., Gordeev, V.V. and Pashkina, V.I. 1994 Dissolved forms of heavy metals in Kara Sea waters. *Oceanology*, **34**(5), 673–680.

Kuznetsov, S.I. 1970 *Lake Microflora and its Geochemical Activity*. Nauka, Leningrad, 440 pp.

Lazarev, K.G. 1957 *Hydrochemical Study of the Lowland Region of the Amu Darya River*. Academy of Science, Moscow, 185 pp.

Martin, J.-M. and Windom, H. 1991 Present and future role of ocean margins in regulating marine bio-geochemical cycles of trace elements. In: R.F.C. Mantoura, J.-M. Martin, R. Wollast [Eds] Proceedings of the Dahlem Conference on Marginal Seas Process in Global Change, Volume 4. Wiley and Sons, 45–67. (In English).

Martin, J.-M., Guan, D.M., Elbaz-Puolichet, F., Thomas, A.J. and Gordeev, V.V. 1993 Preliminary assessment of the distribution of some trace elements (As, Cd, Cu, Fe, Ni, Pb and Zn) in a pristine aquatic environment: the Lena River estuary (Russia). *Marine Chemistry*, **43**, 185–199. (In English).

Nikanorov, A.M. 1989 *Hydrochemistry*. Gidrometeoizdat, Leningrad, 351 pp.

Nikanorov, A.M. and Tsirkunov, V.V. 1991 Hydrochemical regime of USSR rivers. Analysis of long-term data. In: Proceedings of the IV Hydrological Congress, Volume 9. Water Quality and the Scientific Basis for its Protection. Gidrometeoizdat, Leningrad, 336–343.

Raspletina, G.F., Ulanova, D.S. and Sherman, E.E. 1967 Hydrochemistry of Lake Ladoga. In: *Hydrochemistry and Hydro-optics of Lake Ladoga.* Nauka, Leningrad, 60–122.

Shiller, A.M. and Boyle, E.A. 1987 Variability of dissolved trace metals in the Mississippi River. *Geochim. Cosmochim. Acta*, **51**, 3273–3277. (In English).

Skakalsky, B.G. 1983 Hydrochemical zones of local water runoff in the European territory of the USSR. In: *The Problems of Modern Hydrology.* Gidrometeoizdat, Leningrad, 120–130.

Skopintsev, B.A. 1941 Under-ice hydrochemical regime of the Volga river and some rivers of the Volga basin. *Hydrochemical Materials*, **12**, 159–168.

Smirnov, M.P. and Tarasov, M.N. 1977 Maps of long-term average seasonal indices of colour of USSR river waters. *Hydrochemical Materials*, **66**, 3–10.

Smirnov, M.P. and Tarasov, M.N. 1982 Maps showing susceptibility to oxidation by bichromate of USSR river waters. *Hydrochemical Materials*, **89**, 3–12.

Smirnova, L.F. 1974 Hydrological and hydrochemical regimes of Lake Ilmen. In: *Theoretical Basis of Fisheries Regulation in Inland Water Bodies.* Newsletters of the State Research Institute of Fishery Protection (GOSNIIORKh), Leningrad. Volume 86, 26–40.

Stchukarev, S.A. 1924 General review of Georgian waters from a geochemical point of view. Proceedings of the State Central Research Institute for Spa Studies, **5**, 13–18.

Tarasov, M.N. and Beschetnova, E.I. 1987 Hydrochemistry of the Lower Volga river under impoundment (1935–1980). *Hydrochemical Materials*, **101**, 120 pp.

Tsirkunov, V.V. 1985 Evaluation techniques and trends in basic components of hydrochemical regime and ionic river transport. Candidate of Science (Geogr.) Thesis. Hydrochemical Institute, Rostov-on-Don, 260 pp.

USSR Surface Water Resources 1972a The Lena-Indigirka Region. In: *Hydrochemical Characteristics of Surface Waters.* Volume 17. Gidrometeoizdat, Leningrad, 380–435.

USSR Surface Water Resources 1972b Northern Regions. In: *Hydrochemical Characteristics of Surface Waters.* Volume 3. Gidrometeoizdat, Leningrad, 293–351.

USSR Surface Water Resources 1973 Northern Caucasus. In: *Chemical Composition of Waters.* Volume 8. Gidrometeoizdat, Leningrad, 294–334.

Veselovsky, N.V., Danilova, G.N. and Manikhina, R.K. 1975 Major ion regime and water mineralisation in the Don river and its tributaries. *Hydrochemical Materials*, **62**, 18–31.

Voronkov, P.P. 1963 Hydrochemical explanation to distinguish local runoff and the method for dividing its hygrograph. *Meteorology and Hydrology*, **8**, 21–28.

Voronkov, P.P. 1970 *Hydrochemistry of Local Runoff in the European Territory of the USSR*. Gidrometeoizdat, Leningrad, 188 pp.

Votintsev, K.K. and Glazunov, I.V. 1963 Hydrochemical regime of Lake Baikal near Listvennichnoe village. In: *Hydrochemical Research of Lake Baikal*. Nauka, Moscow, 3–56.

Zenin, A.A. 1965 *Hydrochemistry of the Volga River and its Reservoirs*. Gidrometeoizdat, Leningrad, 259 pp.

Zhernovskaya, L.F. 1982 Particular aspects of hydrochemistry of Lake Chany. In: *Pulsing Lake Chany*. Nauka, Leningrad, 198–215.

Appendix 2

Table I   Average annual flow-weighted concentrations of major ions for rivers of the main hydrochemical types of the former USSR

Table II Long-term assessments of median ($X_{50\%}$) and variation ($X_{5\%}$, $X_{95\%}$) in concentrations of nutrients for the main rivers of the former USSR

Appendix 2 Table I Average annual flow-weighted concentrations of major ions for rivers of the main hydrochemical types of the former USSR

River/monitoring station	Catchment area (km²)	Water discharge (km³ a⁻¹)	Type of hydrochemical regime[1]	Period of observation	Chemical composition (mg l⁻¹)						TDS (mg l⁻¹)
					Ca^{2+}	Mg^{2+}	$Na^+ + K^+$	HCO_3^-	SO_4^{2-}	Cl^-	
Barents and White Seas											
Kola/1,429 km	3,780	1.40	EE–C,Cl	1955–59	3.2	0.6	2.6	11.3	2.5	2.8	23.1
Onega/Porog	55,700	16.8	EE–C,S	1955–63	22.3	7.1	2.3	72.3	22.1	3.0	133
Northern Dvina/Ust-Pinega	348,000	101	E–C,S	1947–75	30.0	7.2	6.0	83.0	38.7	5.6	170
Pechora/Ust-Tsilma	248,000	110	EE–C	1939–90	10.4	2.8	4.0	38.9	8.0	4.1	68.8
Mezen/Malonisogorsk	56,400	20.0	EE–C	1963–89	14.9	4.2	6.6	62.3	10.7	3.3	102
Kara Sea											
Ob/Salekhard	2,430,000	400	S–C	1955–90	18.6	5.1	6.3	78.0	8.5	6.5	121
Northern Sosva/Kultbaza	65,200	19.3	S–C	1953–75	5.6	2.5	4.2	21.5	6.7	5.6	46.2
Angara/Boguchany	832,000	102	S–C	1953–62	20.7	4.0	7.7	77.2	10.4	6.2	126
Biryusa/Biryusinsk	184,000	8.46	S–C	1946–75	12.4	2.3	4.8	42.6	4.6	7.1	73.7
Selenga/Mostovoi	440,200	28.1	S–C	1950–60	22.7	3.8	7.5	90.7	8.9	2.6	136
Tobol/Kustanai	28,000	0.43	K–C,Cl	1947–54	55.7	14.8	55.7	138	68.2	90.4	423
Yenisey/Igarka	2,440,000	562	S–C	1952–90	16.5	3.7	6.2	57.3	10.0	9.5	107
Laptev Sea											
Lena/Kusur	2,430,000	523	S–C,Cl	1953–89	14.8	5.0	13.2	47.0	14.0	20.1	116
Namana/Myakinda	16,600	0.96	S–Cl	1956–75	22.5	6.1	59.0	53.6	26.4	90.3	259
Vitim/Bodaibo	186,000	47.1	S–C	1949–75	6.2	1.3	0.9	20.2	3.8	1.1	33.7
Biryuk/Biryuk	9,700	0.69	S–S,C	1955–75	80.5	20.0	19.6	95.9	160	42.0	415
Yana/Dyangky	216,000	28.9	S–C	1962–75	10.0	2.5	1.4	31.1	8.0	2.0	55.1
East-Siberian and Chukchi Seas											
Indigirka/Vorontsovo	305,000	49.7	S–C	1962–75	12.1	2.6	1.7	38.4	7.2	1.6	63.7
Nera/Andyguichan	21,800	3.74	S–S,C	1945–75	6.3	1.5	1.9	14.2	11.3	1.2	36.2
Amguema/area of the River Shumnaya	26,400	8.89	S–C	1961–75	3.0	0.6	2.0	8.6	4.8	1.5	20.4
Kolyma/Srednekolymsk	361,000	67.5	S–C	1955–75	11.0	2.3	2.5	37.0	8.2	1.7	62.7
Kirenga/Shorokhovo	46,500	20.5	S–Cl,C	1960–75	20.5	5.7	24.2	58.9	24.9	34.7	169
Ayan-Yuryakh/Emteguei	9,560	2.13	S–C,S	1953–88	3.4	1.1	1.0	10.5	6.9	1.2	26.8

Continued

Appendix 2 Table I Continued

River/monitoring station	Catchment area (km²)	Water discharge (km³ a⁻¹)	Type of hydrochemical regime[1]	Period of observation	Chemical composition (mg l⁻¹)						TDS (mg l⁻¹)
					Ca^{2+}	Mg^{2+}	Na^++K^+	HCO_3^-	SO_4^{2-}	Cl^-	
Bering Sea											
Anadyr/Snezhnoye	106,000	31.1	F–C	1952–88	3.8	1.5	1.3	19.5	4.8	2.4	33.3
Kamchatka/Klyuchi	45,600	25.5	F–C,S	1957–90	8.9	4.7	5.8	46.0	14.2	4.7	90.0
Seas of Okhotsk and Japan											
Amur/Komsomolsk	1,730,000	316	F–C	1950–58	8.2	1.6	4.0	34.2	5.0	1.4	54.1
Penzhyna/Kamenskoe	71,600	22.7	F–C,S	1961–90	4.3	1.2	1.5	15.8	7.2	1.3	33.2
Taui/Talon	25,100	11.5	F–C	1961–75	3.3	0.7	2.7	11.6	3.7	2.1	24.0
Tumnin/Tumnin	13,900	5.29	F–C	1955–75	5.3	1.3	4.0	22.6	5.1	1.9	40.2
Bureya/Kamenka	67,400	28.0	F–C	1949–75	4.5	1.1	2.7	18.5	3.8	1.4	31.9
Baltic Sea											
Neman/Sovetsk	91,800	18.0	EE–C	1955–62	53.3	12.0	7.0	203	16.8	7.5	301
Venta/Kuldiga	13,200	2.06	EE–C	1947–60	57.8	12.5	4.2	217	13.3	4.5	311
Lielupe/Elgava	12,000	2.30	EE–C,S	1947–60	87.8	13.0	2.5	223	83.6	4.3	445
Gauya/Valmiera	6,150	1.61	EE–C	1947–60	46.7	9.8	1.6	179	11.7	2.9	252
Daugava/Daugavpils	64,500	15.0	EE–C	1947–60	33.9	6.5	1.5	139	6.4	1.2	193
Neva/Novosaratovka	281,000	75.6	EE–C	1946–55	9.3	2.0	2.4	28.0	5.2	4.5	51.4
Black and Azov Seas											
Don/Aksai	420,000	21.5	EE–C,S,Cl	1949–58	59.8	17.6	51.5	182	88.1	52.3	451
Kalaus/Svetlograd	4,540	0.12	EE–S,Cl	1948–54	162	99.7	446	230	1021	325	2301
Derekoika/Yalta	49.7	0.02	B–C	1949–75	66.0	12.3	16.6	221	43.9	15.3	375
Dnieper/Kiev	239,000	40.9	B–C	1946–60	40.7	5.7	4.3	144	10.2	3.1	209
Bzyb/Dyirkhva	1,410	2.95	T–C	1950–86	26.2	4.1	4.51	95.6	12.3	1.8	151
Rioni/Sakochakidze	13,300	13.2	T–C	1963–70	39.1	6.2	9.0	135	21.4	3.6	216
Inguri/Darcheli	3,640	4.03	T–C	1963–80	33.2	10.9	5.9	37.8	9.9	2.2	138
Kodory/Varcha	1,990	4.10	T–C	1966–72	15.8	2.3	4.8	54.3	12.5	1.7	90.5
Southern Bug/Alexandrovka	46,200	2.36	B–C	1949–58	64.8	17.1	32.9	299	26.4	16.4	460
Chamlyk/Voznesenskaya	569	0.04	T–C,S	1954–75	90.8	25.1	36.7	297	131	15.6	599
Kuban/Zaitsevo Koleno	45,900	6.23	T–C,S	1957–61	42.9	10.0	17.8	134	55.5	11.8	272

Continued

Appendix 2 Table I Continued

River/monitoring station	Catchment area (km²)	Water discharge (km³ a⁻¹)	Type of hydrochemical regime[1]	Period of observation	Chemical composition (mg l⁻¹)						TDS (mg l⁻¹)
					Ca^{2+}	Mg^{2+}	$Na^{+}+K^{+}$	HCO_3^{-}	SO_4^{2-}	Cl^{-}	
Caspian Sea											
Volga/Verkhnelebyazhye	1,360,000	238	EE–C,S	1950–58	45.7	9.4	15.4	127	49.7	18.5	265
Oka/Novinki	245,000		EE–C,S	1944–49	54.0	10.3	5.0	141	60.0	5.8	279
Moskva/Zvenigorod	5,000	0.97	EE–C	1948–54	46.5	11.2	2.7	190	6.8	2.8	261
Ural/Orenburg	82,300	3.09	K–C,Cl	1948–58	52.1	14.0	58.1	177	65.8	63.7	431
Belya/Ufa	100,000	18.1	EE–C,S	1950–52	61.2	12.3	6.6	155	73.6	8.9	318
Vyatka/Vyatskiye Polyany	124,000	24.3	EE–C	1950–53	22.9	7.9	4.9	115	19.6	2.3	181
Terek/Kargalinskaya	37,400	9.05	T–C,S	1937–55	65.2	13.9	22.4	176	88.6	19.3	385
Kura/Salyany	187,600	16.9	T–S,C,Cl	1939–55	53.5	18.1	46.7	211	67.5	40.5	437
Akhuryan/Aikadzor	8,140	0.87	T–C	1947–75	35.1	12.1	19.2	163	23.2	10.1	264
Pshavis Aragvi/Magaroskari	736	0.54	T–C	1958–75	32.1	6.2	11.5	119	21.9	5.2	197
Aral Sea											
Syr Darya/Kazalinsk	219,000	16.1	T–S,C,Cl	1950–58	87.9	31.9	88.7	215	262	56.5	719
Amu Darya/KyzylDjar	231,000	38.5	T–S,C,Cl	1955–62	64.9	18.4	72.3	131	136	93.2	518
Naryn/Uchkurgan	58,400	14.4	T–C,S	1950–54	46.1	11.1	20.2	143	61.3	14.1	292
Karadarya/Kampyrravat	12,400	4.52	T–C,S	1950–54	52.1	13.5	19.0	167	73.1	6.9	333
Gunt/Khorog	13,700	3.17	T–C,S	1940–75	21.3	3.03	9.42	65.8	22.4	5.11	127
Internal-drainage regions											
Ily/Ushzharma	129,000	15.1	K–C,S	1950–58	47.4	9.7	24.0	159	48.8	17.7	309
Sarysu/St.189	26,900	0.06	K–Cl,S	1951–86	80.1	60.6	273	189	342	424	1420
Chu/Ulanbel	67,500	0.64	K–S,C,Cl	1956–75	94.2	48.0	155	320	332	91.4	1036
Nura/Romanovskoye	45,100	0.53	K–Cl,S,C	1950–75	62.6	27.7	147	194	174	162	769
Emba/Zhanbike	34,700	0.33	K–Cl,S,C	1964–75	94.1	27.1	211	149	207	314	1003
Tedjen/Pul and Khatun	70,600	1.04	K–S,C,Cl	1951–70	88.8	57.9	318	221	482	306	1476

[1] See text and Alekin, 1970

TDS   Total dissolved solids

Appendix 2 Table II Long-term assessments of median (X₅₀%) and variation (X₅%, X₉₅%) in concentrations of nutrients for the main rivers of the former USSR

River/monitoring station	Period of observation	NO_3-N X5%	NO_3-N X50%	NO_3-N X95%	NH_4-N X5%	NH_4-N X50%	NH_4-N X95%	PO_4-P X5%	PO_4-P X50%	PO_4-P X95%	SiO_2 X5%	SiO_2 X50%	SiO_2 X95%	Fe X5%	Fe X50%	Fe X95%
Barents and White Seas																
Kola/1,429 km	1980–90	0.00	0.04	0.30	0.00	0.03	2.85	0.000	0.000	0.078	0.54	2.30	5.00	0.02	0.14	0.48
Onega/Porog	1980–90	0.02	0.05	0.33	0.00	0.09	0.37	0.000	0.007	0.026	0.80	2.00	3.38	0.10	0.39	0.81
Northern Dvina/Ust-Pinega	1980–90	0.01	0.02	0.41	0.00	0.12	0.41	0.000	0.005	0.037	0.93	2.28	4.84	0.11	0.32	0.52
Pechora/Ust-Tsilma	1980–90	0.00	0.00	0.31	0.00	0.22	0.95	0.000	0.008	0.048	0.48	1.60	4.32	0.00	0.24	0.50
Mezen/Malonisogorsk	1980–90	0.003	0.02	0.17	0.00	0.08	0.44	0.009	0.021	0.046	1.54	2.75	5.62	0.00	0.42	0.87
Kara Sea																
Ob/Salekhard	1980–90	0.002	0.06	0.35	0.09	0.60	1.85	0.020	0.065	0.137	0.62	2.85	8.06	0.20	1.00	2.36
Northern Sosva/Kultbaza	1980–84	0.00	0.03	0.34	0.00	0.91	3.47	0.017	0.086	0.916	0.75	3.90	10.20	0.13	2.15	4.89
Angara/Boguchany	1980–90	0.00	0.03	0.50	0.09	0.14	0.67	0.000	0.010	0.049	0.71	1.55	5.27	0.08	0.23	1.24
Biryusa/Biryusinsk	1980–90	0.00	0.01	0.39	0.00	0.00	0.21	0.000	0.000	0.050	2.28	3.55	4.40	0.003	0.08	0.23
Selenga/Mostovoi	1980–90	0.00	0.01	0.25	0.00	0.00	0.09	0.000	0.000	0.023	2.20	5.40	9.30	0.00	0.33	1.91
Yenisey/Igarka	1980–90	0.00	0.02	0.43	0.03	0.28	1.17	0.000	0.008	4.200	1.20	3.00	4.27	0.00	0.03	0.11
Laptev Sea																
Lena/Kusur	1980–90	0.00	0.03	0.25	0.01	0.08	0.49	0.000	0.004	0.022	0.50	1.50	3.20	0.05	0.37	1.49
Namana/Myakinda	1980–85	0.00	0.00	0.03	0.00	0.02	0.36	0.000	0.006	0.028	0.29	3.05	5.89	0.01	0.34	1.36
Vitim/Bodaibo	1980–90	0.00	0.00	0.22	0.00	0.00	0.09	0.000	0.000	0.012	1.51	3.00	5.08	0.00	0.05	0.29
Biryuk/Biryuk	1980–90	0.00	0.00	0.06	0.00	0.02	0.12	0.000	0.002	0.056	1.56	3.00	4.12	0.00	0.07	0.39
Yana/Dyangky	1980–89	0.00	0.01	0.08	0.00	0.01	0.19	0.000	0.001	0.017	0.13	2.22	4.06	0.00	0.12	1.20
East-Siberian, Chukchi and Bering Seas																
Ayan-Yuryakh/Emteguei	1980–90	0.00	0.02	0.25	0.00	0.17	2.01	0.000	0.010	0.086	0.30	3.30	6.21	0.00	0.11	0.77
Anadyr/Snezhnoye	1980–90	0.00	0.04	0.43	0.00	0.38	1.73	0.000	0.016	0.082	1.19	4.95	9.12	0.77	0.38	0.86
Kamchatka/Klyuchi	1980–90	0.00	0.10	0.37	0.02	0.05	0.12	0.019	0.075	0.105	8.65	12.60	15.80	0.26	1.14	1.72

Continued

Appendix 2 Table II Continued

River/monitoring station	Period of observation	NO₃-N X₅%	X₅₀%	X₉₅%	NH₄-N X₅%	X₅₀%	X₉₅%	PO₄-P X₅%	X₅₀%	X₉₅%	SiO₂ X₅%	X₅₀%	X₉₅%	Fe X₅%	X₅₀%	X₉₅%
Seas of Okhotsk and Japan																
Amur/Komsomolsk	1980–90	0.00	0.02	0.44	0.03	0.43	1.07	0.000	0.021	0.083	0.26	2.15	6.16	0.08	0.37	1.19
Penzhyna/Kamenskoe	1980–86	0.00	0.03	0.14	0.00	0.05	0.06	0.000	0.021	0.124	0.69	5.41	14.40	0.10	0.77	2.14
Taui/Talon	1980–90	0.00	0.02	0.24	0.00	0.24	1.37	0.000	0.011	0.073	0.00	4.20	8.10	0.08	0.43	0.96
Bureya/Novobureisky	1981–90	0.01	0.03	0.69	0.04	0.53	1.15	0.000	0.016	0.075	0.03	3.20	6.25	0.04	0.27	0.82
Baltic Sea																
Neman/Sovetsk	1980–90	0.12	0.37	3.29	0.06	0.42	3.05	0.008	0.046	0.550	0.19	2.06	3.82	0.13	0.30	0.80
Venta/Kuldiga	1947–88	0.01	0.46	6.48	–	–	–	0.001	0.017	0.195	–	–	–	0.20	2.30	6.80
Lielupe/Elgava	1947–88	0.01	0.45	9.80	–	–	–	0.001	0.039	0.600	–	–	–	0.10	2.40	7.20
Gauya/Valmiera	1947–88	0.01	0.45	4.52	–	–	–	0.001	0.016	0.212	–	–	–	0.20	3.00	6.90
Daugava/Daugavpils	1947–88	0.01	0.34	2.47	–	–	–	0.001	0.036	0.170	–	–	–	0.10	2.60	7.00
Neva/Novosaratovka	1980–90	0.13	0.23	0.58	0.00	0.03	0.14	0.000	0.000	0.022	0.00	0.10	0.50	0.00	0.07	0.15
Black and Azov Seas																
Don/Aksai	1980–90	0.00	0.23	1.50	0.00	0.08	0.75	0.003	0.042	0.193	0.00	0.28	1.38	–	–	–
Kalaus/Svetlograd	1986–90	0.31	1.15	2.68	0.00	2.75	7.00	0.000	0.120	0.667	1.39	8.41	4.80	0.19	1.12	3.21
Derekoika/Yalta	1986–90	0.01	0.39	0.89	0.01	0.15	0.35	0.000	0.038	0.096	0.20	0.40	0.79	0.00	0.12	0.48
Dnieper/Kiev	1980–87	0.01	0.11	0.37	0.08	0.48	4.50	0.004	0.063	0.241	0.34	2.30	9.60	0.01	0.10	0.58
Bzyb/Dyirkhva	1980–90	0.10	0.50	0.22	0.00	0.56	1.66	0.000	0.006	0.027	1.00	2.60	6.97	0.00	0.12	0.30
Rioni/Sakochakidze	1980–90	0.11	1.00	4.43	0.02	0.77	2.95	0.005	0.030	0.084	1.84	4.40	7.86	0.01	0.12	1.06
Inguri/Darcheli	1980–90	0.00	1.20	6.40	0.15	1.00	4.85	0.005	0.018	0.055	1.70	4.40	9.20	0.06	0.24	0.71
Kodory/Varcha	1980–90	0.00	0.50	1.75	0.03	0.46	1.58	0.001	0.010	0.034	1.50	2.70	5.74	0.00	0.13	0.34
Southern Bug/Alexandrovka	1980–90	0.02	0.25	1.81	0.05	0.28	1.70	0.018	0.097	0.330	0.20	0.60	7.57	0.02	0.20	0.57
Chamlyk/Voznesenskaya	1980–85	0.64	4.05	15.70	0.00	0.06	1.84	0.000	0.000	0.024	3.36	7.20	16.20	–	–	–
Kuban/Tichovsky	1980–90	0.14	1.06	2.76	0.02	0.34	1.19	0.000	0.013	0.094	0.90	2.50	4.10	0.00	0.09	0.26

Continued

Appendix 2 Table II Continued

River/monitoring station	Period of observation	NO₃-N			NH₄-N			PO₄-P			SiO₂			Fe		
		$X_{5\%}$	$X_{50\%}$	$X_{95\%}$	$X_{5\%}$	$X_{50\%}$	$X_{95\%}$	$X_{5\%}$	$X_{50\%}$	$X_{95\%}$	$X_{5\%}$	$X_{50\%}$	$X_{95\%}$	$X_{5\%}$	$X_{50\%}$	$X_{95\%}$
Caspian Sea																
Volga/Verkhnelebyazhye	1980–90	0.05	0.45	1.66	0.00	0.03	0.40	0.000	0.017	0.093	0.24	2.35	4.58	0.00	0.32	2.06
Oka/Novinki	1980–90	0.69	2.05	4.14	0.10	0.43	3.27	0.028	0.140	0.306	0.21	2.30	6.79	0.05	0.29	0.85
Moskva/Zvenigorod	1980–90	0.01	0.14	0.96	0.10	0.30	1.92	0.018	0.055	0.219	1.14	2.70	5.21	0.73	0.23	1.04
Belya/Aragvi-Pasanaury	1980–90	0.00	0.25	2.10	0.00	0.47	1.83	0.000	0.005	0.037	1.50	4.40	9.14	0.00	0.10	0.41
Ural/Orenburg	1980–90	0.02	0.07	0.77	0.11	0.46	1.47	0.080	0.280	0.988	1.20	4.35	9.76	0.10	0.29	0.76
Terek/Kargalinskaya	1980–90	0.04	0.66	3.29	0.00	0.20	0.82	0.000	0.062	0.189	2.13	6.65	2.13	0.08	0.85	2.96
Kura/Salyany	1980–90	0.79	2.02	3.00	0.01	0.07	0.41	0.000	0.009	0.034	2.15	4.96	8.14	0.01	0.06	0.18
Akhuryan/Aikadzor	1980–90	0.71	2.28	5.65	0.00	0.23	5.33	0.021	0.125	0.390	8.60	13.80	18.80	0.04	0.25	0.56
Aral Sea																
Syr Darya/Kazalinsk	1980–90	0.02	0.99	3.79	0.00	0.05	0.23	0.000	0.014	0.078	2.00	7.65	22.00	0.00	0.00	0.40
Amu Darya/KyzylDjar	1980–90	0.10	0.46	1.56	0.00	0.03	0.19	0.000	0.011	0.037	1.50	3.61	12.70	0.00	0.04	0.14
Naryn/Uchkurgan	1980–90	0.16	0.70	1.71	0.00	0.03	0.21	0.000	0.006	0.048	1.50	2.95	7.15	0.00	0.03	4.20
Gunt/Khorog	1980–90	0.06	0.25	1.10	0.00	0.00	0.11	0.000	0.012	0.050	1.85	4.33	9.13	0.00	0.09	0.50
Internal-drainage regions																
Ily/Ushzharma	1980–90	0.15	0.62	1.43	0.00	0.02	0.10	0.000	0.010	0.035	1.12	3.50	6.22	0.00	0.05	0.25
Sarysu/St.189	1980–90	0.00	0.02	0.80	0.00	0.22	1.76	0.000	0.020	—	1.11	3.60	15.00	0.00	0.09	0.66
Nura/Romanovskoye	1980–90	0.00	0.20	1.02	0.00	0.19	2.86	0.015	0.141	0.714	0.26	3.40	25.60	0.00	0.12	0.46
Murgab/Takhta-Bazar	1981–90	0.23	1.81	10.30	0.00	0.08	0.72	0.000	0.005	0.043	1.74	5.70	11.00	0.05	1.25	4.49
Emba/Zhanbike	1983–90	0.03	0.17	0.22	0.01	0.12	0.14	0.000	0.020	0.051	0.52	1.55	3.83	0.11	0.18	0.79

Chapter 3*

WATER USE AND THE INFLUENCE OF ANTHROPOGENIC ACTIVITY

Water use is determined primarily by the availability of water resources which are unevenly distributed in the former USSR (Table 3.1). The majority of river water (86.5 per cent) flows in northern and eastern areas and in the Arctic Ocean and Pacific Ocean basins. Only 13.5 per cent of river water resources are located in the most populated and economically developed western and southern regions, draining into the Atlantic Ocean and the Aral and Caspian Seas. According to water resource availability, the former USSR can be divided into three zones: zones of high water availability, zones of intermediate water availability, and zones of insufficient water availability. The zones with insufficient water availability comprise more than 25 per cent of the territory, including Southern Ukraine, Moldova, middle and lower reaches of the River Volga, the Caspian lowland, southern parts of West Siberia, Kazakhstan and the Turkmenistan lowland amongst others. Only 2 per cent of total water resources are found in these territories. Among the Republics of the former USSR, the territory of the Russian Federation, Latvia, Estonia and Lithuania have a considerable part of their territory in the zone of high water availability. Economic development in water catchment areas has led to changes in surface runoff, water balance and flow regime. The most noticeable alterations occurred in the 1950s and 1960s.

3.1 Water use and its influence on water balance

The main anthropogenic factors which influence changes in water balance include water abstraction and discharge of wastewaters, deforestation, irrigation and land drainage, urbanisation and water transfer between regions. The greatest changes affect smaller rivers which can completely disappear as a result of economic development.

Water balance can be described in terms of water input and output. Water input includes surface resources and their related groundwaters accessible for exploitation, and water transit between neighbouring

---

*   *This chapter was prepared by G.M. Chernogaeva and A.P. Lvov. A part of section 3.1 was prepared by V.Y. Georgievsky*

Table 3.1   Long-term water resource characteristics of the Republics of the former USSR

Republic	Area (10^3 km^2)	Population (10^3)	Water resources[1] (km^3 a^{-1})
Russia	16,905.0	140,017	4,262.0
Kazakhstan	2,666.8	15,253	123.0
Ukraine	601.0	50,307	216.9
Turkmenistan	488.1	2,970	72.7
Uzbekistan	415.6	16,591	107.5
Byelorussia	207.6	9,744	57.9
Kyrgyzstan	192.3	3,724	48.7
Tajikistan	143.1	4,119	98.2
Azerbaijan	86.6	6,303	24.0
Georgia	69.7	5,100	61.4
Lithuania	65.2	3,474	23.0
Latvia	63.7	2,552	31.9
Estonia	45.1	1,496	15.6
Moldova	33.7	4,025	12.6
Armenia	29.8	3,169	8.6

[1]   Includes freshwater reservoirs and ground-water sources   Source: State Water Register, 1991

territories, including inter-basin transfer. Water output (i.e. total water use) includes irrevocable use by urban populations, industry and thermal power engineering, irrigation and agricultural supply and loss of water due to evaporation from reservoirs. The water balance is also influenced by the regulation of water flow for fisheries, river transport and sanitary purposes. In some territories, water output includes specific types of irrevocable water losses including abstraction of river flow in a foreign part of a river basin and water losses to the river bed. The difference between water input and output of the water management balance determines, to a great extent, the possibilities for optimal development in a region.

The volume of water use in the former USSR has increased annually since 1970 (Figure 3.1). In comparison with 1950, water consumption increased by almost four times to approximately 370 km^3 a^{-1} in 1990 (Avakyan and Shirokov, 1990). This increase in water use was mainly due to an increase in water used for irrigation, especially in Uzbekistan, Turkmenistan and Kazakhstan. Approximately 50 per cent of total water consumption is for irrigation, about 30 per cent for industrial needs, about 10 per cent for municipal and domestic needs and about 10 per cent for other needs.

Water consumption
and transfer
$(10^9\,m^3\,a^{-1})$

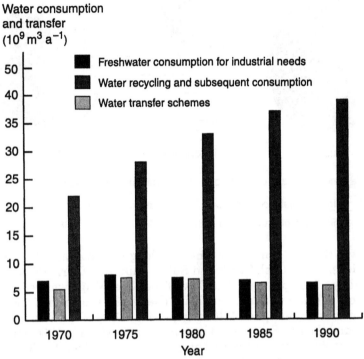

Figure 3.1 Water consumption and water transfer in 1970–90

Although one of the main water-users is industry, most (up to 74 per cent) of the water used by industry is recycled. The greatest proportion of industrial water use goes to the energy industry (54 per cent), then fuel, chemical, petrochemical industries, ferrous metallurgy, timber and mechanical engineering industries. The structure of water use varies from region to region as shown in Figure 3.2 (Main Statistics, 1988).

An excess of water abstraction over water use is caused mainly by the filling of reservoirs and losses during transportation to consumers. The greatest water abstraction occurs in Central Asia, Kazakhstan, Trans Caucasus, southern regions of Russia and Ukraine. The economic development of the water catchments in the above mentioned regions and direct water abstraction from rivers has led to reductions in river flow (Figure 3.3).

According to Vodogretsky (1979), Georgievsky and Moiseenkov (1984), Shiklomanov (1978), Shiklomanov and Markova (1987) and Shiklomanov and Georgievsky (1990), the river flow of the Amu Darya and Syr Darya rivers has been significantly reduced. The Syr Darya water resources were almost exhausted at the beginning of the 1980s, and the Amu Darya flow at the mouth was reduced by 92 per cent. The reduction in the Kuban river inflow into the Azov Sea in the second half of the

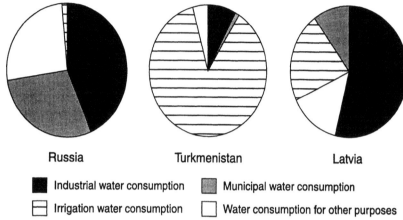

Russia   Turkmenistan   Latvia

■ Industrial water consumption   ▨ Municipal water consumption

▤ Irrigation water consumption   ☐ Water consumption for other purposes

Figure 3.2   Structure of water use in 1987 (After Main Statistics, 1988)

1980s was 6 km^3 a^{-1} or 44 per cent of its long-term average. The flow of the Dnieper, Don, Ural, Terek and Sulak, Kura and Ily rivers has been reduced by 27–31 per cent. The flow of the River Volga was reduced by 10 per cent. There were no significant changes in the flows of the large rivers of the northern part of the European territory of the former USSR, Siberia and the Far East caused by economic activities in 1990. Only the Ob river basin's flow was reduced by 3 per cent, while for other basins the reduction was less than 1 per cent.

On the whole, total river discharge was reduced by 142 km^3 a^{-1} by 1990, which is low (3 per cent) in comparison with the total river flow (4,400 km^3 a^{-1}) in the former USSR. This is due to the major river flow resources (84 per cent) belonging to the Arctic Ocean and the Pacific Ocean basins where economic activities still do not influence the characteristics of the water resources balance due to an abundance of moisture and the cold climate limiting evapotranspiration. Major flow reductions due to anthropogenic factors occur in southern regions where total natural water resources account for 540 km^3 a^{-1} or 12 per cent of total water resources of the country.

3.2   Sources of pollution

Problems associated with the decrease in water resources under the influence of economic activity and changes in river flow regime are exacerbated by the discharge of different types of wastewater.

3.2.1   Point sources of pollution

In general, during 1975–89, there was an increase in the volume of wastewaters discharged into natural waters. By 1989, the total volume of wastewaters was approximately 160 × 10^9 m^3 with over 40 million tonnes of pollutants discharged. Of the total volume of wastewater,

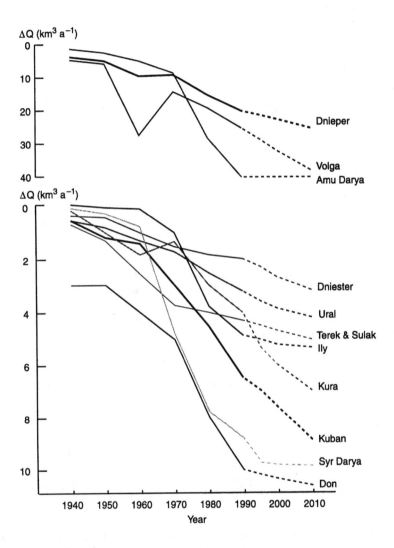

Figure 3.3 Actual and predicted variations in annual river discharge under anthropogenic impact

20–25 per cent are wastewaters that do not meet established standards (Figure 3.4). This situation occurred in Russia, Latvia, Lithuania, Georgia, Azerbaijan and Kazakhstan (State Report, 1990).

In the years before the dissolution of the USSR, there was a tendency for an increase in the volume of wastewaters discharged without treatment or which were insufficiently treated (Table 3.2), although during

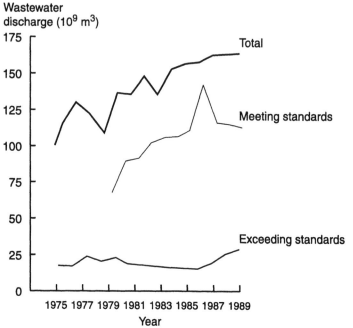

Figure 3.4 Wastewater discharge into natural water bodies (surface and ground) (After USSR National Report, 1991)

Table 3.2   Discharge of polluted wastewaters into water bodies of the former USSR

Wastewaters	1985	1987	1988
Without treatment (10^6 m³)	6,850	6,677	8,062
Insufficiently treated (10^6 m³)	9,046	13,943	20,370
Total (10^6 m³)	**15,896**	**20,620**	**28,432**

this time the capacity of wastewater treatment facilities increased from 25.5 km³ a⁻¹ in 1985 to 28.8 km³ a⁻¹ in 1990, as did the volume of recycled water. The discharge of contaminants varied greatly across the country (Table 3.3).

3.2.2 Diffuse pollutant sources

Urban runoff
The decrease in infiltration capacity of urban areas and the increase in the runoff rate has led to the leaching of many pollutants. Their fluxes to

Table 3.3 Discharge of contaminants into water bodies in the republics of the former USSR, expressed as a percentage of the total discharged, 1989

Republic	BOD	Petroleum products	Suspended substances	Sulphates	Chlorides	Total phosphorus	Ammonium nitrogen	Phenols	Detergents	Cu	Zn	Fe
Total discharged (t)	1,599	74	2,236,000	20,953,000	19,189,000	65,689	240,056	925	15,726	1,002	2,369	3,743
Russia	66	41	75	55	44	80	73	30	80	80	85	92
Ukraine	8	9	7	6.5	5.5	12	12	8	7	9	1	5
Byelorussia	1	1		1.5	6	1	3		1	2	1	
Uzbekistan	6	33	5	1	4		1.5					
Kazakhstan			1.5			1	2		4	2.5	2	
Georgia	1	2	2.5					1				
Azerbaijan	1	5	1.5	9.5	2	1	1	3	3			
Lithuania	4		2			1				3	3	1
Moldova						1						
Latvia	3		1.5			1.5			3			
Kyrgyzstan											5	
Tajikistan	3	8	2	9	10							
Armenia				14	15							
Estonia	3		1.5			1	5	57				

BOD   Biochemical oxygen demand   Source: State Report, 1990

rivers are comparable with wastewaters from industrial enterprises. The influence of urbanisation on hydrological processes depends greatly on the physical and geographical conditions of the area and on the size of town and the history of its development. For small and average towns, with a population < 300,000, the proportion of area exposed to urban runoff is generally < 20 per cent and is from 30–70 per cent in large cities where the population is over a million. The corresponding increase in surface runoff, compared with unurbanised landscapes in small towns, ranges from 25 per cent to 50 per cent which is similar to surface discharge in forest and forest-steppe zones (Chernogaeva, 1989). In towns with an 'impermeable' area of about 30–40 per cent, surface runoff may increase by 100–200 per cent. The maximum increase in surface discharge is noted for arid zones (Pravoschinsky, 1965; Skakalsky, 1973; Ustuzshanin, 1988). On the whole, for small and average sized towns, total surface runoff ranges from 2,000 m^3 a^{-1} to 25 × 10^6 m^3 a^{-1}. For large cities, the volume of surface runoff increases up to 45 × 10^6 m^3 a^{-1} and in the largest cities may reach 200 × 10^6 m^3 a^{-1} (Figure 3.5).

The contribution of surface runoff from large industrial cities entering water bodies, reaches 40–50 per cent of the total flux of contaminants including point sources (Chernogaeva, 1989). The total input of pollutants from urban runoff is greatest compared with other pollutant sources in years of highest runoff because there is a complete flushing of formerly accumulated wastes. In dry years the contribution of urban wastewaters and groundwater is greatest (Morokov, 1987).

Pollution of groundwater is generally less than that of surface water. The total input of pollutants from urban areas to groundwater is 5–15 per cent. However, besides the infiltration of pollutants due to rain and snow melt, sources of groundwater pollution include losses in sewage systems and sewage treatment plants and seepage from solid waste storage and dump sites.

The type and concentration of pollutants, due to runoff from urbanised territories, depends on the size of the town and the type and size of its industrial activity. The volume of suspended substances in surface runoff of large industrial cities in high rainfall regions may reach 100,000 to 400,000 t a^{-1}, organic matter may reach 10,000 to 25,000 t a^{-1} and petroleum products up to 3,000 t a^{-1}.

The influence of towns on the quality of rivers is especially great where the urbanisation in the water catchment area may reach 90 per cent, i.e. for small basins. In larger basins this influence is less prevalent, because the proportion of urbanisation is significantly reduced. For example, in the basin of the River Lena urbanised areas occupy 0.1 per cent, in the basins of the Northern Dvina 0.3 per cent, the Neva 1 per cent, the Volga 1.9 per cent, the Don 2.7 per cent, the Neman 3.2 per cent, the Dnieper 3.8 per cent, the Kura 4.8 per cent and the Prut 3.2 per cent.

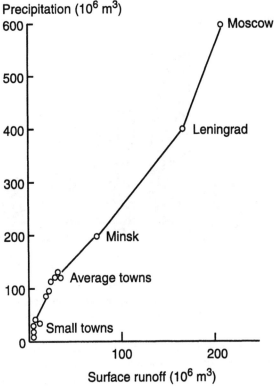

Figure 3.5 Urban runoff in relation to size of town and precipitation (After Chernogaeva, 1989)

In the republics of the former USSR, urbanised areas occupy, on the whole, not more than 2 per cent of the total area.

Agriculture
Different types of agricultural activity in water catchments have led to significant pollution of surface waters with suspended and organic substances, mineral phosphorus, nitrogen and pesticides. In 1989, agricultural land in Russia comprised 37.3 per cent of the total land area in the former USSR, whereas in Kazakhstan it comprised 36.9 per cent, in the republics of Central Asia 14 per cent, in Ukraine and Moldova 7.4 per cent, in Byelorussia and the Baltic Republics 2.8 per cent and in the Caucasian Republics 1.6 per cent (State Report, 1990). In some regions of the former Soviet Union such as Donbass, areas around the Dnieper, the Mid-Volga region, Ural and Kuzbass, agricultural areas reach 70–80 per cent.

Livestock are a significant source of water pollution with organic substances, nitrogen, phosphorus and potassium. The polluting effect

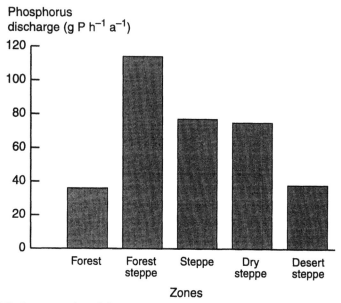

Figure 3.6 Average value of phosphorus (P$_{total}$) discharge from arable land in various zones to water bodies in the European territory (After Nazarov and Kuznetsov, 1991)

occurs both directly as a result of surface flushing of polluted areas and indirectly from the use of livestock wastes as fertiliser in agriculture.

Since 1965, agricultural assets have increased by almost five times, the area of irrigated and drained land by 12 times and the use of fertilisers and pesticides has also increased. These substances enter water bodies via surface runoff from both non-irrigated and irrigated fields, together with water from drained land.

The mechanism of nutrient discharge depends on the amount of discharge, climate conditions (Figure 3.6), type of soil, soil fertility and the amount of fertilisers used. The discharge of nitrogen is one order greater than that of phosphorus (Nazarov and Kuznetsov, 1991). The greatest discharge of nutrients per unit area is from ploughed fields and the minimum is from forests: if the water catchment is covered by 30 per cent forest, the discharge of nitrogen and phosphorus is therefore reduced by more than three times.

From agricultural land, 60–70 per cent of annual phosphorus and nitrogen is discharged during spring floods, 5–10 per cent during summer and 25–30 per cent during autumn. In larger regions and water catchments, about 80 per cent of mineral nitrogen enters a water body as a result of diffuse agricultural sources (Tarasov et al., 1977; Chernogaeva and Mukhamedzshanova, 1987).

The discharge of nutrients, organic matter and suspended substances from agricultural lands consistently decreases from north to south as a

Table 3.4 Leaching of nutrients from agricultural drainage to river basins of different natural vegetation zones

Natural vegetation zone	River sampling station	Area of arable land (%)	Surface runoff (mm a^{-1})	Discharge from arable land into water bodies (kg km^{-2} a^{-1})				
				Nitrogen	Phosphorus	Potassium	Humus	Suspended substances
Mixed forest	Sura-Kadnyevo	52	50	123	5	34	614	5,931
	Klyazma-Kovrov	25	82	75	6	42	1,350	24,116
	Moksha-Shevelevsky Maidan	56	64	141	7	40	656	6,625
	Oka-Murom	50	78	84	4	31	744	11,483
	Desna-Chernigov	52	62	68	3	27	733	9,243
Forest steppe	Seim-Rylsk	69	60	74	3	30	496	5,040
	Medveditsa-Argedinskaya	59	62	75	3	26	636	6,364
	Khoper-Besplemyanovsky	64	38	92	3	28	551	5,509
	Seversky Donets-Belaya Kalitva	67	18	56	2	20	499	4,990

Source: Chernyshev and Ivanova, 1992

Table 3.5   Mineralisation of drainage waters and salt discharge for some irrigation returns in arid zones

Region	TDS ($g\ l^{-1}$)	Discharge of salts ($t\ ha^{-1}\ a^{-1}$)
Tajikistan		
Vakhsh valley	1.4–4.9	25
Uzbekistan		
Golodnaya steppe	3.2–4.9	8.3–20.1
Azerbaijan		
Mugano-Solyan massif	25–30	57
Northern Caucasus		
Gudermes system	0.5–7.2	1.0–6.8
Lower Don system	0.3–11.0	3.1–5.1
Azov system	1.8–6.1	7.4–10.9
Petrov-Anastasyev system	0.5–25	23.1–40.1
Terek-Kuma system	0.9–4.4	3.2–8.8
Dagestan		
Old Terek system	4.0–7.2	200

TDS   Total dissolved solids   Source: Tarasov *et al.*, 1977

Table 3.6   Nitrogen and phosphorus fluxes with surface drainage discharge from meliorated land as a percentage of total quantity used

Agricultural crop	Soil type	Mineral nitrogen	Total phosphorus
Grain crops and	Loam-soil	3.5	0.5
vegetables	Loam-sand	8.5	0.5
Perennial grass	Loam-soil	0.6	0.5
	Loam-sand	0.5	0.5

Source: Report of the Central Research Institute, 1989

result of a reduction, in the same direction, of water runoff and soil runoff (Table 3.4). Water quality of irrigation returns, particularly total dissolved solids, depends on local soil and climate conditions. This leads to a significant variation in water quality in different irrigation systems (Table 3.5).

Drainage and agricultural development of wetland areas lead to a significant increase in discharge of nutrients to the water catchment. Table 3.6 provides results from the monitoring of water quality from drained lands in Byelorussia (Report of the Central Research Institute, 1989). According to this data, soil fertility and composition have little influence on phosphorus discharge compared with nitrogen discharge. However, the amount of nitrogen in drainage water varied according to land use (Velikevitch and Usovitch, 1977). For example, land ploughed for crops yielded 21.8–34.7 kg N $ha^{-1} a^{-1}$, corn fields 15–24.2 kg N $ha^{-1} a^{-1}$ and pasture land 6–6.3 kg N $ha^{-1} a^{-1}$. Phosphorus discharge varied slightly and was generally up to 1 per cent of the total quantity used as fertiliser. In drainage waters, nitrogen was found mainly in the nitrate form comprising 89–99 per cent, ammonia 1–11 per cent, and nitrites 0.5–0.65 per cent. In open drainage networks, the concentration of ammonium nitrogen increases up to 28–60 per cent. In flows from drained agricultural fields during heavy rain, ammonium nitrogen was prevalent: NH_4-N 80–97 per cent, NO_3-N 3–18 per cent and NO_2-N 1–2 per cent.

Special attention should be given to the problem of pollution of surface waters with pesticides used in agriculture. Runoff and discharge of pesticides, according to experiments carried out in different regions of the former USSR, represent only a small percentage of the total amount used in the field (Bobovnikova *et al.*, 1986). The proportion can be greater (above 12 per cent) if herbicides are used which have been formulated as wetting powders; these can be more easily flushed out, especially on steep slopes.

In irrigated agriculture, depending on the type of irrigation and time period between pesticide application and subsequent irrigation, the pesticide flushing rate does not exceed 1 per cent. From 0.3 per cent to 1.3 per cent of pesticides used are flushed out with surface runoff (Korotova *et al.*, 1983). However, pesticides still enter water systems and negatively influence the biota, especially in small rivers where dilution is limited. Low concentrations of DDT and its metabolites, such as DDE, isomers of hexachlorocyclohexane (HCH) and similar products used in industries, as well as polychlorobiphenyls (PCB), are dangerous for aquatic biota because they can be accumulated in lipid-rich tissue. At all trophic levels, there is a trend for an increase in the biological accumulation of fat-soluble compounds which have high coefficients of octanol–water distribution.

Atmospheric fallout
Contamination by atmospheric precipitation in the former USSR is caused both by emissions from local sources and by transboundary global atmospheric transport. On average, there is a total annual atmospheric fallout of approximately 14×10^6 t of sulphur, 7×10^6 t of ammonium nitrogen, 5×10^6 t of nitrate nitrogen, 127 t of benzo(a)pyrene (Atmospheric Load of Pollutants on the Territory of the USSR, 1991).

Atmospheric contamination is found everywhere and corresponds mainly to the density of local pollution sources. The highest levels of atmospheric contamination occur in the European territory, where in central pristine areas in winter the average monthly concentrations of benzo(a)pyrene (0.7–0.9 ng m^{-3}) are close to the maximum allowable concentrations (MAC), and the concentration of sulphur dioxide in some

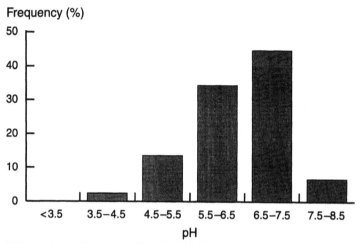

Figure 3.7 pH of precipitation monitored in 118 Russian towns (After State Report, 1994)

latitudes reaches 10–20 mg m^{-3} (Atmospheric Load of Pollutants on the Territory of the USSR, 1991). The Western regions of the European territory suffer considerable anthropogenic load. In addition to existing national sources, there is long-range transboundary transport of contaminated air masses from Western Europe.

Contamination of water catchments from atmospheric sources has a negative impact on surface water quality. This is especially true in areas of high rainfall, where the discharge of contaminants of atmospheric origin in surface runoff is more common than their accumulation in water catchments (Chernogaeva *et al.*, 1990). The contribution of atmospheric components to total river contamination can be significant. For example, in some rivers subject to strong anthropogenic influence, the atmospheric share of sulphate can exceed 25 per cent (Chernogaeva and Moshiashvili, 1995).

Acidification of winter precipitation is, in general, not great. For most of the country, including urbanised and agricultural regions, pH values vary between 5.5 and 5.7. Higher acidity of winter precipitation (pH 4.5–5.0) occurs in the north and north-east of the European territory, in Siberia and the Far East. Increased acidification in internal territories is related to long-range transport mainly from the industrial regions of the Ukraine, Central European and the Ural regions (State Report 1990, 1993). Monitoring of pH in atmospheric precipitation in 118 towns in Russia during 1994 showed that 85 per cent of precipitation was slightly acidic or alkaline with pH > 5.6 (Figure 3.7) and 15 per cent with pH < 5.6 (State Report, 1995).

In pristine regions, DDT and other persistent pesticides may enter surface waters mainly via global atmospheric transport and accumulation in snow. This was noted in most rivers of the temperate zone of the former USSR where persistent pesticides originate mainly (up to 70 per cent) from snow cover.

3.2.3 Sources of surface water pollution along river courses

Transport
Water transport played an important role in the former Soviet Union. The total length of waterways used as transportation routes in the country is 123,200 km, of which 21,000 km are artificial waterways such as reservoirs and canals. In the north of the European territory, in the North-West, the Volga-Vyatka and East Siberian regions, the share of transported loads on rivers is 20 to 40 per cent of the total transported volume. Water transport is also important in the development of the gas-exploring regions of Eastern Siberia, in the Norilsk metallurgical industrial centre and in the mining industry of Yakutia.

Every year during the navigation period, ships of the river fleet produce approximately 3×10^6 m^3 of industrial and domestic waste-water, 550,000 m^3 of oil-containing waters and 26,000 t of solid wastes. These inputs are, however, limited when compared with other sources. For example in Russia, the different types of waste from river ships may be compared with the waste from a town of 250,000 inhabitants (State Report, 1993).

In industrial enterprises associated with the river fleet located on river banks, most polluted wastewaters originate from galvanising works. Only 25 per cent of the enterprises have wastewater treatment facilities. The discharge of polluted wastewaters reaches 5×10^6 m^3. The total volume of water used in recycled water systems comprises one fifth of the total wastewater discharge.

Water reservoirs
In the former USSR, there are more than 1,000 reservoirs with a total volume of approximately 1,070 km^3. The 194 largest water reservoirs (volume exceeding 100×10^6 m^3) correspond to a total volume of 1,040 km^3 and to an exploitable volume of 506 km^3 (State Water Register, 1991). There are 114 large reservoirs (total volume of 365.3 km^3) located in the European territory. In the Asian territory, the total volume of 80 large reservoirs is nearly twice the storage volume of the large European reservoirs. The most intensive building period occurred in 1955–60 (218 km^3), 1965–70 (338 km^3) and 1975–80 (178 km^3) (Figure 3.8).

Most of the large and middle-sized reservoirs of the former USSR are multipurpose reservoirs (e.g. used for energy, irrigation, water transport, water supply). Small reservoirs are often constructed for a particular need. For example, small high-pressure reservoirs are used in energy projects or solely for irrigation purposes.

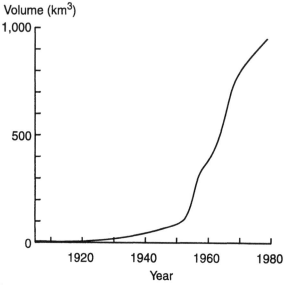

Figure 3.8   Changes in total volume in water reservoirs of the former USSR

The largest reservoir in terms of water volume is the Bratsk reservoir (169.3 km^3 with an exploitable volume of 48.2 km^3). The distribution of large reservoirs within the European and Asian parts of the former USSR is presented in Table 3.7 and their distribution according to economic areas and republics in Table 3.8. The maximum density of large reservoirs (16) is located in the northern and north-western industrial areas of the European territory. The maximum water volume (402.3 km^3) is stored in the largest reservoirs of the West Siberian industrial area.

3.3   Pollution of groundwater

Groundwaters, especially phreatic aquifers, suffer from pollution due to the influence of human activity. Generally, sources of groundwater contamination are oil fields, mining enterprises, filtration fields, sludge collectors and waste tips of metallurgical plants, chemical waste and fertiliser storage areas, dumps and settlements not provided with sewer systems. The principal pollutants are nitrogen compounds, sulphates, chlorides, calcium, magnesium, oil products and phenols (Figure 3.9).

Thus, in Russia, there are 759 detected sites of permanent pollution which have been registered (State Report, 1990). In over 22 per cent of these sites the MAC was exceeded by 100 times for at least one variable; in 34 per cent it was exceeded by 10–100 times and in 44 per cent it was exceeded between 1–10 times (see MAC levels in Table 4.5). Of these 759 major sites of groundwater contamination, 572 are in the European part of Russia. At 320 sites (42 per cent), contamination is related to

Table 3.7   Distribution of water reservoirs with a volume exceeding 100×10^6 m^3 in the European and Asian territories of the former USSR

Volume (km^3)	European territory			Asian territory		
	n	Total volume (km^3)	Exploitable volume (km^3)	n	Total volume (km^3)	Exploitable volume (km^3)
0.1–1	78	27.1	15.1	59	18.6	14.6
1–10	22	79.2	30.1	8	26.1	20.1
10–50	13	201.0	127.0	9	260.0	160.0
> 50	1	58.0	35.0	4	370.0	114.0

Table 3.8   Distribution of large water reservoirs with a volume exceeding 500×10^6 m^3

Economic area or Republic	Number of reservoirs	Volume (km^3)	
		Total	Exploitable
Russia	49	810.5	363.8
Northern and north-western parts of European territory	16	81.3	44.1
Central, Central-Chernozem and Volga-Vyatsk	5	37.1	21.5
Volga	5	116.1	49.8
Northern Caucasus	8	34.0	17.4
Urals	6	26.6	16.6
Western Siberia	1	8.8	4.4
Eastern Siberia	6	402.3	160.0
Far East	2	104.3	50.0
Ukraine	6	43.8	18.4
Byelorussia	1	0.8	0.33
Uzbekistan	6	13.7	10.2
Kazakhstan	8	80.3	44.4
Azerbaijan	3	18.0	9.1
Latvia	1	0.5	0.14
Kyrgyzstan	2	20.0	14.5
Tajikistan	3	15.5	7.71
Turkmenistan	2	1.4	0.96

industrial enterprises. These are mainly located in the areas of Nizhny Novgorod, Tula, Moscow, Perm, Ekaterinburg, Chelyabinsk, Orenburg, Rostov, Samara, Kemerovo, Irkutsk, Krasnoyarsk, Khabarovsk regions

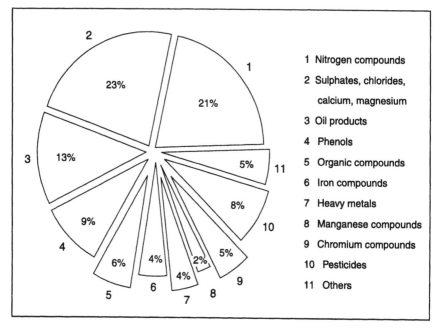

Figure 3.9 Causes of groundwater pollution in the former USSR according to different pollutants, as a percentage of the total causes

and Bashkortostan. Chemical, metallurgical, energy, oil-related, machine-building and medical branches of industry are the main sources of pollution in these territories (Drinking Water, 1994).

At 154 sites (20 per cent), groundwater contamination originates from agriculture, including waste storage sites, livestock and poultry farms and processing plants, and also as a result of water filtration from agricultural land contaminated with toxic chemicals and fertilisers. Most of the agricultural pollution sources are found in the areas of Saratov, Nizhny Novgorod, Vladimir, Lipetsk, Chita, Sakhalin, Stavropol and Krasnodar regions and the Kabarda-Balkar Republic. At 97 sites (13 per cent), pollution is of municipal origin, principally due to disorganised domestic waste dumps and lack of proper sewerage systems. The majority of municipal pollution sources are located in the Komi and Udmurtia Republics, Voronezh and Saratov regions. At 83 sites (11 per cent), pollution originates from naturally poor groundwaters which enter extraction wells due to an inadequate regime of water extraction in the waterworks (mainly in Arkhangelsk, Smolensk, Ivanovo, Samara, Pensa and Orenburg regions). At 93 sites (12 per cent), groundwater pollution is due to a combination of impacts from industrial, municipal and agricultural activities. These sites are found in the Komi Republic, Kaliningrad, Saratov, Samara, Rostov, Chita and Magadan regions. At

12 sites (2 per cent) of groundwater pollution, the actual sources of pollution have not been revealed.

Of the registered 759 sites of pollution, half of them are linked to diffuse sources; 14 cover an area of > 100 km^2, 6 sites cover 50–100 km^2, 28 sites cover 20–50 km^2, 31 sites cover 10–20 km^2, 39 sites cover 5–10 km^2, 132 sites cover 1–5 km^2 and 140 sites cover < 1 km^2. The polluted area for 369 sites (49 per cent) has not been determined because their pollution is characterised by data from a single point (industrial or domestic well).

The greatest sites of pollution are located in the vicinity of the towns of Moscow (Lytkarino 100 km^2, Staraya Kupavna 40 km^2), Vologda (Cherepovets 30 km^2), Chelyabinsk (Magnitogorsk 150 km^2), Kemerovo (Kemerovo 150 km^2 and 500 km^2), Irkutsk (Angarsk 42 km^2) region, in Bashkortostan (Tuimaza district 230 km^2) and the Primorye (Spassk-Dalny 56 km^2).

3.4 Wastewater treatment

The total volume of all wastewater discharged into water bodies from urban, industrial and agricultural sources in the former USSR has increased. In 1976, wastewater discharge amounted to 111.4 km^3 while in 1990 it amounted to 164.4 km^3. Of the water discharged in 1990, 67.2 per cent originated from drainage from irrigation systems and industrial plant cooling waters. This wastewater did not undergo any treatment. It was assumed that water, after passing through a heat exchange apparatus, would only have a higher temperature and not contain polluting substances. However, monitoring of the water has shown contamination by petroleum products and an increase in salt content.

Monitoring of drainage waters from irrigation systems has also shown the presence of different pollutants. Settling ponds have been put into operation in some irrigation systems, especially for rice crops, and these have been shown to lower the content of toxic chemicals and other polluting substances in the drainage water.

In 1990, 33.5 km^3 of wastewaters were treated before being discharged into water bodies. This represents 76.8 per cent of the total wastewater requiring treatment. In 1990, 10.1 km^3 of wastewater was discharged into water bodies without any treatment (compared with 13.2 km^3 in 1979). The total capacity of treatment plants was 44.7 km^3, of which 42.1 km^3 was for secondary treatment (biological treatment) and the remaining 2.6 km^3 for primary and tertiary treatment.

Despite the fact that the treatment capacity exceeded the volume of processed wastewater, the efficiency was low. Only 10 km^3 (29.9 per cent) of wastewaters were treated sufficiently to meet standards, of which 7.8 km^3 were biologically treated, 1.9 km^3 were treated at mechanical installations and 0.2 km^3 at physico-chemical installations. The rate of efficient treatment was 21 per cent for the Baltic Sea,

16 per cent for the Caspian Sea, 52 per cent for the Black Sea and 25 per cent for the Azov Sea basins. The main reasons for this were that the treatment facilities were overloaded by pollutant inputs, the low level exploitation of the installations and the insufficient control of industrial wastewaters discharged into sewerage systems. Of the total volume of wastewater treated adequately to meet standards, 4.1 km^3 were industrial wastewaters and 5.6 km^3 were municipal and domestic wastewaters. In 1990, the total quantity of untreated wastewater discharge was 2.6 $km^3 a^{-1}$ to the Azov and 2.8 $km^3 a^{-1}$ to the Caspian Sea basins.

From 1976 to 1986, there was an annual increase in treated wastewaters meeting standards (15.4 km^3 in 1976, 17.2 km^3 in 1980, 22.9 km^3 in 1986). From 1986, the executive authorities responsible for water management and water protection began to tighten wastewater treatment criteria and the volume of inadequately treated wastewater began to decrease. Such measures became necessary due to the presence of substances in wastewater not being effectively removed during treatment. This was especially important for wastewaters from biological treatment installations due to minimal attention to local treatment of wastewater before its discharge into sewerage systems. An increase in detergents in wastewaters which were not all removed by biological treatment, and the low efficiency of treatment technology in removing nitrogen and phosphorus compounds, led to intensive eutrophication of some water bodies. This was particularly noticeable in the Dnieper and Volga river basins.

Biological treatment facilities were constructed according to a standard scheme: grit traps, primary sedimentation tanks, biofilters or aeration tanks (with pressurised-air aeration in most) and secondary sedimentation tanks. Sludges were disposed of mainly at silt grounds. Mechanical sludge dehydration facilities were constructed at many treatment plants but their efficiency was very low. High energy and reagent consumption were characteristic features of such systems. Disposal of biologically active sludges became a serious problem, especially in large cities. It was also impossible to use the sludges as fertilisers on agricultural land because of their high heavy metal content.

Horizontal sedimentation tanks of different types (in mining and gold-mining industries), slag accumulators (in heat and power plants, enrichment factories), oil traps (in transport enterprises, oil refineries) and fat traps (in meat processing enterprises) were used as mechanical treatment systems. As a rule, their efficiency was not high. In mechanical treatment processes, thinly layered sedimentation mechanisms of different design appeared to be more efficient in removing roughly dispersed (suspended and emulsive) admixtures from water. Microfilters, sedimentation centrifuges and other facilities were highly efficient in the separation of suspended substances. Physico-chemical treatment installations included different neutralisation stations, flotation units, emulsifiers and sorbtion facilities.

Intensive experimental and design research was undertaken by water treatment specialists and scientists into developing high-efficiency wastewater treatment technologies, into improving treatment installations and into the development of water supply circulation systems based on these principles. Certain enterprises had built such systems by 1990, for example the Verkh-Isetsk metallurgical plant, the Selenga pulp and paper mill, the Tobolsk petrochemical plant and the Togliatti motor plant.

The sorption treatment process is one of the physico-chemical treatment methods which has been applied to substances resisting biological treatment, such as refractory substances (xenobiotics) and petroleum products. Electrochemical treatment systems have been installed, especially in galvanising works. Oxidation methods, with ozone and chlorine reagents as oxidisers, are also being developed.

The efficiency of biological treatment systems were further developed to increase productivity and to reduce energy consumption. New biofilters with a high load of plastic fillers were developed. In addition, sand and charcoal biosorbent filters were used to increase the degree of treatment. Considerable attention has been paid to the development and application of methods for removal of both nitrogen and phosphorus compounds from wastewaters. However, all of the methods appeared to be rather expensive and complicated for implementation. The industrial production of compact wastewater treatment facilities for rural areas, recreation camps and sanatoriums started in 1990.

3.5 Water legislation

Water legislation in the former Soviet Union, which aimed at promoting effective and well-founded use of water resources and protecting against pollution and depletion, was the juridical and organisational background for water protection in the country. 'The Fundamentals of Water Legislation of the USSR and the Union Republics' (1970) was adopted by the USSR Supreme Council on 10 December 1970. It served for almost two decades as the principal legal act for water management and protection in the USSR, with only one amendment made in 1980. The act assigned the State full control of water resources as the mutual heritage of the Soviet people and served as a background for water relations in the country. The main objective of the act was the regulation of water to ensure, for the sake of the present and future generations, the rational use of water resources for the needs of the population and economy, for protection of waters from pollution, eutrophication and depletion, and for the prevention and minimisation of impacts on water bodies due to natural disasters and anthropogenic causes.

'The Fundamentals of Water Legislation' defined the responsibilities of authorities in the USSR and Union Republics. The following were considered as the responsibilities of the Union:

- Determination of the basic regulations in the field of water resource use, their protection from pollution, eutrophication and depletion;
- Determination of all-Union standards and objectives for water use, water quality and methods for their control;
- Determination of unified systems relating to State water and water use registration, including keeping of the State Water Register;
- Adoption of plans of multiple use and protection of water bodies under the Union's responsibility;
- Planning of all-Union actions on water use and protection, prevention and cessation of harmful impacts;
- State inspection of water use and protection, and regulation of implementation regimes;
- Determination of water bodies whose water use should be regulated by the USSR authorities.

The following positions were considered as the responsibilities of the Union Republics:

- Distribution of joint State water reserves in the territory of the Republic;
- Determination of regulations for water use and their protection from pollution, eutrophication and depletion;
- Planning of measures for water use and protection, prevention and cessation of impacts;
- Adoption of plans of multiple use and protection of water bodies under the Republic's responsibility;
- Implementation of State inspections regarding water use.

The Water Legislation defined the special authorities which were, together with the government executive bodies, responsible for State management of water use and protection, either directly, or through their basin or territorial boards of water management and state inspection of water use. Therefore, a State system for water resource management and inspection, based on the basin principle, was established in the country. The USSR Water Legislation delegated to these authorities significant rights for the regulation of water.

Because economic activities affect significantly the state of water resources, 'The Fundamentals of Water Legislation' emphasised that sites, where industrial enterprises and other economic activities occurred or were planned, had to be agreed with the water management authorities. For example, it was forbidden to put into operation new or reconstructed plants which were not equipped with water protection facilities.

Water use regulation played an important role in the Water Legislation. It determined types of water use, it assigned an order of issuing permits for different types of water use, and it defined the rights and commitments of water users. Drinking water supply, domestic and other human consumption needs, had formal priority in use of water resources. As a rule, it was forbidden to use groundwater resources for needs which were not directly connected with drinking and domestic supply.

The Water Legislation ordered users, which were using water bodies for industrial purposes, to take measures for the reduction of water consumption and cessation of polluted wastewater discharges by improvements in technologies and water supply systems. Where water consumption by an enterprise exceeded permitted limits, the executive authorities had a right to limit and even totally stop water consumption from communal mains for industrial needs, in order to prioritise water use for drinking and domestic purposes.

Wastewater discharge into natural water bodies was allowed only where it was not followed by an increase in concentration of contaminants over the standards established by water management authorities. Wastewater discharge was determined as a type of water use which is why wastewater discharge permission had to be agreed by sanitary and epidemiological inspectors and fishery authorities.

The Water Legislation assigned responsibility, including administrative and criminal, for cases of violation. Levels of responsibility were determined by the Decree of the Presidium of the USSR Supreme Council 'On Administrative Responsibility for Violation of Water Legislation' (1980). The Criminal Codes of the Union Republics determined what constituted environmental crimes and the levels of responsibility for them. Most of them were formulated at the beginning of the 1960s. During the subsequent period, what constituted environmental crimes remained practically unchanged. This led to numerous gaps in the application of responsibility for environmental crimes and to the legal protection of water resources lagging totally behind practical needs.

'Rules of Surface Water Protection from Wastewater Pollution', which was adopted in 1974, was an important Act for the regulation of wastewater discharge. This document was valid until 1991, when an updated version was adopted. Lists of MACs in water bodies of economic and fishery importance formed the main component of these rules. Maximum allowable concentrations are water quality criteria and define the permitted levels of critical variables which occur as contaminants in wastewater discharges into water bodies. The 'Instruction on the Order of Fixing and Adoption of Permission on Special Water Use', which was issued in 1984, was another regulation valid in the Russian Federation up to the present time.

Taking into account the importance of following the regulations on wastewater discharge for preservation of water quality, the USSR Government adopted, in 1979, 'The Regulations on State Inspection for Water Use and Protection' (1979). This document established the institution of State inspectors for water use and protection and determined their rights and responsibilities. In particular, they had a right to charge enterprises and other organisations for violation of established quotas on wastewater discharge into water bodies, and could even completely close them down.

The legislation established guidelines for managing the State Water Register. Data on water registration and use in the State provided the background information for the preparation of water balances, plans of multiple water use and protection and water management.

Thus, the USSR legislation sufficiently covered problems related to use and protection of water resources. However, the legislation was not put into full scale operation and, despite all the measures, water resources gradually deteriorated. Evaluation of the situation at the time of the dissolution of the Soviet Union showed that the effectiveness of the water legislation was limited. Besides general factors related to the peculiarities of the directive economy and a lack of control for legislation execution, lack of effectiveness was due to a number of specific problems including:

- Vague determination and allocation of responsibilities among the executive bodies responsible for water use and protection;
- Low reliability of data concerning water registration and use;
- Drawbacks in the system of wastewater discharge into water bodies and standards of water quality.

In addition, the implementation of legislation was difficult because there was not an adequate budget to support it. It should be noted that, beginning from 1974, the State plans for economic development had a special section relating to action for environmental protection. Enterprises were required to plan in their budgets for spending on environmental protection. In particular, attention was paid to protection of freshwater as the natural resource suffering from the greatest anthropogenic impact. For example, it was planned to spend around 70 per cent of all financial resources allocated for environmental protection in 1976–90, on water protection although, at the same time, it comprised a negligible part of the State budget (1.5 per cent of State investments). However, as a rule, spending of this budget was unsatisfactory: about 15 per cent of these resources were not used annually.

3.6   References

(All references are in Russian unless otherwise stated).

Atmospheric Load of Pollutants on the Territory of the USSR 1991 Issue 1. Gidrometeoizdat, Moscow, 250 pp.

Avakyan, A.B. and Shirokov, V.M. 1990 *Complex Utilisation and Protection of Water Resources.* Minsk, 240 pp.

Bobovnikova, T.M., Malakhov, S.G. and Virchenko, E.P. 1986 Global and local contamination of river systems by pesticides. Theses of the Reports at the V All-Union Hydrological Congress. Gidrometeoizdat, Leningrad, 40–41.

Chernogaeva, G.M. 1989 Water balance of urban territories and their environmental impact. In: *Hydrological Aspects of Urbanisation.* Moscow, 15–20.

Chernogaeva, G.M. and Moshiashvili, L.D. 1995 Trends in the sulphate flux formation by river flow. Newsletter of the Russian Academy of Science, Geographical Series, No. 2, 40–49.

Chernogaeva, G.M. and Mukhamedzshanova, D.F. 1987 Evaluation of nitrogen cycle components in the Estonian SSR. In: *Background Environmental Pollution Monitoring.* Issue 4. Gidrometeoizdat, Leningrad, 256–265.

Chernogaeva, G.M., Petrukhin, V.A. and Gromov, S.A. 1990 Balance of contaminants in river basins of some pristine areas of the USSR. In: *Background Environmental Pollution Monitoring.* Issue 6. Gidrometeoizdat, Leningrad, 186–193.

Chernyshev, E.D. and Ivanova, N.B. 1992 Geographical and ecological problems of natural waters of central and southern Russia. In: *Proceedings of the International Symposium 'Ecoinformatics Problems',* Zvenigorod, 1992. Russian Academy of Science, Moscow, 116–120.

Drinking Water 1994 State Report, Moscow, 95 pp.

Georgievsky, V.Y. and Moiseenkov, A.I. 1984 Natural hydrographs of large rivers, regulated by water reservoir cascades (example: Volga river). Proceedings of the State Hydrological Institute, **291**, 54–61.

Korotova, L.G., Tarasov, M.N. and Demchenko, A.S. 1983 Study of the migration of some phospho-organic pesticides in soil-water system in the scope of prognoses of water pollution. *Hydrochemical Materials*, **88**, 109–119.

Main Statistics on Water Use in the USSR in 1987 (according to the State Register) 1988 USSR Ministry of Melioration and Water Management, Moscow, 53 pp.

Morokov, V.V. 1987 *Natural and Economic Principles of Regional Planning for River Protection from Pollution.* Gidrometeoizdat, Leningrad, 283 pp.

Nazarov, G.V. and Kuznetsov, V.K. 1991 Zonal trends in nutrient runoff from natural and agricultural areas. Proceedings of the V All-Union Hydrological Congress, Volume 5. Gidrometeoizdat, Leningrad, 184–189.

On Administrative Responsibility for Violation of Water Legislation 1980 News Bulletin of the USSR Supreme Council, No. 41, 845 pp.

Pravoschinsky, N.A. 1965 On the impact of rain and storm waters in cities on river pollution. In: *Water Management of Minsk, Byelorussia.* Science and Technology, 322–329.

Report of the Central Research Institute of Complex Use of Water Resources 1989 To develop and implement the recommendations on protection of surface waters from pollution caused by runoff from agricultural lands and urban territories. NGR81090523, Minsk, 186 pp.

Shiklomanov, I.A. 1978 *Impact of Economic Activities on River Flow.* Gidrometeoizdat, Leningrad, 333 pp.

Shiklomanov, I.A. and Georgievsky, V.Y. 1990 Current and expected changes in river flows of the USSR and their consideration within large-scale water management projects. Proceedings of the V All-Union

Hydrological Congress, Volume 4. Gidrometeoizdat, Leningrad, 243–247.

Shiklomanov, I.A. and Markova, O.L. 1987 *Problems of Water Availability and Flow Transfer into the Sea*. Gidrometeoizdat, Leningrad, 293 pp.

Skakalsky, B.G. 1973 Urbanisation impact on river water quality. Proceedings of the State Hydrological Institute, **206**, 134–144.

State Report 1990 *State of the Environment and Environmental Protection Activities in the USSR in 1989*. Moscow, 360 pp.

State Report 1993 *State of the Environment in the Russian Federation in 1992*. Ministry of Environment, Moscow, 81 pp.

State Report 1994 *State of the Environment in the Russian Federation in 1993*. Ministry of Environment, Moscow, 238 pp.

State Report 1995 *State of the Environment in the Russian Federation in 1994*. Ministry of Environment, Moscow, 257 pp.

State Water Register 1991 Gidrometeoizdat, Leningrad, 301 pp.

Tarasov, M.N., Klimenko, O.A., Semenov, I.V., Brazhnikova, L.V. and Demchenko, A.S. 1977 Issues of river pollution investigations and prognoses. *Hydrochemical Materials*, **67**, 3–112.

The Fundamentals of Water Legislation of the USSR and the Union Republics 1970 News Bulletin of the USSR Supreme Council, No. 50, 566 pp.

The Regulations on State Inspection for Water Use and Protection 1979 Resolutions of the USSR Council of Ministers, No. 17, 14 pp.

USSR National Report to the UN Conference on the Environment and Development 1991 USSR Ministry of Environment, Moscow.

Ustuzshanin, B.S. 1988 Urbanisation impact on the hydrological regime and water quality. Summary Information. Series 3727. Issue 2. Inland Hydrology, Obninsk, 50 pp.

Velikevitch, P.A. and Usovitch, N.A. 1977 On the influence of melioration and agricultural intensification on the concentration of nutrients in Polesie rivers. In: *Development and Implementation of Complex Water Protection Measures*. Kharkov, 72–96.

Vodogretsky, V.E. 1979 *The Impact of Agro-forest Meliorative Measures on Annual Water Discharge*. Gidrometeoizdat, Leningrad, 184 pp.

Chapter 4[*]

WATER QUALITY MONITORING SYSTEMS

Surface water and groundwater resources in the former USSR have been monitored on a regular basis for many years. Monitoring certain aspects of groundwaters started over a century ago and, in 1993, a centenary of observations at Site No. 1 in the Kamennaia steppe, Voronezh region was marked. Regular surface water monitoring began in the mid-1930s based on a network of hydrological observations. From 1936, Hydrological Yearbooks regularly published data on the chemical composition of rivers and lakes. Until World War II, monitoring was carried out at several hundred sites for major ions, pH, colour index, transparency, etc., of surface waters.

After the war, the monitoring network on rivers, lakes and reservoirs increased up to 2,000 sites and observation programmes included monitoring for compounds such as inorganic nitrogen (NH_4^+, NO_2^-, NO_3^-), phosphates, iron, silica, dissolved oxygen, and susceptibility of organic matter to oxidation by permanganate. From 1968, observations were performed regularly to study concentrations of contaminants including petroleum products, surfactants, phenols, heavy metals and, later, pesticides. Details of pesticides and chemical variables analysed, and the methods for their determination, are given in General Appendices I, II and III.

The main principles of the regular national monitoring system, which continue to be used until now, were formulated in the mid-1970s (Izrael et al., 1978). At that time, the National Service of Observation and Control of Environmental Pollution (NSOCEP) was formed on the basis of the monitoring network and the operational bodies of research institutes of the USSR State Committee for Hydrometeorology (Goskomhydromet), called at the time Hydrometeoservice, with the assistance of various ministries including the Ministries of Health, Melioration and Water Management, and Agriculture. When creating NSOCEP, 3,529 surface water quality monitoring stations of the Hydrometeoservice and 148 stations from other ministries and agencies were included.

---

[*]   *This chapter was prepared by V.V. Tsirkunov. In section 4.3 data relating to the monitoring system for groundwater quality were submitted by M. Polkanova and I. Akuz*

This chapter examines the water quality monitoring system in the former USSR in the early 1990s, its advantages and disadvantages, and the requirements (using Russia as an example) of the water quality monitoring system and ways for its improvement.

4.1　General description

The acquisition of water quality information has been regulated mainly by State standards and ministerial guidelines, in particular those set by the State Standard GOST 17.1.3.07-82 (1982), Guidelines on Organising the Monitoring System (1984), Regulations for Surface Water Protection (1991), Manual on Chemical Analysis of Inland Surface Waters (1977) and Manual on Hydrobiological Analysis (1983). All environmental monitoring was financed by the State.

Most water quality monitoring has been performed by the following agencies:

- Ministry for the Protection of the Environment and Natural Resources, responsible for the co-ordination of monitoring activities of different agencies, monitoring sources of anthropogenic impact, organisation of services for analytical inspection and control;
- USSR State Committee of Hydrometeorology and Environmental Control, responsible for monitoring water bodies outside the zone of direct pollution impact and background monitoring, developing and introducing methods of chemical analysis of surface waters and compiling the State Water Register;
- USSR Ministry of Melioration and Water Management, responsible for monitoring artificial waterways and hydro-construction at sites of water intakes and wastewater discharges and compiling the State Water Register (part relating to "Water Use");
- USSR Ministry of Geology, responsible for groundwater monitoring and compiling the State Water Register (part relating to "Groundwaters");
- USSR State Committee of Sanitary and Epidemiological Inspection, responsible for the monitoring of drinking water supply sources and drinking water quality;
- USSR Ministry of Fisheries, responsible for monitoring major water bodies used for commercial fishing as well as for monitoring fish and other aquatic biota, establishing MAC and safety exposure levels of chemicals in water basins where commercial fishing occurs.

All monitoring systems of federal agencies function on similar principles and with a hierarchical structure of several levels. In general, they include a network of fixed monitoring stations and the corresponding departments in the Republics, Autonomous Republics and Regions within the structures of the Ministry for Protection of the Environment and Natural Resources and the USSR State Committee of Sanitary and Epidemiological Inspection, Regions within the structures of the USSR State Committee of Hydrometeorology and the Ministry of Geology, and the major river basins within the structures of the Ministries of Water Management and Fisheries. The federal level is represented by the Agency's headquarters as well as leading Ministerial research institutes providing scientific, methodological and technological guidance. Co-ordination of inter-agency activities in water quality monitoring was conducted by the Ministry for Protection of the Environment and Natural Resources. However, its capacity in this respect was rather limited and every agency actually pursued its own independent policy.

Work on the assessment of water quality was carried out by a number of ministerial and academic research bodies as well as universities and institutes. The leading role belonged to the Goskomhydromet research institutes, namely the Hydrochemical Institute in Rostov-on-Don (water quality monitoring of rivers, lakes and reservoirs); the Institute of Applied Geophysics in Moscow (co-ordination of monitoring in all environments); the State Hydrological Institute in Leningrad (hydrological observations, calculations and forecasts); the State Oceanographic Institute in Moscow (water quality monitoring of estuaries and coastal seas); the Institute of Applied Meteorology ('Taiphun') in Obninsk (monitoring pesticides, some toxic compounds such as dioxins and radioactivity); and the All Union Research Hydrometeorological Institute - World Data Centre (VNIIGMI-MCD) in Obninsk (in charge of the database on environmental monitoring). Besides these, the State Committee of Hydrometeorology and Environmental Control had a number of regional research institutes (e.g. Far Eastern, Trans-Caucasian, West-Siberian, Central Asian, Ukrainian) which monitored the state of water bodies and assessed the water quality of their territories.

Groundwater monitoring and quality assessment were carried out by the Institute VSEGINGEO of the Ministry of Geology in the Moscow Region. Protection of water from pollution, schemes for the complex use of water resources, management and modelling of water quality and monitoring of pollution sources were carried out by the research institutes of the Ministry of Water Management (e.g. VNIIVO in Kharkov, CNIIKIVR in Minsk, Soyuzvodproject in Moscow).

Significant work on establishing MACs and defining the concentration of toxic substances in aquatic ecosystems and evaluating their influence on biota, especially fish, was carried out by specialised institutes of the Ministry of Fisheries (e.g. VNIRO, GosNIORH, CaspNIIRH, AzNIIRH). Affiliations of the USSR State Committee for Standardisation (e.g. the All-Russian Research Institute of Standardisation of General System Technologies) deal with issues of standardisation and metrology in the field of designing techniques, measurement systems, quality control and guarantees.

Research Institutes of the USSR Academy of Sciences, such as the Institute of Water Problems and the Institute of Geography in Moscow, the Institute of Lake Studies in Leningrad, the Limnological Institute in

Irkutsk Region, and the Universities of Moscow, Leningrad, Kiev, Novosibirsk, Rostov, etc., played a major role in studying particular aspects of water quality, such as eutrophication of surface waters and evaluating the influence of sources of pollution when implementing large-scale construction plans.

4.2   Surface water quality

The goal of the surface water quality monitoring system in Russia is to provide "*systematic data acquisition on water quality of water bodies and provision to central administrative and economic authorities as well as to all interested organisations concerned with systematic information and prognoses, of the levels of pollution of water bodies and emergency information on sharp variations in the level of water pollution*" (Guidelines on Organising the Monitoring System, 1984).

The organisation and implementation of monitoring are determined by the following basic principles: comprehensive and systematic observations, co-ordination of the time of observation with particular hydrological phases, and measurement of parameters using standard methods. The system is based on a network of fixed stations of 'regular observations' located at water bodies both in areas subjected to considerable anthropogenic impact and in relatively pristine locations. The sites are selected taking into account the present use of a water body and future development plans.

Monitoring stations include one or more observation sites which were selected taking into consideration hydrometeorological and morphometric features of a water body, location of pollution sources, volume and composition of discharged wastewaters and interests of water users in accordance with the protection of surface water from pollution. At each observation site of the monitoring station, one or several vertical sections are defined on rivers to take into account the mixing of the main water flow with wastewaters or with water from tributaries, and on reservoirs to take into account the width of the contaminated zone.

In 1990, the network of the State Committee for Hydrometeorology and Environmental Control, consisted of 3,282 monitoring stations, 4,556 observation sites and 4,995 vertical sections situated on 2,258 water bodies including 1,970 rivers, 145 lakes and 143 reservoirs (Table 4.1).

Observation sites are sub-divided into four categories determined according to the economic significance of the water body, as well as water quality, size and volume of the water body, river discharge, etc.:

I.   The first-grade stations are located at medium and large water bodies of significant economic importance, i.e. near cities of a population of over 1 million; in spawning and wintering areas of highly valuable commercial aquatic organisms; in areas with recurrent emergency discharges of pollutants and large-scale mortality of

Table 4.1 Regular monitoring network for surface water quality of the USSR State Committee for Hydrometeorology, 1990

Hydrometeorology territorial board	Number of observed water bodies				Number of observation sites	Number of samples taken in different station categories[1]					Total number of individual determinations
	Rivers	Lakes	Reservoirs	Total		I	II	III	IV	Total	
Azerbaijan	45	4	7	56	72	–	216	334	271	821	28,282
Armenia	47	2	3	52	86	10	36	66	550	662	21,676
Byelorussia	58	10	5	73	93	–	–	792	551	1,343	50,897
Georgia	75	6	5	85	113	135	–	340	241	716	17,814
Kazakhstan	119	10	10	139	230	2,255	612	2,188	821	5,876	131,886
Kyrgyzstan	49	1	4	54	80	–	–	424	832	1,256	36,330
Latvia	31	9	4	44	70	–	72	951	156	1,179	31,033
Lithuania	56	5	2	63	84	–	705	986	97	1,788	34,849
Moldova	15	1	3	19	35	–	–	312	122	434	14,447
Tajikistan	47	5	1	53	71	241	–	453	233	927	19,228
Turkmenistan	15	2	3	20	35	396	–	355	30	781	16,110
Uzbekistan	73	1	13	87	126	–	72	969	648	1,689	83,587
Ukraine	136	8	15	159	244	699	429	2,667	1,178	4,973	144,532
Estonia	26	1	1	28	47	–	–	185	292	477	12,814
Upper Volga	82	5	7	94	149	–	901	2,517	80	3,498	81,216
Far East	60	–	1	61	74	–	136	460	479	1,075	28,158
Trans-Baikal	80	2	–	82	105	323	267	411	367	1,368	33,528
West Siberia	73	9	2	84	116	1,416	207	360	614	2,597	58,814
Irkutsk	43	1	4	48	89	348	–	640	566	1,554	43,147
Kamchatka	36	–	–	36	39	–	–	141	268	409	14,805

Continued

Table 4.1   Continued

Hydrometeorology territorial board	Number of observed water bodies				Number of observation sites	Number of samples taken in different station categories[1]					Total number of individual determinations
	Rivers	Lakes	Reservoirs	Total		I	II	III	IV	Total	
Kolyma	37	–	3	40	50	–	74	319	166	559	14,134
Krasnoyarsk	84	8	3	95	123	314	144	420	926	1,804	71,350
Murmansk	33	8	4	45	54	306	–	593	204	1,103	25,038
Omsk	52	8	–	60	93	656	396	575	354	1,981	45,281
Volga	50	–	5	55	84	399	–	1,337	241	1,977	57,344
Primorsk	33	1	1	35	44	–	180	366	134	680	22,642
Sakhalin	45	1	–	46	52	–	180	550	149	879	28,708
Northern	75	4	2	81	123	72	–	1,207	613	1,892	64,342
North Western	102	15	7	124	162	–	252	1,363	912	2,527	70,896
Northern Caucasus	78	–	6	84	150	1,373	90	828	843	3,134	63,302
Ural	61	13	10	84	139	–	–	1,935	330	2,265	57,690
Moscow	19	–	3	22	35	326	144	572	33	1,075	22,095
Central-Chernozem	54	–	5	59	92	–	362	686	803	1,851	59,356
Yakutsk	45	3	1	49	63	–	156	189	400	745	21,634
Tiksi	9	1	–	10	14	–	–	146	38	184	4,887
Bashkiria	27	1	3	31	46	–	–	478	164	642	17,937
Total	1,970	145	143	2,258	3,282	9,269	5,631	27,115	14,706	56,721	1,549,789

[1] See text for details

aquatic organisms; and in areas of authorised discharges of waste-water resulting in high levels of water pollution.

II.   The second-grade stations are located at water bodies near cities with a population of 0.5–1 million; in spawning and wintering areas of valuable commercial aquatic organisms; at dammed river sections important for fisheries; at sites where wastewater from irrigated areas and industrial wastewater are discharged; at rivers crossing the national boundary; and in regions with medium water pollution.

III. The third-grade stations are located at water bodies near cities with a population less than 0.5 million; at the furthest downstream observation site of medium and large rivers; in mouth areas of polluted tributaries of large rivers; and in regions where the discharge of wastewaters cause low pollution levels.

IV.   The fourth-grade stations are located in non-polluted sections of water bodies, although in practice these stations are often subject to pollution incidents.

In 1990, the monitoring network included 22 sites of Category I, 62 of Category II, 1,247 of Category III and 1,951 of Category IV.

Surface water pollution observations were based on physical, chemical and biological variables with simultaneous measurements of hydrological variables. Physical and chemical variables were monitored at all stations and biological variables at 10 per cent of the stations. More than 90 per cent of the stations were provided with hydrological measurements, such as water discharge. The list of variables was compiled taking into account the following:

- The ecological and economic significance of the water body;
- Sources of pollution from adjacent territories, including volume and composition of emissions and discharges into the environment;
- Specific features of the water body; and
- Information needs of consumers.

Monitoring programmes depended on the category of the station (Table 4.2). The monitoring programme at the third-grade station on the Yenisey river can be considered as an example. In terms of sampling sites, two stations were established: the first was situated 7.5 km upstream from Igarka city with samples taken at the surface in the middle of the river; the second was located 1 km downstream of the city with samples taken regularly at four points, in the middle of the river (surface and bottom) and at the surface from two points at a distance of about one tenth of the river width from the right and left banks. Thirty-two samples were taken and 1,413 determinations for 53 variables were performed at this station in 1990 (Table 4.3).

Monitoring of biological variables at the least polluted sites is conducted in periods of vegetation growth, either monthly (first to third grade stations) or quarterly (fourth grade stations). At the sites with higher pollution levels, certain biological variables subjected to serious

Table 4.2   Characteristics of observations and analyses performed for various monitoring types

| Frequency of observations | Monitoring programme for sites in different categories | | | |
	Category I	Category II	Category III	Category IV
Daily	Short programme 1	Visual observations	–	–
Every 10 days	Short programme 2	Short programme I	–	–
Every month	Short programme 3	Short programme 3	Short programme 3	–
During major hydrological events	Obligatory programme	Obligatory programme	Obligatory programme	Obligatory programme

Short programme 1 (SP1): water discharge (for streams) or water level (for lakes and reservoirs); visual observations; temperature; specific conductivity; dissolved oxygen

Short programme 2 (SP2): (SP1) + pH; suspended substances; chemical oxygen demand; biochemical oxygen demand; concentration of 2–3 pollutants typical for the particular site

Short programme 3 (SP3): (SP2) + stream velocity (for reference discharge measurements); concentration of all pollutants for the particular site

Obligatory programme: (SP2) + stream velocity (for reference discharge measurements); colour; transparency; odour; scale; redox-potential (Eh); dissolved gases: oxygen, carbon dioxide; major ions: Cl^-, SO_4^{2-}, HCO_3^-, Ca^{2+}, Mg^{2+}, Na^+, K^+, TDS; nutrients: NH_4^+, NO_2^-, NO_3^-, PO_4^{3-}, Fe_{tot}, Si; widely distributed pollutants: petroleum products, volatile phenols, heavy metals

Source: Leading document, 1992

variation following anthropogenic impact may be measured at intervals of 2 to 3 years. The recommended programme of hydrobiological monitoring is presented in Table 4.4.

One of the weakest points in the inland surface water quality monitoring system has been the sampling and shipping of samples for analysis. In approximately 75 per cent of cases sampling was made by the hydrological service, mainly by the staff of the hydrometric stations, and less often by field expeditions. Sampling the hydrological networks allowed the integration of hydrological and water quality observations, taking into account the main phases of the hydrological regime. However, some samples were mailed to local and regional laboratories and thus 30–35 per cent of the samples took more than 10 days to arrive at laboratories for analysis.

The samples were sent for chemical analysis to 120 laboratories belonging to 36 Hydrometeorology Territorial Boards (listed in Table 4.1). Major laboratories analysed samples for 50–60 variables. The laboratories were equipped mainly with Soviet-made equipment, optical and electrochemical instruments, gas chromatographs, etc. Most analytical instruments were outdated and there was a shortage of spare parts,

Table 4.3 Example of the monitoring of physical and chemical variables at a Category III station on the Yenisey river, near Igarka city on two different dates

Parameters	11 June 1990	01 October 1990
Water discharge ($m^3 s^{-1}$)	78,000	12,100
Transparency (cm)	51	51
Temperature (°C)	7.6	6.0
pH	7.10	7.75
Total hardness (mmol l^{-1})	0.51	1.50
O_2 (mg l^{-1})	10.5	11.2
HCO_3^- (mg l^{-1})	25.0	79.5
SO_4^{2-} (mg l^{-1})	8.6	12.8
Cl^- (mg l^{-1})	4.0	13.0
Ca^{2+} (mg l^{-1})	7.1	21.3
Mg^{2+} (mg l^{-1})	1.9	5.3
Na^+ (mg l^{-1})	2.5	8.6
K^+ (mg l^{-1})	0.4	0.7
Mineralisation (mg l^{-1})	51.2	143
Suspended solids (mg l^{-1})	2.3	2.0
CO_2 (mg l^{-1})	2.4	1.3
Surfactants (mg l^{-1})	0.01	0.02
NO_3^- (mg l^{-1})	–	–
NO_2^- (mg l^{-1})	–	–
NH_4^+ (mg l^{-1})	0.28	0.06
Si (mg l^{-1})	1.1	3.0
PO_4^{3-} (mg l^{-1})	0.014	0.005
P_{tot} (mg l^{-1})	0.28	0.06
COD (mg l^{-1})	24.0	12.1
BOD (mg l^{-1})	1.4	0.5
Phenols (mg l^{-1})	0.009	0.008
Nitro compounds (mg l^{-1})	0	0.26
Fe (µg l^{-1})	100	170
Cu (µg l^{-1})	8.7	3.6
Ti (µg l^{-1})	0	3.6
Al (µg l^{-1})	12	81
Mn (µg l^{-1})	2.4	5.8
DDT (µg l^{-1})	0	0
HCH (µg l^{-1})	0.127	0.053

Measurements made 1 km downstream of the city, middle of the river, surface layer

DDT Dichlorodiphenyltrichloroethane
HCH Hexachlorocyclohexane
–   No data available for given dates

COD Chemical oxygen demand
BOD Biochemical oxygen demand

Source: State Water Register,1991

accessories, reagents and materials, glassware, sampling devices, and other materials. However, there was a system of chemical analysis quality control according to international standards. Every year, laboratories analysed standard samples for individual variables. However, the

Table 4.4   Recommended biological monitoring programme in the former USSR

	Total number	Total number of species	Total biomass	Biomass of main groups	Number in main groups	Number of species in a group
Phytoplankton	+ (10^3 cells/cm^3)	+	+ (mg dm^{-3})	+ (mg dm^{-3})	+ (10^3 cells/cm^3)	+
Zooplankton	+ (No. m^{-3})	+	+ (mg m^{-3})	+ (mg m^{-3})	+ (No. m^{-3})	+
Zoobenthos	+ (No. m^{-3})	+	+ (g m^{-2})	+ (g m^{-2})	+ (No. m^{-2})	+
Periphyton	–	+	–	–	–	–
Macrophytes	–	+	–	–	–	–

It is also recommended to determine the following indices:

Microbiological indices: total number of bacteria (10^6 cells/cm^3), number of saprophytic bacteria (10^3 cells/cm^3), ratio of total bacteria number to saprophyte bacteria number

Phytoplankton photosynthesis intensity and organic matter breakdown: photosynthetic activity (mg dm^{-3} day^{-1}) evaluated in oxygen or carbon; organic matter degradation (mg dm^{-3} day^{-1}) evaluated in oxygen or carbon; chlorophyll content (mg dm^{-3}); enzyme activity of seston (µg dm^{-3} h^{-1} alpha-naphthol)

volume of this work was insufficient and there were no other stages of data quality control.

Data obtained from analytical laboratories were sent to regional computer centres for recording and processing. Copies of magnetic tapes were sent to the Hydrochemical Institute in Rostov-on-Don where the national data bank on surface water quality was located. The Hydrometeorology Territorial Boards prepared, published and disseminated water quality information on the territories they monitored to the consumers. Such information for the whole of the former USSR was prepared by the Hydrochemical Institute (physical and chemical data on surface water quality) and the Institute of Applied Geophysics (hydrobiological data). In addition to routine information, the existing surface water quality monitoring network provided some short-term emergency information on high pollution incidents.

The yearbook, 'Surface Water Quality in the USSR', was the basic information document used for all major water management projects. This document included water quality information for hydrographic regions of the former USSR and individual river basins. Water quality was assessed based on analysis of the frequency of values exceeding the MAC for individual variables as well as information on integrated indices (Table 4.5). Conclusions on improvement or deterioration of water quality were based on comparisons with previous years. The

Table 4.4   Continued

	Dominant and indicator species of saprobic state (name, % of total)	Number of groups	Protective cover in sampled area (100 m^2)	Vegetation distribution	Dominant species[1]
Phytoplankton	+	–	–	–	–
Zooplankton	+	–	–	–	–
Zoobenthos	+	+	–	–	–
Periphyton	+	–	–	–	–
Macrophytes	–	–	+	+	+

o icological indices: e.g. acute and chronic toxicity testing of *Daphnia magna* biotesting algae by monitoring changes in total number of cells

[1] Name, percentage cover, development stage, abnormalities, etc.

Source: Leading Document, 1992

yearbook also provided information on the sources of pollution of water bodies as well as a list of the most heavily polluted water bodies in the former USSR. For example, in 1990 the average annual concentrations of one or several indicators exceeded the MAC by 10 times in 474 water bodies (at 619 stations) and exceeded the MAC by 30 times in 82 water bodies (at 101 stations) (Goskomhydromet, 1991). At the same time, water quality was reported to have improved at 87 monitoring stations and to have deteriorated at 171 monitoring stations, with no significant changes reported at 984 stations.

The information system which combined all information from different agencies was the State Water Register which consisted of three main parts: surface water, groundwater and use of water resources. These were prepared by the Committee for Hydrometeorology, the Ministry of Geology and the Ministry of Melioration and Water Management respectively. At the end of the 1980s, information on water quality from different agencies was compiled by the Ministry of Environmental Protection in the 'State Report on the State of the Environment in the USSR'.

In addition, microbiological water quality monitoring was carried out by laboratories of the Sanitary and Epidemiology Committee to prevent the distribution of water-borne diseases (Table 4.6). There were 2,286 Sanitary and Epidemiology Centres in Russia at different levels, 66 in Republics and Regions, 545 in municipalities and 1,675 in rural districts.

Table 4.5   Percentage of measurements exceeding maximum allowable
concentrations (MAC) for selected variables in surface waters of the
Russian Federation, 1990

Parameters	MAC (mg l⁻¹)	Number of measurements	1 MAC	10 MAC	50 MAC
O_2	≥ 4.0 in winter ≥ 6.0 in summer	46,579	5.50	–	–
BOD_5	3.0	38,165	29.50	–	–
SO_4^{2-}	100	27,269	24.28	3.56	–
Cl^-	300	27,398	5.63	–	–
TDS	1,000	26,626	9.82	–	–
NH_4-N	0.39	35,923	29.68	1.74	–
NO_3-N	9.0	31,208	0.92	–	–
NO_2-N	0.02	34,891	28.08	2.22	–
Fe	0.5	27,928	20.06	–	–
Cu	0.001	29,625	74.85	13.94	1.19
Zn	0.1	25,639	34.44	2.39	0.41
Ni	0.01	11,678	10.36	1.21	0.60
Cr (VI)	0.001	9,920	46.70	8.08	0.94
Pb	0.03	9,755	2.40	–	–
Hg	0.00001	4,614	0.065	–	–
Cd	0.001	2,524	2.73	–	–
As	0.05	1,140	2.81	–	–
Phenols	0.001	33,120	46.14	6.08	–
Petroleum products	0.05	36,513	59.44	7.39	–
Surfactants	0.1	29,839	4.64	–	–
DDT	absent	15,376	9.20	9.20	9.20
HCH	absent	15,258	42.84	42.84	42.84

MAC   Maximum allowable concentration   DDT   Dichlorodiphenyltrichloroethane
TDS   Total dissolved solids   HCH   Hexachlorocyclohexane
BOD   Biochemical oxygen demand

4.3   Monitoring groundwaters

Groundwater monitoring was carried out at a network of observation
sites (wells). The basis of this network was founded about 100 years ago.
Since that time the network has been expanded continuously and, in the
1970s, it included up to 22,000 sites (mostly wells) in the former Soviet
Union. Observations are carried out by 110 stationary hydrogeological
stations. For most of the monitoring sites, the series of observations has
been continuous for more than 30–50 years. Initially, observations were
made at aquifers closest to the surface but in the 1950s and 1960s the
network was expanded to include deeper confined aquifers. The initial
objective of the network was to study trends in natural groundwater
regimes. However, with the increase in anthropogenic impact on the

Table 4.6   Microbiological water quality monitoring in the former USSR

Type of water supply	Type of analysis[1]	Frequency of analysis
Centralised drinking and municipal supply source:		
Surface water	Coliform	1 per month
Groundwater	Coliform	1 every 3 months
Drinking water in the distribution system	Total microbes Coliform	Depending on population numbers: 10,000 – 1 per month 50,000 – 5 per month 100,000 – 10 per month 1 million – 20 per month > 1 million – 60 per month
Recreational waters (at the beach)	Coliform	2 before the bathing season 2 per month during bathing season
Hospitals and veterinary hospitals sewage waters (after disinfection during outflow into reservoir)	Coliform	1 every 3 months

[1]   If intestinal infections of bacterial and viral origin   corresponding pathogenic bacteria and viruses
      increase, analyses are carried out on

quality and quantity of groundwaters, more and more observations were organised at water abstraction sites, irrigated lands, industrial and mining enterprises.

The groundwater monitoring system (Table 4.7) includes both a regional observation network and a specialised network. The regional observation network was designed to study the natural (background) regime and groundwater quality with the purpose of detecting and forecasting changes in resources and trends in their formation over large territories. The specialised observation network was created in areas of special hydrogeological conditions or in economically important areas, such as major groundwater abstraction sites, important areas for drinking water supply, areas of intensive contamination of groundwater, major irrigation systems, areas with developed mining and industrial enterprises and city agglomerations.

At present, the monitoring network of Russia comprises 17,927 observation sites, 6,686 of them are established for the study of natural groundwater conditions and 11,241 for the study of groundwaters affected by human activities or used for economic purposes. In particular, there are 6,925 sites at water abstraction intakes, 859 sites in areas influenced by reservoirs, 1,462 sites in areas affected by land drainage and irrigation and 1,754 sites in urban areas. The observation network has an uneven distribution. Its density varies from 10,000–62,000 km^2 per site (e.g. Tuva, Yakutia, Magadan region) to

Table 4.7   Groundwater monitoring networks in the former USSR

Network	Observation sites		Observations and analyses
	Number	%	
Total sites	43,958	100	Groundwater levels, well and spring discharge, groundwater temperature, water flow into mines, aeration zone humidity and salinity
Under natural conditions	15,512	35	
Under polluted conditions	28,446	65	
Phreatic aquifers	29,534	67	
Confined aquifers	14,424	33	
Regional network	25,578	58	
Specialised network	18,380	42	
Groundwater quality	3,441	8	Brief chemical analysis: total mineralisation, pH, HCO_3^-, Cl^-, CO_2, Ca, Mg, Na+K[1]
Regional network	2,086	61	
Specialised network	1,355	39	

[1]   Full chemical analysis (5% of total sample number): total mineralisation, total hardness, carbonate, salinity, pH, CO_2, oxidation (mg O_2), HCO_3^-, Cl^-, SO_4^{2-}, NO_3^-, NO_2^-, Ca, Mg, K, Fe, NH_4^+, Na, SiO_2, Mn, C, Al, Zn, Br, Be, Mo, As, Pb, S, F, I, U, Rn. In areas of groundwater pollution and depending on polluting substances, the following can be monitored in addition to the above: Bi, Co, Ni, Ti, Cr, methanol, furfurol, formaldehyde, naphthen acids, petroleum products, rodonid, styrol, toluene, phenol, surfactants, caprolactam, pesticides, COD, BOD_5

40–100 km^2 (e.g. central part of the European territory of Russia and Northern Caucasus).

For the regional network, levels, temperature and groundwater quality variables are monitored. The sampling frequency for unpolluted monitoring sites is once per year with levels measured five times per month. At contaminated stations, levels are monitored three times per week, and sampling for chemical analysis is once per year. For the specialised network, observations are carried out taking into account specific features of the area.

Analysis of groundwater quality for the regional network is performed by certified hydrochemical laboratories according to standardised methods (GOST 2761-84, 1984). For all samples, dry residue, chloride, sulphate, water hardness, ammonium, nitrate, nitrite and pH are usually determined.

For the specialised network, monitored variables depend on the objectives of observations and the specific characteristics of the groundwater contamination (e.g. petroleum products, pesticides, heavy metals, other toxic compounds). A general lack of mobile laboratories, prevented the analysis of unstable components directly on site.

The information structure of a groundwater monitoring system has three levels: local, regional and federal (with an account of inter-State objectives and information exchange). At each level, information and corresponding recommendations are forwarded to administrative bodies to be taken into account for economic and environmental protection measures. At the regional level, operational information and tabulated results of the groundwater regime are presented in the Hydrogeological Yearbook and in the Groundwater Regime Assessment Bulletin. Year-books and five-year reports are forwarded to the National and Territorial Geological Information Funds and laboratories of the Research Institute responsible for scientific and methodological supervision of the monitoring system (VSEGINGEO, Moscow). The content and form of the reports are standardised (GOST 41-05-282-87, 1987; GOST 7.63-90, 1990). Data concerning the groundwater regime are published annually in the form of information bulletins and, from the end of the 1980s, were included in the annual USSR reports on the State of the Environment.

4.4   Monitoring pollution sources

Monitoring pollution sources is one of the weakest points in the monitoring system of the former USSR. It was based primarily on wastewater quality and quantity data provided by major water users. Numerous reasons made the data unreliable and the results could deviate by an order of magnitude or more from the true values. Although the monitoring system of the Ministry of Water Management accounted for more than 90,000 water users by the end of the 1980s, many point sources of pollution were not included in this system. Often there was a lack of data on discharge from storm sewers and domestic sewers, wastewaters from large livestock farms, the mining industry, etc. Due to limited financial and technical resources, only 10–15 water quality variables were usually measured and data concerning the quality and concentrations of toxic contaminants and general toxicity of wastewater were scarce.

Data on pollution sources were gathered by the basin departments of the Ministry of Water Management and compiled in special statistics forms. These statistical data were published as a manual 'Main Statistics on Water Use in the USSR'.

The Ministry of Environmental Protection had a service of specialised inspections for analytical control in Republics, Regions, cities and districts. The main function of this service was to obtain data on the composition and properties of wastewater discharges, the efficiency of water protection measures, as well as to check the reliability of data concerning wastewater discharges provided by water users.

4.5   Advantages and disadvantages

As mentioned above, the existing water quality monitoring system was established in the middle of the 1970s. It should be emphasised that at that time it was not sufficiently lagging behind the monitoring systems of developed countries. Compared with a similar system in the USA in the 1970s, some characteristics of the monitoring system in the former USSR were definitely worse (in terms of analytical instrumentation, material and technical supply), but for some characteristics it was better (data for almost all monitoring stations were supported by measured and calculated hydrobiological information, in many regions hydrobiological monitoring was initiated and national reports on water quality were published). However, in more recent years, in spite of a number of modifications, the monitoring system has not undergone significant changes and thus lags behind real information needs.

Advantages of the monitoring system in the former Soviet Union include regular and long-term series of observations which allowed the assessment of long-term trends in water quality variables, sometimes for more than 30 years. Thus, unique data on long-term variations in chemical composition of rivers and groundwaters in different physico-geographical zones exposed to different rates of anthropogenic impact are available for the former Soviet Union. An attempt to present this information, at least partially, is made in other chapters of this book.

Another significant advantage of this monitoring system was the possibility of obtaining information comparable at the national level (but only within each Federal Agency), because data were collected, analysed and processed according to standard principles and methods. There have also been successful attempts to develop and use new analytical methods and monitoring types (e.g. background monitoring, monitoring of stable isotopes, use of ecologically harmless fluorescent tracers, biotests, system of assessment of water ecosystem status).

In addition to the disadvantages mentioned previously, other disadvantages include:

- An insufficiently distinct legal framework of the monitoring system;
- A lack of clearly formulated goals and objectives of the monitoring system which should reflect the close relationship with environmental management systems;
- Insufficiently clear distribution of responsibilities between agencies at the federal level, frequent restructuring, and poor co-ordination at the inter-agency level resulting in the duplication of some functions and gaps in the others;
- Attempts to answer all possible questions within the framework of a single, centrally-managed monitoring system;
- Outdated principles of full monitoring orientated to fixed-station observation and, thus, limited possibilities for estimating the spatial extent of pollution and detection of emerging issues;

- Lack or complete absence of monitoring programmes specific to the basins of major rivers and aquifers and to the assessment of significant pollution problems (e.g. eutrophication, salinisation);
- Insufficient monitoring of bottom sediments and suspended solids, physical parameters of habitats, concentrations of toxic compounds in particles, organs and tissue of biota, eco-toxicological parameters, etc.;
- Outdated instrumentation and equipment and, as a result, limited possibilities for the detection and determination of trace amounts of pesticides, heavy metals and other toxic compounds;
- Inefficient appreciation of quality assurance/quality control procedures and, hence, low reliability of data;
- Low application of modern information technology and the limited possibility for dissemination, processing and presentation of information;
- Water quality assessment based on an outdated and inflexible MAC system;
- Poor accessibility, low reliability and, at times, absence of ancillary information on land or water use, population statistics and economic activities necessary to interpret the information on water resource quality;
- Inadequate financial and logistical support for the monitoring activities.

4.6   References

(All references are in Russian unless otherwise stated).

Goskomhydromet 1991 *Yearbook of Surface Water Quality in the USSR in 1990*. VNIIGMI-MCD, Hydrochemical Institute, Obninsk, 432 pp.

GOST 17.1.3.07-82 1982 Guidelines for Nature Protection. Hydrosphere. Water quality control rules for water bodies and water streams. Introduced 01.01.83, valid until 01.01.91. Gosstandart Publication, Moscow, 12 pp.

GOST 2761-84 1984 Centralised industrial and drinking water supply sources. Hygiene, technical demands and selection rules. Gosstandart Publication, Moscow.

GOST 41-05-282-87 1987 Report on geological research on entrails of the earth. Hydrological and engineering-geological works. Gosstandart Publication, Moscow.

GOST 7.63-90 1990 Report on geological research on entrails of the earth. General demands on content and design. Gosstandart Publication, Moscow.

Guidelines on Organising the Monitoring System and Control of Water Quality in Water Bodies and Streams by Goskomhydromet in the Framework of OGCNK 1984 Gidrometeoizdat, Leningrad, 40 pp.

Izrael, Y.A., Gasilina, N.K., Rovinski, F.Y. and Filippova, L.M. 1978 *The USSR Environment Pollution Monitoring System*. Gidrometeoizdat, Leningrad, 117 pp.

Leading document 1992 Guidelines for Nature Protection. Hydrosphere. Organisation and execution of observations over surface water pollution by RD52.24.309-02, Roskomghydromet, St Petersburg, 68 pp.

Manual on Chemical Analysis of Inland Surface Waters 1977 Gidrometeoizdat, Leningrad, 541 pp.

Manual on Hydrobiological Analysis of Waters and Bottom Sediments 1983 Gidrometeoizdat, Leningrad, 239 pp.

Regulations for Surface Water Protection 1991 Moscow.

State Water Register 1991 Annual data. Krasnoyarsk, 80–201.

Chapter 5*

SALINISATION

With increasing intensity and diversity of anthropogenic impacts on the environment, salinisation of natural waters has become a widespread and important water quality issue. Salinisation is defined as a process whereby there is an increasing concentration of dissolved salts (in most cases major ions) or as the result of this process (Williams, 1987). A number of publications describe salinisation of water in different countries around the world (e.g. Laaksonen and Malin, 1985; Doniec et al., 1986; Van der Weijden and Middelburg, 1989; Lofvendahl, 1990; Smith et al., 1993) and globally (Williams, 1987; Meybeck et al., 1989). It has been noted that as a result of economic activity, mineralisation of water in many water bodies has increased by two to three times or more compared with the natural rate of mineralisation.

In the USSR, salinisation was described (often without reference to the term) in a number of publications, mostly in relation to the study of the influence of irrigation, draining of wetlands and mining activities. Efforts have been undertaken to assess and to describe the reasons for the increase in major ion concentrations in waters of the former USSR Republics (e.g. Chertko, 1982; Chebotareva, 1988; Nikanorov et al., 1989; Peleshenko and Khilchevsky, 1991; Tsirkunov et al., 1992), in large regions (e.g. Leonov, 1979) and in the country as a whole (e.g. Nikanorov and Tsirkunov, 1984, 1991; Maksimova, 1991).

5.1   Main sources of salinisation of surface waters

As was noted in Chapter 2, the natural mineralisation of surface waters in the USSR varies within two to three orders of magnitude, reaching 3–4 g l^{-1} or more in arid and semi-arid areas (see Figure 2.4). In this chapter, only the anthropogenic factors which cause changes in major ion concentrations are described. Taking into consideration the great variety of natural conditions and types of anthropogenic impact, all factors causing salinisation are found in the former USSR.

Long-term trends in salinisation of river waters and the main factors causing salinisation are presented in Table 5.1 and Figure 5.1. The

---

*   This chapter was prepared by V.V. Tsirkunov. Section 5.3 was prepared by V.F. Baron and L.A. Ostrovsky

Table 5.1 Long-term changes in concentration of major ions and mineralisation of some rivers of the former USSR

Numbers refer to Fig. 5.1	River	Location	Water catchment area (km²)	Period of observation	TDS[1] (mg l⁻¹)	Relative long-term changes (%)[2]							
						Ca^{2+}	Mg^{2+}	$Na^{+}+K^{+}$	HCO_3^{-}	SO_4^{2-}	Cl^{-}	TDS	Q_j[3]
Prevailing source of salinisation: draining of wetlands and the use of fertilisers													
1.	Neva	Novosaratovka	281,000	1946–90	72.0	18.1	67.5	266	0	229	82.7	50.6	0[4]
2.	Shelon	Zapolye	6,820	1946–74	320	50.8	99.3	150	10.7	178	140	66.0	–45.8
3.	Velikaya	Pyatonovo	20,000	1946–74	228	19.4	19.9	75.3	0	228	128	19.0	–30.8
4.	Daugava	Daugavpils	64,500	1947–88	295	25.9	43.9	649	7.6	358	962	36.2	0
5.	Lielupe	Elgava	12,000	1947–88	693	38.7	93.4	766	35.4	74	777	51.0	0
6.	Neman	Sovetsk	91,800	1955–90	428	35.0	83.1	165	24.5	321	376	52.1	21.4
7.	Dnieper	Kiev	239,000	1946–75	259	0	65.0	204	0	225	503	24.7	0
8.	Desna	Chernigov	81,400	1947–75	309	0	30.0	175	–15.2	167	472	10.1	0
9.	Berezina	Bobruysk	20,300	1949–91	306	48.0	44.1	245	23.1	332	468	55.7	0
10.	Styr	Lutsk	7,200	1947–75	405	–8.7	31.1	164	–18.3	118	785	0	0
Diverse sources of salinisation: industry, agriculture, urbanisation, mining, etc.													
11.	Kola	1,429 km	3,780	1947–75	31.3	0	17.1	39.2	0	51.8	34.4	15.0	0
12.	Umba	Poyalka	6,470	1950–75	35.7	0	0	102	0	43.9	102	21.0	0
13.	Onega	Porog	55,700	1955–89	148	0	–13.2	180	–5.5	16.2	42.8	0	0
14.	Northern Dvina	Ust-Pinega	348,000	1947–90	156	0	0	176	5.4	8.9	27.5	7.4	0
15.	Volga	Verhnelebyazhye	1,360,000	1951–89	295	0	93.8	154	31.3	58.8	130	28.8	25.4
16.	Volga	Nizhny Novgorod	479,000	1945–75	185	47.1	73.7	253	9.1	115	627	54.9	–29.1
17.	Oka	Novinky	245,000	1938–75	346	17.8	33.0	265	9.1	40.8	285	31.0	0
18.	Moskva	Zvenigorod	5,000	1948–75	286	0	0	285	0	188	403	0	–24.2
19.	Vyatka	Vyatski Polyany	124,000	1939–75	367	12.5	14.4	203	0	63.9	323	20.5	0
20.	Belaya	Ufa	100,000	1938–75	382	44.8	0	369	0	37.3	611	50.9	0
21.	Dniester	Bendery	66,100	1955–90	494	0	78.9	360	0	90.0	135	35.8	0
22.	Southern Bug	Alexandrovka	46,200	1949–90	623	30.8	85.5	0	6.5	191	187	33.5	66.5
23.	Biuk-Karasu	Belogorsk	275	1951–75	378	0	15.5	164	0	135	112	23.2	0
24.	Volchya	Vasilkovka	11,600	1960–91	2,750	17.3	0	0	0	112	0	38.1	0
25.	Kalchik	Mariupol	1,250	1946–91	3,020	57.6	0	156	0	54.8	247	69.5	631
26.	Kuban	Zaitsevo Koleno	45,900	1966–85	401	0	32.3	68.8	0	50.8	43.8	23.4	0

Continued

Table 5.1 Continued

Numbers refer to Fig. 5.1	River	Location	Water catchment area (km²)	Period of observation	TDS[1] (mg l⁻¹)	Relative long-term changes (%)[2]							
						Ca²⁺	Mg²⁺	Na⁺+K⁺	HCO₃⁻	SO₄²⁻	Cl⁻	TDS	Q_j[3]
Diverse sources of salinisation: industry, agriculture, urbanisation, mining, etc. (continued)													
27.	Don	Aksai	420,000	1949–90	777	44.4	132.0	229	16.5	178	202	89.8	0
28.	Ural	Orenburg	82,300	1948–92	691	47.4	67.5	0	19.4	118	85.6	64.8	46.5
29.	Tura	Turinsk	29,900	1947–75	187	24.0	39.8	271	13.1	121	125	49.3	0
30.	Tobol	Kurgan	98,800	1948–92	997	0	-5.7	34.9	-5.3	21.7	0	9.8	-59.2
31.	Irtysh	Omsk	321,000	1945–75	177	50.7	0	93.0	0	135	41.3	22.1	0
32.	Nura	Romanovka	45,100	1950–75	883	0	38.1	73.8	0	48.2	90.7	42.3	0
33.	Tom	Tomsk	57,000	1952–75	106	0	0	214	0	107	138	12.4	0
34.	Yenisey	Krasnoyarsk	300,000	1950–75	105	0	0	47.5	-7.9	51.3	119	0	0
35.	Angara	Boguchany	832,000	1953–92	140	0	27.0	0	0	72.9	25.4	9.1	0
36.	Selenga	Mostovoi	440,200	1950–92	159	0	51.8	0	0	89.3	0	5.1	0
37.	Amur	Bogorodskoye	1,730,000	1950–90	58.3	0	0	65.6	-17.2	40.2	199	0	0
38.	Khasyn	Kolymskoe Highway	682	1957–75	41.5	0	0	162	13.0	41.3	80.9	24.1	0
Prevailing source of salinisation: irrigation													
39.	Terek	Kargalinskaya	37,400	1946–90	613	17.1	20.3	133	9.8	70.1	170	42.7	0
40.	Kura	Salyany	187,600	1939–91	730	51.1	96.4	271	2.4	386	270	120	0
41.	Murgab	Takhta–Bazar	34,700	1950–70	521	0	59.0	68.7	-22.1	80.4	77.9	27.3	0
42.	Amu Darya	Kzyl Djar	231,000	1955–90	1,070	80.9	305.0	207	-6.2	294	214	149	-123
43.	Syr Darya	Kazalinsk	219,000	1950–90	1,280	58.5	250.0	344	-21.4	229	278	140	-121
44.	Chu	Ulanbel	67,500	1956–75	1,084	0	24.1	46.8	-14.0	92.4	0	23.2	-43.1
45.	Ily	Uchzharma	129,000	1950–91	390	-13.2	83.4	139	0	83.2	104	29.9	0

TDS   Total dissolved solids

1   Median value for the last five years of observation

2   Using the formula: (a × n)/MeF8, %, where:

a = non-parametric estimate of changes per year (Seasonal Kendal slope estimate);
n = number of years of observation;
MeF8 = median value of concentration in the first eight years of observation

3   Q_j = average monthly discharge

4   Absence of statistically significant (90%) long-term changes

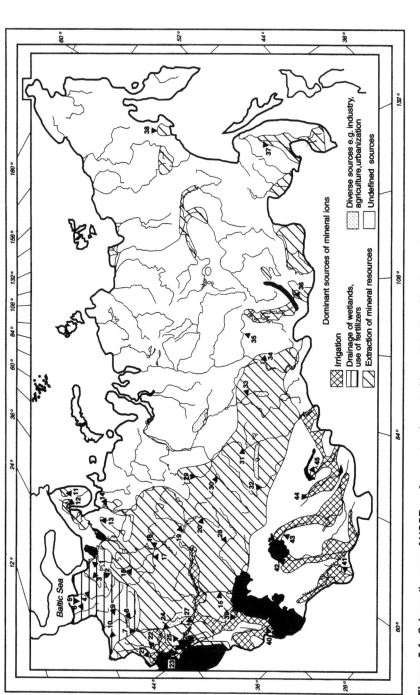

Figure 5.1 Schematic map of USSR surface waters impacted by salinisation. Numbers refer to rivers listed in Table 5.1

relative values of long-term changes in time-series of mean quarterly concentrations were calculated using non-parametric methods. In the absence of significant long-term changes in water discharge, trends were calculated from concentration residuals with corrections for the dependence of concentration on water discharge (Hirsch *et al.*, 1982; Helsel, 1993). For the estimated background values, median values of the first eight years of observation were used. Given the general lack of data concerning anthropogenic inputs of various substances into water bodies, the data on prevailing sources of salinisation given in Table 5.1 and Figure 5.1 should be regarded as approximate.

5.1.1   Wetland draining and fertiliser input

The main factor causing salinisation of surface waters in the Baltic states, Byelorussia, Northern Ukraine, and western and north-western parts of Russia is the use of fertilisers in agriculture and the draining of wetlands (Figure 5.1).

The main mineral fertilisers used are KCl, NH_4Cl, $(NH_4)_2SO_4$ and superphosphate containing up to 38 per cent of sulphates. A considerable proportion of these fertilisers is flushed out into water bodies, especially if the fertilisers are applied on fields in winter by direct spreading on snow. In particular, Na^+ and Cl^- in fertilisers are flushed out easily (85 per cent and 92 per cent of the total amount applied respectively) (Kuptsov and Khoretskaya, 1987).

Total mineralisation of waters from drained lands increases to ≥ 1 g l^{-1}; Cl^-, $Na^+ + K^+$ and SO_4^{2-} concentrations increase significantly and the chemical composition of water changes from bicarbonate to sulphate dominance. A specific feature of bog waters is abnormally high concentrations of Fe^{2+} which may reach concentrations of 30 mg l^{-1}. As a result of oxidising conditions created in the process of draining, a reduction in total iron concentrations has been noted (Lukashev *et al.*, 1976).

Such changes have also been observed in the ultra-fresh waters of Karelia, where the total area of drained lands is more than 60,000 ha and the mineralisation of water due to drained lands has increased 4–10 times (Kuraptseva, 1990). In Estonia, drained lands comprise about 40 per cent of all agricultural land used, and mineralisation of waters in reclaimed catchment areas has increased from 70 mg l^{-1} to 420 mg l^{-1} (Kuptsov and Khoretskaya, 1987). In the Ukrainian Pripyat Polesie region, with a total area of 7,095,000 ha, the area of drained land was 465,000 ha in 1966, and reached 2,343,000 ha by 1979. During the same period, the proportion of drained land in catchment areas for most rivers of this territory increased from 1 per cent to 12 per cent (Garasevich, 1990). This has led to the fact that, during the last 20 years, the flow-weighted mean of mineralisation of waters in the Pripyat river basin increased 1.5–4 times. The dynamics of intra-annual fluctuations of Cl^- concentrations at the Kiev monitoring station on the Dnieper river

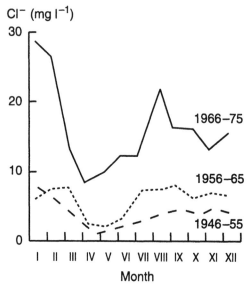

Figure 5.2 Dynamics of seasonal changes in chloride concentrations in three 10-year periods on the River Dnieper at Kiev

(situated downstream of the main drained areas) show an intensification in salinisation in the 1960s and 1970s during the period of intensive land reclamation in the territory (Figure 5.2).

A comparison of monitoring data for small rivers in the basin of the Neman river showed that, where reclamation and agricultural use of drained lands occurred in the periods 1956–64 and 1965–75, there was an increase in chloride flux of 93–123 per cent in spring and 85–329 per cent during summer low-flows; the sulphate flux increased by 12–55 per cent and 63–131 per cent respectively (Krasutskaya, 1981). Studies of Latvian rivers showed that such non-point sources of anthropogenic impact, such as drainage runoff and fertiliser flushed from agricultural lands, may be the reason for the significant increase in concentrations of major ions in some regions. As a result of this, in the 1980s, the anthropogenic flux of salts into rivers of Latvia exceeded the natural level by 2–4.5 times for chlorides, sulphates and the sum of sodium and potassium (Tsirkunov *et al.*, 1992).

An increase in mineralisation of large and medium-sized rivers, has occured due to the influence of land reclamation. The increase in SO_4^{2-} may be 1.5 times or more, the increase in $Na^+ + K^+$ may be 3–4 times or more, and for Cl^- it may be 5–7 times or more (Table 5.1). The smallest increase in concentration has been noted for bicarbonates and is usually related to an increase in groundwater discharge on drained lands. In some cases (e.g. Desna and Styr rivers) a decrease in HCO_3^- concentrations occurred which may be explained by specific processes in drained peat bogs (Zhukhovitskaya and Kadatskaya, 1981). Similar observations have been made for the Turya and Goryn rivers (Zakrevsky, 1991). For most rivers in drained areas, water mineralisation and concentrations of major ions did not reach MACs and were not obstacles to water use. Nevertheless, the present transformation of the chemical composition of water may have an unfavourable impact on aquatic ecosystems and therefore must be studied and controlled.

5.1.2   Diverse sources of salinisation

For most of the former Soviet Union, salinisation of surface waters occurs due to diverse inputs including industrial, agricultural and municipal wastewaters, runoff from urban areas, livestock farms and an increase in atmospheric precipitation of salts. The influence of these sources depends on the volume and mineral content of wastewaters in relation to the discharge and mineralisation of river waters. The annual average volume of industrial and municipal wastewater discharge may reach 10–50 per cent of the annual average volume of discharge of small rivers (Leonov, 1979).

The concentration of dissolved salts in wastewaters of different industries varies greatly. For example, in food, pulp and paper, textile and leather industries, effluent mineralisation varies from 2–25 g l^{-1}. Municipal sewage treatment plants and livestock farms discharge wastewaters with a mineral content of 0.5–2 g l^{-1} and 5–6 g l^{-1} respectively (Hygienic Assessment, 1976). Water quality monitoring data in Estonia, upstream and downstream of wastewater input sites to rivers, has shown that industrial and urban sources cause an increase in river water mineralisation by 11–88 mg l^{-1} (Kuptsov and Khoretskaya, 1987). In the former Soviet Union, 115 km^3 of wastewaters were discharged annually into natural water bodies (Main Statistics, 1990).

Information on the influence of urbanised areas on salinisation of surface waters is limited. In order to fill this gap to at least some extent, approximate calculations of inputs of salts into rivers upstream and downstream of six major cities, with a population of over one million people, have been made (Table 5.2). In Alma-Ata, Ekaterinburg and Tashkent, other rivers, besides those listed in Table 5.2, also receive part of the cities' wastewaters. For these cities (e.g. Minsk and Ekaterinburg), estimates show that more than 80,000–100,000 t a^{-1} of salts are discharged into the rivers as a result of all types of wastewaters, storm runoff, etc. The higher values obtained for the town of Yerevan may be explained by the discharge of drainage waters from irrigated lands of the Ararat valley. More than 0.25×10^6 t of salts flow annually into the River Moskva from Moscow with its 9 million inhabitants.

An additional factor causing salinisation of surface and groundwaters is the ongoing use of de-icing salts, particularly NaCl, in winter. Due to the underdevelopment of road networks, de-icing salts are used a lot less

Table 5.2 The influence of six major cities on changes in concentrations of major ions and ion discharge of rivers (average for 1986–90)

River	City	Average annual concentrations (mg l^{-1})								Average long-term water discharge ($m^3 s^{-1}$)	Average increase in salt transport ($10^3 t a^{-1}$)
		HCO$_3^-$		SO$_4^{2-}$		Cl$^-$		TDS			
		(U/S)	(D/S)	(U/S)	(D/S)	(U/S)	(D/S)	(U/S)	(D/S)		
Razdan	Yerevan	268	230	44.8	112	90.3	644	578	1,586	7.3	232.0
Svisloch	Minsk	184	225	22.6	51.3	11.2	49.6	290	446	16.8	82.1
Malaya Alma-Atinka	Alma-Ata	77.3	266	16.3	84.9	2.2	33.0	131	533	2.1	27.0
Iset	Ekaterinburg	54.7	135	51.8	98.4	12.2	61.2	167	442	9.7	84.3
Moskva	Moscow	188	195	24.7	51.6	18.3	62.9	310	436	67.6	269.0
Salar	Tashkent	129	174	62.0	130	9.7	41.6	273	470	–	–

TDS   Total dissolved solids   U/S   Upstream of city   D/S   Downstream of city

than, for example, in the USA, although the local impact from cities such as Moscow and Leningrad may be considerable.

The construction of water reservoirs also has a certain influence on the salinisation of surface waters, especially in arid conditions. Factors which lead to an increase in mineralisation of water in reservoirs include the high content of soluble minerals in the bedrock of reservoirs, an increase in evaporation and ice formation (Tarasov and Kriventsov, 1976). However, the influence of these factors, especially on reservoirs with high flow-through, is very small. Impoundment of rivers may also lead to changes in seasonal variations of concentrations of major ions and mineralisation. An example of the transformation of a hydro-chemical regime, accompanied by a general increase in mineralisation, is given for the Volga river (Figure 5.3). Following impoundment, the amplitude of seasonal fluctuations of mineralisation decreased, the mineralisation value (given as TDS) at low-flow also decreased to a certain extent but the minimum TDS increased significantly and the time of its appearance shifted from May/June to September. Such changes have also been observed for other large impounded rivers (see Chapter 14).

An additional flux of salts into surface waters arises from atmospheric fallout. This is related to increased emissions of sulphur oxides and soluble aerosols. Data on the chemical composition of polluted atmospheric precipitation has shown that concentrations of SO_4^{2-} and Cl^- may reach tens of milligrams per litre with general mineralisation up to 200 mg l^{-1}. The contribution of the atmospheric component is particularly great in industrially developed regions (e.g. suburbs of Moscow, Donbass, Kuzbass, Ural) and regions affected by long-range transboundary atmospheric transport (e.g. Baltic States, north-west Russia). According to Izrael (1983), the average annual concentration of sulphate in precipitation falling in some regions of the Baltic states increased two to three times from 1958 to 1976 (see also Chapter 11).

5.1.3 Irrigation

Irrigation, as the most water-consuming type of economic activity, has a great influence on the quantity and quality of water resources.

In the USSR in 1989, 132 km^3 of freshwaters were used for irrigation (Main Statistics, 1990) mainly in Central Asia and Kazakhstan, in the basins of the Rivers Amu Darya, Syr Darya, Chu, Ily, Talas and others, in the Trans-Caucasian republics, in the basin of Kura, in Moldova, the southern part of the Ukraine and Russia and in the basins of the Dniester, Southern Bug, Dnieper, Don, Kuban, Terek, Volga and Ural (Figure 5.1).

The volume of drainage waters discharged into natural water bodies was 34.9 km^3 in 1989 and approximately 9 km^3 were diverted into storage pools and ponds. As a result, total mineralisation of surface

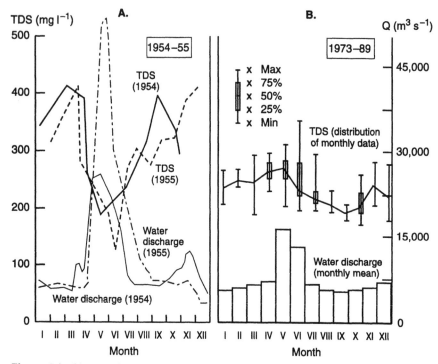

Figure 5.3 Changes in seasonal variation of water mineralisation (TDS) and discharge (Q) of the River Volga as a result of impoundment. **A.** River Volga at Volgograd, 1954–55; **B.** River Volga at Verhnelebyazhye, 1973–89 (After Zenin, 1965)

waters and concentrations of some major ions may exceed, by several times, the MACs and water becomes unsuitable for use, not only for drinking supply, but also for irrigation. Long-term river chemistry data showing the increase in salt concentrations in regions with intensive irrigation are given in Table 5.1.

Considering the importance of this problem in the former Soviet Union, it is not surprising that numerous publications were devoted to the study of irrigation and its effects on surface water quality (e.g. Tarasov *et al.*, 1966; Kirsta, 1975; Tarasov and Kriventsov, 1976; Stepanov and Chembarisov, 1978; Rubinova, 1979; Chembarisov and Bakhritdinov, 1989). Studies have shown how irrigation waters change in mineral and chemical composition in the process of flowing through drains and following infiltration. Mineralisation of drainage waters varies greatly depending on salinisation of soils and rocks in the irrigation zone, drainage conditions, level of groundwaters, methods of irrigation, age of irrigation schemes and other natural and anthropogenic factors.

The level of salinisation of surface waters depends on the ratio between discharge and mineralisation of river water and the inflow of irrigation return waters. For example, the increase in mineralisation of the River Don, due to inflows from the Lower Don and Azov irrigation systems, was 7–10 per cent in an average year (Tarasov et al., 1966). Mineralisation of the Kuma river, which has a much smaller discharge, increased 10–15 per cent due to one small irrigated area.

One of the most typical uses of water bodies is the combination of river impoundment with the development of irrigation in the basin. Lake Balkhash is such an example. After impoundment of the River Ily (the main river in the lake's basin) in 1970 and the formation of the Kapchagai reservoir, the lake level began to fall and mineralisation of water increased from 2.28 g l^{-1} to 3.04 g l^{-1} TDS in 1983 (Pozdnyakova, 1991). The increase in mineralisation of the Ily (Table 5.1) is related both to the irrigation returns and to the additional weathering of soils and rocks forming the bed and banks of the Kapchagai reservoir. The discharge of irrigation returns had a considerable influence on the Eastern Balkhash tributaries. For example, the development of rice growing in the basin of the River Karatal increased its ionic flux 1.5 times (Pozdnyakova, 1991).

Another example is the River Syr Darya which is considered here in more detail because the influence of irrigation on its salinisation is profound. The River Syr Darya is the longest river (2,137 km) and the second largest river in Central Asia in terms of water discharge (37.8 km^3). At present, the basin drains four states, Kazakhstan, Kyrgyzstan, Tajikistan and Uzbekistan where the principal crops are cotton and rice, mostly cultivated in the lower basin. Water consumption by irrigation is more than 90 per cent of the total water use and the potential water resources are used up completely, and in some parts of the basin use has exceeded availability several times. During the late 1980s, there was limited water for use in irrigation and this demanded strict control over operational water distribution. In 1989, 60 km^3 of water were abstracted from the basin of the River Syr Darya, of which 47 km^3 were used in irrigation and 13 km^3 for inter-basin transfer (Main Statistics, 1990). In 1985, in the basin of the River Syr Darya a total of 2.7×10^6 ha were irrigated. The largest irrigation area, 1.3×10^6 ha, is situated in the upper part of the basin in the Fergana valley and, in the middle section of the river (Golodnaya steppe and Tashkent oasis), 1×10^6 ha or more are irrigated. The total discharge of irrigation returns and sewage waters in the basin was 14 km^3 a^{-1} and approximately 2 km^3 a^{-1} of this was discharged beyond the limits of the basin. Mineralisation of these waters was 2.4 g l^{-1} TDS for the Fergana valley and 6.0 g l^{-1} TDS for the Central Golodnaya steppe and some other regions (Chembarisov and Bakhritdinov, 1989).

Under natural conditions in the Syr Darya, there is a slight increase in mineralisation downstream (Stepanov and Chembarisov, 1978;

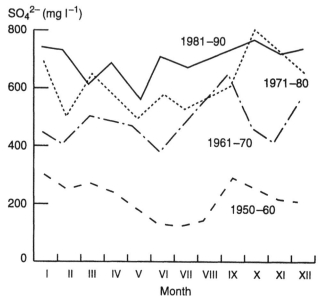

Figure 5.4 Trends in seasonal variation of sulphate concentrations over 10-year periods from 1950 for the River Syr Darya at Kazalinsk

Rubinova, 1979). At present the increasing influence of drainage waters is leading to the rapid increase in dissolved salts in the lower reaches of the river. In 1912, mineralisation of the Syr Darya varied seasonally from 0.2 g l^{-1} to 0.4 g l^{-1} TDS (Stepanov and Chembarisov, 1978), whereas in the second half of the 1980s the average mineralisation value reached 1.3 g l^{-1} TDS. Seasonal trends in sulphate concentration from the 1950s are shown in Figure 5.4.

There have also been changes in the salt transport of the Syr Darya and other Central Asian rivers. Formerly, most salts which formed in the upper reaches were transported to the Aral Sea. Now these salts stay, almost completely, on irrigated lands in the middle and lower reaches of the river causing salinisation.

The intensive salinisation of water and the high concentration of pesticides in water has led to the water in the lower parts of the Rivers Syr Darya and Amu Darya (see Chapter 15) becoming practically unsuitable for drinking. The problem is exacerbated by the diversion of water for irrigation (very often with further discharge of irrigation returns into depressions) which, in the 1980s, led to the near cessation of natural water discharge into the Aral Sea. This, in turn, led to the well-known sharp decrease in lake level, a reduction in its surface area and consequently an intensive process of desertification of vast surrounding areas. An additional important factor of salinisation now is the atmospheric salt input as a result of aeolian transfer from the dried bed of the

Aral Sea to the eastern basins of the Amu Darya and Syr Darya. The Aral Sea is the most dramatic example of the consequences of uncontrolled development of irrigation without proper consideration of all the possible ecological and socio-economic consequences.

5.1.4 Mining activities

Strong local and sometimes regional influences on salinisation of natural waters are produced by the extraction of mineral resources, especially of coal and salt. The total volume of water discharged from various mines in the USSR in 1989 was 3.2 km^3 (Main Statistics, 1990).

The greatest influence of this factor was found in Donbass, the largest coal-mining and industrial centre of the former Soviet Union. Of the total amount of wastewater entering the surface water network of Donbass and particularly small rivers, the greatest proportion was occupied by mine waters with an annual volume of discharge of 400–500 × 10^6 m^3 (Ivanova et al., 1980; Peltikhin et al., 1981). Thus, for the Lozovaya and Kamyshevakha rivers (tributaries of the River Lugan) and for the Krepenkaya and Nagolnaya rivers (tributaries of the River Mius) the share of mine waters in the annual runoff of the river was 60–70 per cent (Peltikhin et al., 1981). During periods of low flow, the proportion of mine waters in many rivers reached 70–90 per cent (Peltikhin, 1985). Considering mineralisation of mine waters varies from 1 g l^{-1} TDS to 35 g l^{-1} TDS for different mines (Kroik and Charun, 1986), with mean values varying between 1.1–5.7 g l^{-1} TDS or more in different areas (Karpushin et al., 1986), many small rivers of Donbass can clearly no longer serve as a water source for domestic and even industrial needs. For example, the average mineralisation of mine waters discharged into the basin of the River Volchya is 3.7 g l^{-1} TDS, and the median value of mineralisation of water at the Vasilkovka cross-section reached 2.75 g l^{-1} TDS in the second half of the 1980s (Table 5.1). This increase is caused by more than a two-fold increase in sulphate concentration in the discharged mine waters. The total ion flux in mine waters discharged into the Donbass rivers in 1975 was 1.15 × 10^6 tons, an increase of 40 per cent compared with 1963 (Peltikhin et al., 1981). Sulphate ions comprise the largest share (40 per cent) of the total ion flux and amount to 435,000 t, causing the widespread occurrence of sulphate-type waters in most rivers of Donbass.

In phreatic groundwaters used for water supply, a significant source of salinisation is settling ponds. The total discharge of NaCl-rich waters into settling ponds in Western Donbass in 1981 was 75,000 m^3 (Antroptsev et al., 1987). Mineralisation of water in these ponds varied from 2.6 g l^{-1} to 12 g l^{-1} TDS. The increase in Na$^+$ and Cl$^-$ ion concentrations was detected up to a depth of around 30 m when mineralisation of groundwaters increased to 2 g l^{-1} TDS.

Water salinisation in the reservoirs of Donbass has also occurred. During the 20-year period of operation of six typical reservoirs in the central part of Donbass, water mineralisation in flood periods increased 1.5–2.5 times (i.e. by 600–900 mg l^{-1}), and during low-flow periods 1.3–1.8 times. The chemical composition of water also changed (Peltikhin et al., 1980).

During the last 15–20 years, there have been significant changes in the quantitative and qualitative composition of mine waters, in particular the volume of acidic waters with a high concentration of sulphate has decreased to a great extent and the standard of water purification in mine wastewater treatment facilities has improved.

Mining of oil shale deposits and the release of associated mining drainage has resulted in the salinisation of surface and groundwaters in Estonia (Kuptsov and Khoretskaya, 1987). Mining works have also resulted in lowering groundwater levels 20–60 m below the mining face. Average long-term water drainage in mines is approximately 300,000 m^3 d^{-1}. The development of large depressions, collapsed roofs and subsidence of lower borders of unsaturated zones has led to changes in conditions of groundwater formation and their hydrochemistry. The natural mineralisation of waters of the Ordovician aquifer is 400–600 mg l^{-1} TDS with some wells having high sulphate and iron concentrations of up to 1,000 mg l^{-1} and 2 mg l^{-1} respectively. The total increase in mineralisation of surface waters due to discharge from mines has been exacerbated by the influence of wastewater discharge from oil shale processing. As an integrated result of these activities, mineralisation of surface waters in north-east Estonia increased from 250 mg l^{-1} TDS in the 1950s to 1,200 mg l^{-1} TDS in the 1980s.

The negative impact of the mining industry is especially marked on the vulnerable ecosystems of Siberia and the far North. This is illustrated by diamond extraction processes in the basin of the Vilyui river (a tributary of the Lena river). The open pit 'Mir' discharges 5×10^6 m^3 a^{-1} of hydrosulphide waters containing up to 95–100 g l^{-1} NaCl into the Irelyakh river; this is equivalent to 0.5×10^6 t a^{-1} of salts (Shpeizer et al., 1986). As a result, the River Irelyakh has turned into a 30–35 km wastewater canal. The River Malaya Boutobia, into which the Irelyakh river flows, has also lost its fishery potential. At the confluence, water mineralisation in the flood period, when the main discharge from the settling pond usually occurs, was 2–13 g l^{-1} TDS and in the low-flow period it was 2–3 g l^{-1} TDS. The plume of highly mineralised waters of the diamond extracting industry can be traced along the Vilyui river over a distance of 150 km (Shpeizer et al., 1986).

In the Ob basin, the influence on salinisation of natural waters may occur due to the inflow of brines during the development of gas and oil industries in Western Siberia. Unfortunately, because of the lack of permanent monitoring networks in areas of gas and oil extraction there are no reliable data on this effect. As an indirect proof of the extent of this impact, data on changes in Cl^- concentrations at most downstream

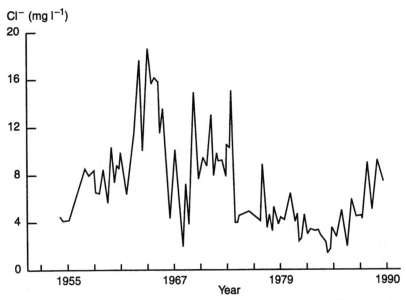

Figure 5.5 Long-term changes in chloride concentrations of the River Ob at Salekhard

stations of the River Ob at Salekhard have been considered (Figure 5.5). Peak concentrations in the mid-1960s are probably due to the massive discharge of brines into the River Ob during the exploration of oil and gas fields.

Accidents at mills and mines are also a serious threat to aquatic ecosystems. An accidental discharge of liquid waste from the Stebnikovsky potassium plant is an example of serious accidental saline pollution. As a result of dam failure at a sludge reservoir on 15 September 1983, 500 km upstream of Novodnestrovsk town, the Dniester river received $4.5 \times 10^6 \, m^3$ of brine containing about 250 g l^{-1} of salts (sodium, potassium and magnesium chlorides and sulphates) (Present State, 1986). The brine was moving with the Dniester flow to the town of Novodnestrovsk at a speed of 30 km d^{-1}. Maximum mineralisation was reduced along the flow at a rate of 16.5 per cent per 100 km. By the end of September, the main mass of salts (approximately 1×10^6 t) had accumulated as a bottom layer behind the dam of the Novodnestrovsk water reservoir. Close to the dam, the thickness of the salt layer reached 8 m. To deal with this problem, a separate regime of discharge for surface and bottom water layers was operated and floating pumps were installed for pumping out the brine. By the end of 1984, the brine which remained in localised hollows in the reservoir bed had a volume of about $1.6 \times 10^6 \, m^3$. To remove the brine completely, each hollow had to be treated separately. The Novodnestrovsk reservoir dam and the control of water discharge prevented strong saline pollution downstream that may have been lethal to biota in the lower reaches of the river. However, benthic fauna in the upper reaches of the river were all killed and serious disruptions in the functioning of the Novodnestrovsk water reservoir ecosystem were also observed (Nikanorov et al., 1988).

5.2 General salinisation trends in surface water

The previous section described the principal factors involved in salinisation processes, and Table 5.1 and Figure 5.1 reflect the geographic distribution and scope of salinisation in the former Soviet Union. These showed that in the vast territories of the Siberian and Far Eastern river basins and in the north-east of the European territory, river salinisation only occurs locally and mainly in regions where natural resources are extracted. Since the 1950s, regular monitoring of major ion concentrations in many large rivers (e.g. Ob, Yenisey, Lena, Indigirka, Kolyma, Anadyr, Kamchatka) has not shown any signs of regional salinisation (see Chapter 2).

At the same time in western, central and especially southern parts of the country, the salinisation process has become more widespread and regional salinisation in some rivers and river basins has occurred to such an extent that water use has become restricted or totally impossible. The range of mineralisation in the rivers affected by salinisation has varied from 30 mg l^{-1} to 3,000 mg l^{-1} TDS. For most rivers, in spite of a significant relative increase in salt concentrations, their absolute values did not exceed MACs.

The greatest long-term increase in major ion concentrations in surface waters has been noted for Cl$^-$ ions (up to 8–10 times), SO$_4^{2-}$ and Na$^+$+ K$^+$ (up to 4–6 times). A more limited increase has been observed for Mg^{2+} (up to 2–3 times) and Ca^{2+} (up to 1.5–2 times). For many rivers of southern Russia (e.g. Don, Terek, Ural, Kuma), Central Asia and Kazakhstan, as well as in the lower basin of the River Kura, values exceeding the MAC for sulphates (100 mg l^{-1}) and, less frequently, for chloride (300 mg l^{-1}), sodium (120 mg l^{-1}) and magnesium (40 mg l^{-1}) ions has been recorded. This impairs the quality of river water as a source for drinking water and leads to additional expenses in water treatment. For most rivers, long-term changes in HCO$_3^-$ concentrations have been statistically insignificant. For some rivers, where an increase in HCO$_3^-$ concentrations have been recorded, the increase has not exceeded 35 per cent. At the same time, in some rivers, a noticeable decrease in HCO$_3^-$ concentrations has been observed. In the Rivers Syr Darya, Amu Darya and Chu (Table 5.1) where average annual mineralisation has already exceeded 1 g l^{-1} TDS, the decrease in HCO$_3^-$ may be related to the precipitation of calcium carbonate. In rivers with relatively low mineral concentrations found in areas of major land drainage (e.g. Desna, Styr), the decrease could

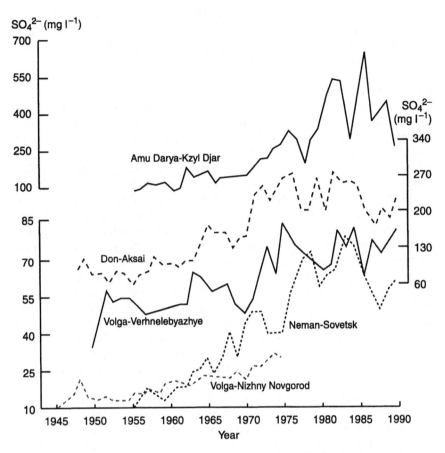

Figure 5.6 Long-term changes in sulphate concentrations of some large rivers in the former USSR

probably be explained by the process of draining peat bogs (Zhuk-hovitskaya and Kadatskaya, 1981).

The long-term changes in major ion concentrations, described by Nikanorov and Tsirkunov (1984, 1991) for the former USSR, show that the greatest increase in concentrations occurred from the 1950s to the mid-to-end of the 1970s. In the 1980s, concentrations were stabilised in most rivers. An example of long-term changes in sulphates for large rivers of the former USSR is shown in Figure 5.6.

The selection of the main factors discussed above which have led to salinisation in certain regions and river basins may be rather arbitrary because of the lack of reliable data for some pollution sources. In each region and river basin simultaneously various factors may influence sali-nisation. This could clearly be seen in the example of Estonia, where salinisation due to land drainage and the use of fertilisers has had an intensive regional impact, in addition to factors such as transboundary atmospheric inputs and local discharge of domestic and industrial waste-waters as well as mining effluents.

5.3   The impact of irrigation on groundwater quality

The principal factor influencing salinisation of groundwater in the former Soviet Union is irrigation. Satisfying the needs of irrigated farming has led to irrevocable water withdrawal from surface and groundwater sources. Water withdrawal exceeding the critical limit has led to a reduction in volume of water bodies and the degradation of delta areas. A reduction or cessation of flooding causes deltas to dry out, causes an increase in mineralisation and contamination of groundwater and a general reduction in groundwater resources. In conditions of insufficient natural drainage, irrigation causes a rise in groundwater levels. In arid zones, the rise in groundwater levels leads to secondary soil salinisation, the creation of salt marshes and to salinisation of groundwaters.

The drainage network, used for management of water and salt regimes of soils, has caused problems for the use of irrigation returns enriched by salts and anthropogenic pollutants (e.g. pesticides, herbi-cides, nitrates). This has led to pollution of both river waters, as a result of mixing with drainage discharge, and groundwaters due to filtration directly from the river bed and from irrigated fields. The changes in mineralisation and level of groundwater are illustrated by the example of the Kislovodsk irrigation system in the Volgograd region (Figure 5.7).

The principal changes in groundwater salinisation recorded in the different Republics of the former Soviet Union are as follows:

- *Russian Federation*. The main irrigation areas are concentrated in the southern parts of the Russian Federation. As a result of groundwaters rising in the valleys of the Rivers Don (in middle and lower reaches) and Western Manych up to critical levels, their mineralisation has increased from 3–5 g l^{-1} to 20–30 g l^{-1} TDS. In the Volga-Akhtuba flood plain, groundwater mineralisation has increased from 0.3 g l^{-1} to 9 g l^{-1} TDS and in the Volga valley mineralisation increased from 0.6 g l^{-1} to 6 g l^{-1} TDS. These changes were caused not only by irriga-tion itself, but also by over-regulation of the River Volga and by a reduction in the groundwater discharged into it. In the plains of North Caucasus, the formation of highly mineralised groundwaters has occurred in formerly dry alluvial sediments (Left-Yegorlyk and Right-Yegorlyk irrigation systems). In the vicinity of the Kuma-Manych canal, an increase in groundwater level from 5–10 m to 2–5 m has led to an increase in mineralisation from 1–10 g l^{-1} to 20 g l^{-1} TDS. The groundwater quality has deteriorated slightly in the deltas of the Rivers Terek and Sulak and on the lower reaches of the River Podku-mok because of an increase in mineralisation up to 3–5 g l^{-1} TDS.

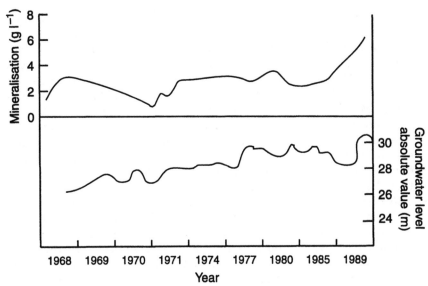

Figure 5.7 Long-term changes in mineralisation and groundwater levels in the Kislovodsk irrigation system, Volgograd region

- *Ukraine.* Under the influence of irrigation in the southern regions of the Ukraine (Odessa, Nikolaev, Kherson Regions), the formation of a groundwater horizon within the zone of influence of irrigation has occured at a rate of 0.2–1.3 m a^{-1}. In flood plains of small rivers (Large and Small Kuyalnik, Yalpug, Kirgnizh and others) the levels of groundwater mineralisation have increased. Within the Crimea Peninsula, groundwaters withdrawn for irrigation had led to a decrease in levels in the main aquifer of > 30 m and to water quality deterioration due to increased mineralisation. Further degradation due to the penetration of sea waters for 1–3 km within the peninsula has also occurred.
- *Moldova.* Substantial changes in the quality of groundwaters have been observed within the flood plain of the Rivers Dniester and Prut. In areas of Pliocene-Quaternary age, there has been a decrease in salinisation of groundwaters. In the vast area between the Rivers Dniester and Prut, there has been a reduction in groundwater mineralisation by 0.2–0.5 g l^{-1} TDS.
- *Georgia.* Irrigation is carried out mainly by overhead sprinklers which use smaller amounts of water. In the north-east part of Georgia, a slight increase in mineralisation has been noted.
- *Armenia.* Under the influence of irrigation, groundwaters have risen within the Ararat, Sevan and Shirak basins followed by an increase in mineralisation of the groundwater.
- *Azerbaijan.* The main irrigation areas are concentrated within the Kura-Araks lowland. A rise in groundwater to critical levels and an increase in

groundwater mineralisation up to 10–20 g l^{-1} have been typical for most irrigated areas in the Mugan, Shirvan, Mil steppes and others.

- *Kazakhstan*. The greatest changes in groundwaters occurred in the Kzyl Orda region, in the present and ancient delta and lower reaches of the River Syr Darya, where mineralisation of groundwaters has increased up to 10 g l^{-1} TDS. Within the Ural region, on irrigated land between the Rivers Maloe Uzen and Zauralny Syrt, there has been an increase in groundwater levels of 2–3 m and mineralisation has also increased. Similar changes are characteristic for the Ural-Kushum irrigation system. In the rest of the country, significant changes in groundwater have occurred in the valleys of the Rivers Uil, Irgiz and others and where there has been an increase in groundwater mineralisation.
- *Uzbekistan*. In the Golodnaya, Dji-Zak and Karshin steppes, the Bukharo-Karakul oasis, Surkhandarya, Fergana and Zeravshan valleys and new areas of irrigation, there has been an increase in groundwater level by approximately 20 m. The depth of groundwater is controlled by drainage at a level of 2–3 m from the surface. The decrease in mineralisation of groundwater to 1–3 g l^{-1} TDS is widespread. Exceptions to this are the lower reaches of the Amu Darya where, against a background of natural mineralisation of groundwater (up to 3 g l^{-1} TDS), there is strong pollution of anthropogenic origin.
- *Tajikistan*. Similarly to Uzbekistan, the main areas of irrigation are located in endorheic depressions, river valleys and alluvial fans. Under the influence of irrigation, the level of groundwater in the Dalversin steppe has risen by 15–20 m, in the Golodnaya steppe by 10–50 m and in Lyakkat basin by 10–15 m. Natural mineralisation of groundwaters varies from 1–3 g l^{-1} TDS but, in some areas, pesticide and nitrate pollution have been observed. In the Vakhsh, Yavan and Dangirin valleys, between 30 and 60 per cent of the fertilisers used are flushed out into groundwater. For example, mineralisation of groundwater used for water supply at the Tuya-Bugazsai alluvial fan has increased since 1963 from 0.8 g l^{-1} to 1.7 g l^{-1} TDS, its hardness has reached 19.6 meq l^{-1} and the nitrate concentration has increased to 50 g l^{-1}.
- *Kyrgyzstan*. Deterioration has occurred in the Chuy valley and in the Osh basin in areas of heavy clay, where there is highly mineralised groundwater.
- *Turkmenistan*. Irrigation in Turkmenistan has a 1,000-year history and therefore this section deals with relatively newly irrigated lands (i.e. irrigated in the last 30–40 years). In the Kopetdag plain, the level of groundwater at Kaakhin (which is irrigated) has increased by 10 m, at Gyaur by 27 m and at Ashkhabad by 25 m. Mineralisation has increased to 3–5 g l^{-1} TDS. For deltas of the Rivers Murgab and Tedjen, the difference in groundwater level is 3–30 m and mineralisation is 1–3 g l^{-1} TDS. In the Sarykamysh delta region of the River Amu Darya, under the influence of irrigation, the groundwater level

has risen by 20–50 m and is now at a depth of 0–3 m. The rate of the rise was 0.8–1.1 m a^{-1}. Phreatic groundwaters are characterised by higher than normal mineralisation and by pollution from agricultural activities. The information presented in this chapter shows that salinisation of surface and groundwaters in the former Soviet Union is a serious problem. It is especially crucial in the basin of the Aral Sea in Central Asia and Kazakhstan. Here it has become necessary to undertake complex measures to improve living conditions for the population and, primarily, to supply safe drinking water. Solutions to this problem should include the transfer of economic activity from extensive forms to sustainable ones, including the implementation of efficient water consumption strategies, the reduction of irrigated land area and the ban of monoculture in agricultural practices.

5.4   References

(All references are in Russian unless otherwise stated).

Antroptsev, A.M., Antonov, U.I. and Kudryavtsev, T.D. 1987 The influence of industrial mining activity in the Western Donbass on natural water mineralisation. *Hydrochemical Materials*, **15**, 120–127.

Chebotareva, A.G. 1988 Mineralisation of rivers in Moldova. Shtiinza, Kishinev, 131 pp.

Chembarisov, E.I. and Bakhritdinov, B.A. 1989 Hydrochemistry of river and drainage waters. Ukituvchi, Tashkent, 232 pp.

Chertko, N.K. 1982 Fluxes of elements by small and average-sized rivers in Byelorussia exposed to drainage and chemical melioration impact. Newsletter of the Byelorussian University, 2(3), 46–51.

Doniec, M., Grela, J. and Madej, P. 1986 Perspektywy zasolenia gornej Wisly wodami kopalnianymi przy uwzglednieniu zmian zagospodarowania zasobow wodnych. Gosp. wod., **46**(7), 165–169. (In Polish).

Garasevich, I.G. 1990 The basic features of the hydrochemical regime of rivers in the Pripyat Polesie Region under land-reclamation. *Hydrochemical Materials*, **108**, 13–30.

Helsel, D.R. 1993 Statistical analysis of water quality data. In: *US Geological Survey, National Water Summary 1990–1991 – Hydrologic Events and Stream Water Quality*, US Geological Survey Water Supply Paper 2400, 93–100. (In English).

Hirsch, R.M., Slack, J.R. and Smith, R.A. 1982 Techniques of trend analysis for monthly water quality data. *Water Resources Research*, **18**, 107–121. (In English).

Hygienic Assessment of Sewage Water Purification by Soil 1976 Meditsyna, Moscow, 184 pp.

Ivanova, A.A., Lebedeva, E.M. and Klepeshnev, A.M. 1980 Mine water influence on the composition and properties of surface waters in Donbass. *Hydrochemical Materials*, **8**, 43–53.

Izrael, Y.A. 1983 *Acid Rain*. Gidrometeoizdat, Leningrad, 269 pp.

Karpushin, N.M., Rodin, D.P., Sharikov, A.P. and Krasnyanskaya, O.A. 1986 The influence of anthropogenic factors on the hydrochemical regime of water bodies in the Donets-Makeevka industrial region. *Hydrochemical Materials*, **12**, 21–26.

Kirsta, B.T. 1975 Water mineralisation and chemical discharge of the Turkmenistan rivers and methods for their calculation. Ylym, Ashkhabad, 171 pp.

Krasutskaya, O.V. 1981 Ionic flux in rivers of the Neman basin under conditions of increased economic activity in the catchment area. In: *Economic and Geochemical Research of the Byelorussian Landscapes*. Nauka i Tekhnika, Minsk, 60–67.

Kroik, A.A. and Charun, I.E. 1986 Changes in chemical composition of surface waters of the Western Donbass under impact from the coal mining industry. *Hydrochemical Materials*, **12**, 88–96.

Kuptsov, A.N. and Khoretskaya, A.S. 1987 Mineralisation changes in river waters of Northern Estonia under increased economic activity. State Hydrological Institute, **19**, 122–131. Gidrometeoizdat, Leningrad.

Kuraptseva, S.V. 1990 Changes in the hydrochemical regime of some small rivers in Karelia following agricultural land-reclamation. *Hydrochemical Materials*, 65–78.

Laaksonen, R. and Malin, V. 1985 Regional water quality in Finland. *Aqua Fennica*, **15**(2), 201–209. (In English).

Leonov, E.A. 1979 Assessment of changes in water mineralisation of large rivers in the European part of the USSR under increased economic activity. State Hydrological Institute, **15**, 172–183. Gidrometeoizdat, Leningrad.

Lofvendahl, R. 1990 Changes in the flux of some major dissolved components in Swedish rivers during the present century. *Ambio*, **19**(4), 210–219. (In English).

Lukashev, K.I., Kovaleva, V.A. and Zhukhovitskaya, A.L. 1976 Conditions for the formation of chemical composition in marsh waters and their changes after land reclamation. In: *Water Quality and the Scientific Basis for its Protection*. Proceedings of the IV All-Union Hydrological Congress, Volume 9. Gidrometeoizdat, Leningrad, 208–291.

Main Statistics on Water Use in the USSR in 1989 1990 The Central Scientific Research Institute of Complex Use of Water Resources, Minsk, 45 pp.

Maksimova, M.P. 1991 Anthropogenic changes in ionic flux of large rivers in the USSR. *Water Resources*, **5**, 65–69.

Meybeck, M., Chapman, D. and Helmer, R. [Eds] 1989 *Global Freshwater Quality: A First Assessment*. Blackwell Reference, Oxford, 306 pp. (In English).

Nikanorov, A.M. and Tsirkunov, V.V. 1984 *Study of the Hydrochemical Regime and its Long-term Variations in the Case of Some Rivers in the USSR*. IAHS Publication, No. 150, 287–294. (In English).

Nikanorov, A.M. and Tsirkunov, V.V. 1991 Hydrochemical regime of USSR rivers. Analysis of long-term data. In: *Water Quality and the Scientific Basis for its Protection*. Proceedings of the V All-Union Hydrological Congress, Volume 5. Gidrometeoizdat, Leningrad, 336–344.

Nikanorov, A.M., Bryzgalo, V.A., Kosmenko, L.S., Tkachenko, T.B. and Slutskaya, N.V. 1988 Restoration of a freshwater ecosystem following severe salinisation. Gidrometeoizdat, Leningrad, 97 pp.

Nikanorov, A.M., Tsirkunov, V.V. and Laznik, M.M. 1989 Changes in hydrochemical regime and ionic flux of rivers in Latvia (long-term data). *Hydrochemical Materials*, **95**, 153–163.

Peleshenko, V.I. and Khilchevsky, V.K. 1991 Assessment of anthropogenic impact on chemical composition of river water in the Ukraine. In: *Water Quality and the Scientific Basis for its Protection*. Proceedings of the V All-Union Hydrological Congress, Volume 5. Gidrometeoizdat, Leningrad, 225–230.

Peltikhin, A.S. 1985 Mine water impact on the composition of river waters in the Azov region. *Hydrochemical Materials*, **93**, 11–17.

Peltikhin, A.S., Bogogosian, A.T. and Kaplin, V.T. 1980 Changes in mineralisation and ion content in Donbass reservoirs following a long period of operation. *Hydrochemical Materials*, **8**, 93–96.

Peltikhin, A.S., Bogogosian, A.T. and Kaplin, V.T. 1981 The influence of coal-mining wastewaters on river waters of the Ukrainian Donbass. *Hydrochemical Materials*, **9**, 25–28.

Pozdnyakova, G.V. 1991 Salt balance of Lake Balkhash and its anthropogenic changes after 1970. In: *Water Quality and the Scientific Basis for its Protection*. Proceedings of the V All-Union Hydrological Congress, Volume 5. Gidrometeoizdat, Leningrad, 383–389.

Present State of River and Reservoir Ecosystems in the Dniester Basin 1986 Shtiinza, Kishinev, 174 pp.

Rubinova, F.E. 1979 Changes in Syr Darya river discharge following water-management and building activities in its basin. Proceedings of SARNIGMI, **58**(139), Gidrometeoizdat, Moscow, 138.

Shpeizer, G.M., Rodionova, V.A. and Mazurova, T.M. 1986 Anthropogenic influence on the hydrochemical regime of the Vilyui river and its tributaries. *Hydrochemical Materials*, **13**, 62–71.

Smith, R.A., Alexander, R.B. and Lanfear, K.J. 1993 Stream water quality in the conterminous United States – Status and trends of selected indicators during the 1980s. In: *US Geological Survey, National Water Summary 1990–1991 – Hydrologic Events and Stream Water Quality*, US Geological Survey Water Supply Paper 2400, 111–140. (In English).

Stepanov, I.N. and Chembarisov, E.I. 1978 The influence of irrigation on river water mineralisation. Nauka, Moscow, 119 pp.

Tarasov, M.N. and Kriventsov, M.I. 1976 Changes in the hydrochemical regime of water bodies as a result of river impoundment and the development of irrigation. In: *Water Quality and the Scientific Basis for its Protection.* Proceedings of the IV All-Union Hydrological Congress, Volume 9. Gidrometeoizdat, Leningrad, 227–235.

Tarasov, M.N., Korolyov, I.A., Lapshina, T.P., Kobileva, E.A. and Zakharova, V.V. 1966 Chemistry of drainage waters from irrigated territories of the Northern Caucasus. Gidrometeoizdat, Leningrad, 118 pp.

Tsirkunov, V.V., Nikanorov, A.M., Laznik, M.M. and Dongwei, Z. 1992 Analysis of long-term and seasonal river water quality changes in Latvia. *Water Research*, **26**(9), 1203–1216. (In English).

Van der Weijden, C.H. and Middelburg, J.J. 1989 Hydrochemistry of the River Rhine: long-term and seasonal variability, elemental budgets, base levels and pollution. *Water Research*, **23**(10), 1247–1266. (In English).

Williams, W.D. 1987 Salinisation of rivers and streams: An important environmental hazard. *Ambio*, **16**(4), 180–185. (In English).

Zakrevsky, D.V. 1991 Drainage and land reclamation impact on the chemical composition of river waters in Pripyat Polesie in Ukraine. *Water Resources*, **6**, 50–59.

Zenin, A.A. 1965 Hydrochemistry of the Volga river and its reservoirs. Gidrometeoizdat, Leningrad, 259 pp.

Zhukhovitskaya, A.L. and Kadatskaya, O.V. 1981 The influence of land reclamation on hydrochemical conditions. In: *Materials on Anthropogenic-Geochemical Research of the Byelorussian Landscapes.* Nauka i Tekhnika, Minsk, 121–134.

Chapter 6[*]

EUTROPHICATION OF LAKES AND RESERVOIRS

The initial effects of anthropogenic eutrophication were first noted at the end of the last century although intensive study of this process only began in the 1960s. One of the principal problems of eutrophication is related to the enrichment of water bodies with nutrients which leads to an increase in productivity and an increase in organic matter — one of the most important factors which defines water quality.

6.1   The study of eutrophication

The problem of anthropogenic eutrophication of water bodies in the former Soviet Union has been addressed relatively recently compared with other countries. The first research on anthropogenic eutrophication of aquatic ecosystems started in 1963 on Lake Plestcheevo (Pereslavl) by the Institute of Geography of the USSR Academy of Science (Fedorova, 1967). Studies on Lake Seliger (Tver region) and the Valdai lakes (Novgorod region) were also carried out by the Institute in 1964. The slow change from initial oligotrophic and mesotrophic status probably occurred while the watershed was underdeveloped (Pokrovskaya and Rossolimo, 1967; Shilkrot, 1967).

In the 1960s, attention was focused on eutrophication of complex aquatic ecosystems including huge reservoirs (e.g. Dnieper cascade) and those reservoirs situated within the steppe zone. Along with natural and geographical factors promoting an increase in a reservoirs' trophic level, sources of nutrients related to human activities became more significant (Denisova et al., 1974). The 1970s witnessed the highest level of interest in anthropogenic eutrophication within the former USSR. During that time, anthropogenic eutrophication was detected in nearly all lakes and reservoirs in areas of intensive economic activity.

The rate and intensity of lake eutrophication is most clearly seen in lakes which have been studied regularly over numerous years, such as the following examples:

- *Lake Beloye* (Moscow region). The study of this lake began in 1910–12 and it was studied continuously from 1923 to 1940. From 1967, a sharp increase in eutrophication was noted (Shilkrot, 1968).

---

[*]   *This chapter was prepared by V.G. Drabkova*

- *The Narochan lakes* e.g. Naroch, Myastro and Batorino lakes (Byelorussia). The study of these lakes began in 1946 and they are still being studied up to the present. More detailed research is carried out on Lake Naroch than on the other lakes. These lakes have been subjected to a continuous increase in nutrient load of anthropogenic origin mainly caused by recreational activities. However, the ratio of catchment area to lake surface area is low, and large cities and industries are absent from the catchment which is covered mostly by woods and marshes. These features limit the flow of nutrients and pollutants into the lakes, especially into mesotrophic Lake Naroch, which helps to preserve the natural state of this water body (Ecosystem of the Narochan Lakes, 1985).

- *Lake Krasnoye* (Leningrad region). This lake has been studied since 1964. The study originally aimed to assess changes in the lake's regime under the influence of anthropogenic factors. It was found that the initial process of eutrophication within the lake was related to intensification of agriculture (Eutrophication of a Mesotrophic Lake, 1980).

- *Trakai Lakes* (Lithuania). The study of these lakes, situated in a developed area, has been carried out periodically since 1952. The studies have shown strong enrichment of the lakes with phosphorus and nitrogen and an expansion of the oxygen deficient zone (Klimkaite, 1977).

- *Lake Ladoga* (Russia). This long-term study commenced in 1956. Since the 1960s, the concentration of phosphorus in Lake Ladoga has increased three times, and this has affected the whole structure of its ecosystem. The lake status changed from oligotrophic to mesotrophic (Anthropogenic Eutrophication of Lake Ladoga, 1982).

- *Lake Sevan* (Armenia). The study of Lake Sevan began in 1923. Water resource use of Lake Sevan since 1939, mainly for irrigation, has led to a significant fall in lake level. The surface area diminished by 12 per cent or 170 km^2. The main changes in morphology and chemistry which occurred in the lake are shown in Table 6.1. During the last 30 years, the lake was affected by intensive processes of anthropogenic eutrophication. Until the 1950s, the lake had all the features characteristic of oligotrophic water bodies. Simultaneously, with the fall in water level and the morphometric changes in the lake, a particular hydrochemical and hydrobiological regime was established which was followed by quantitative and qualitative changes in the characteristic biocoenosis of its ecosystem (Parparov *et al.*, 1977). A shortage of oxygen in the hypolimnion and an increase in phytoplankton primary production are considered as the most negative consequences of eutrophication in all parts of Lake Sevan (Figure 6.1).

- *Lake Plestcheevo*. This lake has been studied periodically since 1921. Since 1978, general limnological studies have been performed which have detected signs of water quality and fisheries deterioration (Factors and Processes of Eutrophication of Lake Plestcheevo, 1992).

Table 6.1   Some limnological characteristics of Small and Great Sevan

	Small Sevan		Great Sevan	
	Period 1	Period 2	Period 1	Period 2
Surface area (km^2)	384	328	1,032	916
Volume (km^3)	19.5	12.8	39.0	20.9
Mean depth (m)	50.9	39.2	37.7	23.6
Transparency (m)	14.3	3.5	14.3	2.8
pH	9.2	8.6–8.8	9.2	8.6–8.8
NO$_3$-N (mg l^{-1})	0	0.10	0	0.10
PO$_4$-P (mg l^{-1})	0.32	0.025	0.32	0.025

Period 1   Natural condition before water abstraction (up to 1939)

Period 2 Recent condition (1977–82)
Source: Parparov, 1987

For nearly all large reservoirs, long-term ecosystem studies have been undertaken since their filling which allow for the determination of trophic state evolution. In addition to long-term studies, some short-term surveys have been carried out in regions where economic activities are most intensive. The goal of these studies has been to determine changes in aquatic ecosystem structure and functioning in lakes impacted by anthropogenic influence of varying intensities. Such surveys were performed in lake areas of Estonia, East Latvia, Lithuania, Karelia Isthmus (Leningrad region), Bolshezemelskaya tundra, Southern Aral (Uzbekistan), South Urals, Byelorussia (Braslav lakes), Trans-Baikal (Ivan-Arakhley lakes), and Karelia.

6.2   Origins of lake eutrophication

Eutrophication of water bodies is due to an excessive input of nutrients, particularly phosphorus and nitrogen, creating the possibility for rapid development of biological production processes when other favourable conditions are present. Hence, the concentrations of nutrients in water are the primary variables influencing potential eutrophication. Phosphorus is the main chemical element which controls eutrophication. Phosphorus in natural conditions limits the development of biological communities in most water bodies. Most oligotrophic and mesotrophic lakes are characterised by a phosphorus to nitrogen ratio (µg P:µg N) of 1:30 or 1:40 respectively. In eutrophic lakes the ratio varies from 1:15 to 1:25; in hypereutrophic lakes the ratio is 1:12 to 1:18; in lakes with a high content of humic substances, the ratio is 1:135 (Alekin *et al.*, 1985). In algal cells, the nitrogen content is only 6–17 times greater than that of phosphorus. Therefore, nitrogen in natural water bodies is practically always present in excess. In the assessment of eutrophication, due to the ratio of phosphorus and nitrogen, attention generally should be given to

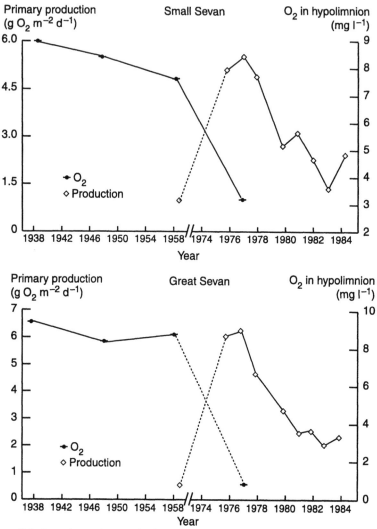

Figure 6.1 Long-term changes in the values of phytoplankton primary production (mean values for the vegetation growth period) and oxygen content in the hypolimnion (mean values for August) of Small Sevan and Great Sevan (After Ghezalian, 1983; Parparov, 1987)

phosphorus as the main element responsible for eutrophication (although it should be remembered that this is not always the case).

6.2.1 Input of nutrients from different sources
The proportion of diffuse phosphorus flux which cannot be controlled (e.g. runoff from agricultural land and atmospheric precipitation) is

approximately 20 per cent of all anthropogenic inputs into water, while 80 per cent is related to untreated or poorly-treated point sources from cities, industry and farming.

The total population growth of the country cannot account alone for the increase in phosphorus fluxes. In 200 years (1780–1980), the population has grown 6.3 times from 42 million to 264 million, while the total phosphorus content in domestic sewage has increased 395 times. This is due to an increase in urban populations (from 2 per cent to 63 per cent of the total population) and due to an increase in the use of phosphorus detergents (Alekin *et al.*, 1985).

A similar situation has been observed for livestock. The input of nutrients into water from livestock farms has grown faster than the number of cattle. This is explained by changes in farming technology. During the period 1916–81, the number of cattle increased from 58.4 million to 115.1 million (National Economy, 1981) while the phosphorus input into water through livestock effluent increased 60–70 times.

The flux of nutrients from agricultural lands is higher than from uncultivated land. In comparison with coniferous forests, phosphorus outflow from agricultural fields is 2–10 times greater and nitrogen 1.5–5 times greater. Ploughing an area without changing other conditions leads to an increase in the outflow of nutrients. Moreover, ploughing in the basin of lakes with relatively small catchment areas greatly affects the phosphorus flux. An increase in ploughed area from 10 per cent to 45 per cent increased the phosphorous load on small lakes in Estonia by approximately four times. The higher the catchment to lake ratio, the smaller the influence of ploughing (Nazarov, 1988).

The phosphorus flux directly from agricultural fields due to the use of phosphorus mineral fertilisers is not greater than 2 per cent of the total anthropogenic phosphorus flux into water. However, it should be noted that in the former Soviet Union, the loss of mineral fertiliser at the stages of production, storage and transportation accounted for up to 55 per cent of the total phosphorus flux into water.

The origin of anthropogenic phosphorus fluxes into aquatic ecosystems in different regions varies. In the European part of the former USSR the following zones were distinguished (Razumikhina and Monina, 1991):

- Zones with prevailing phosphorus flux of municipal origin (e.g. Moscow and Leningrad regions);
- Zones with prevailing phosphorus flux of livestock origin (e.g. mainly the northern part of non-chernozem regions);
- Zones with prevailing phosphorus flux of crop-growing origin (e.g. central chernozem regions, Ukraine).

The correlation of phosphorus fluxes from different sources for large and small lakes also varies (e.g. Lake Ladoga and small lakes of the Pskov region (Figure 6.2 and see Chapter 19)). The structure of phosphorus input into the smaller lakes of the Pskov region is relatively

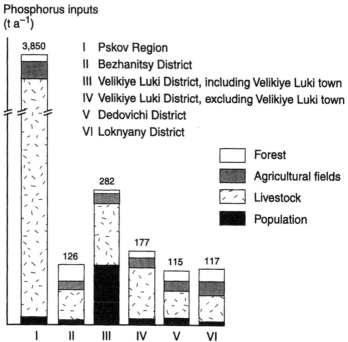

Figure 6.2 Phosphorus inputs into lakes of the Pskov region (After Alekseev and Golubetskaya, 1993)

similar. The greatest danger is from livestock; the phosphorus load from agricultural land is 1.7–3 times lower than that from livestock. The load from municipal wastes in comparison with livestock is 10 times lower (with the exception of the town of Velikiye Luki) (Figure 6.2). For Lake Ladoga, the main source of phosphorus is also livestock (up to 60 per cent of the total input) although industrial and municipal inputs are increasing (see Chapter 19). In Estonia, the input of phosphorus to Lake Pskov-Chud comes from various sources: 22 per cent from waters, 28 per cent livestock, 39 per cent crop-growing, 3 per cent precipitation and 8 per cent natural load (K.A. Simm, Personal communication, Institute of Zoology and Botany, Estonia). The percentage of phosphorus and nitrogen input from various sources into reservoirs has been illustrated by the Ivankovo reservoir, Moscow region (Table 6.2).

6.2.2   External nutrient load

The relationship between phosphorus load and bioproductivity has been well studied. This section considers only the widely accepted correlations between (i) external nutrient load (ratio of the nutrient flux from the external source (e.g. atmosphere, catchment area) per unit of lake surface area) and (ii) nutrient concentrations in the water body and

Table 6.2   The percentage contribution of different sources of nutrient input into the Ivankovo reservoir

Sources	Nitrogen (%)	Inorganic phosphorus (%)
Towns	17.5	47.0
Mineral fertilisers	60.4	34.5
Livestock	6.5	13.8
Production of peat fertilisers	15.5	4.2
Recreation	0.1	0.5

Source: Avakyan *et al.*, 1977

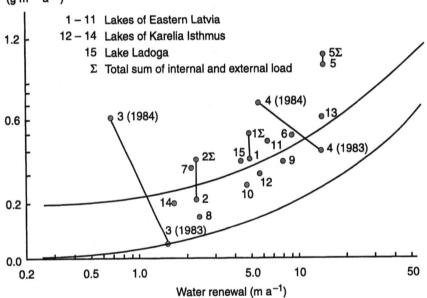

Phosphorus load
$(g\ m^{-2}\ a^{-1})$

1 – 11 Lakes of Eastern Latvia
12 – 14 Lakes of Karelia Isthmus
15 Lake Ladoga
Σ Total sum of internal and external load

Water renewal $(m\ a^{-1})$

Figure 6.3 Evaluation of the trophic status of lakes of the north-west by phosphorus load (Lp), average depth (z) and water renewal (ρ) using Vollenweider's (1968) method (After Drabkova and Stravinskaya, 1989). For lake number see Table 6.3

primary production, described by Vollenweider (1968, 1974, 1976), Dillon and Rigler (1974) and their numerous successors.

The results of all-year-round studies of phosphorus fluxes from all sources and the simultaneous study of water balance characteristics have allowed the determination of the total phosphorus loads in lakes of the north-west (Table 6.3). Load values were incorporated into Vollenweider's scale where they were plotted against average depth, and water residence time, of lakes (Figure 6.3). The upper line on Figure 6.3 is the

Table 6.3   External phosphorus load (L), renewal rate (ρ), phosphorus retention coefficient (R), and phosphorus load (P) according to Dillon (1975), in the lakes of the North West of the former USSR

Lake	Years	L ($g\ m^{-2}\ a^{-1}$)	ρ (a^{-1})	R	P ($g\ m^{-2}\ a^{-1}$)
Eastern Latvia[1,2]					
1. Rudush	1981–82	0.41	0.93	0.26	0.33
	1982–83	0.43	1.09	0.21	0.31
	1983–84	0.28	0.66	0.31	0.29
2. Grizhan	1981–82	0.22	0.53	0.27	0.31
	1982–83	0.27	0.58	0.26	0.34
	1983–84	0.16	0.36	0.39	0.26
3. Ilzes	1982–83	0.12	0.61	0.67	0.066
	1983–84	0.66	0.26	0.95	0.12
4. Lapiitis	1982–83	0.42	7.4	0.29	0.04
	1983–84	0.81	3.0	0.74	0.07
5. Laborzh	1981–82	1.44	5.1	0.31	0.19
	1982–83	1.38	6.1	0.33	0.15
6. Udrinka	1981–82	0.58	0.94	0.59	0.25
	1982–83	0.48	1.04	0.56	0.20
7. Vorkalu	1977–78	0.53	1.11	0.54	0.22
	1978–79	0.36	0.73	0.41	0.28
	1979–80	0.14	0.51	0.42	0.16
8. Sleinovas	1977–78	0.18	1.57	0.48	0.06
	1978–79	0.29	1.81	0.42	0.09
	1979–80	0.08	0.95	0.34	0.055
9. Snidzinyas	1977–78	0.40	2.24	0.25	0.13
	1978–79	0.52	2.89	0.39	0.109
	1979–80	0.15	1.08	0.20	0.11
10. Becheru	1977–78	0.35	1.28	0.49	0.14
	1978–79	0.30	1.35	0.44	0.12
	1979–80	0.13	1.00	0.40	0.08
11. Rogaizhu	1977–78	0.62	1.68	0.47	0.20
	1978–79	0.59	1.71	0.36	0.22
	1979–80	0.25	0.91	0.44	0.15
Karelia Isthmus					
12. Krasnoye[3]	1973	0.15	0.82	0.03	0.18
	1974	0.30	–	0.22	–
	1976	0.27	0.76	0.36	0.23
	1977	0.27	0.76	0.37	0.22
13. Maloe Lugovoye[3]	1977–78	0.69	7.5	0.38	0.06
14. Nachimov[3]	1979–80	0.21	0.28	0.64	0.27
15. Ladoga[4]	1959–62	0.14	0.081	0.702	0.51
	1976–79	0.39	0.083	0.701	1.38

Phosphorus load $P = L(1-R)/\rho$
[1]   Sorokin, 1983
[2]   Kalinina and Prytkova, 1988
[3]   Stravinskaya and Ulyanova, 1984
[4]   Raspletina and Gusakov, 1982

critical phosphorus load, above which lakes are classified as eutrophic. The lower line corresponds to a permissible load and defines oligotrophic and mesotrophic water bodies.

The estimate of the phosphorus load in lakes of the Narochan group (Byelorussia), using average phosphorus concentration in precipitation, main inputs and average long-term water balance, was 0.04 g m^{-2} a^{-1} for Lake Naroch, 0.12 g m^{-2} a^{-1} for Lake Myastro and 0.25 g m^{-2} a^{-1} for Lake Batorino (Zhukova, 1991). For the Pskov-Chud lake, the values were 0.8 g m^{-2} a^{-1} for Pskov Lake and 0.1 g m^{-2} a^{-1} for Chud Lake, while the estimated critical load was lower at 0.21 g m^{-2} a^{-1}.

In the initial stages of eutrophication, studies showed a direct dependence of phytoplankton primary production and degradation processes and nutrient input from nutrient load. However, it became clear that the concept of external phosphorus load as the main factor controlling eutrophication could not explain all the changes in lake development. This was especially true of eutrophic lakes which were subjected to continuous nutrient input. Thus, nutrient input does not always cause a proportional change in the level of biological productivity of water bodies. This depends, to a great extent, on internal processes in lakes, on the accumulation of nutrients by sediments and on their turnover rate.

6.2.3 Internal nutrient load

An internal load is the result of the diffusion of nutrients from the interstitial waters of the sediment into the overlying water. It characterises the partial return into the water of nutrients which entered the water body earlier, mostly in organic form, and were ultimately deposited in bottom sediments. While the external load reflects the degree of impact on a water body, the internal load reflects the intensity of the cycle and redistribution of nutrients within the water body itself.

As a rule, the absolute values of annual internal phosphorus load increase with an increase in the trophic level of a water body from zero for ultra-oligotrophic lakes, to > 10 g P m^{-2} a^{-1} for hypereutrophic water bodies (Martynova, 1993). Unfortunately, there are few data on internal loads for water bodies in the former Soviet Union. For the six lakes of Eastern Latvia which were studied in detail, the internal phosphorus loads measured were 0.01–0.16 g m^{-2} a^{-1} (Stravinskaya, 1989). It was shown that if the internal phosphorus load did not exceed 1 g m^{-2} a^{-1}, the internal load depended weakly on the external load and was not more than 0.1 g m^{-2} a^{-1}.

In non-carbonate lakes, there is a strong relationship between internal and external loads. When the external load increases from 1 g m^{-2} a^{-1} to 3 g m^{-2} a^{-1}, the ratio between external to internal loads decreases from 14 to 2. In reservoirs with extremely high external loads, up to 5–6 g m^{-2} a^{-1}, the phosphorus flux from bottom sediments does not exceed 0.3 g m^{-2} a^{-1} (e.g. Kiev reservoir). Similarly for carbonate lakes,

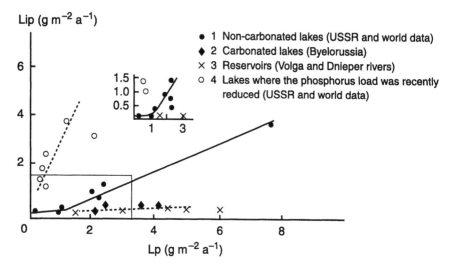

Figure 6.4 The relationship between internal phosphorus load (Lip) and external load (Lp) for various water bodies (After Martynova, 1993)

the ratio of external and internal loads changes from 8–9 to 20–40 (Figure 6.4).

Mobile forms of iron in lake sediments may reduce greatly the internal phosphorus load. This is illustrated by the example of Narochan lakes of Byelorussia (Zhukova *et al.*, 1990). In the highly productive Batorino lake, the quantity of mobile forms of iron in the sediment was twice that of Lake Naroch and the internal phosphorus load in Lake Batorino was only 0.19 g m^{-2} a^{-1}. In mesotrophic Lake Naroch, the internal phosphorus load was about 0.4 g m^{-2} a^{-1}, which is 10 times greater than the external load.

Comparison of external and internal loads for a water body allows the determination of the principal factors which regulate the nutrient content of the water body.

6.3   Biological indicators of anthropogenic eutrophication

Changes in nutrient regime may influence the structure of aquatic ecosystems. If the level of bioproductivity increases, the composition of biological communities changes and the stability of trophic links is disturbed. Structural changes are important, because their study allows a better understanding of the ecology of different species, their adaptation characteristics and the specific features of their physiology. It should be stressed that structural changes cannot always be detected clearly. However, even when changes in specific composition of aquatic communities are not observed, anthropogenic eutrophication is followed by considerable variations in quantitative variables.

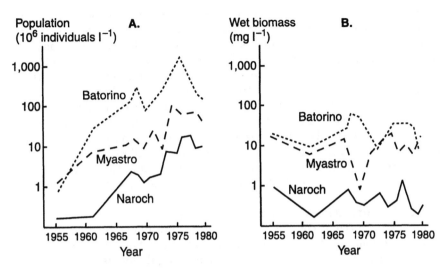

Figure 6.5 Phytoplankton in Byelorussian lakes in August. **A**. Population numbers;
B. Wet biomass (After Mikheeva, 1985)

6.3.1 Qualitative changes in population structure
Structural changes in biological communities subjected to anthropogenic eutrophication have been widely studied. Some results have been universally accepted and data from water bodies of the former Soviet Union have also contributed to this understanding.

The lakes of Eastern Latvia showed that, as the process of anthropogenic eutrophication developed and the concentration of nutrients increased, the phytoplankton composition changed considerably towards an increase in blue-green and green algae which are consistently dominant in hypereutrophic shallow lakes. In deeper, stratified lakes, also with high nutrient concentrations, mobile peridinium and chrysomonad species are dominant and cryptomonad and euglenid algae increase in the biomass (Trifonova *et al.*, 1988).

In the Narochan lakes (Byelorussia), the process of eutrophication led to the substitution of large-cell species with small-cell species. This occurred at the expense of the development of small blue-green algae, green and other algae and the decrease in larger diatoms and dinoflagellates (Mikheeva, 1985). As a result, phytoplankton numbers increased rapidly but the biomass remained relatively stable (Figure 6.5).

The leading role of nutrients in phytoplankton succession has been shown by experimental lake fertilisation. When the original phytoplankton composition was modified, large algal forms with complex shapes were substituted by small algae of simple form (Lavrenteva, 1982).

The structure of heterotrophic communities also changes with an increase in the level of eutrophication. Zooplankton respond to an increase in trophic status by an increase in the proportion of smaller species, such as rotifers. This leads to an increase in the total size of the community, but also to a decrease in total zooplankton biomass in hypereutrophic lakes. In lakes of Eastern Latvia which are similar in terms of morphometry, the proportion of small zooplankton species (i.e. protozoa and rotifers) increased from 16 per cent to 65–70 per cent of the total zooplankton biomass with increasing eutrophication. The increase in trophic level was followed by a decrease in the number of zooplankton species which formed the main part of the biomass, and they changed the diversity index from 3.2 to 0.76 (Lavrentev *et al.*, 1988).

Unlike plankton communities, which react rapidly to changes in the aquatic environment, benthic communities are much slower to react.

6.3.2 Quantitative changes in biomass

Changes in nutrient composition are not generally followed by changes in aquatic community structure but are invariably followed by quantitative changes in biological indicators, such as algal biomass.

Phytoplankton biomass can be correlated with phosphorus concentration. The maximum concentration of phosphate-phosphorus, at which active growth of various algae is possible, is 1 mg l^{-1} P; above this level, there is a decline in phytoplankton development (Chu, 1943). Such concentrations are rare in lakes, even in lakes under strong anthropogenic impact. Data from Swedish lakes have confirmed these results (Forsberg and Ryding, 1980) with concentrations of chlorophyll *a* reaching their maximum when the total phosphorus concentration was 1 mg l^{-1}. A direct relationship between chlorophyll *a* and total phosphorus concentration in these lakes was observed only up to phosphorus concentrations of 0.1 mg l^{-1}. This relationship had an exponential character. The study of lakes in Eastern Latvia, the Karelian Isthmus and Bolshezemelskaya tundra, where the total phosphorus concentration did not exceed 0.15 mg l^{-1}, also showed a direct relationship between concentrations of phosphorus and chlorophyll *a* (Figure 6.6). The smaller lakes of Estonia also showed a clear relationship between these two indicators (Figure 6.7).

Low transparency of lake water can limit phytoplankton development even in the presence of extremely high (8–9 mg l^{-1} P) phosphorus concentrations (e.g. Dautkul lakes in the Aral region) (Konstantinova, 1987).

The biomass of heterotrophic organisms increases to a certain level and then decreases. This is illustrated by the Eastern Latvian lakes with different levels of anthropogenic impact. High phosphorus concentration sharply decreases the activity of heterotrophic organisms, but their numbers and biomass do not change significantly. Thus, in the deep lakes of this region which show marked stratification, the total number of bacteria increased simultaneously with an increase in phosphorus concentration up to 0.13 mg l^{-1}, and then bacterial numbers decreased. At the same time, bacterial biomass reached a maximum when the

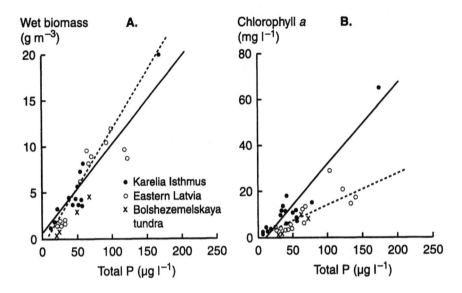

Figure 6.6 The relationship between phytoplankton (open water season) and average annual concentration of total phosphorus in various lakes. **A.** Wet biomass; **B.** Chlorophyll *a* (After Trifonova, 1990)

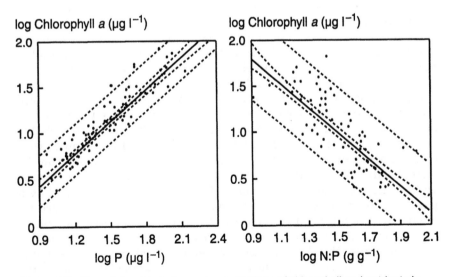

Figure 6.7 The relationship between concentrations of chlorophyll and nutrients in small lakes in Estonia (After Milius *et al.*, 1991)

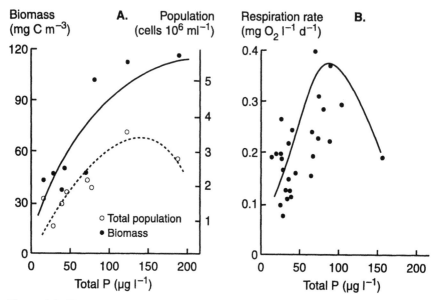

Figure 6.8 The dependence of bacterioplankton development on total phosphorus concentration in Eastern Latvian lakes. **A**. Population and biomass of bacteria; **B**. Bacteria respiration rate

phosphorus concentration in these lakes was also at a maximum. The activity of bacterioplankton, characterised by their respiration rate and their production, generally decreased when phosphorus concentrations exceeded 0.1 mg l^{-1} (Figure 6.8).

Other heterotrophic communities behaved in a similar way. While eutrophication of lakes of the Latgal Hills increased, zooplankton numbers increased, but overall biomass decreased due to the dominance of small forms (Makartseva and Lavrentev, 1989). Zooplankton (meso- and protozooplankton) activity also decreased in lakes with high total phosphorus concentrations. Phosphorus concentrations > 0.16 mg l^{-1} do not stimulate the development of invertebrates (Tsimdin *et al.*, 1982).

Studies of the lakes of Eastern Latvia (Latgal Hills) showed that the rate of organic matter degradation by zoobenthos depended on the phosphorus concentration of the water (Belyakov and Skvortsov, 1989). The trend for littoral zones of deep lakes was described by a parabolic function (Figure 6.9). The maximum allowable values for organic matter degradation corresponded to a phosphorus concentration of 0.05 mg l^{-1} P, after which there was a sharp decrease in values of this indicator. For the littoral zone of shallow lakes, the maximum corresponded to the higher phosphorus concentration of 0.11 mg l^{-1} P.

These examples show that an increase in phytoplankton activity corresponds to the increase in anthropogenic eutrophication within the

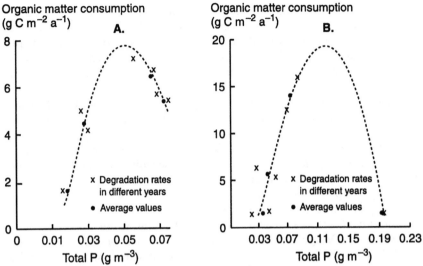

Figure 6.9 The relationship between organic matter consumption by zoobenthos and total phosphorus concentration in lakes of Eastern Latvia. **A**. Littoral zone of deep lakes; **B**. Littoral zone of shallow lakes (After Belyakov and Skvortsov, 1989)

limits of the phosphorus concentrations characteristic for water bodies in the north-west of the European territory.

As for macrophyte communities, an increase in organic carbon per square unit area of vegetation has been observed (Denisova, 1989). However, this trend is not always clearly seen, because water transparency and morphological features of water bodies can influence macrophyte production. The activity of heterotrophic forms, e.g. zooplankton and bacterioplankton, increases with increasing eutrophication up to concentrations of total P of approximately 0.1 mg l^{-1} and then decreases.

6.3.3 Reservoir eutrophication

Eutrophication processes in reservoirs differ from those in lake ecosystems. These differences are most clearly seen in the first years of reservoir operation. A major change occurs when flows are altered and riverine attributes are replaced by lacustrine attributes. In addition, the influx of large amounts of nutrients from flooded territories leads to a significant increase in productivity. The position of a reservoir in the river system and the features of the hydrological regime also influence the reservoir ecosystems.

Large river reservoirs in the former Soviet Union are often situated in cascades. In upstream reservoirs, the nutrient content depends directly on the influx from the inflowing rivers. In downstream reservoirs, nutrients are controlled by the influx from upstream reservoirs and the related nutrient transformation processes within the water body.

During the initial stage of reservoir filling the nutrient content in a reservoir is significantly influenced by the flooded areas. Enrichment with nutrients and organic matter, due to extraction from soil and the decay of flooded vegetation, is especially intensive in the first two or three years of reservoir operation. This is generally accompanied by intensive growth of blue-green algae. This effect decreases in time and is reflected by phytoplankton growth. The reservoir system on the Dnieper river provides a good example of this is. In the summers of 1965–67, the Kiev reservoir (situated in a forest zone) had an average wet phytoplankton biomass of 37–97 g m^{-3}. Since 1968, the average biomass has decreased to 2–24 g m^{-3}. In the Kakhovka reservoir, situated in a steppe zone, extensive phytoplankton blooms were observed in 1956–57. The average phytoplankton biomass in the summer of those years was 112–260 g m^{-3} but in 1958–68 it was only 18–39 g m^{-3} (Primachenko-Shevchenko *et al.*, 1973; Denisova, 1979). The same situation was observed in the Volga river system. In the first year of operation of the Kuybyshev reservoir in 1956, average phytoplankton biomass along the reservoir was 8 g m^{-3} in the vegetation growth period (reaching 44.3 g m^{-3} near the dam) due to growth of blue-green algae. In subsequent years, the intensity of phytoplankton development decreased to 1.5 g m^{-3} in 1958 and then increased slightly to 2.8 g m^{-3} in 1970 and 5.2 g m^{-3} in 1972 (Kuzmin, 1978).

Following reservoir filling, general trends can be observed: the variability of phytoplankton decreases and its biomass increases. Algal blooms in the initial years of reservoir operation occur practically everywhere, but in the following years their intensity and rate of occurrence decrease (Sirenko and Gavrilenko, 1978). In this respect, eutrophication in reservoirs differs from that of lake ecosystems. In addition to these general trends it is necessary to take into consideration the complicated hydrodynamic structure of reservoirs which facilitates the development of a large ecological variety of biotopes, especially in shallow areas.

Thus, initially, reservoirs are generally characterised by intensive eutrophication processes. The stabilisation phase is marked, as a rule, by a decrease in the level of eutrophication where the response to anthropogenic impacts becomes similar to that of lakes. The latter can be illustrated using the example of the Rybinsk reservoir on the Volga river which has been intensively studied since 1954 (Figure 6.10, see also Chapter 16) and where an increase in phytoplankton biomass has been observed since the beginning of the 1970s as a result of anthropogenic eutrophication. This period was characterised by an increase in phytoplankton biomass and an increase in heterotrophic communities (zooplankton and bacterioplankton).

Reservoirs in the European territory suffer from the most intensive processes of anthropogenic eutrophication whereas Siberian reservoirs are less affected (Table 6.4). As a rule, an increase in the eutrophication

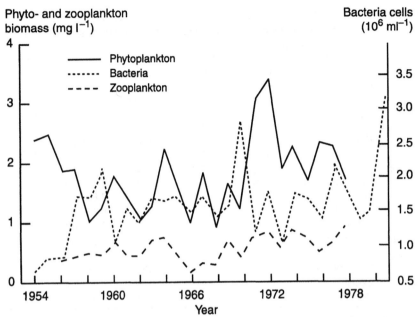

Figure 6.10 Changes in biomass of phytoplankton and zooplankton and the total number of bacteria in the Rybinsk reservoir on the Volga river (After Romanenko, 1985)

of reservoirs subjected to anthropogenic impact leads to a decrease in the self-purification capacity of the water and decreases the activity of heterotrophic communities (as occurs in lakes). This leads, in turn, to a silting-up of reservoirs, eventually turning them into bogs. This has already occurred in some parts of the Dnieper reservoirs, such as Kanev, Kakhovka and Kiev (Sirenko, 1989).

6.4   Rate of eutrophication

6.4.1   Changes in trophic status of water bodies

In the last few decades, the increase in phosphorus fluxes has influenced the trophic status of water bodies. The trophic status of lakes can be determined by the Carlson Index (Carlson, 1977). This index uses water transparency according to Secchi disc transparency (Z_s) and chlorophyll a concentration and express lake trophic status on a continuous scale (Table 6.5). As a result of its use, it can be shown that in regions of intensive economic activity practically no oligotrophic lakes remain. Most lakes have acquired the status of eutrophic and even hypereutrophic.

In order to analyse the spatial distribution of lakes by trophic level, a rapid survey was carried out in Byelorussia, Latvia, Lithuania, Estonia, Pskov region, Novgorod region, Leningrad region and Karelia (Table 6.6). Studies were performed using the same methods over an annual

Table 6.4   Concentration of nutrients, average annual primary production and average phytoplankton biomass during the period of vegetation growth in selected reservoirs

River	Reservoir	P_{total} (mg l^{-1})	N_{total} (mg l^{-1})	Production (g C m^{-2} a^{-1})	Biomass (mg l^{-1})	Sources
Volga	Rybinsk	0.049	1.3	76	0.7–6.5	1, 2
	Gorky	0.055	1.29	112	0.7–11.7	1, 2
	Kuybyshev	0.076	1.76	100	1.5–5.2	1, 2
	Saratov	0.057	1.5	100	0.9–1.3	1, 2
Dnieper	Kiev	0.072	2.2	144	2.1–9.6	2, 3
	Dnieprodzerzhinsk	0.124	–	231	1.4–25.0	2, 3
	Zaporozhye	0.081	–	205	7.8–120.0	2, 3
	Kakhovka	0.094	2.1	271	0.1–19.2	2, 3
Don	Tsimlyansk	–	–	418	2.8–21.8	2
Dniester	Dubossary	0.003–0.41	0.5–4.7	436	0.9–3.6	4, 5
Yenisey	Khantaisk	0.005–0.028	–	–	0.13–1.56	6
	Krasnoyarsk	–	–	150	0.28–0.38	2, 7
Angara	Irkutsk	–	–	115	0.09–0.2	2, 8
	Bratsk	–	–	101	1.8–5.0	2, 8

Sources:
1	Kuzmin, 1978	5	Gorbatenky et al., 1977
2	Romanenko, 1985	6	Domysheva, 1986
3	Stcherbak, 1989	7	Chaikovskaya, 1975
4	Shalar, 1971	8	Kozhova, 1973

Table 6.5   Chlorophyll a daily primary production rate at optimal light and Secchi disc transparency in lakes of different trophic status based on the Carlson Trophic State Index (TSI)

TSI[1]	Chlorophyll a (μg l^{-1})	A_{opt} (μg C l^{-1} d^{-1})	Z_s (m)
0	0.01	–	–
20	0.1	3	16
40	1	30	6
60	10	300	2
80	100	3,000	0.75
100	1,000	–	–

[1]
0–20	Ultraoligotrophic lakes	A_{opt}	Primary production rate at optimal light
20–40	Oligotrophic lakes	Z_s	Secchi disc transparency
40–60	Mesotrophic lakes		
60–80	Eutrophic lakes		Source: Bulion, 1993
80–100	Hypereutrophic lakes		

Table 6.6   Frequency distribution of total phosphorus concentrations in lakes of different regions, as a percentage of the total number of lakes

P_{total} (µg l^{-1})	Estonia	Karelia	Lithuania	Byelorussia	Latvia	Novgorod region	Pskov region	Leningrad region	Finland[1]
< 20	43	23	15	6	25	8	3	3	16
20–40	19	38	74	75	42	44	32	18	55
40–60	19	19	9	19	25	32	31	27	26
60–80	12	12	2	–	8	12	25	18	–
> 80	7	8	–	–	–	4	9	34	3
n	26	26	34	16	12	26	34	33	–

n   Number of lakes   Source: Alekseev, 1993
[1]   Data on Finland are provided for comparison

Table 6.7   Comparison of total phosphorus and chlorophyll *a* in waters of small lakes of Karelia and Estonia

Region	P_{total} (µg l^{-1})	Chlorophyll *a* (µg l^{-1})	Number of lakes studied
Karelia[1]	6–12	0.3–1.0	39
	13–33	1.0–10.0	83
	38–77	0.0–30.0	31
Estonia[2]	9–17	3–6	26
	17–50	6–23	35
	50–132	23–60	15

[1]   Average annual data for 1986–89 (Lozovik *et al.*, 1991)   [2]   Average data for May–August in 1985–88 (Milius *et al.*, 1991)

cycle (Koplan-Diks *et al.*, 1993). With the increase in anthropogenic load, an increase in dispersion and greater asymmetry in distribution of total P concentrations occurred (Table 6.6). The proportion of lakes not suitable for economic use (P_{total} > 0.1 mg l^{-1}) also increased. The Leningrad region had the largest number of such lakes. Of the territories studied, Estonia and Karelia, where most lakes are still mesotrophic, had the most favourable situation (Table 6.7).

Studies carried out by the Limnological Research Laboratory of the Byelorussian University on lakes of the Western Dvina (Daugava) basin, showed a high level of anthropogenic transformation of lake ecosystems. Of 465 lakes studied, 75 per cent were affected by agricultural development in catchment areas and 43 per cent were affected by livestock practices. The majority of these lakes (76 per cent) can be considered as

Table 6.8   Lake trophic level in the Western Dvina (Daugava) river basin, Byelorussia

Administrative districts	Number of lakes	Percentage of lakes that are:			
		Mesotrophic	Eutrophic	Dystrophic	Hypereutrophic
Beshenkovichi	17	18	70	12	–
Braslav	75	11	77	11	1
Verkhnedvinsk	8	–	50	50	–
Vitebsk	15	–	93	7	–
Glubokoye	19	32	58	5	5
Gorodok	34	9	79	12	–
Lepel	46	17	76	7	–
Miory	32	6	66	28	–
Polotsk	63	13	74	13	–
Postavy	20	5	85	10	–
Rossony	29	7	83	10	–
Senno	11	9	64	18	9
Ushachi	66	22	75	3	–
Chashniki	22	–	91	9	–
Shumilin	8	25	62	13	–
Total	**465**	**12**	**76**	**11**	**1**

Source: Yakushko, 1989

eutrophic lakes, of which 67 per cent are small and shallow (Table 6.8). In regions with intensive economic activity, such as the Leningrad region and Byelorussia, the total productivity of lakes is increasing.

6.4.2   Organic sedimentation and filling rate of lakes

The increase of lake primary production is accompanied by an accumulation of organic matter, which can lead to the complete filling of a lake. Therefore eutrophication, together with other types of anthropogenic impact, accelerates the disappearance of lakes by increasing the intensity of sedimentation. Unfortunately, there are not enough data to quantify this rate of degradation in the former Soviet Union. However, it is known that many small lakes of Byelorussia, the Karelian Isthmus and the Valdai hills have lost 50 per cent of their volume due to silt formation. The main reasons for this are:

- Internal processes related to an increase in primary production and organic matter;
- External processes related to activities such as wood cutting and land ploughing in the lake basin and, as a consequence of these activities, an increase in runoff of particulate matter into water bodies.

Estimating sedimentation rate is complicated. Mean sediment accumulation rates for 25–50 years were evaluated for the lakes of Eastern

Latvia on the basis of paleolimnological data. The sedimentation velocity was 0.3–0.4 mm a^{-1} and the sedimentation rate was 8–46.4 g m^{-2} a^{-1} with a terrigenous matter component of 3–32 g m^{-2} a^{-1}. For the majority of eutrophic lakes in this region, total sedimentation rates were 55–68 g m^{-2} a^{-1} with a terrigenous matter component of 43–56 g m^{-2} a^{-1} (Yakovleva, 1983; Sergheeva et al., 1988).

Based on data obtained with sediment traps, sedimentation rates in the Trakai lakes situated in developed areas of Lithuania were significantly higher and were in the range 28.8–112.4 g m^{-2} per month (Martinkenene and Tamoshaitis, 1977). The importance of internal processes in sedimentation was emphasised for these lakes.

In Latvia, 750 lakes (25 per cent of the total number) disappeared during 1930–73 (Glazycheva, 1974). In the woodlands of Eastern Ukraine, approximately 186 lakes marked on maps in the 1950s have disappeared in the last 30 years (Antonov et al., 1989).

This evolution can be compared with data from Northern Poland (Churski, 1983). After the last glaciation (c. 9,000 years ago) the total lake area had been reduced by two thirds. In the last 40 years, the rate of lake filling has increased greatly. The results of a study of 100 lakes showed that, of 50 small shallow lakes (area < 1 ha), 29 lakes disappeared and 21 were about to disappear; of 50 deep lakes, 6 had disappeared and 3 were about to do so.

6.5 Consequences of eutrophication

The most serious consequences of an excessive increase in lake and reservoir production levels are: algal blooms and an associated change in the macrophyte role in aquatic ecosystems; and a change in oxygen regime. These phenomena are closely linked and, in addition to the resultant deterioration in water quality, they also influence fish productivity.

6.5.1 Algal blooms

Algal blooms are associated with significant changes in the main phytoplankton composition and are characterised by the mass development of a few species. As early as 1937, Dolgov (1937) noted that out of 53 reservoirs in the USSR, 32 had algal blooms. These blooms were generally caused by blue-green and green algae. Blue-green algae tended to be dominant because they are able to fix atmospheric nitrogen dissolved in water and therefore are not nitrogen limited. They can also survive with lower levels of carbon dioxide dissolved in water than required by other algae. Blooms started in the deeper water and rose to the surface due to an expansion of the phytoplankton gas vacuoles, eventually replacing completely the less-buoyant algae.

One of the main consequences of algal blooms is their toxic effect on some species. In addition to animal deaths (Vinberg, 1954), a decrease in

Table 6.9   Biomass of submerged macrophytes and phytoplankton and general phosphorus concentration in different types of lakes in Byelorussia

Type of lake	Phosphorus concentration (mg l^{-1})	Biomass of submerged macrophytes (g m^{-2})	Biomass of phytoplankton (g m^{-3})
Oligotrophic and dystrophic	0.01–0.03	2.0	0.8
Mesotrophic	0.02–0.05	2.5–8.0	0.8–2.0
Eutrophic	0.05–0.15	1.0–2.5	2.5–10.5
Hypereutrophic	0.15	1.0	10.5–75.5

Source: Ghighevich, 1991

livestock production has been reported when animals drink from water bodies with algal blooms. In the former Soviet Union, cases of Haff disease were registered in the area around Lake Yuks, Leningrad region and Lake Sartlan, Novosibirsk region. Although certain trophic levels create the potential for algal blooms and their possible toxic effects, they do not predetermine their annual appearance.

Algal blooms cause a decrease in water transparency and, as a result, certain species of macrophytes disappear because of a decrease in light penetration. A complete reduction in vegetated areas can be observed if the transparency of water is < 1 m. These effects have been noted in water bodies of the basins of the Don, in Byelorussian woodlands and Eastern Latvia (Denisova, 1988; Ghighevich, 1991; Khmelev and Khlyzova, 1991). In the Narochan lakes (Byelorussia), the maximum depth of macrophyte occurrence was reduced over the period 1959–81. The maximum depth at which macrophytes were found decreased in Lake Naroch from 9 m to 7.5 m, in Lake Myastro from 6 m to 4.2 m and in Lake Batorino from 9 m to 7.5 m (Ghighevich, 1985). The overall area and biomass of submerged macrophytes decreased when the trophic level of reservoirs increased (Table 6.9). Macrophyte degradation leads to a reduction in their role as a protecting barrier.

6.5.2   Changes in oxygen regime

Deterioration of the oxygen regime is another negative consequence of anthropogenic eutrophication of water bodies. High levels of organic matter accumulate in deeper water following high primary production and the cessation in degradation of organic matter. The dissolved oxygen in the hypolimnion (and in the entire water column during the ice cover period) is used up during oxidation processes and anaerobic conditions are created. In these conditions, fermentation is the main mechanism of energy production, as a result of which there is discharge into the water of ammonia, hydrogen sulphide and methane. These gases not only spoil the taste of water, but also may be toxic with detrimental consequences for the aquatic ecosystem.

A regular decrease in the oxygen content of large deep lakes is a reliable indicator of the onset of eutrophication, even if this change does not lead to the appearance of anaerobic zones. For example, in Lake Ladoga, a slight decrease in oxygen concentration in the hypolimnion occurs. At mid-depth (i.e. between 52–89 m) the dissolved oxygen saturation in summer was 100–102 per cent in 1962–81, and 90–96 per cent in 1986–89. In the deepest zone (depth > 89 m) these values were 100–103 per cent in 1962–81 and 90–97 per cent in 1986–89 (Tregubova and Kulish, 1992).

6.6   References

(All references are in Russian unless otherwise stated).

Alekin, O.A., Drabkova, V.G. and Koplan-Diks, I.S. 1985 Eutrophication of continental waters. In: *Anthropogenic Eutrophication of Natural Waters*. Chernogolovka, 25–34.

Alekseev, V.L. 1993 The analysis of spatial heterogeneity of trophic levels of lakes on local and regional scales. In: *Anthropogenic Redistribution of Organic Matter in the Biosphere*. Nauka, St Petersburg, 136–142.

Alekseev, V.L. and Golubetskaya, N.P. 1993 Combined analysis of field data and indirect assessment of phosphorus load in a region. In: *Anthropogenic Redistribution of Organic Matter in the Biosphere*. Nauka, St Petersburg, 142–145.

Anthropogenic Eutrophication of Lake Ladoga 1982 Nauka, Leningrad, 304 pp.

Antonov, A.D., Korbutyak, M.V. and Korotun, I.N. 1989 Research of lake systems in western forest regions of Ukraine. In: *Lake History. Rational Use and Protection of Lakes*. VIII All-Union Symposium, Abstracts. Minsk, 138–139.

Avakyan, A.B., Kaminski, V.S. and Falkovskaya-Chernysheva, L.N. 1977 Nutrient sources in large water bodies (example: Ivankovo reservoir). In: *Anthropogenic Eutrophication of Natural Waters*. Volume 2. II All-Union Conference Report. Chernogolovka, 259–264.

Belyakov, V.P. and Skvortsov, V.V. 1989 Degradation of organic matter by zoobenthos. In: *Transformation of Organic Matter and Nutrients Under Anthropogenic Eutrophication of Lakes*. Nauka, Leningrad, 186–188.

Bulion, V.V. 1993 Primary production and trophic classification of water bodies. In: *Research Methods of Plankton Primary Production in Inland Water Bodies*. Gidrometeoizdat, St Petersburg, 147–157.

Carlson, R.E. 1977 A trophic state index for lakes. *Limnology and Oceanography*, 22(2), 361–369. (In English).

Chaikovskaya, T.S. 1975 Phytoplankton in the Yenisey river and in the Krasnoyarsk reservoir. In: *Biological Research of Krasnoyarsk Reservoir*. Novosibirsk, 46–91.

Chu, S.P. 1943 The influence of the mineral composition of the medium on the growth of planktonic algae. II. The influence of the concentration of inorganic nitrogen and phosphate phosphorus. *Journal of Ecology*, **31**, 109–148. (In English).

Churski, Z. 1983 Eutrophication and the disappearance of lakes in the Brosnica Lake District, Northern Poland, as a result of human interference. *Hydrobiologia*, **103**, 165–168. (In English).

Denisova, A.I. 1979 Formation of the hydrochemical regime of Dnieper reservoirs and methods for its forecast. Daukova Dumka, Kiev, 230 pp.

Denisova, A.I., Nakhshina, E.P., Zhuravleva, L.A. and Palamarchuk, I.K. 1974 The factors influencing the formation of the hydrochemical regime of the Dnieper river and its eutrophication in impounded flow conditions. In: *Anthropogenic Eutrophication of Water Bodies*. I All-Union Conference Report. Chernogolovka, 85–91.

Denisova, I.A. 1988 Trends in the development of aquatic macrophytes in lakes with different levels of eutrophication. In: *Structural Changes of Lakes Ecosystems Under Increasing Nutrient Load*. Nauka, Leningrad, 119–132.

Denisova, I.A. 1989 The production of macrophyte vegetation. In: *Transformation of Organic Matter and Nutrients Under Anthropogenic Eutrophication of Lakes*. Nauka, Leningrad, 72–77.

Dillon, P.J. 1975 The phosphorus budget of Cameron Lake, Ontario: The importance of flushing in lakes. *Limnology and Oceanography*, **20**, 28–39. (In English).

Dillon, P.J. and Rigler, F.H. 1974 A test of a simple nutrient budget model predicting the phosphorus concentration in lake water. *Journal of the Fisheries Research Board of Canada*, **31**(11), 1771–1778. (In English).

Dolgov, G.I. 1937 Reservoir exploitation problems. *Hygiene and Sanitation*, **2**, 1–12.

Domysheva, V.M. 1986 Hydrochemical research. In: *Hydrochemical and Hydrobiological Studies of Khantai Reservoir*. Novosibirsk, 10–27.

Drabkova, V.G. and Stravinskaya, E.A. 1989 Intensity of phosphorus and carbon cycles in lakes of different trophic levels. In: *Transformation of Organic Matter and Nutrients under Anthropogenic Eutrophication of Lakes*. Nauka, Leningrad, 243–251.

Ecosystem of the Narochan Lakes 1985 Minsk University, Minsk, 302 pp.

Eutrophication of a Mesotrophic Lake 1980 Nauka, Leningrad, 245 pp.

Factors and Processes of Eutrophication of Lake Plestcheevo 1992 Yaroslavl, 200 pp.

Fedorova, E.I. 1967 Hydrochemical changes in Pereslavl (Plestcheevo) lake under the impact of pollution. In: *Lake Typology*. Nauka, Moscow, 53–79.

Forsberg, C. and Ryding, S.-O. 1980 Eutrophication parameters and trophic state indices in 30 waste-receiving Swedish lakes. *Archiv für Hydrobiologie*, **89**, 189–207. (In English).

Ghezalian, M.G. 1983 Oxygen deficit in Lake Sevan. In: *Production Processes in the Sevan Lake Ecosystem*. Armenian Academy, Yerevan, 95–108.

Ghighevich, G.S. 1985 Macrophytes. In: *Ecosystem of the Narochan Lakes.* Minsk University, Minsk, 116–123.

Ghighevich, G.S. 1991 Role of macrophytes as bioindicators under anthropogenic impact. (Example: Byelorussian lakes). In: *Anthropogenic Changes in Ecosystems of Small Lakes.* Gidrometeoizdat, St Petersburg, 204–206.

Glazycheva, L.I. 1974 Experiences of level regime assessment in lakes of the Latvian SSR in connection with their economic use. Latvian INTI, Riga, 123 pp.

Gorbatenky, G.G., Byzgu, S.E. and Zubkova, E.I. 1977 The influence of anthropogenic eutrophication on water composition and quality in Dubossary reservoir, a water body of multiple use. In: *Anthropogenic Eutrophication of Natural Water.* Volume I. Chernogolovka, 110–116.

Kalinina, L.A. and Prytkova, M.Y. 1988 Total phosphorus balance and nutrient load in lakes. In: *Structural Changes of Lake Ecosystems Under Increasing Nutrient Load.* Nauka, Leningrad, 42–54.

Khmelev, K.F. and Khlyzova, N.U. 1991 Anthropogenic changes in vegetation in ecosystems of flood-land lakes in the Don river basin. In: *Anthropogenic Changes in Ecosystems in Small Lakes.* Gidrometeoizdat, St Petersburg, 217–220.

Klimkaite, I.N. 1977 Trends in lake eutrophication in the Lithuanian SSR (Example: Trakai lakes). In: *Anthropogenic Eutrophication of Natural Waters.* II All-Union Conference Report. Chernogolovka, 139–143.

Konstantinova, L.G. 1987 Abiotic conditions of ecosystems. In: *Reactions of Limnological Systems to Anthropogenic Impact.* FAN, Tashkent, 9–20.

Koplan-Diks, I.S., Krylenkova, N.L., Milius, A.U. and Stravinskaya, E.A. 1993 The possibility of quantitative assessment of the spatial heterogeneity of trophic levels of lakes. In: *Anthropogenic Redistribution of Organic Matter in the Biosphere.* Nauka, Leningrad, 132–136.

Kozhova, O.M. 1973 Hydrobiological characteristics of Bratsk and Irkutsk reservoirs. In: *The Biological Regime of Bratsk Reservoir.* Irkutsk, 10–40.

Kuzmin, G.V. 1978 Phytoplankton. In: *The Volga River and its Life.* Leningrad, 122–140.

Lavrentev, P.V., Makartseva, E.S. and Maslevtsov, V.V. 1988 Ratio of quantitative indices of micro and meso-plankton in lakes of different trophic levels. In: *Structural Changes of Lake Ecosystems Under Increasing Nutrient Load.* Nauka, Leningrad, 241–244.

Lavrenteva, G.M. 1982 Phytoplankton in small lakes in the North West of USSR and the problem of ecosystem management in water bodies used for fisheries. Synopsis of a Doctor of Science Thesis in Biology. Leningrad, 44 pp.

Lozovik, P.A., Sabylina, A.V., Kovalenko, V.N., Basov, M.I. and Kharkevich, N.S. 1991 Hydrochemical characteristic of small lakes in Karelia. In: *Anthropogenic Changes in Ecosystems of Small Lakes.* Gidrometeoizdat, St Petersburg, 34–37.

Makartseva, E.S. and Lavrentev, P.V. 1989 Degradation of organic matter by zooplankton. In: *Transformation of Organic Matter and Nutrients Under Anthropogenic Eutrophication of Lakes*. Nauka, Leningrad, 184–186.

Martinkenene, F. and Tamoshaitis, U. 1977 Seasonal changes in chemical composition of bottom sediment in Trakai lakes. In: *Anthropogenic Eutrophication of Natural Waters*. Chernogolovka, 149–153.

Martynova, M.V. 1993 Internal nutrient load. In: *Anthropogenic Redistribution of Organic Matter in the Biosphere*. Nauka, Leningrad, 93–102.

Mikheeva, T.M. 1985 Annual and long-term trends in number and biomass of phytoplankton. In: *Ecosystem of the Narochan Lakes*. Minsk University, Minsk, 70–86.

Milius, A.U., Starast, K.A. and Lindpere, A.V. 1991 Relationship between chlorophyll and nutrients in small lakes in Estonia. In: *Anthropogenic Changes in Ecosystems of Small Lakes*. Volume 2. Gidrometeoizdat, St Petersburg, 234–238.

National Economy of the USSR 1981 Statistics, Moscow, 235 pp.

Nazarov, G.V. 1988 Influence of catchment area on nutrient discharge. In: *Evolution of the Phosphorus Cycle and Eutrophication of Natural Waters*. Nauka, Leningrad, 33–38.

Parparov, A.S. 1987 Phytoplankton production characteristics in Lake Sevan and their connection with processes of oxygen consumption in the hypolimnion. In: *Productive Hydrobiological Research of Water Ecosystems*. Nauka, Leningrad, 82–90.

Parparov, A.S., Parparova, R.M. and Simonyan, A.A. 1977 Some characteristics of eutrophication intensity in Lake Sevan. In: *Anthropogenic Eutrophication of Natural Waters*. Volume 2. II All-Union Conference Report. Chernogolovka, 160–162.

Pokrovskaya, T.N. and Rossolimo, L.L. 1967 Features of Seliger lake eutrophication. In: *Lake Typology*. Nauka, Moscow, 27–52.

Primachenko-Shevchenko, A.D., Denisova, A.I. and Tulupchuk, U.M. 1973 Long-term trends in nutrients and phytoplankton in Dnieper reservoirs. In: *Organic Matter and Energy Cycle in Lakes and Reservoirs*. No. 2, 113–117.

Raspletina, G.F. and Gusakov, B.L. 1982. Application of direct and indirect methods for calculation of nutrient load and concentrations in Lake Ladoga. In: *Anthropogenic Eutrophication of Ladoga Lake*. Nauka, Leningrad, 222–242.

Razumikhina, V.N. and Monina, T.A. 1991 Mapping of phosphorus inflow into water bodies of European territory USSR from different sources. In: *Anthropogenic Changes in Ecosystems in Small Lakes*. Gidrometeoizdat, St Petersburg, 115–118.

Romanenko, V.I. 1985 Microbiological processes of production and degradation of organic matter in inland water bodies. Nauka, Leningrad, 294 pp.

Sergheeva, L.V., Trifonova, I.S. and Khomutova, V.I. 1988 Spore-pollen, lithologic and pigment analysis of bottom sediment. In: *Structural Changes of Lake Ecosystems Under Increasing Nutrient Load*. Nauka, Leningrad, 14–19.

Shalar, V.M. 1971 Phytoplankton in Moldovian reservoirs. Kishinev, 204 pp.

Shilkrot, G.S. 1967 Features of typological changes in Valdai lake. In: *Anthropogenic Factors in Lake Development*. Nauka, Moscow, 58–90.

Shilkrot, G.S. 1968 Hydrochemical regime of the lake in the latest stage of anthropogenic eutrophication (Example: Beloye lake). *Hydrobiology Journal*, 4(4), 20–27.

Sirenko, L.A. 1989 Conclusion. In: *Vegetation and Bacterial Inhabitants in the Dnieper River and its Reservoirs*. Daukova Dumka, Kiev, 219–223.

Sirenko, L.A. and Gavrilenko, M.Y. 1978 *Algal blooms and eutrophication*. Kiev, 231 pp.

Sorokin, I.N. 1983 Morphometry of lakes and their external water exchange. Incoming substances and their absorption in lakes. In: *Changes in the 'Catchment-Lake' System Under the Influence of Anthropogenic Factors*. Nauka, Leningrad, 69–78.

Stcherbak, V.I. 1989 Formation features, condition and primary production of phytoplankton. In: *Vegetation and Bacterial Inhabitants in the Dnieper River and its Reservoirs*. Daukova Dumka, Kiev, 81–97.

Stravinskaya, E.A. 1989 The influence of external and internal loads on nitrogen and phosphorus regimes in lakes. In: *Transformation of Organic Matter and Nutrients Under Anthropogenic Eutrophication of Lakes*. Nauka, Leningrad, 61–78.

Stravinskaya, E.A. and Ulyanova, D.S. 1984 Phosphorus load and nutrient concentration in lake water. In: *Trends in the Formation of Water Quality in Different Types of Lakes in the Karelian Isthmus*. Nauka, Leningrad, 61–78.

Tregubova, T.M. and Kulish, T.P. 1992 Oxygen regime. In: *Lake Ladoga - Criteria of Ecosystem Condition*. Nauka, Leningrad, 35–239.

Trifonova, I.S. 1990 *Ecology and Succession of Lake Phytoplankton*. Nauka, Leningrad, 180 pp.

Trifonova, I.S., Stanislavskaya, E.A. and Petrova, A.L. 1988 Species and characteristics of algal flora of lakes of different trophic levels. In: *Structural Changes of Lake Ecosystems Under Increasing Nutrient Load*. Nauka, Leningrad, 133–163.

Tsimdin, P.A., Druvietis, I.U., Kachalova, O.L., Liepa, R.A., Matisone, M.N., Melberga, A.G., Pareghe, E.A., Rodionov, V.I. and Rudzroga, A.I. 1982 Typological changes of a small lake in Latvia. *Proceedings of the Latvian Academy of Science*, 3, 71–78.

Vinberg, G.G. 1954 Toxic phytoplankton. *The Achievements of Modern Biology*, 38(2), 216–226.

Vollenweider, R. 1968 Scientific fundamentals of the eutrophication of lakes and flowing waters with special reference to nitrogen and phosphorus as factors in eutrophication. Organization for Economic Co-operation and Development Technical Report, DA5/SCI/68.27, Paris, 250 pp. (In English).

Vollenweider, R. 1974 Eutrophication: consideration of some basic factor affecting productivity of lakes. In: *Abstracts XIX. Coagr. Internat. Assoc. of Limnol.* Winnipeg, Canada, 219 pp. (In English).

Vollenweider, R. 1976 Advances in defining critical levels for phosphorus in lake eutrophication. *Mem. Ist. Ital. Idrobiol.* **33**, 53–83. (In English).

Yakovleva, L.V. 1983 Sedimentation and sediment accumulation in lakes. In: *Changes in the 'Catchment-Lake' System Under the Influence of Anthropogenic Factors.* Nauka, Leningrad, 155–178.

Yakushko, O.F. 1989 Changes in aquatic systems under intensive economic activity. In: *Landscapes in Byelorussia.* Minsk, 214–222.

Zhukova, T.V. 1991 Ratio of external and internal nutrient loads in different lakes of the Narochan group. In: *Anthropogenic Changes in Small Lakes Ecosystems.* Gidrometeoizdat, St Petersburg, 170–173.

Zhukova, T.V., Martynova, M.V. and Zhukov, E.P. 1990 Bottom sediment in ecosystems of the Narochan lakes. I. Internal nutrient load. *Water Resources,* **2**, 130–138.

Chapter 7[*]

GROUNDWATER CONTAMINATION BY NITROGEN AND PHOSPHORUS COMPOUNDS

Nitrates (NO_3^-), nitrites (NO_2^-), and ammonium (NH_4^+) ions are widespread but are toxic in high concentrations. Related to this, NO_3^-, NO_2^- and NH_4^+ are subject to MACs in waters for drinking and industrial purposes. The World Health Organization (WHO) recommends a MAC of nitrate nitrogen (NO_3-N) in drinking water of 10 mg l^{-1}, which corresponds to the standards of the European Community (EC) and the former USSR (50 and 45 mg l^{-1} NO_3^- respectively). No MACs for nitrites and ammonium have been set by WHO, although these compounds have greater toxicity than nitrates. The MACs defined by the EC for NO_2^- and NH_4^+ in drinking water are 0.1 mg l^{-1} and 0.5 mg l^{-1} respectively (Chapman, 1992). In the former Soviet Union, MACs for NO_2^- and NH_4^+ in water bodies should not exceed 0.02 mg l^{-1} of NO_2^--N and 0.39 mg l^{-1} for NH_4^+-N.

Research has shown an increase in nitrogen compounds in surface waters (see Chapter 6) which can be related to various economic activities, primarily the use of fertilisers in agriculture. Considering the velocity of surface water circulation, the residence time of nitrogen compounds in surface water varies from less than one year for major rivers without reservoirs or small lakes, to a few years for river systems with few reservoirs and medium-sized lakes, and to decades for very large lakes such as Ladoga and Baikal. Nitrogen compounds are also directly taken up by the aquatic biota and are key factors in the process of eutrophication. In addition, there is the accumulation of nitrogen compounds in groundwater where the process of water circulation is slow and the nitrogen cycle is limited by low biological activity. This chapter is concerned mainly with the origins and migration of the nitrogenous compounds in ground water, gases (NH_3, N_2, NO_2, NO, N_2O_3), inorganic ions (NO_3^-, NO_2^-, NH_4^+) and organic compounds (e.g. amino acids), which are linked to complex reactions such as ammonification, nitrification and denitrification.

---

* *This chapter was prepared by V.P. Zakutin and M.P. Polkanov*

7.1   Occurrence and control of nitrogen compounds

The main geochemical factor influencing changes in nitrate, nitrite and ammonium concentrations and their ratios is the redox potential of waters (Eh), which determines the direction and intensity of chemical and biochemical transformations of these compounds (Krainov *et al.*, 1989). The results of long-term geochemical research of groundwater contamination by nitrogen compounds in the former Soviet Union, showing the relationship between Eh–pH and the stability of various nitrogen forms in water, are presented in Figure 7.1. The research has shown that nitrates have thermodynamic and biochemical stability at higher values of the redox potential, whereas ammonium is more stable at lower values. Nitrite is an ephemeral compound which is formed as a result of differently directed processes of nitrification and nitrate reduction (or denitrification). There are regular trends in nitrate, nitrite and ammonium content in groundwater depending on the value of their redox potential (Figure 7.2). These specific hydrogeochemical features of nitrogen compounds determine their occurrence and distribution in groundwater.

Under natural conditions, the levels of nitrogen compounds in groundwater (i.e. background concentrations) are very low: 3.41 mg l^{-1} NO_3^- and 0.56 mg l^{-1} NH_4^+ according to Shvartsev (1978) for groundwater of upper aquifers. The usual sources of nitrogen inputs to ground water are mainly gas and dust emissions from industrial and domestic activities including fossil fuel combustion, industrial effluents, mineral and organic fertilisers, livestock effluents, municipal wastes and silo effluents. The concentrations of nitrogen compounds in these sources may exceed hundreds and even thousands of milligrams per litre (Krainov and Zakutin, 1993). Inputs from these sources can affect groundwater quality not only at a local level but also at a regional level. In addition, the formation of higher concentrations of NH_4^+ in groundwater (up to several mg l^{-1}) occurs under certain biochemical conditions as a result of ammonification processes (Zakutin and Chugunova, 1992).

7.2   Intensity of pollution with nitrogen compounds

Water pollution and its intensity can be characterised by the ratio of average concentration to MAC, referred to here as the pollution index (Goldberg and Gazda, 1984). The MACs are 45 mg l^{-1} NO_3^- and 0.5 mg l^{-1} NH_4^+. In wet areas, and for groundwaters with natural background levels of NO_3^-, the pollution index is < 0.02 for NO_3^- whereas the pollution index for arid areas is 0.02–0.2. Polluted waters with higher concentrations of NO_3^- resulting from anthropogenic activities are characterised by three ratios depending on the level of pollution. A pollution index from 0.2–0.5 reflects low pollution, an index from 0.5–1 is characteristic of moderately polluted water and an index > 1 characterises highly polluted water (Figure 7.3). A similar approach has been used to

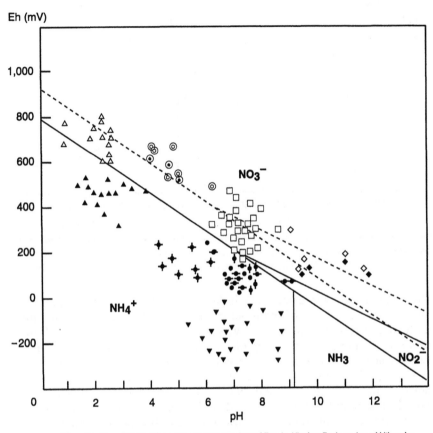

Eh (mV)

NO₃⁻

NH₄⁺

NH₃

NO₂⁻

pH

□ Mineral fertilisers (NO₃⁻ – N) used in agricultural regions of Russia, Ukraine, Byelorussia and Lithuania;
• Industrial effluents of Russia with a high concentration of unoxidised substances (NH₄⁺); △ Acid wastes
of industrial enterprises in the Urals with NO₃⁻ in wastewaters; ▲ Acid wastes of industrial enterprises in the
Urals with NH₄⁺ in wastewaters; ✛ Water from deep aquifers originating from West Siberian gas and oil
fields (NH₄⁺); ✦ Organic fertilisers used in agricultural regions of Lithuania (NH₄⁺); ➤ Effluents from
livestock farms in Lithuania (organic N and NH₄⁺); ▼ Sulphide effluents of industrial enterprises in Ukraine
(NH₄⁺); ◇ Alkali effluents of Lithuanian industries (NO₃⁻); ◆ Alkali effluents of Lithuanian industries (NH₄⁺);
⊚ Atmospheric fallout of NO₃⁻ in the Ukraine; ⊙ Atmospheric fallout of NH₄⁺ in the Ukraine.

Figure 7.1 The relationship between the redox potential of water (Eh) and pH and the
different forms of nitrogen present under different conditions with data points showing
polluted groundwaters with a high concentration of nitrogen compounds (After Krainov
et al., 1991)

assess nitrogen contamination of groundwaters in the USA (Madison
and Brunett, 1984).

Waters with minimum background levels of NH₄⁺, formed as a result
of natural processes under aerobic conditions, have a pollution index < 1.

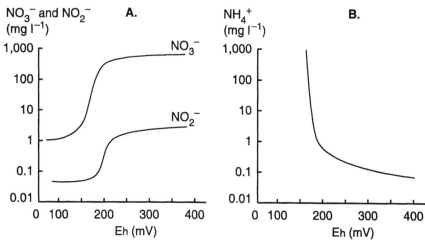

Figure 7.2   Changes in maximum concentrations of nitrates, nitrites and ammonium in polluted groundwaters of agricultural regions in the European part of the former USSR (n ~ 1,000). **A.** Nitrates and nitrites; **B.** Ammonium (After Krainov and Zakutin, 1993)

Under anaerobic conditions, the index ranges from 1–2 at low pollution levels, from 2–3 for moderate pollution, and the index is 3–4 for high pollution levels due to the activity of ammonifying bacteria. Waters severely affected by anthropogenic activities and high bacterial ammonifying processes, have the highest pollution index > 4 (Figure 7.4).

Figures 7.3 and 7.4 are schematic maps based on administrative borders, not hydrogeological divisions, and should only be considered as an initial attempt to estimate the distribution of nitrate and ammonium in groundwaters of the former USSR. Schematic maps based on hydrogeological divisions are currently being prepared. These results should serve as a basis for assessing the spatial distribution of nitrogen pollution of water and form the basis for further research.

Statistical assessment of nitrate and ammonium distribution in groundwater of the former USSR subject to intensive pollution showed that these waters had high concentrations of NO_3^- and NH_4^+, with a high probability of samples exceeding MACs (Tables 7.1 and 7.2). Particularly unfavourable concentrations occurred in areas where there is significant anthropogenic impact and where there is little protection of aquifers from contamination.

The intensity of regional nitrate pollution was greatest for well-oxygenated waters of alluvial aquifers (e.g. Kama-Vyatka, Caspian and North East Caucasian basins), for aquifers of crystalline rock regolith (e.g. Ural, Borzin basins) and for karstic aquifers (e.g. Leningrad, Kama-Vyatka and Moldovian basins). All these areas receive significant amounts of fertilisers annually. Nitrate-containing waters occur up to a depth of 100 m.

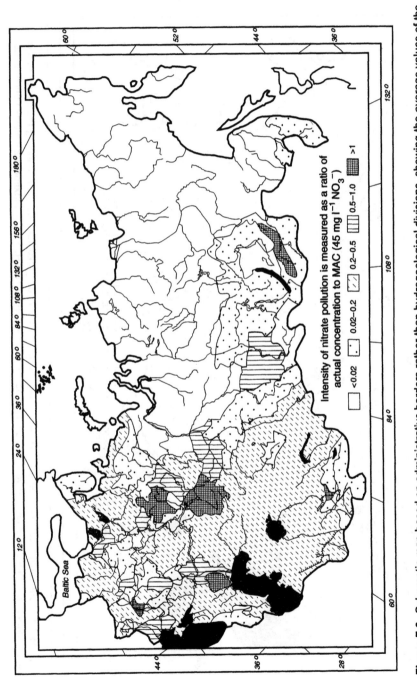

Figure 7.3 Schematic map based on administrative borders rather than hydrogeological divisions, showing the average values of the nitrate pollution index of groundwaters of phreatic aquifers in the former USSR (After Zakutin *et al.*, 1994). MAC = maximum allowable concentration

Intensity of nitrate pollution is measured as a ratio of actual concentration to MAC (45 mg l^{-1} NO$_3^-$)

<0.02   0.02–0.2   0.2–0.5   0.5–1.0   >1

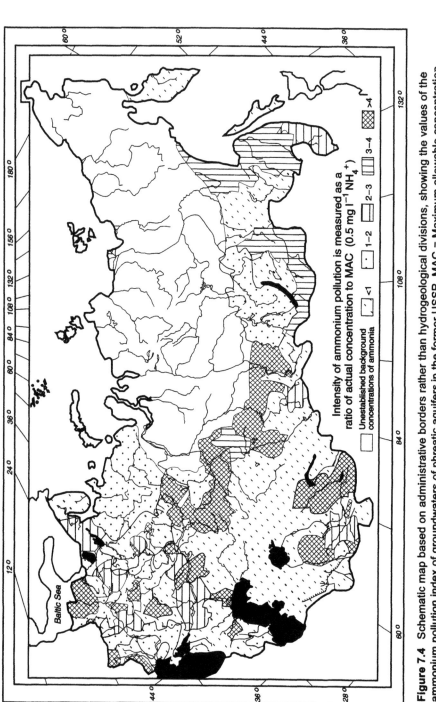

Figure 7.4 Schematic map based on administrative borders rather than hydrogeological divisions, showing the values of the ammonium pollution index of groundwaters of phreatic aquifers in the former USSR. MAC = Maximum allowable concentration

Intensity of ammonium pollution is measured as a ratio of actual concentration to MAC (0.5 mg l^{-1} NH$_4^+$)

Unestablished background concentrations of ammonia

<1   1–2   2–3   3–4   >4

Table 7.1   Nitrate distribution in phreatic aquifers

Region (basin)	Water-bearing rock type	Number of samples	NO_3^- (mg l^{-1})			Pollution index[1]	Samples exceeding MAC (%)[2]
			Min.	Max.	Average		
Leningrad	Devonian sandy/ clay sediment	49	1.2	100	23.3	0.5	18
	Ordovician limestone	87	1.9	100	22.3	0.5	10
Kama-Vyatka	Neogene-quaternary sand	83	0.1	701	54.5	1.2	28
	Permian limestone	662	0.1	456	46.6	1.0	32
Caspian	Quaternary sand	74	0.2	882	36.2	0.8	19
Moldovian	Neogene limestone	137	3.2	1,200	37.0	0.8	18
North East Caucasus	Quaternary sand	30	0.5	550	81.9	1.8	53
Ural	Palaeozoic crystalline rock	64	1.2	556	80.1	1.8	40
Borzin	Palaeozoic crystalline rock	50	0.3	87	48.6	1.1	27

MAC   Maximum allowable concentration
[1]   Ratio of the actual average concentration and the MAC (45 mg l^{-1} NO_3^-)

[2]   Percentage of samples exceeding 45 mg l^{-1} (MAC)

Table 7.2   Ammonium distribution in phreatic aquifers

Region (basin)	Water-bearing rock type	Number of samples	NH_4^+ (mg l^{-1})			Pollution index[1]	Samples exceeding MAC (%)[2]
			Min.	Max.	Average		
Moscow	Devonian sandy/ clay sediment	39	0.1	10.5	1.6	3.2	54
Baltic	Devonian sandstone	176	0.1	9.9	2.1	4.1	80
Black Sea	Neogene sandy/ clay sediment	151	0.1	10.0	1.4	2.8	28
West Siberian	Palogene sandy/ clay sediment	216	0.1	8.0	2.6	5.1	69

MAC   Maximum allowable concentration
[1]   Ratio of the actual average concentration and the MAC (0.5 mg l^{-1} NH_4^+)

[2]   Percentage of samples exceeding 0.5 mg l^{-1} (MAC)

The intensity of regional ammonium pollution is greatest for anoxic phreatic aquifers containing high concentrations of organic compounds (up to tens of milligrams of organic carbon per litre) such as those contaminated by municipal sewage or livestock wastes (e.g. Moscow, Baltic and Black Sea basins) or when there is a natural occurrence of organic compounds in the aquifer (e.g. oil and gas fields in the West Siberian basin).

7.3   Spatial distribution of nitrogen

The distribution of nitrates and ammonium in groundwater depends to a great extent on the natural geochemical properties of the water systems.

Phreatic aquifers are characterised by geochemical zones of nitrate-containing waters related to climate and soil conditions. Spatial distribution is due to a combination of the oxygenation state of the groundwater and regional nitrate contamination of the total environment. Under similar conditions (i.e. the same levels of anthropogenic input of nitrates into aquifers and similar geological, hydrogeological and agro-biochemical conditions) the natural trend for groundwaters from wet areas to dry areas is an increase in concentration of NO_3^- as well as an increase in Cl^- and in mineralisation. Field experiments in desert conditions of Turkmenistan have shown that NO_3^- concentrations may increase by almost 25 times (compared with wet areas) and reach 1 g l^{-1} (Zakutin et al., 1994).

There is a stable correlation (r = 0.573, n = 633) between the average quantity of nitrogen in mineral fertilisers applied to land (kg N ha^{-1} a^{-1}) and the average concentration of NO_3^--N in groundwater. There is also a positive correlation for NO_3^- with Cl^- and a negative correlation for NO_3^- with HCO_3^- in groundwater (Enikeev et al., 1989; Jygar, 1990).

Point sources of agricultural contamination of phreatic aquifers (e.g. areas of livestock farming, mineral or organic fertiliser storage) account for the localised occurrence of higher ammonium concentrations in groundwater. This is reflected by gradual changes in concentrations of NO_3^- and NH_4^+ in waters as they filter along a horizontal transect.

There are several types of distribution of nitrogen compounds in groundwater which differ from one another by their differences in redox states (Zabulis, 1988). One such type belongs to conditions under which redox states, as expressed by Eh values, are invariably high, e.g. aerobic environments which promote the development of nitrifying bacteria and exclude microbiological and chemical denitrification. Under these conditions, and based on the relationship between redox potential and pH observed in the former USSR (Figure 7.1), the nitrogen compounds are located in the field of biochemical and thermodynamic stability of NO_3^- ions. In groundwaters with high Eh values, nitrates do not take part in the nitrogen biogeochemical cycle. The pattern of NO_3^- concentration is

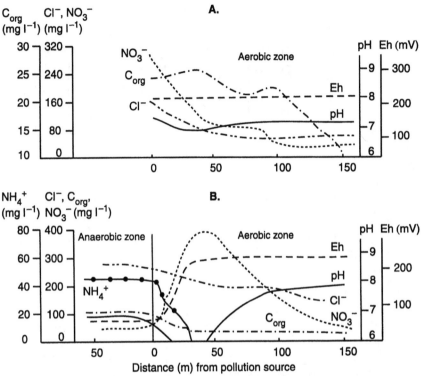

Figure 7.5 Changes in nitrate and ammonium concentrations in polluted groundwater in areas around livestock farms in Lithuania. **A.** Under aerobic conditions; **B.** Transition from anaerobic to aerobic conditions (After Zabulis, 1988)

affected by physical processes such as dispersion in water and dilution of nitrate-containing water by precipitation (Figure 7.5).

This type of distribution is found not only at local, but also at regional, levels. Thus, consistently high values of redox potential are character-istic of groundwaters of alluvial aquifers intersected by a river, for aquifers in alluvial fans and piedmont loops and for the recharge zone of artesian aquifers. Typically, when diffuse nitrate sources of pollution occur (e.g. nitrate fertiliser application over large areas), nitrates may migrate freely in the water. Such nitrate distributions were studied in the Alma-Ata (Kazakhstan), Ashkhabad (Turkmenistan) and Bishkek (Kyrgyzstan) deposits of groundwaters of alluvial fans (Krainov and Zakutin, 1993). Within the boundaries of these deposits, there is a constant decrease downstream in Eh values towards the piedmont zone but which do not reach the level at which processes of denitrification and nitrate-reduction commence.

The second type of distribution occurs in conditions where the redox state varies in different directions, both in the vicinity of the pollution

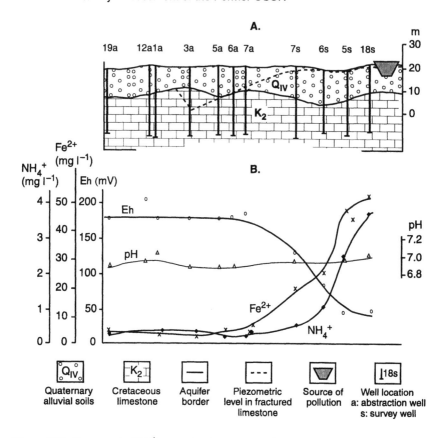

Figure 7.6   Changes in NH_4^+ groundwater concentrations in the Donbass region, Russia, polluted by industrial effluents with a high content of organic substances (COD = 100 mg l^{-1} O_2). **A.** Schematic hydrogeological profile; **B.** Hydrogeochemical profile of a limestone aquifer (After Krainov et al., 1989)

source and further along the groundwater flow. This causes the transformation of nitrogen compounds as nitrification, denitrification, nitrate-reduction and ammonification occur.

In polluted areas, nitrogen compounds are distributed according to changes in the redox state of groundwater, which in turn determines the process of microbiological transformation of nitrogen compounds. In phreatic aquifers usually near the source of pollution where anaerobic conditions with low Eh values are formed, NH_4^+ is the prevailing form of nitrogen (Figure 7.6). The further from the contamination source, the higher the Eh value. Correspondingly, there is an increase in redox potential and a decrease in denitrification and nitrate-reduction processes and an increase in the activity of nitrificating bacteria causing an increase in nitrate concentration. While groundwater moves from a recharge area to a transit area, a redox trend occurs reflected by an O_2 gradient. Along the direction of water flow, there is a decrease in dissolved O_2 and NO_3^- concentrations and an increase in NH_4^+ (Figure 7.7).

7.4   Vertical variation of nitrogen compounds

The vertical variation of nitrogen compounds in groundwater is related to the piezometric level and to their redox status (Krainov *et al.*, 1989). Usually, this is a straightforward vertical variation where a decrease in Eh values occurs with aquifer depth. When waters are well oxygenated, as in karsts or in the upper parts of aquifers (Eh values > 200 mV), the dominant form of nitrogen is NO_3^-. Thus in the Donbass, concentrations of NO_3^- up to 200 mg l^{-1} are present in water of karst limestone aquifers to depths of about 100 m. A decrease in Eh values with depth causes a decrease in NO_3^- concentrations and is often accompanied by the process of ammonification leading to an increase in NH_4^+ in waters.

Contamination of groundwater in agricultural regions can lead to an inverse relationship in terms of the NO_3^- profile due to the occurrence of anaerobic conditions caused by the input of organic substances. Consequently, in the surface aquifer oxygen concentrations decrease by one order of magnitude, from 1 to 10 m l^{-1} O_2; the typical concentration in unpolluted waters of the upper section of an aquifer is $0–0.1$ mg l^{-1} O_2. In an agricultural field in Lithuania which was irrigated with livestock waste, phreatic groundwaters were characterised by a minimal amount of oxygen and by an ammonium concentration reaching 100 mg l^{-1}. However, in deeper aquifers in this area where dissolved oxygen (up to 3 mg l^{-1} O_2) is still present, the ammonium level was very low (< 0.1 mg l^{-1}) probably due to an inflow of unpolluted water.

Although the above example relates to agricultural activity in a local area (1 km^2) and for aquifers of quaternary depositions in the Baltic region, this phenomenon also occurs at a regional level. Thus, in the karst region of Lithuania (c. 1,000 km^2), which serves as a recharge zone for the Upper Devonian aquifers, the long-term use of liquid manure as fertiliser has led to anaerobic conditions and high concentrations of ammonium, up to 150 mg l^{-1} (Zakutin and Paoukchtis, 1993).

7.5   Phosphorus

Alongside the large-scale contamination of natural waters with nitrogen compounds, the increasing concentrations of phosphorus compounds in groundwater has also become ecologically significant. In the former USSR, many publications are devoted to the study of phosphorus in surface waters but there are practically no publications concerning phosphorus in groundwaters.

Under natural conditions, average concentrations of phosphorus compounds in groundwater with a low mineral content and pH values

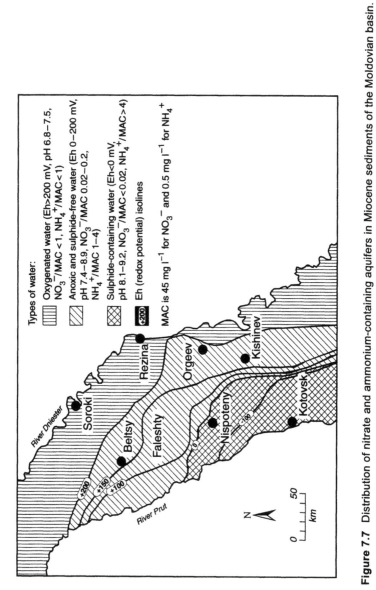

Figure 7.7 Distribution of nitrate and ammonium-containing aquifers in Miocene sediments of the Moldovian basin.

close to neutral do not exceed 0.1 mg l^{-1} P (Driescher and Gelbrecht, 1988). This is also characteristic for groundwaters of the former Soviet Union. However, recently, in areas of industrial and agricultural pollution a recorded increase in phosphorus concentrations in groundwater has been found. For example, in quaternary aquifers of the Kopetdag (Turkmenistan), Baltic (Latvia) and Black Sea (Ukraine) basins concentrations have reached 1 mg l^{-1} P. Even higher concentrations of phosphorus up to hundreds and thousands of milligrams per litre were recorded in industrially developed regions of the Kola Peninsula, the Urals, Lithuania and Byelorussia (Krainov and Zakutin, 1993).

The main form of phosphorus found in water is derived from the dissociation of phosphorus acid, H_3PO_4. Under natural conditions, phosphate levels are controlled by the presence of iron and calcium. Hence, the highest levels of phosphate contamination are observed for sodium-containing groundwaters. This is due to kinetic processes of secondary mineral formation in polluted groundwater and the formation of soluble organic forms of phosphorus, calcium and iron (Krainov *et al.*, 1993).

7.6   Conclusions

The principal aim of studying the pollution of groundwater with nitrogen and phosphorus compounds in the former USSR, has been to establish general trends in the distribution and migration of nitrates, nitrites, ammonium ions and phosphates in groundwater. The main objectives of future research are:

- The large-scale mapping of the distribution of nitrogen and phosphorus compounds in ground waters used for drinking and industrial purposes;
- Computer modelling of the migration of these compounds in aquifers, combining hydrodynamic and geochemical factors of mass transfer;
- Studying the kinetics of microbiological transformations of nitrogen and phosphorus compounds under varying conditions;
- Studying temporal changes in inorganic nutrients in groundwater;
- Studying the distribution of organic forms of nitrogen and phosphorus in groundwater.

7.7   References

(All references are in Russian unless otherwise stated).

Chapman, D. [Ed.] 1992 *Water Quality Assessments. A Guide to the Use of Biota, Sediments and Water in Environmental Monitoring.* Chapman & Hall, London, 585 pp. (In English).

Driescher, E. and Gelbrecht, J. 1988 Phosphat im unterirdischen Wasser. 1. Mitteilungen zum Vorkommen von Phosphat im Grundwasser – eine Literaturübersicht. *Acta Hydrophys*, **32**(4), 213–235. (In German).

Enikeev, N.I., Grigorova, G.L. and Pryadunenko, T.I. 1989 *Study of Groundwater Quality Under Anthropogenic Conditions (Example: Chirchik-Akhangaran Region)*. FAN, Tashkent, 200 pp.

Goldberg, V.M. and Gazda, C. 1984 *Hydrogeological Fundamentals for Protection of Groundwater Pollution*. Nedra, Moscow, 262 pp.

Jygar, P. 1990 Research and approbation of introduced indices of agricultural pollution of ground waters by field modelling. (Example: Northern Estonia). *Estonian Academy of Science Newsletter Geology Series*, **39**(3), 108–114.

Krainov, S.R. and Zakutin, V.P. 1993 *Groundwater Pollution in Agricultural Areas*. Geoinformmark, Moscow, 86 pp.

Krainov, S.R., Zakutin, V.P. and Solomin, G.A. 1989 *Nitrogen Compounds in Groundwater Used for Industrial and Drinking Purposes*. VIEMS, Moscow, 66 pp.

Krainov, S.R., Solomin, G.A. and Zakutin, V.P. 1991 Redox conditions of nitrogen transformations in ground waters (in relation to geochemical-ecological problems). *Geochemistry*, **6**, 822–831.

Krainov, S.R., Matveeva, L.I., Zakutin, V.P., Chugunova, N.N. and Solomin, G.A. 1993 Phosphorus geochemistry in ground waters with relatively neutral reaction (in relation to geochemical-ecological problems). *Geochemistry*, **4**, 549–565.

Madison, R.J. and Brunett, J.O. 1984 Overview of the occurrence of nitrate in ground water of the United States. US Geological Survey Water-Supply Paper No. 2275, 93–105. (In English).

Shvartsev, S.L. 1978 *Hydro-geochemistry of Hypergenesis Zones*. Nedra, Moscow, 288 pp.

Zabulis, R.M. 1988 *Groundwater Protection Against Pollution in the Areas of Large Cattle-breeding Facilities in the Lithuanian SSR*. Vilnius, 71 pp.

Zakutin, V.P. and Chugunova, N.N. 1992 Ammonium in ground waters in areas with free water exchange. *Reports of the Academy of Science*, **326**(3), 535–540.

Zakutin, V.P. and Paoukchtis, B.P. 1993 Effects geochimiques de la pollution agricole des eaux souterraines dans la region karstique de la Lituanie. *Hydrogeologie*, **1**, 65–70. (In French).

Zakutin, V.P., Fetisenko, D.A., Panteleeva, Z.N., Bogomolova, A.A. and Chugunova, N.N. 1994 Nitrate pollution of ground waters in the CIS and contiguous countries. *Water Resources*, **3**, 374–380.

Chapter 8[*]

HEAVY METALS, NATURAL VARIABILITY AND ANTHROPOGENIC IMPACTS

The former Soviet Union is characterised by a natural geochemical heterogeneity which leads to variable background concentrations of heavy metals. As a result of the naturally high metal concentrations, local populations and livestock are, in some places, exposed to endemic diseases. In addition to the naturally variable background concentrations of metals, anthropogenic metal contamination leads to extremely high heavy metal concentrations in some sites.

Anthropogenic heavy metal pollution is one of the major factors affecting water quality in the former Soviet Union. In heavily industrialised areas and around oil refineries, the heavy metal loads on water and wetland surfaces due to atmospheric fallout were estimated at around $1,000$ kg km^{-2} a^{-1}. The amount of wastewater discharged into Russian water courses in 1992 was 70.6 km^3, of which 38 per cent was untreated or only partially treated (State Report, 1993). The volume of solid waste from the metallurgical industry is $50–80 \times 10^6$ t a^{-1}. The wastes from galvanic works in Russia are approximately 500,000 t a^{-1}. Wastewaters from these works are discharged either into surface waters or to city sewerage, sometimes without prior treatment. Eighty per cent of solid industrial waste in Russia is still stored without adequate technical precautions.

8.1 Methods of sampling, sample preparation and analysis

Numerous sites have been sampled for heavy metals across the former Soviet Union (Appendix 8 Table I) and the authors would like to express their gratitude to colleagues and friends who took part in joint field surveys. Non-metallic materials were used for sampling in natural water bodies. Water was filtered through 0.4 µm Nucleopore filters which had been treated with purified acid. Cellulose-nitrate filters of 0.7 µm pore diameter (Mytischy), polyethyleneterephthalate filters of 0.5 µm pore diameter (Dubna), cellulose acetate-nitrate Millipore filters of 0.45 µm pore diameter and other filters, including paper filters, were also frequently used. The use of some of these filters resulted in unreasonably high results for metal concentrations in water. Where this occurred, the

---

*   *This chapter was prepared by A.V. Zhulidov and V.M. Emetz*

results obtained were used only for general comments on the pollution of ecosystems. These data were obtained from the regular monitoring network of surface waters of the State Service for Observations and Control of Environmental Pollution (OGSNK).

After filtration, water samples were preserved by acidification with nitric acid to pH < 2. For Hg determinations, if transportation of samples to laboratories took more than six days, 5 per cent $K_2Cr_2O_7$ was added to the samples, up to 0.05 per cent $K_2Cr_2O_7$ concentration.

Solubilisation of bottom sediment, seston, soil and biota samples was conducted by nitric acid attack (Nikanorov *et al.*, 1985, 1993; Nikanorov and Zhulidov, 1991). Analysis of metals was performed using atomic absorption and chemical spectrography as well as colourimetry and neutron activation methods (APHA, 1985; Nikanorov *et al.*, 1985; Nikanorov and Zhulidov, 1991). For additional details on methodology in more recent studies, see Appendix 8.

8.2   Trends in river and lake waters and bottom sediments

Routine measurements of heavy metal concentrations in water commenced in the 1950s, alongside the development of methods for the analysis of metal concentrations (Konovalov, 1959; Konovalov *et al.*, 1966a,b, 1982a,b; Stradomsky and Konovalov, 1968; Konovalov and Koreneva, 1980; etc.). Long-term average concentrations of Hg, Cd, Pb, Zn and Cu in water and bottom sediments of rivers and lakes are presented in Tables 8.1 and 8.2.

Analysis of published data and data from the Hydrochemical Institute and OGSNK for the period 1976–85 shows there were four types of long-term trends in heavy metal concentrations in waters of the former Soviet Union (Lobtchenko *et al.*, 1988, 1991):

- Variable metal concentrations with sharp variations. An example of this was the inter-annual trends in Pb concentration in the Nura river (Kazakhstan) and Chu river (Kyrgyzstan);
- Stable inter-annual metal concentrations. This occurred in a few water bodies, for example, trends in Zn concentration in rivers within the Lena river basin (Russia) and Cu concentration in the Volga river (Russia), and in rivers within the Western Bug river basin (Ukraine);
- Positive trends, i.e. long-term increases in metal concentrations. An example is the inter-annual trends in Cu concentration in the Lopan river at Kharkov, Ukraine from 1979 to 1988;
- Negative trends, i.e. long-term decreases in metal concentrations. An example is the Zn concentration in the Volga river and rivers within the Amu Darya river basin (Central Asia).

The last type of trend was also observed for Cd, Pb and Zn from 1975 to 1993 in the Usman river, Don basin, in the Voronezh Biosphere Reserve (Figure 8.1). A similar variation was observed in dissolved Zn concentrations in the Vashutkiny lakes in the Bolshezemelskaya tundra, Russia

Table 8.1   Mean annual concentrations of dissolved ($\mu g\ l^{-1}$) and suspended ($\mu g\ g^{-1}$, dry weight) heavy metals in rivers and lakes in the former USSR, 1973–93

Metal		Background levels			Anthropogenic pollution			
	n	Mean ± S.E.	Min.		n	Mean ± S.E.	Max.	Extremely high values[1]
Hg	Dissolved	94	0.09 ± 0.03	0.01	103	10 ± 2	22.8	65.1
	Suspended	234	0.23 ± 0.03	< 0.01	182	6 ± 1	18.4	34.6
Cd	Dissolved	298	0.044 ± 0.007	0.001	115	0.9 ± 0.1	2.3	5.8
	Suspended	182	0.16 ± 0.03	0.05	64	5.7 ± 0.9	15.4	97.4
Pb	Dissolved	535	0.13 ± 0.02	0.02	238	2.1 ± 0.5	16.9	74 [640][2]
	Suspended	741	7.4 ± 0.9	1.5	315	31 ± 7	84.2	232
Zn	Dissolved	1,537	0.6 ± 0.1	0.04	2,721	31 ± 8	181	646 [2,160][3]
	Suspended	1,247	30 ± 5	3.2	2,149	343 ± 51	1,230	5,430
Cu	Dissolved	1,221	2.1 ± 0.4	0.21	1,124	51 ± 11	229	443 [3,310][4]
	Suspended	1,135	12 ± 2	1.0	995	140 ± 24	438	985

Data were obtained from the following sites (numbers correspond to those in Appendix 8 Table I): 1–20, 23–36, 38, 39, 41–48, 50–69, 72, 74, 75, 77, 80–82, 86, 89–99, 101–114, 116–118 Areas with excessive metal concentrations were excluded

n   Number of samples
S.E.   Standard error

[1] Values not included in calculation of mean
[2] Based on OGSNK data for the Nura river in 1984 (Lobtchenko et al., 1991)
[3] Based on OGSNK data for the Ulba river close to the Tushinskogo ore mine (5.6 km downstream of the discharge of mine waters) in 1984 (Lobtchenko et al., 1991)
[4] Based on OGSNK data for the Schuchia river, in the city of Norilsk, 14 km upstream of the estuary (Lobtchenko et al., 1991)

Table 8.2   Mean annual concentrations ($\mu g\ g^{-1}$) of heavy metals in bottom sediments of rivers and lakes in the former USSR, 1973–93

Metal	Background levels			Anthropogenic pollution			
	n	Mean ± S.E.	Min.	n	Mean ± S.E.	Max.	Extremely high values[1]
Hg	437	0.15 ± 0.04	0.01	385	13 ± 4	51	164
Cd	481	0.39 ± 0.05	0.05	542	38 ± 7	124	930
Pb	875	6.3 ± 0.8	1.80	1,356	60 ± 14	456	4,250
Zn	1,834	31 ± 7	4.00	1,508	800 ± 111	2,800	12,200
Cu	1,715	18 ± 3	2.70	2,482	360 ± 81	1,280	5,380

Data were obtained from the following sites (numbers correspond to those in Appendix 8 Table I): 1–20, 22–36, 38, 39, 41–69, 71, 74, 75, 77, 80–82, 86, 89–99, 101–118 Biogeochemical provinces with excessive metal concentrations were excluded

n   Number of samples
S.E.   Standard error

[1] Values not included in calculation of mean

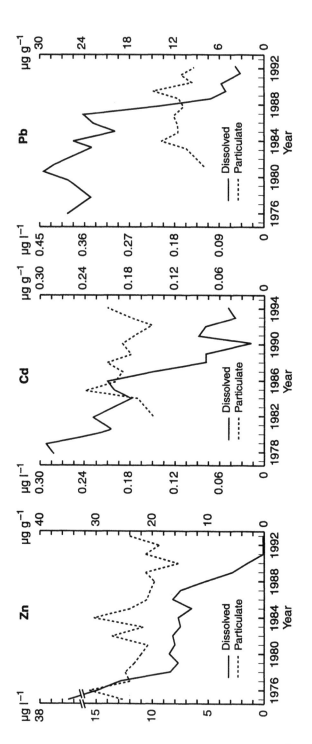

Figure 8.1 Variations in dissolved and particulate concentrations data for Zn, Cd and Pb in the Usman river, Voronezh Biosphere Reserve showing biased trends for dissolved metals due to improved techniques for sampling and analysis

where Zn concentrations were 4.2 µg l^{-1} in 1978, 2.1 µg l^{-1} in 1983, 1.5 µg l^{-1} in 1987 and 0.04 µg l^{-1} in 1992. The decrease in heavy metal concentrations in the Usman river and the Vashutkiny lakes was an arte-fact due to a decrease in sample contamination during sampling, filtration, storage and analysis processes. This is supported by the lack of negative trends in Cd, Pb and Zn concentrations in suspension in the Usman river from 1975 to 1993 (Figure 8.1).

Similar reasons for negative trends in dissolved metal concentrations have been noted in other rivers from 1958 to 1988 (Meybeck *et al.*, 1989) and in Lake Huron from 1965 to 1980 for Pb (Sturgeon and Bergman, 1987). The data on long-term dynamics of metals in the Usman river and the Vashutkiny lakes illustrate the magnitude of determination errors of dissolved metal concentrations in surface waters. Thus, it is more appro-priate to use the data on metal concentrations in bottom sediments, hydromorphic soils and biota to estimate levels of metal contamination of water bodies of the former Soviet Union (Nikanorov *et al.*, 1985, 1993; Zhulidov, 1990; Nikanorov and Zhulidov, 1991).

8.3   Natural geographic variability in freshwater bodies and wetlands

Biogeochemical provinces of the former Soviet Union were defined by Kowalsky (1974, 1978) on the basis of water, sediment, soil and plant analyses (Figure 8.2). Within the basic biogeochemical provinces, there are sub-provinces where the biogeochemical parameters of the ecosystem do not correspond to the zonal characteristics of the province (non-zonal sub-provinces). These areas are usually found over large ore massives and in regions of volcanic activity.

8.3.1   Concentrations in biogeochemical provinces and the related health implications

Forest steppe and chernozem steppe (medial zone)
This zone is situated in the middle of the former Soviet Union (Figure 8.2) and comprises approximately 11 per cent of the total area of the country. Metal concentrations in surface waters, soils, floodplain grass and pastures are considered as being close to the physiological optimum for people and animals (Kurazhkovsky, 1969; Kowalsky, 1974, 1978). This province may be considered as the most healthy with regard to metal concentrations (Co, Cu, Mn, Zn, Mo, Sr, etc.) in the natural envi-ronment (Table 8.3) and is considered as a standard against which other provinces are compared.

Taiga (humid zone)
This zone (together with transitional sub-zones) occupies the northern part of the former Soviet Union (40 per cent of the country) (Figure 8.2).

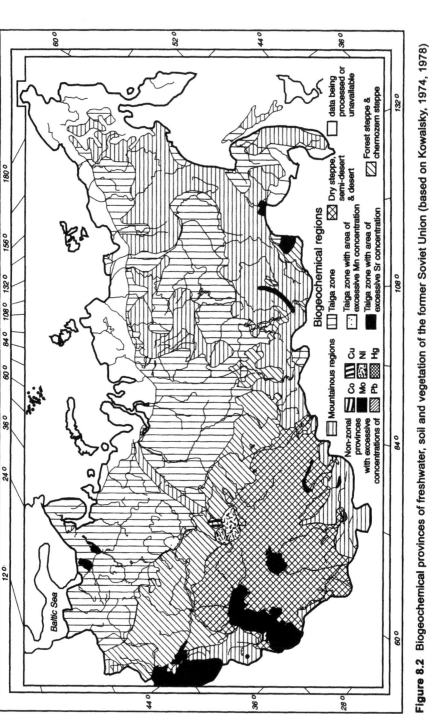

Figure 8.2 Biogeochemical provinces of freshwater, soil and vegetation of the former Soviet Union (based on Kowalsky, 1974, 1978)

Table 8.3 Heavy metal concentrations (mean ± S.E.) in rivers and wetlands in different zones of the former USSR, 1973–93

Sample	n	Co	Cu	Mn	Zn	Mo	Sr
Forest steppe and chernozem steppe[1]							
Water ($\mu g\ l^{-1}$)	231	0.15 ± 0.03	1.3 ± 0.3	9 ± 1	0.8 ± 0.2	0.8 ± 0.1	90 ± 11
Seston ($\mu g\ g^{-1}$ d.w.)	231	1.1 ± 0.2	3.4 ± 0.4	240 ± 37	15 ± 3.2	0.25 ± 0.06	98 ± 8.4
Bottom sediments ($\mu g\ g^{-1}$ d.w.)	1,545	2.3 ± 0.3	16 ± 1.6	430 ± 52	33 ± 4.1	0.42 ± 0.06	140 ± 12
Periphyton ($\mu g\ g^{-1}$ d.w.)	547	0.74 ± 0.08	5.8 ± 0.7	180 ± 22	22 ± 3.4	0.29 ± 0.04	120 ± 10
Hydromorphic soils ($\mu g\ g^{-1}$ d.w.)	2,534	13 ± 1.7	35 ± 3.9	630 ± 51	110 ± 12	1.6 ± 0.2	220 ± 27
Floodplain grass ($\mu g\ g^{-1}$ d.w.)	831	0.28 ± 0.03	6.3 ± 0.9	100 ± 11	30 ± 3.8	0.39 ± 0.03	47 ± 3.8
Taiga zone[2]							
Water ($\mu g\ l^{-1}$)	198	0.05 ± 0.01	0.48 ± 0.09	30 ± 9	0.21 ± 0.05	0.42 ± 0.08	230 ± 48
Seston ($\mu g\ g^{-1}$ d.w.)	198	0.32 ± 0.04	1.8 ± 0.3	630 ± 92	6.1 ± 0.9	0.10 ± 0.02	210 ± 39
Bottom sediment ($\mu g\ g^{-1}$ d.w.)	821	0.74 ± 0.08	4.2 ± 0.3	1,100 ± 245	14 ± 2.1	0.16 ± 0.03	290 ± 54
Periphyton ($\mu g\ g^{-1}$ d.w.)	103	0.12 ± 0.02	2.7 ± 0.4	420 ± 89	10 ± 1.5	0.09 ± 0.01	260 ± 42
Hydromorphic soils ($\mu g\ g^{-1}$ d.w.)	1,231	8.0 ± 1.3	10 ± 1.4	2,400 ± 394	27 ± 3.1	0.53 ± 0.04	840 ± 153
Floodplain grass ($\mu g\ g^{-1}$ d.w.)	543	0.09 ± 0.02	4.2 ± 0.7	420 ± 72	19 ± 2.4	0.60 ± 0.09	210 ± 27
Dry steppe, semi desert, and desert zones[3]							
Water ($\mu g\ l^{-1}$)	95	0.08 ± 0.017	0.61 ± 0.08	4.2 ± 0.6	14 ± 3	3.1 ± 0.5	530 ± 87
Seston ($\mu g\ g^{-1}$ d.w.)	95	0.42 ± 0.05	1.3 ± 0.2	95 ± 11	41 ± 6	0.83 ± 0.07	270 ± 32
Bottom sediments ($\mu g\ g^{-1}$ d.w.)	634	2.8 ± 0.32	6.4 ± 0.9	80 ± 12	80 ± 9	1.3 ± 0.18	390 ± 41
Periphyton ($\mu g\ g^{-1}$ d.w.)	115	1.3 ± 0.15	2.3 ± 0.4	52 ± 7	59 ± 9	0.75 ± 0.09	400 ± 33
Hydromorphic soils ($\mu g\ g^{-1}$ d.w.)	2,667	16 ± 1.3	28 ± 3.3	105 ± 14	210 ± 19	5.6 ± 0.4	460 ± 41
Floodplain grass ($\mu g\ g^{-1}$ d.w.)	833	0.67 ± 0.05	5.5 ± 0.7	27 ± 2	15 ± 2	3.1 ± 0.4	290 ± 21

S.E.   Standard error
n   Number of samples
d.w.   Dry weight

For the location of sample sites see Appendix 8 Table I

1   Data were obtained from the following sites: 19, 50–53, 77, 111

2   Data were obtained from the following sites: 3, 5, 6, 31, 33, 34, 38, 39, 56, 62, 114

3   Data were obtained from the following sites: 67, 80, 92–95, 97, 99

It is characterised by low Co, Cu, Mo and Zn concentrations in surface waters, soils, floodplain grass and pastures (Table 8.3; Kowalsky, 1974, 1978) causing physiological deficiencies. Concentrations of Mn and Sr are often higher than the concentrations found in steppe zones. Various farm animal diseases have been recorded in this zone including Co deficiency, vitamin B12 hypo- and avitaminosis, anaemia and anomalous skeletal developments.

Within this zone, there are zonal provinces with increased concentrations of some metals. An area with excessive Mn concentrations occurs in the south-east of the Republic of Mari and the north of Tatarstan, Russia (Figure 8.2). Natural Mn concentrations in aquatic ecosystems and wetlands of this territory exceed 2–53 times those documented in other parts of the taiga zone. This excess is usually coupled with a lack of iodine (Kowalsky, 1974) and epidemic diseases of endocrine (thyroid) glands have been noted among the local population (Kowalsky, 1974, 1978).

Zonal provinces with excessive Sr concentrations are located in the east of the Chita region (East Siberia) and the Zeisk area of the Amur region, Russia (Figure 8.2). Natural Sr concentrations in the aquatic ecosystem and wetlands in these provinces are 2–10 times as high as those outside these provinces in the taiga zone. Excessive Sr concentration in the environment is coupled with a lack of Ca and uricolysis disease has been recorded among the local population and livestock (Kowalsky, 1974, 1978; Kowalsky et al., 1978).

Dry steppe, semi-desert and desert (arid) zones
This zone occupies about 14 per cent of the former Soviet Union in its southern part (Figure 8.2) and is characterised by low concentrations of Co, Cu and Mn and high concentrations of Zn, Mo and Sr in aquatic and wetland ecosystems compared with the steppe zone (Table 8.3). The disproportionate amount of Cu, Mo and Co concentrations in the diet of farm animals (the ratio of Zn to Cu in pasture plants is < 3.0 and the ratio of Mo to Cu is > 0.4) results in endemic ataxia, especially in sheep (Kowalsky, 1974).

Mountainous regions
Mountainous regions are numerous in the former Soviet Union, including Carpathian, Crimea, Caucasus, Urals, Kopetdag, Saur, Tarbagatay, Djungar Alatau, Tian-Shan, Pamir-Alai, Altai, Sayan, the mountains of the Baikal area, Trans-Baikal area, north-east Siberia and Sikhote-Alin (Figure 8.2). The total area of mountainous regions occupies almost one-third of the former Soviet Union. In these regions, there are various biogeochemical types with either low concentrations of Co, Cu and Zn or with high concentrations of Co, Cu, Zn, Mo, and Sr compared with the steppe zone (Appendix 8 Table II). A great diversity of biogeochemical parameters occurs within each mountain range, which is

Table 8.4   Cobalt concentrations in rivers and wetlands in provinces with excessive cobalt concentrations in Azerbaijan

Sample	Sumgait River[1] Mean ± S.E.	n	Pirsagat River[1] Mean ± S.E.	n
Water ($\mu g \, l^{-1}$)	9 ± 1	34	14 ± 2	28
Seston ($\mu g \, g^{-1}$ d.w.)	29 ± 4	34	40 ± 5	28
Bottom sediments ($\mu g \, g^{-1}$ d.w.)	53 ± 6	85	59 ± 5	72
Periphyton ($\mu g \, g^{-1}$ d.w.)	29 ± 4	29	55 ± 4	24
Hydromorphic soils ($\mu g \, g^{-1}$ d.w.)	64 ± 7	187	79 ± 5	81
Floodplain grass ($\mu g \, g^{-1}$ d.w.)	15 ± 2	89	19 ± 2	54

S.E.   Standard error
n   Number of samples
d.w.   Dry weight

[1]   Correspond to sampling sites Nos. 87 and 88 in Appendix 8 Table I

typical of mountainous region. For example, in the Carpathians, natural concentrations of Co, Mn, Zn, Mo and Sr in the Goryn river are 2–10 times higher than those in the Sluch river, whereas the concentrations of Cu in the Goryn river are 2–4 times lower than those in the Sluch river (Appendix 8 Table II).

In mountainous biogeochemical provinces where very low concentrations of Co, Cu and Zn (along with low concentrations of iodine) occur, the local population suffers from endemic goitre and vitamin B12 hypo- and avitaminosis (Kurazhkovsky, 1969; Kowalsky, 1974, 1978).

8.3.2   Non-zonal biogeochemical provinces, excess natural metals and consequent health effects

The most well known province with excessive Co concentrations (Figure 8.2) is located in the Sumgait and Pirsagat river basins in Azerbaijan. Concentrations of Co in freshwater ecosystems and wetlands of the area are 10 to 100 times higher than that in the medial zone (Table 8.4). Suppressed synthesis of vitamin B_{12} have been recorded among the local population and livestock (Kowalsky, 1974, 1978).

The most intensively studied area with excessive Mo concentrations (Figure 8.2) is located north of Lake Sevan and in the Razdan river basin (Armenia). Concentrations of Mo in freshwater and wetland ecosystems in the area are 9–100 times higher than those in medial zones (Table 8.5). Endemic molybdenum podagra has been recorded among the local population and disrupted purine exchange has been recorded among sheep and cattle (Kowalsky, 1974, 1978).

The most well known province with excessive Pb concentrations is located in the north of Armenia, in the Debet river basin (Figure 8.2). Concentrations of Pb in freshwater ecosystems of the area are 5–200

Table 8.5   Molybdenum concentrations in aquatic ecosystems in provinces with excessive molybdenum concentrations in Armenia

Sample	Lake Sevan[1] Mean ± S.E.	n	Razdan River[1] Mean ± S.E.	n
Water (µg l^{-1})	–	–	56 ± 7	39
Seston (µg g^{-1} d.w.)	8 ± 1	45	15 ± 2	39
Bottom sediments (µg g^{-1} d.w.)	14 ± 1	52	23 ± 3	47
Periphyton (µg g^{-1} d.w.)	10 ± 1	32	14 ± 1	39
Hydromorphic soils (µg g^{-1} d.w.)	21 ± 3	87	35 ± 3	145
Floodplain grass (µg g^{-1} d.w.)	18 ± 2	143	24 ± 2	171

S.E.   Standard error
n   Number of samples
d.w.   Dry weight

[1]   Correspond to sampling sites Nos. 84 and 85 in Appendix 8 Table I

Table 8.6   Lead concentrations in the Debet river and wetlands in provinces with excessive lead concentrations in Armenia

Sample	Tumanyan village[1] Mean ± S.E.	n	Akhtala village[1] Mean ± S.E.	n
Water (µg l^{-1})	18 ± 2	41	15 ± 2	34
Seston (µg g^{-1} d.w.)	54 ± 7	41	45 ± 6	34
Bottom sediments (µg g^{-1} d.w.)	145 ± 18	86	110 ± 19	57
Periphyton (µg g^{-1} d.w.)	58 ± 7	42	65 ± 6	37
Hydromorphic soils (µg g^{-1} d.w.)	350 ± 41	155	380 ± 52	141
Floodplain grass (µg g^{-1} d.w.)	120 ± 20	94	190 ± 28	84

S.E.   Standard error
n   Number of samples
d.w.   Dry weight

[1]   Corresponds to sampling site No. 83 in Appendix 8 Table I

times higher than those in the medial biogeochemical zone where conditions are taken as optimal (Table 8.6). Endemic diseases of the nervous system (e.g. cephalalgia, myalgia, ischialgia, and gastralgia), gingivitis, and hypermenorrhea have been observed (Kowalsky, 1974, 1978).

The largest area with high Cu concentrations is located in the east of Bashkiria and south-west of the Chelyabinsk region, Russia (Figure 8.2). Concentrations of Cu in this area are 5–200 times higher than those in the medial biogeochemical zone (Table 8.7). Endemic normochromic and hypochromic anaemia occurs among the local population. Farm animals suffer endemic anaemia, hepatitis and liver cirrhosis (Kowalsky, 1974, 1978).

Table 8.7   Copper concentrations in rivers and wetlands in provinces with excessive copper concentrations in Russia

Sample	Skamara River[1] Mean ± S.E.	n	Zingeyka River[1] Mean ± S.E.	n
Water (µg l^{-1})	80 ± 9	56	90 ± 14	45
Seston (µg g^{-1} d.w.)	190 ± 18	56	240 ± 28	45
Bottom sediments (µg g^{-1} d.w.)	320 ± 37	63	350 ± 42	74
Periphyton (µg g^{-1} d.w.)	180 ± 21	56	210 ± 25	49
Hydromorphic soils (µg g^{-1} d.w.)	380 ± 42	231	410 ± 52	187
Floodplain grass (µg g^{-1} d.w.)	230 ± 20	91	190 ± 15	103

S.E.   Standard error
n   Number of samples
d.w.   Dry weight

[1]   Correspond to sampling sites Nos. 72 and 73 in Appendix 8 Table I

The most well known areas with excessive Ni concentrations are located in Kazakhstan (north of Akhtubinsk and Kustanai regions) and in Russia (south of Orenburg region, south-west of Chelyabinsk region and south of Bashkiria) (Figure 8.2). Concentrations of Ni in freshwater ecosystems of the area are 4–1,400 times higher than those in the medial biogeochemical zones (Table 8.8). Specific skin diseases have been recorded in the local population and osteodystrophy occurs in cattle (Kowalsky, 1974, 1978).

The largest area with excessive Hg concentrations is located in Uzbekistan within the boundaries of the southern Fergana mercury-antimony belt (Figure 8.2). The area has a long history of Hg mining. There is evidence of ancient mining (2nd to 4th century BC to 8th to 9th century AD). Traditional Hg mining ceased in the fourteenth century due to the Tamerlan wars. Today, Hg is still mined in this area. The most studied areas are the basins of the Chauvay, Sokh and Shakhimardan rivers. Concentrations of Hg in aquatic ecosystems of the area are 90–2,000 times higher than those in the medial zone (Table 8.9). At the same time, climatic and geographic aspects of the Sokh and Shakhimardan river valleys have encouraged the construction of recreation areas for the southern Fergana residents in the river valleys.

8.4   Water bodies exposed to extreme anthropogenic pollution

The examples presented below for the different hydrographic regions were chosen out of a large number of water bodies in the former Soviet Union exposed to anthropogenic pollution by heavy metals. For each river discussed, the main sources of pollution and the principal heavy metal pollutants are noted. The information is based mostly on data obtained by the OGSNK network (Lobtchenko *et al.*, 1988, 1991).

Table 8.8 Nickel concentrations in rivers and wetlands in provinces with excessive nickel concentrations in Kazakhstan and Russia

Sample	Ilek River[1] Mean ± S.E.	n	Tobol River[1] Mean ± S.E.	n	Ural River[1] Mean ± S.E.	n	Zingeyka River[1] Mean ± S.E.	n	Bolshoylk River[1] Mean ± S.E.	n
Water ($\mu g\ l^{-1}$)	280 ± 25	23	240 ± 27	27	420 ± 51	32	320 ± 29	29	250 ± 31	33
Seston ($\mu g\ g^{-1}$ d.w.)	95 ± 10	23	120 ± 13	27	240 ± 31	32	170 ± 20	29	73 ± 9	33
Bottom sediments ($\mu g\ g^{-1}$ d.w.)	105 ± 9	45	230 ± 21	42	680 ± 87	96	450 ± 64	87	110 ± 10	87
Periphyton ($\mu g\ g^{-1}$ d.w.)	110 ± 12	25	150 ± 17	37	200 ± 21	54	160 ± 18	34	90 ± 10	42
Hydromorphic soils ($\mu g\ g^{-1}$ d.w.)	180 ± 20	98	420 ± 51	88	790 ± 83	234	670 ± 81	178	160 ± 19	91
Floodplain grass ($\mu g\ g^{-1}$ d.w.)	250 ± 21	54	200 ± 18	92	240 ± 21	131	220 ± 19	165	90 ± 10	97

S.E.   Standard error
n   Number of samples
d.w.   Dry weight

[1] Corresponds to sampling sites Nos. 78, 79, 40, 76, 73, and 70 respectively in Appendix 8 Table I

Table 8.9   Mercury concentrations in rivers and wetlands in provinces with excessive mercury concentrations in Uzbekistan

Sample	Sokh River[1] Mean ± S.E.	n	Shakhimardan River[1] Mean ± S.E.	n	Chauvay River[1] Mean ± S.E.	n
Water (µg l⁻¹)	8 ± 1	47	4.2± 0.7	28	4.7± 0.6	35
Seston (µg g⁻¹ d.w.)	21 ± 4	47	14 ± 2	28	16 ± 2	35
Bottom sediments (µg g⁻¹ d.w.)	19 ± 2	86	32 ± 4	31	16 ± 2	52
Periphyton (µg g⁻¹ d.w.)	15 ± 2	25	19 ± 2	33	25 ± 3	35
Hydromorphic soils (µg g⁻¹ d.w.)	34 ± 4	86	45 ± 3	95	92 ± 8	78
Floodplain grass (µg g⁻¹ d.w.)	20 ± 2	42	19 ± 2	31	28 ± 3	62

S.E.   Standard error
n   Number of samples
d.w.   Dry weight

[1]   Corresponds to sampling site No. 100 in Appendix 8 Table I

8.4.1   The Baltic Sea basin

The Pltava river in the city of Lvov, Ukraine, is the most heavily polluted tributary of the Western Bug river, receiving wastewaters from the cities of Busk and Lvov. The principal metal pollutant is Cr which originates mainly from industrial sources (Table 8.10).

For the Pregolia river, near the city of Kaliningrad, Russia, the major pollution sources are steel industries, pulp and paper mills, ship building, and the electronic and machine building factories of the cities of Cherniakhovsk, Gvardeysk, and Kaliningrad. The principal metal pollutant in this river is Zn (Table 8.10).

The Kulpe river, near the city of Shaulyay, Lithuania is one of the most heavily polluted tributaries of the Lielupe river. The major sources of pollution for the Kulpe river are industrial and municipal wastewaters from Shaulyay. The principal metal pollutant is Cu (Table 8.10).

8.4.2   The Black Sea basin

The Svisloch river near Minsk, Byelorussia, receives copper-rich effluents from the city's pharmaceutical industry. Copper concentrations in the Svisloch river are given in Table 8.10. For the Seim river in the city of Kursk, Russia, heavy metal concentrations in the river are due to industrial effluents from the city. Zinc is the principal metal pollutant (Table 8.10). The Byk river, near the city of Kishinev, Moldova, is influenced by both industrial effluents and agricultural runoff where Cu is abundant. Copper concentrations in the Byk river are presented in Table 8.10.

8.4.3   The Azov Sea basin

In the Temernik river, Rostov-on-Don, Russia, heavy metal concentrations in the river are due to the discharge of city effluents. Zinc is the principal metal pollutant (Table 8.10).

Table 8.10   Heavy metal concentrations in rivers subjected to extreme anthropogenic pollution

Site[1]	River	Basin	Republic	Metal	Water (μg l^{-1})	Seston (μg g^{-1})	Bottom sediment (μg g^{-1})	Periphyton (μg g^{-1})
							Concentration	
7	Pltava	Baltic Sea	Ukraine	Cr	78	460	800	1,000
8	Pregolia	Baltic Sea	Russia	Zn	75	360	1,600	950
2	Kulpe	Baltic Sea	Lithuania	Cu	95	200	500	490
16	Svisloch	Black Sea	Byelorussia	Cu	60	150	400	300
17	Seim	Black Sea	Russia	Zn	21	140	800	400
18	Byk	Black Sea	Moldova	Cu	45	120	370	240
25	Temernik	Azov Sea	Russia	Zn	80	450	1,600	80
26	Seversky Donets	Azov Sea	Ukraine	Cu	52	130	470	245
32	Nuduay	Barents Sea	Russia	Cu	82	210	730	400
				Ni	300	120	540	300
47	Ulba	Kara Sea	Kazakhstan	Zn	150	900	2,300	900
49	Schuchia	Lake Piasino, Kara Sea	Russia	Cu	–	270	900	800
74	Chusovaya	Caspian Sea	Russia	Cr	220	540	970	1,200
				Cu	43	120	750	150
100	Chauvay	Central Asian Region	Uzbekistan	Hg	5.5	20	95	56
115	Silinka	Pacific Ocean	Russia	Cu	–	85	290	110

Average arithmetic error in all cases fluctuates within 10–20 per cent

[1] Sampling site numbers correspond to those in Appendix 8 Table I

The Seversky Donets river near Rubezhnoye city, Ukraine, is the most heavily polluted tributary of the River Don. It receives two-thirds of the total volume of wastewater introduced to the Don river basin originating mainly from the chemical industry. Chromium and Cu are the principal heavy metal pollutants. Copper concentrations in the Seversky Donets river are given in Table 8.10.

8.4.4   The Barents Sea basin

The Nuduay river near Monchegorsk, Russia, is one of the most heavily polluted rivers of the Kola Peninsula. The major source of pollution is the 'Severonikel' smelter effluent, and Cu and Ni are the principal metal pollutants (Table 8.10). The mean annual concentration from samples near the city of Monchegorsk was 144 μg l^{-1} for Cu in 1985 and 13,200 μg l^{-1} for Ni in 1982.

8.4.5   The Kara Sea basin

The River Ulba in the mining district of Tishinsky, Kazakhstan, is one of the most heavily polluted tributaries of the River Irtysh. The major sources of pollution are ore mine discharge, effluents from the Ust-Kamenogorsk titanium-magnesium and lead-zinc combined smelters and the Breks river which receives effluents from the Irtysh Polymetal smelter. The principal metal pollutants are Zn and Cu. Zinc concentrations in the Ulba river are given in Table 8.10. In 1984 and 1985, the mean annual concentrations of Zn in the river, 5.6 km downstream of the Tishinsky ore mine discharge, were 2,160 µg l^{-1} and 1,960 µg l^{-1} respectively.

The Schuchia river (Lake Piasino basin) in the Krasnoyarsk region, Russia, is probably one of the most heavily contaminated rivers in the former Soviet Union for Cu and Ni. The major source of pollution is the Norilsk combined smelter. Copper concentrations in the Schuchia river are given in Table 8.10. Based on OGSNK data, the mean annual Cu concentration in the river close to the city of Norilsk in 1981 was 3,310 µg l^{-1}.

8.4.6   The East Siberian Sea basin

Water bodies in this area rank amongst the cleanest in the former Soviet Union. Where sources of pollution occur, these are due to navigation, municipal wastes, effluents from mills, and other major industries. However, owing to a sparse population (< 1 person per km^2) the impact of these sources on water quality is not significant.

8.4.7   The Caspian Sea basin

In the Caspian Sea basin, the Chusovaya river (in the Kama river basin) downstream of the city of Pervouralsk, Russia, receives effluents from the sludge accumulation ponds of a potassium bichromate factory, Cu smelter and metallurgic works. These effluents are rich in Cr, Cu and Zn. Chromium and Cu concentrations in the Chusovaya river are given in Table 8.10. Based on OGSNK data for 1981–85, mean annual Cu and Zn concentrations in the Chusovaya river downstream of Pervouralsk were 67–388 µg l^{-1} and 180–680 µg l^{-1} respectively.

8.4.8   Central Asian region

In the Central Asian region, the Chauvay river, Fergana, Uzbekistan, receives effluents from a Hg ore mine in a region where background Hg concentrations have always been high (Table 8.10).

8.4.9   The Pacific Ocean basin

The Silinka river (in the Amur river basin), near the city of Gorny, Russia, is exposed to the impact of the 'Solnetchny' and other mining effluents rich in Cu (Table 8.10). Based on OGSNK data for 1981–85, the mean annual Cu concentration in the Silinka river downstream of Gorny was 37–87 µg l^{-1}.

8.5   Heavy metals in the aquatic food web

Heavy metals affect all aspects of trophic food webs, although the level of impact varies. In May 1982, the downstream section of the Usman river (Voronezh region, Don basin) and the adjacent wetlands were exposed to an industrial effluent spill containing heavy metals. An example of the impact and long-term dynamics (1981–89) of Cd in the aquatic food web following the pollution incidence are presented below (Figure 8.3).

At the trophic level of filter feeding bivalves, *Sphaerium corneum* (Mollusca, Bivalvia), the Cd concentration in the soft tissue increased following the pollutant discharge to the river. Maximum concentrations of Cd were observed in particulate matter immediately after the wastes were discharged to the river in June and July, 1982. The maximum Cd concentrations in the soft tissues of molluscs were recorded one year later (lag effect). After the effluent spill, and until the end of investigations in 1989, metal concentrations in *S. corneum* tissue were higher than those in particulate matter (Figure 8.3).

At the trophic level of benthic invertebrates, such as *Limnodrilus hoffmeisteri* (Oligochaeta, Tubificidae), Cd concentrations measured in tissue also increased after the spill. As in the case of particulate matter, maximum Cd concentrations in bottom sediments were observed immediately after the spill (June–July 1982) whereas in oligochaetes, as in molluscs, maximum Cd concentrations were recorded a year later (lag effect). Before the pollution incident and immediately after it, the Cd concentration in bottom sediments was higher than that in the oligochaete tissue. In 1983, as well as in 1987–89, Cd concentration in oligochaetes appeared to be higher than that in bottom sediments. Between 1984 and 1986, Cd concentrations in bottom sediments and oligochaete body tissue did not differ significantly.

At a higher trophic level, predators of the benthic invertebrates were examined. Cadmium concentrations remained relatively low and constant in fish (*Perca fluviatilis*) muscle and fledgling duck (*Anas platyrhynchos*) muscle. The reaction of the vertebrates studied to the pollution incident was less pronounced than that of the invertebrates studied and the environmental samples (Figure 8.3).

8.6   Heavy metal trends in freshwater biota

Trends in heavy metals in biota have been studied by analysis of samples kept in museums and similar collections. Samples included insects (Zhulidov and Emetz, 1979; Nikanorov *et al.*, 1982; Zhulidov *et al.*, 1985), molluscs (Nikanorov *et al.*, 1985; Nikanorov and Zhulidov, 1986, 1991), bone tissue of roe deer (*Capreolus capreolus*), moose (*Alces alces*) and European red deer (*Cervus elaphus*) (Zhulidov *et al.*, 1985), and various plant species (Zhulidov and Skalon, 1979; Nikanorov and Zhulidov, 1981, 1991; Nikanorov *et al.*, 1985 amongst others).

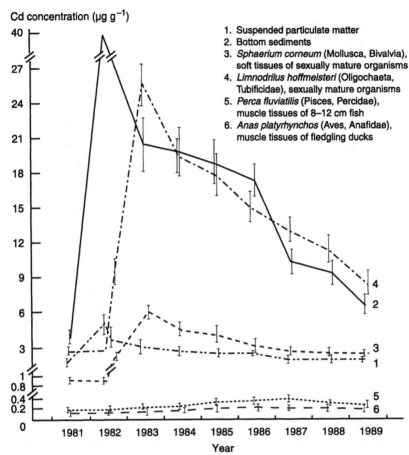

Cd concentration (μg g^{-1})

1. Suspended particulate matter
2. Bottom sediments
3. *Sphaerium corneum* (Mollusca, Bivalvia),
   soft tissues of sexually mature organisms
4. *Limnodrilus hoffmeisteri* (Oligochaeta,
   Tubificidae), sexually mature organisms
5. *Perca fluviatilis* (Pisces, Percidae),
   muscle tissues of 8–12 cm fish
6. *Anas platyrhynchos* (Aves, Anafidae),
   muscle tissues of fledgling ducks

Figure 8.3 Long-term trends in Cd concentration (mean ± S.E., dry weight) in the trophic food web of the Usman river, Voronezh Biosphere Reserve, following an industrial spill in May 1982

8.6.1 Plants

Long-term dynamics of Pb and Hg were studied in two species of vascular plants, *Nuphar lutea* and *Menyanthes trifoliata* (Nikanorov and Zhulidov, 1981, 1991; Nikanorov *et al.*, 1985). Specimen plants were collected in 1913–15, 1939, 1947 and 1958 in the upper reaches of the Usman river and in 1977–80 in the Usman river in the Voronezh Biosphere Reserve. Rhizomes of *N. lutea* and leaves of *M. trifoliata* were analysed.

Between 1913 and 1980 Hg concentrations in the plants gradually increased from 0.03–0.04 μg g^{-1} to 0.12–0.15 μg g^{-1} dry weight. Lead concentrations in the plants also increased (Figure 8.4). From 1913 to 1939, Pb concentrations in the samples of both plant species did not

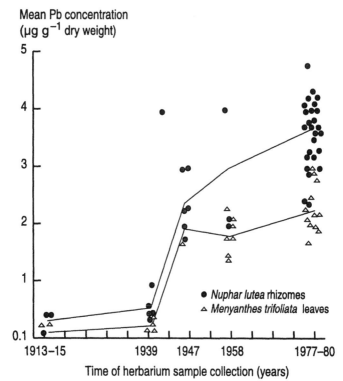

Figure 8.4 Long-term trends in mean Pb concentration in samples of *Nuphar lutea* (L.) rhizomes and *Menyanthes trifoliata* L. leaves, collected from 1913 to 1980 in the Usman river (Nikanorov and Zhulidov, 1991)

increase significantly (P > 0.05), but from 1939 to 1980 there was an increase in Pb concentrations in both species. A notable increase occurred between 1939 and 1947 when highly significant (P < 0.001) Pb concentrations occurred (Figure 8.4). This was probably related to military actions in the region during World War II. In a similar study (1932–52) in an adjoining area (i.e. the Sysola river, near Syktyvkar, Komi Republic, Russia) where military action had not taken place, Pb concentration in the same plant species did not increase significantly.

8.6.2 Animals

Levels of Pb analysed in samples of molluscs and insects collected from wetlands adjoining the Usman river in the Voronezh Biosphere Reserve in 1930–39 were 2–7 times lower than those in molluscs and insects of the same species collected from wetlands adjoining the Usman river at the same reserve in 1977–80 (Table 8.11).

The results presented for heavy metal concentrations in plants and animals showed clearly that the compartments of the Voronezh

Table 8.11   Lead concentrations ($\mu g\ g^{-1}$ dry weight) in molluscs and insects collected in the Usman river wetlands in the Voronezh Biosphere Reserve, 1930–39 and 1977–80

Species	1930–39 Mean ± S.E.	n	1977–80 Mean ± S.E.	n
Mollusca				
Lymnaea stagnalis	1.8 ± 0.1	6	6.4 ± 0.4	12
Coleoptera (imago)				
Dytiscidae (males)				
Dytiscus marginalis	0.53 ± 0.03	7	3.2 ± 0.2	23
Carabidae:				
Carabus cancellatus	0.52 ± 0.02	5	1.8 ± 0.1	13
C. glabratus	0.62 ± 0.03	5	1.8 ± 0.1	13
C. marginalis	0.68 ± 0.02	6	1.6 ± 0.1	15
Scarabaeidae:				
Anomala dubia	0.38 ± 0.04	7	1.6 ± 0.1	15
Cerambicidae:				
Acmaeops collaris	0.42 ± 0.04	7	1.1 ± 0.1	13
Silphidae:				
Silpha obscura	0.53 ± 0.03	10	1.9 ± 0.1	25
S. carinata	0.56 ± 0.04	12	2.3 ± 0.1	32
Heteroptera (imago)				
Ilyocoris cimicoides	0.73 ± 0.02	8	5.4 ± 0.3	51
Notonecta glauca	0.84 ± 0.02	5	3.7 ± 0.2	46

S.E.   Standard error   Source: Nikanorov and Zhulidov, 1991
n   Number of samples

Biosphere Reserve environment are exposed to anthropogenic pollution with metals, in spite of the fact that there are no direct sources of contamination in this area.

8.6.3   Plant–insect food web

Similar trends in heavy metal concentrations have also occurred in plants and animals which are closely associated. Figure 8.5 shows the relationship between Pb concentrations in imago beetles, *Donacia crassipes* (Coleoptera, Chrysomelidae) and those in *Nuphar lutea* (Nymphaeaceae) leaves in the Usman river during different periods.

In 1957–60 and 1978–80, Pb concentrations in the beetles were higher than those in the *Nuphar* leaves, but in 1939–40 this ratio was opposite. From 1913 to 1939, there was a slight but insignificant increase in Pb concentrations in the two species. From 1939 to 1980 there was a marked significant increase ($P < 0.05$) in Pb concentration in the *Nuphar* leaves and in the beetles. Because there is a direct relationship between Pb

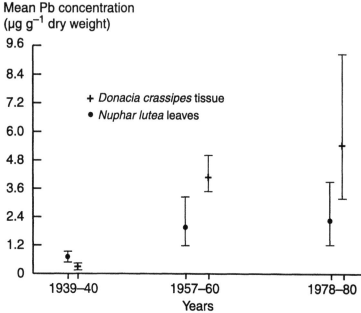

Figure 8.5 Mean Pb concentration in *Nuphar lutea* (L.) leaves and *Donacia crassipes* (Coleoptera) tissues from 1913 to 1980 in the Usman river, Voronezh Biosphere Reserve (After Nikanorov and Zhulidov, 1991)

concentration in the beetles and the *Nuphar* leaves (Nikanorov and Zhulidov, 1991), it may be assumed that Pb accumulation in the beetles is related to the pollution level of the ecosystem.

8.7   References

(All references are in Russian unless otherwise stated).

APHA 1985 *Standard Methods for the Examination of Water and Wastewater.* 16th Edition, American Public Health Association, Washington DC, 1,286 pp. (In English).

Danielsson, L.G., Magnusson, B., Westerlund, S. and Zhang, K. 1982 Trace metal determination in estuarine waters by electrothermal AAS after extraction of dithiocarbamate complex into Freon. *Analytica Chimica Acta*, **144**, 183–188. (In English).

Elbaz-Poulichet, F. and Martin, J.-M. 1987 Dissolved Cd behaviour in some selected French and Chinese estuaries. Consequences on Cd supply to the ocean. *Marine Chemistry*, **22**, 125–136. (In English).

Konovalov, G.S. 1959 Transport of micro-elements in major USSR rivers. *Reports of the USSR Academy of Sciences*, **129**(4), 912–915.

Konovalov, G.S. and Koreneva, V.I. 1980 Concentrations of sodium, potassium, and micro-elements in rivers within the basins of the Azov, Black and Caspian Seas. *Water Resources*, **2**, 188–191.

Konovalov, G.S., Ivanova, A.A. and Kolesnikova, T.K. 1966a Micro-elements in water and particulate matter in rivers of the European part of the USSR. *Hydrochemical Materials*, **42**, 94–111.

Konovalov, G.S., Ivanova, A.A. and Kolesnikova, T.K. 1966b Rare elements in water and particulate matter in rivers of the Asian part of the USSR. *Hydrochemical Materials*, **42**, 112–123.

Konovalov, G.S., Koreneva, V.I., Korenev, A.P. and Dobrovolskaya, N.I. 1982a Micro-elements in the most downstream observation sites of rivers in the European part of the USSR (1976–1978). *Hydrochemical Materials*, **80**, 18–24.

Konovalov, G.S., Koreneva, V.I., Korenev, A.P. and Dobrovolskaya, N.I. 1982b Distribution of micro-element concentrations in rivers of the Asian part of the USSR (1976–1978). *Hydrochemical Materials*, **81**, 3–9.

Kowalsky, V.V. 1974 *Geochemical Ecology*. Nauka, Moscow, 300 pp.

Kowalsky, V.V. 1978 Geochemical ecology – the basis of the system of biogeochemical regionalisation. *Transactions of the Biogeochemical Laboratory*, **15**, 3–21.

Kowalsky, V.V., Blokhina, R.I., Zasorina, E.F., Samarina, I.A. and Khobotiev, V.G. 1978 Strontium-calcium sub-provinces of the biosphere and biogeochemical provinces. *Transactions of the Biogeochemical Laboratory*, **15**, 121–130.

Kurazhkovsky, Y.N. 1969 *Essays on Use of Nature*. Mysl, Moscow, 268 pp.

Lobtchenko, E.E., Kolesnikova, T.K., Dolzhenko, A.S., Levina, M.E., Kuzhekova, N.I., Lyampert, N.A., Shaposhnik, E.G., Sorokina, E.F. and Pervysheva, O.A. 1988 Dynamics of surface water quality in the USSR in 1976–1980. Gidrometeoizdat, Leningrad, 350 pp.

Lobtchenko, E.E., Tsirkunov, V.V., Kuzhekova, N.I., Levina, M.E., Lyampert, N.A., Shaposhnik, E.G., Sorokina, E.F. and Pervysheva, O.A. 1991 Dynamics of surface water quality in the USSR in 1981–1985. Gidrometeoizdat, Leningrad, 335 pp.

Martin, J.-M., Guan, D.M., Elbaz-Puolichet, F., Thomas, A.J. and Gordeev, V.V. 1993 Preliminary assessment of the distribution of some trace elements (As, Cd, Cu, Fe, Ni, Pb and Zn) in a pristine aquatic environment: the Lena River estuary (Russia). *Marine Chemistry*, **43**, 185–199. (In English).

Meybeck, M., Chapman, D. and Helmer, R. [Eds] 1989 *Global Freshwater Quality: A First Assessment*. Blackwell Reference, Oxford, 306 pp. (In English).

Nikanorov, A.M. and Zhulidov, A.V. 1981 Historical biomonitoring of river water pollution with lead based on the study of metal accumulation in plants. *Reports of the USSR Academy of Sciences*, **258**(4), 1019–1021.

Nikanorov, A.M. and Zhulidov, A.V. 1986 Background mercury concentrations in aquatic biota in the lakes of the East European sub-Arctic. In: *Monitoring of Environmental Background Pollution*. Issue 3, Gidrometeoizdat, Leningrad, 186–190.

Nikanorov, A.M. and Zhulidov, A.V. 1991 *Biomonitoring of Metals in Freshwater Ecosystems.* Gidrometeoizdat, Leningrad, 312 pp.

Nikanorov, A.M., Zhulidov, A.V., Emetz, V.M., Kulmatov, R.A. and Kist, A.A. 1982 Specific features of mercury accumulation in tissues of aquatic invertebrates and bottom sediments at the initial stage of metal accumulation in riverine ecosystems. *Reports of the USSR Academy of Sciences,* **264**(1), 1022–1024.

Nikanorov, A.M., Zhulidov, A.V. and Pokarzhevsky, A.D. 1985 *Biomonitoring of Heavy Metals in Freshwater Ecosystems.* Gidrometeoizdat, Leningrad, 144 pp.

Nikanorov, A.M., Zhulidov, A.V. and Emetz, V.M. 1993 *Heavy Metals in Organisms of Russian Wetlands.* Gidrometeoizdat, St Petersburg, 294 pp.

State Report 1993 *State of the Environment in the Russian Federation in 1992.* Russian Federation Ministry for Protection of the Environmental and Natural Resources, Zelenyi Mir, Moscow, 20–22.

Stradomsky, V.B. and Konovalov, G.S. 1968 Stable strontium in some inland surface waters. *Hydrochemical Materials,* **47**, 51–62.

Sturgeon, S. and Bergman, S.S. 1987 Sampling and storage of natural water for trace metals. *CRC Critical Reviews in Analytical Chemistry,* **18**, 209–244. (In English).

Zhulidov, A.V. 1990 Analysis of the factors determining the concentration of heavy metals in animal bodies as bioindicators of environmental condition. In: D.A. Krivolutsky [Ed.] *Bioindicators of Chemical and Radioactive Pollution.* Mir Publishers, Moscow-Boca Raton, 251–303. (In English).

Zhulidov, A.V. and Emetz, V.M. 1979 Lead accumulation in beetle tissue through pollution by automobile exhaust gases. *Reports of the USSR Academy of Sciences,* **244**(6), 1515–1516.

Zhulidov, A.V. and Skalon, N.V. 1979 Variation in heavy metal concentrations in rare and endangered plants of Yakutia and Buryatia during the twentieth century. In: *Problems of Ecology and Environmental Protection.* Kemerovo University, Kemerovo, 176-179.

Zhulidov, A.V., Bogatch, Y., Pokarzhevsky, A.D. and Gusev, A.A. 1985 Historical biomonitoring of heavy metals in natural ecosystems. In: *Radioecology of Soil Animals.* Nauka, Moscow, 186–198.

Appendix 8

Procedures, sampling sites and results for metal determination in water bodies of the former USSR

Sampling and analytical procedures in recent studies

For water sampling an extendible pole to hold the sample container (glass, polyethylene, polypropylene and Teflon-coated) was used. Pre-cleaned (with acid) Teflon-coated equipment was used and samples were stored in acid-cleaned polyethylene and polypropylene bottles. Polyethylene bottles were previously soaked for 2–3 days in hot HCl (30 per cent), rinsed with demineralised water and kept filled with HCl (5 per cent, Merck Suprapur) for 5–7 days. The bottles were then rinsed three times with demineralised water and transferred to pure double polyethylene bags. For Teflon bottles the same procedure was used but HCl was replaced by nitric acid (Merck Suprapur) (Elbaz-Poulichet and Martin, 1987). A volume of up to 2 litres of water was filtered within 0.5–3 hours of collection using acid-cleaned polycarbonate Nucleopore filters (0.4 µm pore diameter, 47 mm filter diameter).

Sampling and analysis of Cd, Pb, Zn, Cd and other metals

The filtered water samples were acidified to pH 2 with ultrapure HCl (or nitric acid), Merck Suprapur grade. Transportation of samples to the laboratory was carried out in double plastic containers using hermetic film, at a temperature of 4–10 °C. The samples (except water) were kept in pure double hermetic soldered plastic containers up to 1991–94. Total dissolved Cd, Pb, Zn and Cu were pre-concentrated using solvent extraction of the chelates formed by the metals with a mixture of complexes (ammonium pyrrolidine ditiocarbamate and diethylammonium diethyl dithiocarbamate) using 1,1,2-trichloro-1,2,2-trifluoroethane, followed by back-extraction into nitric acid (Danielsson *et al.*, 1982; Martin *et al.*, 1993). Total dissolved metal (Cd, Pb, Zn and Cu) concentrations were measured using graphite-furnace atomic-absorption spectrophotometry (GFAASI; Perkin Elmer 3030, HGA-500). Filters containing particulate matter, bottom sediments, soils and biota were dried at 80 °C. For these latter filter samples, the total trace metal concentrations were also determined by GFAAS after digesting with a mixture of HNO_3-HF-$HClO_4$ (Nikanorov *et al.*, 1985; Nikanorov and Zhulidov, 1991). For some determinations, colourimetry and neutron activation were employed (Nikanorov *et al.*, 1985; Nikanorov and Zhulidov, 1991). Some samples of water from pristine regions were analysed in a Class-100 clean room.

Sampling and analysis of mercury
Total Hg concentrations were determined by specific atomic absorption spectrometry on a Perkin-Elmer Coleman Mas-50 Mercury Analyzer, after mineralisation (Nikanorov and Zhulidov, 1991; Nikanorov *et al.*, 1993). Flameless atomic absorption spectrometry on a Spectrophotometer VARIAN AA-475 was also used.

Quality accuracy / quality control
The accuracy and precision of the measurements were assessed using reference standards (Riverine Water Reference Material for Trace Metals, National Research Council of Canada; National Bureau of Standards, USA; International Atomic Energy Agency, Monaco).

Appendix 8 Table I List of water bodies and heavy metal sampling sites

No.	Name of water body	Sampling sites	Year sampled
Baltic catchment area			
1	Kulpe River	Myashkuytchay, Latvia	1980, 1984, 1988
2	Kulpe River	Shaulyay, Latvia	1988
3	Liela-Karpa River	Grini Reserve, Latvia	1987
4	Ignalina lakes	Ignalina, Lithuania	1985, 1989
5	Lake Endla	Endla Reserve, Estonia	1987
6	Kem River	Yuma, Karelia, Russia	1974
7	Pltava River	Lvov (downstream), Ukraine	1988
8	Pregolia River	Kaliningrad, Russia	1988
Black Sea catchment area			
9	Sluch River	Bazalia, Ukraine	1980, 1989
10	Goryn River	Vyshnevetz, Ukraine	1980, 1989
11	Dniester River	Galitch (upstream), Ukraine	1983–86
12	Zolotaya Lipa River	Berezhany (upstream), Ukraine	1983, 1989
13	Salgir River	Belogorsk (upstream), Crimea, Ukraine	1973
14	Alma River	Potchovy (upstream), Crimea, Ukraine	1981
15	Mzymta River	Krasnaya Polyana, Russia	1973–79
16	Svisloch River	Minsk, Byelorussia	1988, 1991
17	Seim River	Kursk, Russia	1981–93
18	Byk River	Kishinev, Moldova	1988, 1989
Azov catchment area			
19	Ivnitza River	Voronezh Biosphere Reserve, Russia	1973, 1976–93
20	Usman River	Voronezh Biosphere Reserve, Russia	1930–39, 1975–93
21	Usman River	Usman, Russia	1913–15
22	Usman River	Estuary area, Russia	1981–89
23	Voronezh River	Voronezh, Russia	1976–90
24	Bakhmut River	Artemovsk, Ukraine	1986, 1989
25	Temernik River	Rostov-on-Don, Russia	1983–87
26	Seversky Donets River	Rubezhnoye, Ukraine	1989–91
27	Lopan River	Kharkov, Ukraine	1984–90
28	Don River	Kotchetovskaya, Russia	1990–92
29	Don River	Veshenskaya, Russia	1989
Barents Sea catchment area			
30	Vashutkiny lakes	Bolshezemelskaya tundra, Russia	1903–04, 1954, 1978, 1983, 1987, 1992
31	Varzuga River	Varzuga, Russia	1974, 1987
32	Nuduay River	Monchegorsk, Russia	1982, 1985, 1987
33	Pinega River	Kerga, Russia	1974, 1987, 1989
34	Kosyu River	Kosyu, Russia	1980, 1987, 1991
35	Pechora River	Kurya (upstream), Russia	1977–80, 1988
36	Pechora River	Oksino and Andeg, Russia	1988
37	Sysola River	Syktyvkar, Russia	1938–52, 1979
Kara Sea catchment area			
38	Bolshoy Yugan River	Larlomkiny, Russia	1986
39	Chitchkayul River	Proletarka, Russia	1987
40	Tobol River	Oktyabrsky, Kazakhstan	1983

Continued

Appendix 8 Table I   Continued

No.	Name of water body	Sampling sites	Year sampled
Kara Sea catchment area continued			
41	Ob River	Barnaul (upstream), Russia	1981, 1985
42	Chuya River	Tashanta, Russia	1983
43	Charysh River	Ust-Kan, Russia	1983
44	Chumysh River	Yeltzovka, Russia	1983
45	Glubochanka River	Belousovka, Kazakhstan	1983, 1987
46	Breks River	Leninogorsk, Kazakhstan	1983, 1987
47	Ulba River	Tishinsky ore mine, Kazakhstan	1984, 1987
48	Salda River	Prokopievskaya, Russia	1987
49	Schuchia River	Norilsk, Russia	1981, 1987
50	Ishim River	Sergeevka (upstream), Kazakhstan	1984, 1987
51	Osha River	Kolosovka, Russia	1973, 1988, 1990
52	Berd River	Legostayevo, Russia	1973, 1988, 1990
53	Bolshaya Ket River	Ilinka, Russia	1987, 1990
54	Yenisey River	Kyzyl (upstream), Tuva, Russia	1981, 1985
55	Osa River	Osa, Russia	1987, 1990
56	Chuna River	Chunoyar, Russia	1980, 1987, 1990
East Siberian catchment area			
57	Ulkan River	Tarasovo, Russia	1988
58	Barguzin River	Taza, Buryatia, Russia	1980
59	Bolshoy Amalat River	Baisa, Buryatia, Russia	1978, 1980, 1988
60	Vostotchnaya		1978, 1980, 1988,
	Khadynga River	Khadynga, Yakutia, Russia	1990
61	Zyryanka River	Zyryanka, Yakutia, Russia	1978
62	Amga River	Olekminsk Reserve, Yakutia, Russia	1991
63	Lena River	Olekminsk, Yakutia, Russia	1991
64	Lena River	Yakutsk (upstream), Russia	1980, 1991
65	Taskan River	Taskan, Russia	1988
Caspian Sea catchment area			
66	Volga River	Rzhiev (upstream), Russia	1981, 1983
67	Volga River	Astrakhan Biosphere Reserve, Russia	1965, 1972, 1976, 1989
68	Ilet River	Paranga, Russia	1987
69	Lisva River	Krutoy Log, Russia	1990
70	Bolshoy Ik River	Mrakovo, Russia	1988
71	Zilim River	Zigazinsky, Russia	1988
72	Skamara River	Yuldybaevo, Russia	1988
73	Zingeyka River	Katzbakhsky, Russia	1983, 1989
74	Chusovaya River	Pervouralsk, Russia	1981–85, 1989
75	Mesha River	Tyulyachi, Russia	1986, 1988
76	Ural River	Ilinka, Russia	1987
77	Buzuluk River	Andreevka, Russia	1987
78	Ilek River	Alga (upstream), Kazakhstan	1984, 1987
79	Ilek River	Akhtubinsk, Kazakhstan	1980, 1984, 1987
80	Kuma River	Prikumsky, Russia	1973, 1984
81	Tseydon River	North Osetia Reserve, Russia	1983–90
82	Ardon River	Alagir (upstream), Russia	1985–89
83	Debet River	Tumanyan and Akhtala, Alaverdi, Armenia	1978–81, 1986, 1989

Continued

Appendix 8 Table I   Continued

No.	Name of water body	Sampling sites	Year sampled
Caspian Sea catchment area continued			
84	Lake Sevan	Lake Sevan, northern part, Armenia	1978–81, 1986
85	Razdan River	Ankavan, Armenia	1978–81
86	Vorotan River	Sisian, Armenia	1978–81
87	Sumgait River	Khilmili, Azerbaijan	1979–81, 1986
88	Pirsagat River	Shemakha, Azerbaijan	1979–81, 1986
89	Chandyr and Sumbar Rivers	Shunt-Khasardag Reserve, Turkmenistan	1981
90	Mashavera River	Kazreti, Georgia	1984
91	Belaya River	Sterlitamak, Russia	1986, 1989
Central Asian catchment area			
92	Chu River	Tokmak (upstream), Kyrgyzstan	1981, 1991
93	Chu River	Furmanovka, Kazakhstan	1984
94	Emba River	Kulsary, Kazakhstan	1984
95	Turgai River	Kuymys, Kazakhstan	1978, 1984
96	Nura River	Malinovka, Kazakhstan	1984
97	Tokrau River	Aktogay, Kazakhstan	1984
98	Lake Balkhash	Saryshagan, Kazakhstan	1984
99	Ily River	Bakanas, Kazakhstan	1980, 1991
100	Sokh, Shakhimardan and Chauvay Rivers	Sokh, Khamzabad and Chauvay, respectively, Uzbekistan	1980, 1984, 1986
101	Sokh River	Mercury ore mine, Gallery No. 39, Uzbekistan	1980, 1984, 1986, 1990
102	Kulsay River	Zaaminsky Reserve, Uzbekistan	1983
103	Tashkesken, Terakli, Serkeli and Akbulak Rivers	Chatkala Biosphere Reserve, Uzbekistan	1980–84, 1986, 1987, 1990
104	Sarychelek River	Sarytchelek Reserve, Uzbekistan	1980–84, 1986, 1987, 1990
105	Atbashi River	Bosogo, Kyrgyzstan	1983
106	Malyi Naryn River	Archaly, Kyrgyzstan	1983
107	Aksai River	Chatyrtash, Kyrgyzstan	1983
Pacific catchment area			
108	Kuenga River	Chernyshevsk, Russia	1983
109	Gazimur River	Gazimur Zavod (upstream), Russia	1983
110	Unda River	Baley (upstream), Russia	1983
111	Aga River	Aga, Russia	1983
112	Urkan River	Zarechnaya (upstream), Russia	1986
113	Zeya River	Verkhnaya Nikolaevka, Russia	1986
114	Lake Chlya	Nikolaevsk-on-Amur, Russia	1986
115	Silinka River	Gorny, Russia	1986
116	Khar River	Bichevaya (upstream), Russia	1986
117	Samarga River	Adimi, Russia	1986
118	Iman River	Kartun, Russia	1986

Appendix 8 Table II Heavy metal concentrations (mean ± S.E.) in rivers and wetlands in mountainous regions, 1973–91

Sample	n	Co	Cu	Mn	Zn	Mo	Sr
Carpathian							
Sluch River (No. 9 in Table I)							
Water (µg l⁻¹)	54	0.09 ± 0.02	1.9 ± 0.4	10 ± 2	0.4 ± 0.1	0.74 ± 0.13	110 ± 9
Seston (µg g⁻¹ d.w.)	54	0.54 ± 0.09	4.1 ± 0.5	280 ± 43	7 ± 0.8	0.28 ± 0.06	120 ± 11
Bottom sediments (µg g⁻¹ d.w.)	123	1.2 ± 0.2	21 ± 2.1	520 ± 63	24 ± 3.5	0.66 ± 0.08	160 ± 14
Periphyton (µg g⁻¹ d.w.)	134	0.24 ± 0.04	6.1 ± 0.9	160 ± 19	15 ± 2.4	0.11 ± 0.02	140 ± 15
Hydromorphic soils (µg g⁻¹ d.w.)	237	5.4 ± 0.4	43 ± 5.2	690 ± 75	40 ± 5	0.64 ± 0.09	240 ± 22
Floodplain grass (µg g⁻¹ d.w.)	215	0.11 ± 0.02	8.6 ± 0.9	110 ± 15	22 ± 3	0.12 ± 0.02	53 ± 4
Goryn River (No. 10 in Table I)							
Water (µg l⁻¹)	46	0.19 ± 0.03	0.61 ± 0.09	21 ± 2	2.1 ± 0.4	2.3 ± 0.35	240 ± 39
Seston (µg g⁻¹ d.w.)	46	1.6 ± 0.2	1.5 ± 0.19	570 ± 54	51 ± 7.4	0.58 ± 0.11	260 ± 30
Bottom sediments (µg g⁻¹ d.w.)	87	4.9 ± 0.5	3.5 ± 0.4	1,200 ± 132	74 ± 10	1.5 ± 0.19	350 ± 33
Periphyton (µg g⁻¹ d.w.)	67	2.2 ± 0.4	2.8 ± 0.4	490 ± 42	33 ± 5.3	0.35 ± 0.06	430 ± 51
Hydromorphic soils (µg g⁻¹ d.w.)	176	27 ± 3.3	21 ± 3.2	2,900 ± 250	290 ± 34	5.3 ± 0.72	500 ± 46
Floodplain grass (µg g⁻¹ d.w.)	124	0.84 ± 0.1	1.4 ± 0.2	560 ± 49	88 ± 11	1.8 ± 0.2	270 ± 24
Caucasus							
Tseydon River (No. 81 in Table I)							
Water (µg l⁻¹)	86	0.53 ± 0.09	2.4 ± 0.4	7 ± 0.8	35 ± 5.6	0.85 ± 0.11	180 ± 19
Seston (µg g⁻¹ d.w.)	86	3.2 ± 0.4	5.7 ± 0.8	170 ± 19	76 ± 10	0.31 ± 0.05	240 ± 31
Bottom sediments (µg g⁻¹ d.w.)	225	5.7 ± 0.7	28 ± 2.4	100 ± 11	140 ± 21	0.41 ± 0.05	290 ± 34
Periphyton (µg g⁻¹ d.w.)	96	1.5 ± 0.2	8.5 ± 0.9	160 ± 18	80 ± 11	0.61 ± 0.03	180 ± 12
Hydromorphic soils (µg g⁻¹ d.w.)	386	20 ± 1.5	64 ± 5.7	520 ± 60	280 ± 31	1.9 ± 0.23	380 ± 28
Floodplain grass (µg g⁻¹ d.w.)	188	0.42 ± 0.05	6.8 ± 0.5	130 ± 14	110 ± 12	0.53 ± 0.04	81 ± 7
Vorotan River (No. 86 in Table I)							
Water (µg l⁻¹)	35	0.06 ± 0.01	0.60 ± 0.09	15 ± 1.8	0.32 ± 0.05	0.41 ± 0.07	89 ± 11
Seston (µg g⁻¹ d.w.)	35	0.35 ± 0.05	1.9 ± 0.3	880 ± 93	6.8 ± 0.9	0.15 ± 0.02	110 ± 14
Bottom sediments (µg g⁻¹ d.w.)	92	1.1 ± 0.2	4.6 ± 0.6	900 ± 118	19 ± 2.6	0.19 ± 0.03	130 ± 18
Periphyton (µg g⁻¹ d.w.)	45	0.16 ± 0.03	2.8 ± 0.4	240 ± 27	15 ± 2.5	0.10 ± 0.02	120 ± 15
Hydromorphic soils (µg g⁻¹ d.w.)	146	3.1 ± 0.35	5.8 ± 0.7	940 ± 84	34 ± 4.3	0.55 ± 0.07	180 ± 22
Floodplain grass (µg g⁻¹ d.w.)	117	0.13 ± 0.02	0.12 ± 0.02	290 ± 25	16 ± 2.2	0.13 ± 0.02	39 ± 4

Continued

Appendix 8 Table II Continued

Sample	n	Co	Cu	Mn	Zn	Mo	Sr
Urals							
Zilim River (No. 71 in Table I)							
Water (µg l⁻¹)	45	0.28 ± 0.05	0.53 ± 0.09	37 ± 4	4.2 ± 0.7	2.5 ± 0.4	320 ± 29
Seston (µg g⁻¹ d.w.)	45	1.9 ± 0.3	1.4 ± 0.17	780 ± 68	8.5 ± 1.5	0.56 ± 0.09	250 ± 31
Bottom sediments (µg g⁻¹ d.w.)	122	4.1 ± 0.5	15 ± 1.6	1,200 ± 115	45 ± 5.8	1.4 ± 0.15	330 ± 29
Periphyton (µg g⁻¹ d.w.)	72	1.7 ± 0.3	6.2 ± 0.7	530 ± 63	29 ± 3.7	0.51 ± 0.09	180 ± 24
Hydromorphic soils (µg g⁻¹ d.w.)	226	34 ± 5.3	12 ± 1.6	2,800 ± 410	145 ± 17	5.3 ± 0.6	760 ± 83
Floodplain grass (µg g⁻¹ d.w.)	165	0.75 ± 0.08	1.5 ± 0.2	320 ± 28	10 ± 1.3	1.7 ± 0.3	130 ± 11
Pechora River (No. 35 and 36 in Table I)							
Water (µg l⁻¹)	76	0.07 ± 0.01	0.51 ± 0.07	5.3 ± 0.6	0.15 ± 0.02	0.51 ± 0.07	100 ± 13
Seston (µg g⁻¹ d.w.)	76	0.38 ± 0.06	5.8 ± 0.62	380 ± 37	6.7 ± 0.8	0.14 ± 0.02	120 ± 11
Bottom sediments (µg g⁻¹ d.w.)	95	0.80 ± 0.13	4.9 ± 0.5	590 ± 62	12 ± 1.1	0.24 ± 0.04	160 ± 15
Periphyton (µg g⁻¹ d.w.)	88	0.15 ± 0.03	3.0 ± 0.4	270 ± 25	8.0 ± 0.9	0.12 ± 0.02	160 ± 19
Hydromorphic soils (µg g⁻¹ d.w.)	324	4.1 ± 0.6	44 ± 3.5	1300 ± 165	40 ± 4.4	0.59 ± 0.09	250 ± 21
Floodplain grass (µg g⁻¹ d.w.)	122	0.13 ± 0.02	6.2 ± 0.4	150 ± 14	51 ± 4.8	0.14 ± 0.02	43 ± 4
Pamir Alai							
Akbulak River (No. 103 in Table I)							
Water (µg l⁻¹)	32	0.34 ± 0.06	1.9 ± 0.3	3.9 ± 0.6	23 ± 4.4	2.1 ± 0.3	180 ± 19
Seston (µg g⁻¹ d.w.)	32	1.5 ± 0.3	4.8 ± 0.8	130 ± 12	65 ± 9.5	0.84 ± 0.12	210 ± 23
Bottom sediments (µg g⁻¹ d.w.)	92	5.8 ± 0.8	21 ± 3.5	195 ± 20	110 ± 15	1.9 ± 0.3	210 ± 18
Periphyton (µg g⁻¹ d.w.)	75	2.2 ± 0.4	9 ± 1.5	143 ± 18	84 ± 11	0.87 ± 0.14	89 ± 10
Hydromorphic soils (µg g⁻¹ d.w.)	127	43 ± 6.6	51 ± 7.4	650 ± 59	320 ± 51	7.7 ± 0.9	280 ± 25
Floodplain grass (µg g⁻¹ d.w.)	101	0.61 ± 0.1	10 ± 1.3	160 ± 14	110 ± 19	1.4 ± 0.15	39 ± 7
Kulsay River (No. 102 in Table I)							
Water (µg l⁻¹)	34	0.06 ± 0.01	0.63 ± 0.14	8.1 ± 1.1	0.46 ± 0.07	0.7 ± 0.1	82 ± 9
Seston (µg g⁻¹ d.w.)	34	0.50 ± 0.09	1.6 ± 0.2	270 ± 33	7.2 ± 1.1	0.17 ± 0.03	98 ± 8
Bottom sediments (µg g⁻¹ d.w.)	75	0.64 ± 0.08	5.2 ± 0.7	290 ± 42	22 ± 3.5	0.21 ± 0.03	180 ± 16
Periphyton (µg g⁻¹ d.w.)	56	0.20 ± 0.04	2.1 ± 0.3	190 ± 17	14 ± 2.2	0.17 ± 0.02	95 ± 15
Hydromorphic soils (µg g⁻¹ d.w.)	181	3.5 ± 0.5	13 ± 2.0	460 ± 55	40 ± 6.5	0.7 ± 0.1	250 ± 21
Floodplain grass (µg g⁻¹ d.w.)	117	0.12 ± 0.02	2.2 ± 0.4	220 ± 24	14 ± 1.9	0.23 ± 0.03	44 ± 3.7

Continued

Appendix 8 Table II Continued

Sample	n	Co	Cu	Mn	Zn	Mo	Sr
Tian-Shan							
Atbashi River (No. 105 in Table I)							
Water (µg l⁻¹)	43	0.38 ± 0.08	0.74 ± 0.12	4.4 ± 0.6	0.55 ± 0.09	1.8 ± 0.2	90 ± 10
Seston (µg g⁻¹ d.w.)	43	2.1 ± 0.4	1.4 ± 0.2	120 ± 15	6.8 ± 0.9	0.95 ± 0.15	110 ± 14
Bottom sediments (µg g⁻¹ d.w.)	112	1.0 ± 0.18	16 ± 2.5	1000 ± 130	100 ± 16	0.3 ± 0.07	480 ± 39
Periphyton (µg g⁻¹ d.w.)	39	0.3 ± 0.04	7.6 ± 0.9	260 ± 31	43 ± 2.8	0.24 ± 0.03	230 ± 35
Hydromorphic soils (µg g⁻¹ d.w.)	223	5.8 ± 0.9	50 ± 7.7	1300 ± 180	160 ± 2.7	5.4 ± 0.8	600 ± 79
Floodplain grass (µg g⁻¹ d.w.)	312	0.11 ± 0.01	7.8 ± 0.8	230 ± 32	35 ± 2.3	0.16 ± 0.02	130 ± 18
Aksai River (No. 107 in Table I)							
Water (µg l⁻¹)	54	0.10 ± 0.02	2.2 ± 0.4	13 ± 1.5	12 ± 2.2	1.6 ± 0.2	280 ± 33
Seston (µg g⁻¹ d.w.)	54	0.70 ± 0.11	4.9 ± 0.8	490 ± 59	55 ± 8.7	0.15 ± 0.02	350 ± 43
Bottom sediments (µg g⁻¹ d.w.)	112	1.0 ± 0.18	16 ± 2.5	1,000 ± 130	100 ± 16	0.3 ± 0.07	480 ± 39
Periphyton (µg g⁻¹ d.w.)	39	0.30 ± 0.04	7.6 ± 0.9	260 ± 31	43 ± 2.8	0.24 ± 0.03	230 ± 35
Hydromorphic soils (µg g⁻¹ d.w.)	223	5.8 ± 0.9	50 ± 7.7	1,300 ± 180	160 ± 27	6.0 ± 0.9	600 ± 79
Floodplain grass (µg g⁻¹ d.w.)	312	0.11 ± 0.01	7.8 ± 0.8	230 ± 32	35 ± 2.3	0.16 ± 0.02	130 ± 18
Baikal and Transbaikal area							
Ulkan River (No. 57 in Table I)							
Water (µg l⁻¹)	2	0.42	0.23	10	0.23	2.6	450
Seston (µg g⁻¹ d.w.)	29	3.8 ± 0.6	3.1 ± 0.41	510 ± 43	7.3 ± 0.9	0.96 ± 0.13	310 ± 33
Bottom sediments (µg g⁻¹ d.w.)	92	6.7 ± 0.9	21 ± 1.3	830 ± 89	24 ± 3.5	1.6 ± 0.19	390 ± 30
Periphyton (µg g⁻¹ d.w.)	43	1.8 ± 0.3	6.7 ± 0.5	230 ± 33	8.5 ± 0.8	0.51 ± 0.07	190 ± 11
Hydromorphic soils (µg g⁻¹ d.w.)	88	53 ± 7.8	44 ± 6.5	960 ± 85	56 ± 8.5	10 ± 1.3	620 ± 55
Floodplain grass (µg g⁻¹ d.w.)	67	0.8 ± 0.12	6.7 ± 0.8	380 ± 22	9.7 ± 1.2	1.5 ± 0.2	260 ± 22
Kuenga River (No. 108 in Table I)							
Water (µg l⁻¹)	18	0.08 ± 0.015	0.38 ± 0.07	4.8 ± 0.6	17 ± 3.1	0.63 ± 0.10	140 ± 18
Seston (µg g⁻¹ d.w.)	18	0.49 ± 0.08	1.6 ± 0.28	130 ± 12	58 ± 8.7	0.19 ± 0.04	100 ± 13
Bottom sediments (µg g⁻¹ d.w.)	67	0.84 ± 0.12	6.9 ± 0.9	170 ± 16	120 ± 19	0.21 ± 0.04	180 ± 16
Periphyton (µg g⁻¹ d.w.)	23	0.20 ± 0.04	2.0 ± 0.3	180 ± 24	80 ± 11	0.17 ± 0.03	91 ± 8
Hydromorphic soils (µg g⁻¹ d.w.)	89	4.8 ± 0.7	21 ± 3.4	690 ± 76	290 ± 42	0.50 ± 0.09	280 ± 41
Floodplain grass (µg g⁻¹ d.w.)	87	0.16 ± 0.03	1.7 ± 0.3	145 ± 14	61 ± 9.5	0.19 ± 0.04	124 ± 9

Continued

Appendix 8 Table II Continued

Sample	n	Co	Cu	Mn	Zn	Mo	Sr
North-East Siberia							
Vostotchnaya Khadynga River (No. 60 in Table I)							
Water (µg l⁻¹)	29	0.32 ± 0.06	0.53 ± 0.11	36 ± 5.6	0.27 ± 0.05	0.38 ± 0.06	150 ± 12
Seston (µg g⁻¹ d.w.)	29	1.9 ± 0.3	2.3 ± 0.4	520 ± 62	6.9 ± 1.1	0.12 ± 0.02	190 ± 21
Bottom sediments (µg g⁻¹ d.w.)	78	3.8 ± 0.7	4.8 ± 0.7	840 ± 93	19 ± 4	0.20 ± 0.02	310 ± 29
Periphyton (µg g⁻¹ d.w.)	34	0.9 ± 0.15	3.0 ± 0.5	280 ± 29	9.5 ± 1.7	0.08 ± 0.02	210 ± 19
Hydromorphic soils (µg g⁻¹ d.w.)	122	21 ± 3.7	16 ± 2.7	1,800 ± 253	33 ± 5.8	0.61 ± 0.09	750 ± 65
Floodplain grass (µg g⁻¹ d.w.)	88	0.82 ± 0.09	1.5 ± 0.2	280 ± 33	11 ± 1.7	0.10 ± 0.02	190 ± 15
Taskan River (No. 65 in Table I)							
Water (µg l⁻¹)	18	0.06 ± 0.01	1.6 ± 0.3	11 ± 1.6	1.8 ± 0.27	0.81 ± 0.14	72 ± 8
Seston (µg g⁻¹ d.w.)	18	0.61 ± 0.11	5.3 ± 0.9	130 ± 16	22 ± 4.5	0.25 ± 0.04	81 ± 10
Bottom sediments (µg g⁻¹ d.w.)	56	0.74 ± 0.13	20 ± 1.7	170 ± 21	40 ± 7.6	0.83 ± 0.07	98 ± 7
Periphyton (µg g⁻¹ d.w.)	23	0.15 ± 0.04	7.3 ± 1.1	168 ± 16	20 ± 3.8	0.34 ± 0.06	65 ± 5
Hydromorphic soils (µg g⁻¹ d.w.)	76	4.1 ± 0.6	43 ± 7.7	760 ± 82	170 ± 29	1.9 ± 0.27	360 ± 44
Floodplain grass (µg g⁻¹ d.w.)	56	0.12 ± 0.03	8 ± 1.2	135 ± 21	41 ± 3.9	2.8 ± 0.35	50 ± 4.3
Sikhote-Alin							
Khar River (No. 116 in Table I)							
Water (µg l⁻¹)	22	0.08 ± 0.07	0.59 ± 0.09	15 ± 2	0.12 ± 0.02	0.33 ± 0.04	170 ± 22
Seston (µg g⁻¹ d.w.)	22	0.27 ± 0.04	1.5 ± 0.3	280 ± 33	4.4 ± 0.7	0.08 ± 0.01	150 ± 18
Bottom sediments (µg g⁻¹ d.w.)	34	0.66 ± 0.09	5.7 ± 0.8	450 ± 54	17 ± 2.8	0.13 ± 0.02	320 ± 30
Periphyton (µg g⁻¹ d.w.)	15	0.09 ± 0.02	1.9 ± 0.3	220 ± 35	8.4 ± 1.2	0.07 ± 0.01	120 ± 13
Hydromorphic soils (µg g⁻¹ d.w.)	154	4.1 ± 0.7	15 ± 2.3	780 ± 85	21 ± 3.7	0.59 ± 0.08	780 ± 85
Floodplain grass (µg g⁻¹ d.w.)	185	0.05 ± 0.01	1.7 ± 0.3	120 ± 14	10 ± 1.7	0.10 ± 0.02	150 ± 12
Samarga River (No. 117 in Table I)							
Water (µg l⁻¹)	2	0.25	1.8	33	11	1.7	180
Seston (µg g⁻¹ d.w.)	22	1.4 ± 0.2	3.6 ± 0.6	580 ± 65	48 ± 8.3	0.5 ± 0.09	140 ± 15
Bottom sediments (µg g⁻¹ d.w.)	32	3.3 ± 0.6	23 ± 3.7	950 ± 86	90 ± 15	0.9 ± 0.17	380 ± 32
Periphyton (µg g⁻¹ d.w.)	27	0.54 ± 0.09	5.7 ± 1.0	370 ± 45	60 ± 11	0.3 ± 0.05	150 ± 16
Hydromorphic soils (µg g⁻¹ d.w.)	67	16 ± 2.5	47 ± 7.5	1,600 ± 140	170 ± 31	4.8 ± 0.7	820 ± 74
Floodplain grass (µg g⁻¹ d.w.)	121	0.35 ± 0.05	6.6 ± 0.6	260 ± 22	33 ± 5.7	4.4 ± 0.3	120 ± 10

n   Number of samples   S.E.   Standard error   d.w.   Dry weight

Chapter 9[*]

ORGANIC POLLUTANTS

Economic activity has led to the production and application of a broad range of organic compounds most of which enter the environment, including water bodies, in some way. Although not all of them are highly toxic, each xenobiotic compound influences, to a certain degree, the natural equilibrium and may have a negative impact on the aquatic ecosystem.

Monitoring organic pollution is a complicated task for several reasons. The actual composition of natural water, particularly its organic matrix, is usually not fully known, and when establishing an observation programme, a researcher may be guided more by the analytical methods available and the technical capabilities of the laboratory than by the indices to be controlled. Substances accounted for are usually those which can be discharged into a water body without regard for organic pollutant transformation in water. Monitoring for organic contamination is also complicated by the chemical and biochemical instability of many organic compounds and their possible degradation after sampling, making the most reliable analytical methods ineffective. Finally, monitoring is also complicated by a need to identify and quantify low or trace concentrations of organic contaminants from a background mixture of compounds of markedly higher concentrations.

Two approaches are traditionally used for assessing organic contamination of surface waters. The first is an approach based on quantitative indices characterising total content of organic matter in water by measuring the chemical oxygen demand (COD), total organic carbon (TOC) and its relatively easily transformable component by determination of the biochemical oxygen demand (BOD). The second is an approach based on determination of different specific groups of organic compounds considered as the most toxic and hazardous for aquatic ecosystems and human health.

In the former USSR, the State Monitoring System monitored a number of indicators of organic pollution (Review of USSR Surface Water Quality, 1976–1982; Hydrochemical Institute, 1983–1992, 1993, 1994). In particular, there are long-term monitoring records for COD and

---

*   *This chapter was prepared by V.S. Petrosyan, Y.Y. Vinnikov, V.A. Kimstach, L.V. Boeva and T.A. Sergeeva*

BOD (Appendix 9 Tables I and II). However, TOC monitoring was carried out only in specific studies and is therefore not discussed in the present chapter.

Data are available for the following groups of organic contaminants which have been monitored regularly: pesticides (e.g. organohalogens, triazines, organophosphorus), petroleum hydrocarbons and heavy fractions of petroleum products (e.g. resins and asphaltenes), anionic synthetic surfactants and volatile phenols (phenol index).

Some non-ionic detergents have been monitored in various water bodies for certain special projects. The assessment of water pollution by polychlorinated biphenyls (PCB) was performed only occasionally, and these data are scattered and difficult to access. More data are available on polycyclic aromatic hydrocarbons (PAH). However, in order to interpret the available data, they need to be thoroughly verified. For this reason the present chapter contains only some illustrative examples on PCB and PAH.

The following groups of specific organic contaminants were monitored over several years in water bodies mainly in regions with chemical, petrochemical, pulp and paper, and metallurgy industries: aniline, dithiophosphates, fats, xanthogenates, lignin and lignosulfonate acids, methanol, lower mercaptans, organic acids (primarily volatile), polyacrylamide, carbon disulphide, turpentine, tallic oil, formaldehyde, furfurol, cyclohexanol, cyclohexane and cyclohexanoxime.

Over the last seven to eight years, several research teams have performed complex surveys of water bodies in large industrial centres. They have used modern analytical methods (e.g. chromato-mass-spectrometry, gas, liquid and thin-layer chromatography, spectrophotometry) and obtained reliable data on a broad spectrum of organic compounds in wastewaters and natural water bodies into which contaminants are discharged.

9.1   Chemical and biochemical oxygen demand

Data on COD and BOD values in the former USSR are discussed below in terms of different hydrographic regions. Appendix 9 Tables I and II show summarised data for COD and BOD in various water bodies over numerous years.

9.1.1   The Baltic hydrographic region

The Neman river is considerably polluted with an obvious trend for COD and BOD showing increased pollution from the headwaters to the mouth. No obvious temporal trend has been reported for the basin as a whole. In the Neris river, the COD and BOD values generally increased in the late 1980s and 1990s compared with the late 1970s. An increase in COD was reported downstream in the Venta and Western Dvina (Daugava) rivers, although no obvious trend was observed in spatial and temporal variations of BOD. In respect of COD, the Western Bug river is the most polluted, although a downward trend has been noted in later years. At the same time, a BOD increase occurred in the late 1980s, showing an increase in non-persistent organic compounds. In the Neva river, BOD and COD values decreased slightly by 1991 at some stations although pollution reduction was not reported in the Neva basin as a whole.

9.1.2 The Barents Sea hydrographic region

Heavy metals are the major pollutants of rivers in the Kola Peninsula, and BOD and COD values are not high in most water bodies and there are no temporal variations (e.g. Pechenga river). In contrast, the Northern Dvina, and its basin as a whole, contains significant amounts of organic compounds. However, it is noteworthy that a marked BOD decrease has been reported recently in one of its tributaries, the Sukhona river.

9.1.3 The Azov-Black Sea hydrographic region

This region has a high concentration of industries, highly developed agriculture and a high population density. There is therefore a high input of both industrial and domestic wastewaters which have a negative impact on regional water bodies.

The pollution rate of the River Danube in the last 15 years has been sustained at the same level, with rather high BOD values. The River Prut, one of the major tributaries of the Danube, is one of the most highly polluted rivers showing a marked increase in pollution downstream. There were no obvious temporal trends in BOD and COD values for this river.

In the River Dniester, there is an increase in pollution in middle reaches and a slight reduction in BOD and COD values towards the mouth, possibly due to the dilution and absence of large-scale pollution sources. Average values for COD in the Dniester as a whole, have been reduced by approximately 1.5 times in the last 15 years, although BOD values in the river and its basin over this time did not change and have remained relatively high. A tributary of the Dniester, the River Reut, is extremely polluted in its middle and downstream sections although pollution levels have remained stable in recent years.

In the Southern Bug, BOD values are relatively high along the river with a constant increase in BOD over time. Values for COD decreased over 1985–89, particularly in the mouth. An insignificant BOD increase in 1990–91 was reported in the river basin as a whole.

Values for COD in the River Dnieper and its reservoirs gradually increase from the upper reaches to the mouth. In recent years, there has been a slight decrease in average COD values in the river. There were no temporal trends in BOD values in the Dnieper and its basin, although BOD values varied within a broad range in different sections of the river.

One of the largest Dnieper tributaries, the River Berezina, is characterised by relatively constant BOD and COD values throughout its flow and no trends in pollution level have been reported. The pollution level of the River Svisloch, a tributary of the River Berezina, had decreased markedly by 1989 in a section downstream of Minsk, although it still remained relatively high. The BOD value in the mouth of this river was low, possibly due to significant self-purification facilitated by a higher water discharge in the mouth section. The COD of the River Desna, another Dnieper tributary, remained practically unchanged whereas an increase in BOD occurred.

In the River Don, COD values were similar in the upstream and middle sections of the river, but increased sharply in the mouth (downstream of the city of Rostov-on-Don). A marked increase was reported in recent years compared with points upstream of the Tsimlyansk reservoir. The River Seversky Donets, the largest tributary of the Don, has high pollution levels in its middle section (in the Ukraine) due to the large-scale discharge of industrial wastewater. Despite some self-purification in the lower reaches, the pollution level there remained high. Values for COD in the downstream section decreased markedly from 1975 to 1989 but some increase was reported again in 1990–93. Values for BOD are high and homogeneous along the length of the river with no marked temporal trends.

The River Kuban is most polluted in the middle reaches due to industrial wastewater discharges and runoff from agricultural land. No obvious temporal trends in COD and BOD values have been reported, their average levels being stable both in the river and the basin as a whole. From 1980, COD average values decreased markedly in the River Belaya, a large Kuban tributary. A decrease in BOD values was also observed at the mouth of this river.

The level of pollution River Rioni by easily oxidisable compounds is slightly above average although it remains constant along the river and over time. Values for COD in the River Rioni are low and are also constant along its course. Markedly higher pollution by organic compounds was reported in the middle reach of River Inguri prior to 1990 but subsequently COD and BOD values have decreased sharply. The pollution level also decreases markedly towards the mouth.

9.1.4   The Caspian Sea hydrographic region

As in the Azov and Black Sea regions, the Caspian hydrographic region is characterised by a high level of industrial and agricultural development, as well as having a high population density. These factors have triggered negative environmental impacts in the region, including impacts on water bodies.

The River Kura and its Araks tributary, are the largest Trans-Caucasian rivers. The upstream stretch of the Kura is unpolluted, but BOD and COD values reach a maximum in the middle stretch. The highest pollution rate was reported in 1980–85, although in subsequent years BOD and COD values were lower. An increase in COD was reported in 1985–91 in the River Araks (particularly near the border), whereas BOD values remained practically constant throughout the river over time. The most heavily polluted Araks tributary is Razdan downstream of Yerevan. In the whole Kura basin in 1977–84, the BOD level decreased; it then remained relatively constant in the following years.

Pollution of the River Terek could be considered average but with minor improvements reported in 1979–84. Values for BOD and COD in the river remained relatively constant in the years following 1984.

The River Volga is the region's largest water body. The river flow is regulated by a series of reservoirs which markedly aggravate the poor environmental condition of the river. Lower COD values in the mouth, in contrast to upstream sections, are possibly due to natural reasons, namely a lower input of refractory organic compounds from the steppe zone soils. At the same time, the BOD values in the mouth were markedly higher, although they have decreased slightly in more recent years. Of the Volga reservoirs, the most unfavourable situation occurs in the Kuybyshev reservoir, although the COD and BOD in all the reservoirs were at approximately the same level. An exception is the Volgograd reservoir where the BOD and COD levels were slightly lower.

One of the largest Volga tributaries, the River Oka, is affected by numerous wastewater discharges from industrial enterprises and polluted tributaries (e.g. the Moskva river). An increasing trend in COD and BOD mean values has been reported both in the river and its basin as a whole.

Pollution of the River Kama and its basin has remained relatively constant. The most heavily polluted areas were the middle and downstream sections with the pollution level increasing towards the mouth. The Belaya river is significantly polluted, particularly in the middle section from the city of Salavat to Blagoveschensk. Over the period 1975–93, the COD and BOD values decreased markedly in the Vyatka river, another Kama tributary.

In the entire Volga basin, the level of pollution by easily oxidisable organic compounds has remained stable over the observation period.

The River Ural is most heavily polluted in its middle reach. No significant changes in COD and BOD values have been reported, either in the river or its basin.

9.1.5   The Central Asia hydrographic region

The River Amu Darya and its tributaries are characterised by low COD and BOD values typical of rivers in desert zones. Therefore, the average BOD value gradually decreases from upstream to downstream reaches in the Amu Darya basin.

The second largest river in the region, the Syr Darya is, in contrast to the Amu Darya, markedly contaminated by organic compounds with their content increasing from upstream sections to the mouth. The COD values in the middle reach and the mouth section have increased twofold, and BOD values have increased almost threefold compared with the upstream section. The Syr Darya tributaries are unevenly polluted. For example, the River Naryn has low BOD and COD values throughout its sections, while pollution levels in the River Chirchik markedly increase from the upstream section to the mouth, as in the Syr Darya. The level of pollution in the river basin by easily oxidisable organic compounds has remained unchanged in recent years.

9.1.6 The Kara Sea hydrographic region

The Ob river section down to Novosibirsk is characterised by low COD values that increase markedly as polluted tributaries flow into the river. The BOD values are high throughout the river particularly below Novosibirsk, yet a considerable decrease in BOD values has been reported in the mouth section, especially in the last decade.

The largest Ob tributary, the River Irtysh and its largest tributary, the River Tobol, are polluted rivers where organic pollution increases markedly downstream. No obvious trends towards reduced pollution levels have been reported in either river.

The River Tom is one of the most heavily polluted Ob tributaries. The COD and BOD values indicate that no significant changes in the state of the river have been reported in the last decade. The most heavily polluted section is the middle section downstream of the city of Kemerovo.

The River Yenisey is another large river in the Kara Sea hydrographic region. The COD and BOD values indicate that the organic pollution of this river is low. The COD values steadily increase downstream which may be related to natural factors. The BOD values are slightly higher in the middle section (below Krasnoyarsk) and decrease towards the mouth.

The Lower Tunguska river, one of the Yenisey's largest tributaries, has high COD values increasing markedly towards the mouth. However, lower BOD values have been reported in the mouth, particularly in recent years.

Another large tributary of the Yenisey is the River Angara, the outlet of Lake Baikal. From its source to the mouth, the Angara receives several tributaries characterised by markedly higher pollution levels than the main river, such as the Vikhoreva tributary. The latter is characterised by extremely high COD and BOD values, whose reduction in 1990–93 is not due to environmental protection but rather to industrial recession and the reduced volume and pollution level of wastewater discharges. Major, direct pollution sources from industrial enterprises are also found along the banks of the Angara. As a result, COD and BOD values increase over twofold from the source to its confluence with the Yenisey. A general trend towards lower pollution by easily oxidisable organic compounds has been reported in the Rivers Yenisey and Angara in the period 1977–89. The same situation has occurred in the River Selenga and its basin.

9.1.7   The East Siberian Sea hydrographic region

Despite the lower anthropogenic impact in this region compared with others, the environmental situation here cannot be considered as favourable. Typical of the region are low mean annual temperatures and permafrost, which slow down self-purification processes. Therefore factors which may pose little danger for ecosystems in other regions could cause serious impacts here.

High COD and BOD values suggest that the region's largest river, the River Lena, is most heavily polluted by organic compounds in its middle reach. However, these levels are reduced markedly towards the mouth. No obvious decrease in organic pollution trends have been reported over time. The same applies to two large tributaries of the Lena, the Rivers Vilyuy and Aldan. Both in the Indigirka and Kolyma river basins, COD and BOD have remained relatively constant.

9.1.8   The Pacific Ocean hydrographic region

The rate of organic pollution in the River Amur, the largest river in the Pacific Ocean region, and its basin is not high. No obvious trends in BOD has been reported. Values for COD decreased from 1975 to 1989, but have stabilised since then. From 1975, BOD and COD values decreased markedly in the Rivers Ussuri and Ingoda, two tributaries of the River Amur.

9.2   Petroleum products

Due to their broad use, crude oil and refinery products are the most common pollutants found in surface waters. Oil is a complex, versatile and variable mixture of different hydrocarbons and unsaturated heterocyclic compounds forming resins and asphaltenes. Today, the term 'petroleum products' in a water pollution context is limited to a hydrocarbon constituent composed of aliphatic, naphthenic and aromatic hydrocarbons including the most hazardous PAH.

Analytical methods used in the former USSR for determination of the petroleum content in water were based on extraction by an organic solvent, hydrocarbon separation by thin-layer chromatography, transfer to a solution by the appropriate solvent and measurement of IR and UV absorption or luminescence. It was assumed that different analytical procedures were expected to yield similar results and this was achieved in most cases (i.e. approximately 75 per cent of samples). However, deviations in the remaining cases could be significant due to discrepancies between standards and analysed petroleum fractions. This in turn, complicated the interpretation of available data on the concentration of petroleum products in water bodies. Extreme changes in the data obtained for concentrations of petroleum products in water are due either to errors in analysis or to changes in analytical techniques.

Data on petroleum products in various water bodies sampled since the 1970s generally showed a decrease in petroleum product concentrations over time (Appendix 9 Table III). This was due either to a reduced discharge of heavily polluted waters and more stringent discharge controls, or a transition to alternative assay techniques. Petroleum product concentrations are rather high in certain regions, especially where large-scale navigation or the exploitation of large oil and gas deposits occurs. Evaluating the pollution levels of water bodies for petroleum products based on a hydrocarbon component is inadequate, especially in cases of chronic pollution, because hydrocarbon transformations result in accumulating resins and asphaltenes compared with the initial composition. Resin and asphaltene concentrations are usually markedly below those of hydrocarbon concentrations, although occasionally the former may exceed the latter by two to three times.

The most hazardous components of oil and coal processing products are PAHs which are extremely carcinogenic, e.g. benzo(a)pyrene. Unfortunately, PAHs are not monitored regularly in water bodies because of the complicated and expensive analytical procedures. Data obtained as a result of specialised studies are usually scattered and not appropriately handled. Bottom sediment is a more representative medium for monitoring petroleum products, including PAH, because these compounds generally accumulate in bottom sediments due to their poor solubility in water.

9.3   Synthetic surfactants

Despite their wide occurrence, particularly in households detergents, surfactants are usually detected at low concentrations in natural waters. This is probably due to their relatively low biochemical stability. Higher concentrations of surfactants usually result from the discharge of large amounts of untreated effluents in minor water bodies (e.g. the Svisloch river downstream of Minsk) especially in the cold season. High levels of dilution in large water bodies usually result in concentrations below 0.2 mg l^{-1} (Appendix 9 Table III). Both maximum and average concentrations of surfactants have decreased in numerous water bodies possibly due to more efficient operation of treatment plants.

The State monitoring system has not performed routine observations for the non-ionic detergents which are widely used at present. However, they have not been measured in several water bodies during specialised research programmes in 1989–93 (Table 9.1).

9.4   Phenols

Until recently, regular monitoring of phenol concentrations in natural waters has been performed in terms of total concentration of volatile

Table 9.1 Non-ionic detergent concentrations in various water bodies, 1989–93

Water body	Sample site	Concentration ($\mu g\ l^{-1}$)
Gorky Reservoir	Downstream of Yaroslavl city	Not detected
Belaya River	Upstream of Salavat	Not detected
	Downstream of Salavat	19
	Downstream of Sterlitamak	Not detected
	Downstream of Ufa	33
	Downstream of Blagoveschensk	44
Sutoloka River	River mouth	435
Sheksna River	Cherepovets, water intake	24
	Cherepovets, downstream of treatment facilities	89
Yagorba River	River mouth	52
Serovka River	River mouth	100
Iset River	Upstream of Ekaterinburg	Not detected
	Downstream of Ekaterinburg	60
Don River	At Rostov-on-Don	24
	Downstream of Rostov-on-Don	55

phenols, known as the 'phenol index'. This index is of minimal information value but determination of individual phenolic compounds has not been broadly practised because of analytical difficulties.

Data on phenol concentrations in surface waters of the former USSR are shown in Appendix 9 Table IV. These data should be considered with some caution because a concentration < 2 $\mu g\ l^{-1}$ estimated by a standard technique is not reliable, while at higher phenol concentrations data reliability is related to storage time and the organic matrix of the samples. In the presence of a preserving agent (e.g. sodium hydroxide) a sample may produce additional phenols which distort the results.

The available data show that both average and maximum phenol concentrations decreased (the latter being most important) in various water bodies from 1980 to the early 1990s. This is due to the smaller volume of heavily polluted wastewaters, to the improved efficiency of treatment plants, and to the more precise adherence to guidelines on sampling, sample storage and analysis. Nevertheless, average phenol concentrations remain high in several water bodies including the rivers Kura, Araks, Danube, Ob, Yenisey, Lena, Indigirka, Tom and Irtysh.

The actual concentration of phenols in natural water can only be estimated by specialised studies using more representative analytical techniques (e.g. gas or liquid chromatography). At present, such studies are scarce and usually do not confirm high concentrations of phenols and their derivatives. This is illustrated by the estimation of phenol in the Moskva river in June 1987 (Table 9.2).

Table 9.2   Concentrations of phenols in the River Moskva, June 1987

Compound	Concentration ($\mu g\ l^{-1}$)	MAC ($\mu g\ l^{-1}$)
Phenol	2.9	1
4-Chlorophenol	7.7	1
4-Nitrophenol	1.4	20
2,4-Dinitrophenol	10.7	30
Pentachlorophenol	103.9	10

MAC   Maximum allowable concentration   Source: Petrosyan, 1992

9.5   Pesticides

Monitoring of pesticide concentrations in surface waters of the former USSR was initiated in 1964–69. Pilot studies were carried out in regions with intensive application of pesticides during that period (i.e. the Ukraine, Moldova, Trans-Caucasus, Central Asia). Initially, the concentrations of persistent organochlorine pesticides (OCP), namely the HCH isomers, DDT and its metabolites DDD and DDE, were estimated. It should be noted that the use of DDT was banned in the former USSR in 1970 although the compound still persists in water bodies. From 1972, the number of pesticides monitored was expanded. Organophosphorus pesticides (OPP) (e.g. parathion-methyl, malathion, phosalone, dimethoate and butiphos) were subsequently measured in the Dnieper, Northern Dvina, Syr Darya and Amu Darya rivers. The chemical nomenclature of the commercial names of chemicals is provided in Appendix I, at the end of this book.

Specific routine monitoring of OCPs in the country's water bodies was launched in 1973. The number of monitoring stations varied from 90 to 100 throughout the observation period. Stations were situated at downstream sections of major rivers, as well as in regions of OCP production and extensive application. The monitoring data have been summarised and published in numerous books and reviews (Yearbook of Residual Pesticides Content, 1989, 1990, 1991; Yearbook of Pesticide Monitoring, 1992; Pesticides, 1993). At present, over 20 pesticides are included in a priority list recommended for water monitoring, although just over 10 chemicals are actually monitored. This list is periodically updated and amended to reflect the variations in pesticide use. Summarised data on pesticide concentrations in surface waters starting from 1970 collected either by various institutions or the State monitoring system are shown in Tables 9.3 and 9.4 respectively.

The survey performed on the mid-Dnieper and Western Polesie in 1969–70 detected OCPs in 67 out of 882 samples (Vrochinsky, 1977). The frequency of OCP detection in the samples was 17 per cent for DDT, 9 per cent for γ-HCH and 8 per cent for DDD and DDE. The maximum

Table 9.3   Pesticide concentrations in surface waters

Water body, region	Pesticide[1]	Observation period	Concentration (μg l^{-1})	Frequency of detection (%)	Source(s)
Dnieper River, Western Polesie rivers, Ukraine	γ-HCH DDT & metabolites	1969–70	0.5–5 50	9 8–17	Vrochinsky, 1977
Rivers and reservoirs, Ukraine	γ-HCH DDT & metabolites	1973–75	0.1–3.0 0.2	– –	Vrochinsky, 1975
Southern Bug river basin, Ukraine	γ-HCH DDT & metabolites	1977	0.08–3.6 0.08–3	40 50	Vrochinsky, 1977
Rivers, lakes and reservoirs, Ukraine	γ-HCH DDT & metabolites	1978–83	Trace–0.015 Trace–6	– –	Maslova et al., 1990
Dnieper River basin, Ukraine	γ-HCH DDT & metabolites Malathion Parathion-methyl Trichlorfon Zineb	1979–83	0.013–0.080 0.016–0.384 8–50 25 10 18–27	33–100 29–66 – 100 100 0.01	Motuzinsky, 1984
Rivers, reservoirs and lakes, Ukraine	γ-HCH DDT & metabolites	1979–88	0.01–0.7 0.002–1.5	100 100	Zatylny, 1989
Surface waters, Latvia	γ-HCH DDT & metabolites	1968–69	Trace–50 Trace–450	23 25	Krasilschikov, 1971
Main rivers, Byelorussia	γ-HCH DDT & metabolites 2,4-D	1976–79	0.338–0.015 0.075–0.005 Trace	82 83 –	Zholudeva, 1979
Surface waters, Armenia	γ-HCH DDT & metabolites	1970–75	10–50 21–50	25 61	Gevorkian et al., 1971
Lake Sevan basin, Armenia	γ-HCH DDT & metabolites	1981–83	0.016–0.037 1–10	– –	Sakhalian, 1983
Surface waters, Trans-Caucasus	γ-HCH DDT & metabolites Granozan	1976–78	0.006–0.22 0.012–0.15 0.16–1.2	58–96 15–77 86–100	Gevorkian, 1978
Surface waters, Uzbekistan	γ-HCH DDT & metabolites Butiphos Parathion-methyl Phosalone Dimethoate	1979–85	0.016–0.076 0.08–0.119 0.9–22 0.15–0.5 0.01–0.42 0.01–71.4	75–100 50–100 – – – 60–79	Orlova and Yaroshenko, 1976 Dzhuraev, 1986 Toryanikova et al., 1984 Toryanikova and Karaseva, 1992

Continued

Table 9.3   Continued

Water body, region	Pesticide[1]	Observation period	Concentration (μg l^{-1})	Frequency of detection (%)	Source(s)
Dniester River, Moldova	α-HCH	1976–85	0.11–0.27	47	Gorbatenky and
	γ-HCH		0.13–0.37	65	Sinelnikova, 1987
	DDT		0.08–0.72	51	Sinelnikova, 1987
	DDD		0.4–0.67	58	
	DDE		0.12–0.53	66	
Reservoirs and lakes, Moldova	α-HCH	1987	0.098–0.224	70–80	Sinelnikova et al., 1990
	γ-HCH		0.085–0.3	70–80	
	DDT		0.194–0.413	70–80	
	DDD		0.157–0.226	70–80	
	DDE		0.111–0.216	70–80	
	Simazine		1.28–3.3	25–30	
Northern Dvina river basin, Arkhangelsk region	γ-HCH	1979–83	0.006–0.036	–	Motuzinsky, 1984
	DDT & metabolites		0.001–0.022	–	
Ponds, Northern Caucasus	γ-HCH	1972–80	0.002–1.7	29–100	Korotova et al., 1989
	Parathion-methyl		0.05–12	14–89	
	Trichlorfon		0.1–72	25–78	
Irrigation systems, water reservoirs, USSR	γ-HCH		0.02–0.05	–	Asin et al., 1986
	Molinate		1–481	–	Demchenko et al., 1977
	Propanil		1–278	–	
	Thiobencarb		3–34	–	Dogadina et al., 1976
	Diuron		110	–	Zholudeva, 1979
	2,4-D		10–1,400	–	Zatula et al., 1987 Likhovidova, 1986
Surface waters monitoring stations for OCP, USSR	γ-HCH	1973–92	0–1.36	40–100	Yearbook of residual pesticide content, 1989, 1990, 1991 Yearbook, 1992 Pesticides, 1993
	DDT & metabolites		0–2.61	9–100	
Chapaevka River, Russia	γ-HCH	1973–92	0–80	–	
	DDT & metabolites		0–3.17	–	

[1]   For commercial names see General Appendix I

concentration of γ-HCH was 5 μg l^{-1} and for DDT and its metabolites it was 50 μg l^{-1}. The latter concentration is rather high, possibly due to the analytical techniques used at the time.

The OCP concentrations estimated in the Ukraine in 1973–75 were < 3 μg l^{-1} (Vrochinsky, 1975). Similar pollution levels of surface waters were recorded in 1977 when the OCP detection frequency in the Southern Bug varied within 40–50 per cent with measured

Table 9.4 Pesticide concentrations in surface waters of the former USSR (1988–92) and in the Russian Federation (1991–92)

Pesticide[1]	Observation period	No. of sampling stations	Stations where pesticides detected (%)	No. of samples analysed	Samples in which pesticides detected (%)	Range of concentrations ($\mu g\ l^{-1}$)
α–HCH	1988–1990	5,883	55.7	51,936	30.2	0–8.7
	1991–1992	1,740	43.8	12,491	24.4	0–3.8
β-HCH	1991–1992	123	9	545	5.3	0–3.7
γ-HCH	1988–1990	5,816	56.5	50,955	30.4	0–11.9
	1991–1992	1,731	45	12,476	26.9	0–7.3
p,p'-DDT	1988–1990	5,994	14.6	53,183	5.2	0–8.3
	1991–1992	1,743	8.8	12,472	3.3	0–3.4
p,p'-DDE	1988–1990	4,865	3.2	50,374	0.4	0–1.9
	1991–1992	1,687	3.8	12,066	0.9	0–0.3
p,p'-DDD	1988–1990	2,587	4.5	23,221	1.1	0–9.4
	1991–1992	215	0.5	1,085	0.1	0–0.02
HCB	1991–1992	46	6.5	177	1.7	0–0.17
Dihydro-heptachlor	1991–1992	66	42.4	599	9.2	0–0.08
Parathion-methyl	1988–1990	343	12.8	2,694	2.3	0–8.1
	1991–1992	78	5.1	444	1.1	0–0.7
Carbophos	1988–1990	104	69.2	874	55.4	0–54
	1991–1992	40	5	203	1	0–2.1
Trichlorfon	1988–1990	3	0	36	0	0
Dimethoate	1988–1990	326	29.8	2,652	12.8	0–1,250
	1991–1992	39	0	196	0	0
Butiphos	1988–1990	128	0	919	0	0
Phosalone	1991–1992	47	0	212	0	0
Prometryne	1991–1992	28	0	164	0	0
2,4-D	1991–1992	116	3.4	499	0.8	0–29
Molinate	1988–1990	6	0	45	0	0
Thiobencarb	1988–1990	9	55.6	92	7.4	0–1
Propanil	1988–1990	16	18.8	27	4.1	0–14
Trifluralin	1988–1990	80	18.8	440	6.3	0–9
	1991–1992	123	37.4	1,102	27	0–1.4
Fluometuron	1988–1990	73	0	607	0	0
TCA	1991–1992	43	0	293	0	0

[1] For alternative commercial names see General Appendix I

concentrations varying from 0.08–3.6 µg l^{-1}. Lower OCP concentrations in the Ukrainian water bodies were reported from 1979 to 1983 (Motuzinsky, 1984). Comparative studies in two regions of the country (the Dnieper and Northern Dvina river basins) showed that the DDT and

HCH concentrations were generally < 0.1 µg l^{-1}. Concentrations of OCP in the Dnieper river and its tributaries were higher.

In 1979–88, monitoring data on pesticide concentrations in aquatic biota of the Dnieper-Bug, Berezan, Tiligul, Khodzhibey, Kuyalnik and Dniester gulfs, the Sasyk water reservoir and the mouth sections of the Danube, Dnieper, Southern Bug and Ingul showed that OCPs were present in water and bottom sediments (Medovar et al., 1987; Zatylny, 1989; Maslova et al., 1990). Average concentrations of DDT and its metabolites in water varied from 0.1 to 0.6 µg l^{-1} with a maximum concentration of 1.5 µg l^{-1}. The average γ-HCH concentrations ranged between 0.01 and 0.1 µg l^{-1}, with a maximum concentration of 0.7 µg l^{-1}.

Several water bodies in the Ukraine were studied for organophosphorus pesticides (carbophos, parathion-methyl, trichlorfon) and the thiocarbamate fungicide, zineb. Although concentrations of OPPs and zineb were recorded at levels up to 50 µg l^{-1} (Motuzinsky, 1984), the reliability of the data is questionable.

High pesticide pollution was reported in Moldovian surface waters. The pesticide γ-HCH was present throughout the Dniester river flow in 1967–70, with a maximum concentration of 5 µg l^{-1} (Gontovaya and Akselrod, 1973). In the following years (1976–87), in addition to OCPs, OPPs and triazine herbicides were detected in Moldovian surface waters including the Dniester river, Dubossary, Kuchurgan and Kongaz reservoirs and some ponds (Gorbatenky and Sinelnikova, 1987; Sinelnikova, 1987; Sinelnikova et al., 1990). The highest detection rate was 47–80 per cent for OCPs, 25–30 per cent for atrazine and simazine, while OPPs were rarely reported. The OCP concentration in Moldovian surface waters generally varied from 0.01–0.1 µg l^{-1}, triazine concentrations varied from 0.1–3.3 µg l^{-1}; and OPPs were detected in trace amounts.

In Latvia, water pollution with OCP has occurred (Krasilschikov, 1971) with DDT recorded in 25 per cent of the total number of samples collected in 1968–69 (concentrations varied from trace amounts to 450 µg l^{-1}) and γ-HCH recorded in 23 per cent of samples.

In Byelorussia in 1976–79, OCPs were detected in all basins of the republic's main rivers. DDT and HCH were detected in 83 per cent of the 960 samples and in all seasons. By the end of 1979, a decrease in concentration of γ-HCH and DDT to 0.015 µg l^{-1} and 0.005 µg l^{-1} respectively was reported. Other pesticides rarely occurred in surface waters.

Long-term monitoring of pesticide concentrations in 56 ponds within an area of intensive agricultural activity in the Northern Caucasus showed a broad variation in pesticide concentrations (Korotova et al., 1989). The most common concentrations of pesticides were: γ-HCH — 0.1 µg l^{-1} (in 60 per cent of samples), parathion-methyl — 1 µg l^{-1} (in 78 per cent of samples) and trichlorfon — 2.5 µg l^{-1} (in 63 per cent of samples). Further analysis showed that γ-HCH was present in ponds all year round, whereas OPPs presented an occasional danger for water bodies for several weeks following application.

Long-term monitoring of water bodies in Armenia, Georgia and Azerbaijan revealed an overall trend of decreasing OCPs. High pesticide pollution of surface waters was reported over the entire Armenian region in 1970–75 (Gevorkian *et al.*, 1971). More recent studies (1976–78) in Trans-Caucasian republics has shown significantly lower OCP concentrations in water bodies. Surveys in Razdan, Vorotan and Debet rivers and Lake Sevan in Armenia, at Rioni, Inguri and Kura rivers in Georgia and the Kura river and Mingechaur reservoir in Azerbaijan showed OCP present in most water bodies. However, maximum concentrations had decreased by approximately an order of two, compared with the previous period (Gevorkian, 1978). Typical for Trans-Caucasian surface waters in the same period was that pollution by granozan was detected in almost all samples. Concentrations of OCP decreased in the Lake Sevan basin in 1981–83 as a result of a cessation in the use of DDT and HCH in the catchment area (Sakhalian, 1983).

Organochlorine pesticides were also reported in surface waters over the entire Uzbekistan region, which ranked amongst the highest in the country in terms of pesticide use. Organochlorine pesticide concentrations in the Amu Darya and Syr Darya rivers in 1975–80 varied from 0.01 to 0.1 $\mu g\ l^{-1}$. The average concentration of γ-HCH was 0.016–0.076 $\mu g\ l^{-1}$ and of DDT was 0.08–0.12 $\mu g\ l^{-1}$. Organophosphorus pesticides (dimethoate and phosalone) were reported in the spring and summer in some water samples collected at river mouths near the Aral Sea. Their concentrations did not exceed 0.01–0.015 $\mu g\ l^{-1}$ (Dzhuraev, 1986). According to other authors (Orlova and Yaroshenko, 1976; Toryanikova *et al.*, 1984; Toryanikova and Karaseva, 1992) the phosalone and dimethoate concentrations in drainage waters in the Chirchik-Angren basin exceeded 70 $\mu g\ l^{-1}$. Parathion-methyl and butiphos were rarely reported in Uzbekistan in water bodies whereas dimethoate was detected in approximately 60–70 per cent of samples. The maximum water pollution by this pesticide was reported in Zeravshan.

Generally, the results of long-term monitoring showed a presence of DDT and its metabolites and α-HCH and γ-HCH in practically all water bodies sampled; this highlights the global distribution of such pesticides (Table 9.4). Where relatively high OCP concentrations were reported at other sites (e.g. downstream section of the Ob river), they were possibly due to long-range atmospheric pesticide transport and condensation under the cold Arctic climate. However, the overall trend in the concentration of these pesticides during the observation period was for lower concentrations over time.

The highest pollution levels occurred in regions around the town of Chapaevsk on the Chapaevka river (mid-Volga basin) due to the presence of industries manufacturing pesticide components. The maximum values of γ-HCH in the region varied within 21–80 µg l^{-1} in 1974–80 and within 0.8–8.7 µg l^{-1} in 1981–92. The maximum concentration of DDT and its metabolites in 1974–82 varied within 1–3.2 µg l^{-1} (except 1979–80) and within 0–0.28 µg l^{-1} in 1983–92. No DDT was reported in the Chapaevka river in 1989–93, except for trace amounts in 1985, 1987 and 1989.

Despite a gradual decrease in levels detected in the OCP determinations and an increase for other classes of pesticides (e.g. OPP, triazine, 2,4-D, thiocarbamate) by the monitoring network, the data series of concentrations in water are too short to show any trends. The surveys carried out on irrigated lands in the former USSR (e.g. Krasnodar, Rostov, Lower Volga, Kherson, Kazakhstan, Kalmykia and Primorsk Regions) showed that surface water pollution occurred by herbicides which are widely used in rice production (Dogadina *et al.*, 1976; Demchenko *et al.*, 1977; Zholudeva, 1979; Asin *et al.*, 1986; Likhovidova, 1986; Zatula *et al.*, 1987). Molinate, propanil and thiobencarb concentrations in drainage waters reached 100 µg l^{-1}. Thiobencarb was reported to migrate through drainage waters to ponds and to the Veseloye reservoir (Lower Don basin) with concentrations varying between 4 and 7 µg l^{-1}. Several days after rice paddies were treated with propanil, it was detected in reservoirs and lakes at concentrations of 2–6 µg l^{-1}. The herbicides 2,4-D and diuron were reported to migrate beyond irrigation networks. Maximum reported concentrations in these studies were 8.1 µg l^{-1} for parathion-methyl, 54.2 µg l^{-1} for malathion, 1.4 µg l^{-1} for trifluralin, 1,253 µg l^{-1} for dimethoate, 1 µg l^{-1} for thiobencarb, 14 µg l^{-1} for propanil and 29 µg l^{-1} for 2,4-D.

9.6   Specific pollutants

As noted above, analytical techniques for most organic compounds used in the monitoring network are not sufficiently sensitive and selective. Their application in surface water analysis usually yields either zero concentrations or random numbers because of sample matrix impacts. Introducing more reliable techniques could radically change the results. For example, xanthogenates are highly unstable compounds whose presence in natural waters at noticeable concentrations is not probable even when large amounts of effluents are discharged. Prior to 1987, xanthogenates were determined using an insufficiently selective analytical procedure and, for example, they were estimated at concentrations up to 1 mg l^{-1} in the Nuduay river, Monchegorsk. Following the introduction of a more selective and sensitive technique that excluded the sample matrix impact, no xanthogenates were detected in 300 samples.

The most reliable (yet still not adequately sensitive) techniques for assessing the chemicals mentioned in the introduction are those currently used to detect methanol, formaldehyde and furfurol. Table 9.5

Table 9.5 Methanol and formaldehyde concentrations in various water bodies

Water body, region	Observation period	Formaldehyde (mg l⁻¹) Maximum	Average	Methanol (mg l⁻¹) Maximum	Average
Northern Dvina River,	1976–1979	–	–	0.84	0.24
Arkhangelsk town	1980–1984	0.20	0.06	0.80	0.18
	1985–1989	0.10	0.05	0.40	0.10
	1990–1993	0.08	0.05	0.26	< 0.10
Vychegda River,	1980–1984	0.33	0.07	1.2	0.24
Syktyvkar town	1985–1989	0.16	0.06	0.60	0.19
	1990–1993	0.08	< 0.03	–	–
Vychegda River,	1980–1984	0.09	0.04	0.60	0.25
Koryazhma town	1985–1989	0.11	< 0.03	0.34	0.16
Seversky Donets River,	1985–1989	0.20	0.06	–	–
Lisichansk town	1990–1991	0.16	0.04	–	–
Rybinsk Reservoir	1990–1993	0.27	0.04	–	–
Cheboksary Reservoir	1989	–	–	0.25	0.14
	1990–1993	0.07	0.04	0.32	0.11
Kuybyshev Reservoir	1990–1993	0.58	0.11	–	–
Oka River,	1975–1978	0.20	0.06	–	–
Kolomna town	1980–1984	0.07	0.05	–	–
	1985–1988	0.05	0.04	–	–
	1990–1993	0.18	0.11	–	–
Oka River,	1987–1989	0.08	0.03	0.40	0.18
Dzerzhinsk town	1990–1993	0.08	0.03	0.36	0.16
Moskva River,	1975–1979	0.20	0.06	–	–
Kolomna town	1980–1984	0.07	0.04	–	–
	1990–1993	0.05	< 0.03	0	0
Vikhoreva River,	1975–1979	0.19	0.08	0.60	0.32
Koblyakovo	1980–1984	0.10	0.04	0.70	0.40
	1985–1988	0.11	0.03	–	–
	1990–1993	0.36	0.14	–	–

shows summarised data on methanol and formaldehyde in selected water bodies.

In the Zeravshan river (Amu Darya tributary) downstream of the city of Navoi, furfurol was detected at three sampling points in 1978–82 (C_{max} = 0.61 mg l⁻¹; C_{av} < 0.1 mg l⁻¹). It was also detected in the Syr Darya river downstream of the city of Namangan in 1978–84 (C_{max} = 0.22 mg l⁻¹; C_{av} < 0.1 mg l⁻¹) and in the Vychegda river in 1980–84 (C_{max} = 1 mg l⁻¹; C_{av} = 0.37 mg l⁻¹).

There is no doubt that specific pollutants require specialised in-depth studies using more reliable techniques which could confirm or override data especially at relatively high pollution levels, such as in the Volga reservoirs and other water bodies.

9.7   Background contamination of surface waters and bottom sediments

Monitoring background levels of organic contaminants in pristine regions was focused on persistent organochlorine pesticides (γ-HCH, DDT) and polyaromatic hydrocarbons (3,4-benzo(a)pyrene (3,4-BP)) (Analytical Review, 1990). Pollution of water bodies in continental pristine regions by these chemicals is mostly due to surface runoff from drained areas, as well as dry and wet atmospheric deposition. On entering water bodies, OCPs and 3,4-BP are relatively quickly redistributed between water and suspended matter and are deposited in bottom sediments as a result of sedimentation. The low solubility of OCPs and 3,4-BP in water on the one hand, and high rates of adsorption by suspended particles on the other hand, largely determine their concentration in surface waters. The hydrological regime of water bodies also affects the surface water pollution and intra-annual variations in concentrations. Regular monitoring of OCPs and 3,4-BP pollution of river and lake waters indicated a low variability in their baseline concentrations in water in recent years (Tables 9.6 and 9.7). There is a clear trend towards a gradual decrease in DDT and HCH in the European part of the former USSR which is less obvious in the Asian part of the country.

Monitoring in pristine regions also showed that lowland water bodies were more polluted than those in mountains, with maximum pollution levels in surface waters reported during spring and autumn floods. No marked qualitative changes were reported in OCP and 3,4-BP concentrations in bottom sediments.

Monitoring in 1988–89 showed relatively high total HCH concentrations in the silt sediments of the Volga delta. This possibly resulted from large-scale use of HCH on flooded agricultural lands in the Volga delta, in place of the banned DDT. Monitoring also showed slow organochlorine decay in bottom sediments due to prevalent anaerobic conditions. More persistent isomers (α-HCH, o,p-DDT and p,p-DDT) were detected in most samples of bottom sediments and accounted for 70–80 per cent of the total organochlorine concentration.

9.8   Case studies of organic pollution in selected water bodies

9.8.1   The Belaya river basin, Republic of Bashkortostan (Volga basin)

In 1989–91, a detailed survey of specific contaminants in effluents discharged into the Belaya river and its tributaries was performed. Three cities, Salavat, Sterlitamak and Ufa, with large-scale chemical and petrochemical industries, are located in the region.

The organic compounds typically found in effluents of Salavat included alkylbenzenes of various structures, phenols, aliphatic alcohols, nitro-derivatives and particularly phthalic acid ethers. In effluents

Table 9.6 Concentrations of organochlorinated pesticides and
3,4-benzo(a)pyrene in river and lake surface waters of various
Biosphere Reserves, 1983–89

Region	HCH (µg l⁻¹) Range	Average	DDT (µg l⁻¹) Range	Average	3,4-BP (µg l⁻¹) Range	Average
European territory of the USSR						
Astrakhan Biosphere Reserve (Volga delta)	6–63	24	20–57	34	2.1–3.8	3.3
Berezina Biosphere Reserve (Berezina River)	1–84	18	15–70	48	2.0–10.5	5.0
Caucasian Biosphere Reserve (Achipse River)	12–104	53	22–123	64	2.8–11.3	6.1
Prioksko-Terrace Biosphere Reserve (Ponikovka River)	12–77	41	30–82	47	3.3–6.8	4.9
Central-Forest Biosphere Reserve (Mezha River)	1–14	7	4.3–118	57	4.0–5.3	4.6
Asian territory of the USSR						
Barguzin Biosphere Reserve (Lake Baikal)	6–64	38	45–79	59	1.1–2.7	1.8
Borovoe (Lake Borovoe)	9–81	45	8–57	39	3.0–4.0	3.5
Glacier Abramova (Kok-su River)	2.5–49	22	4–20	13	0.3–2.7	1.5
Sarychelek Biosphere Reserve (Lake Sarychelek)	15–85	42	4–20	11	4–10	5
Sikhote-Alin Biosphere Reserve (Serebryanka River)	1–2	2	7.5–28	18	2.3–10.2	6.3
Chatkala Biosphere Reserve (Bashkyzylkai River)	6–48	20	4–20	8	0.8–2.6	1.5

HCH, DDT Organochlorinated pesticides   3,4-BP   3,4-benzo(a)pyrene

of Sterlitamak, a complex set of chlorinated organic compounds were found. In Ufa, petroleum products, volatile chlorinated compounds and phenols were typically present in the effluent (Table 9.8).

Upon discharge to the Belaya river, concentrations of the organic contaminants decreased with increasing distance from the pollution source as a result of dilution and degradation. However, the most persistent chemicals (chlorine derivatives and phthalates) were found over 100 km downstream, particularly in winter. Although no chlorine derivatives were reported in the Belaya river from Sterlitamak to Ufa, they were detected in water samples collected upstream of Ufa.

Table 9.7 Concentrations of organochlorinated pesticides and 3,4-benzo(a)pyrene in river sediment of various Biosphere Reserves, 1983–89

	HCH (ng g⁻¹)		DDT (ng g⁻¹)		3,4-BP (ng g⁻¹)	
Region	Range	Average	Range	Average	Range	Average
Astrakhan Biosphere Reserve (Volga delta)	75–118	96.5	5–7	6	–	0.8
Berezina Biosphere Reserve (Berezina River)	2.7–20	8	1–11.5	4	1.8–7	5
Caucasian Biosphere Reserve (Achipse River)	6.3–20	13	3.2–16	9	–	–
Priokso-Terrace Biosphere Reserve (Ponikovka River)	5–6.5	6	1–11	6	0.9–1.9	1.5
Central-Forest Biosphere Reserve (Mezha River)	–	9	–	12	–	1.4
Borovoe (Borovoe River)	0.1–3.4	1.7	0.2–0.8	0.5	0.9–1.9	1.5

HCH, DDT   Organochlorinated pesticides   3,4-BP   3,4-benzo(a)pyrene

Concentrations of some chlorinated compounds increased downstream of Ufa due to the additional discharge of effluents by an industrial complex at Ufa.

The rate of self-purification processes markedly increased in July–September as a result of higher water temperatures (18–23 °C). This, in turn, led to a notable decrease in the concentrations of organic contaminants compared with spring values. For example, dichloro-ethane concentrations were in the order of 10 µg l⁻¹, trichloroethane 3 µg l⁻¹ and total chlorinated compounds C_4-C_8 7 µg l⁻¹. Concentrations of other pollutants decreased to < 1 µg l⁻¹.

Water samples from the Belaya river basin and from the Ufa water supply network were analysed and found to contain polychlorinated dioxins (PCDD) and furanes (PCDF) (Tables 9.9 and 9.10). The survey of 1993–94 indicated that these pollutants were present in practically all waters analysed in the basin (Final report, 1994).

The relatively high concentrations of dioxins and furanes in the Belaya river basin are possibly due to runoff from fields treated with 2,4-D as well as to effluents from Sterlitamak and Ufa where large-scale production of organochlorine products occur. This is supported by the fact that PCDD and PCDF concentrations in the Belaya river down-stream of Sterlitamak increased by 2.5 times. It should be noted that the congener ratio markedly differed in samples from the Belaya river upstream and downstream of Ufa. The analysis of samples from Ufa water supply sources and mains showed that the total content of PCDD

Organic pollutants   231

Table 9.8   Concentrations of organic pollutants at selected sites in the Belaya River, March 1992

Pollutant (µg l^{-1})	8 km downstream of Salavat	8 km downstream of Sterlitamak	10 km upstream of Ufa	10 km downstream of Ufa
Toluene	< 1	53		
Xylene (total isomers)		13		
Styrene		3		
Diethylbenzene (total isomers)	0.5			
Trimethyldiphenyl	0.5	1		
Chloroform	3	14	1	5
Carbon tetrachloride		37		
Dichloroethane	1	30	2	16
Trichloroethylene		32	< 1	2
Trichloropropane		71	8	16
Dichlorobutane				3
Total chlorinated compounds with C$_4$–C$_8$	2	111	16	6
Phenol		1.5	< 1	< 1
Para-crezol		1	< 1	< 1
Other alkylphenols (total)		7		
Dibutylphthalate	4	6	1	2
Dioctylphthalate		4	2	3
Other alkylphthalates (total)	17	21	3	3

Where no data are given, the compound was not found in a sample

and PCDF differed significantly before and after chlorination (Table 9.11). Overall, the pollution level was higher in the autumn than in the spring due to a decrease in surface runoff in the autumn.

9.8.2   The Tom river (Ob basin)

A survey of Tom river pollution was carried out between the cities of Mezhdurechensk, Kemerovo and Tomsk in 1987–90. It focused on organic pollutant inputs via effluents from the region's industries, through tributaries and snow cover.

Organic contaminants in the Tom river and its tributaries included n-alkanes C$_{16-26}$ (concentrations of particular compounds varied from 0.1 µg l^{-1} to 1.0 µg l^{-1}), toluene, xylol, lower alkylbenzenes (0.5–10 µg l^{-1} for each compound), phthalates, primarily dibutyl-, dioctyl- and bis (2-ethylhexyl) phthalate (1–24 µg l^{-1}), naphthalene and alkylnaphthalenes (0.2–2 µg l^{-1}), polycyclic hydrocarbons with 3–5 cycles (< 1 µg l^{-1}) and phenol and alkylphenols (0.2–2 µg l^{-1}). Simultaneously, hard-to-identify oxygen-containing derivatives of the above compounds (alcohols,

Table 9.9   Polychlorinated dioxin and furane concentrations in the Belaya River basin, 1993–94

Water body	Concentration (ET, µg l^{-1})
Upstream of Sterlitamak city	2.3
Downstream of Sterlitamak city	5.7
Near Chesnokovka village	17.4
Near Blagoveschensk city	6.0
Southern water intake, Ufa city	1.95
Northern water intake, Ufa city	0.58
River Dema	0.40
River Shugurovka	1.64
River Sutoloka	21.8
Krasny Klutch spring	5.6
River Inzer	1.8
River Zilim	0.2

PCDD   Polychlorinated dibenzo-dioxins
PCDF   Polychlorinated dibenzo-furanes

ET   Equivalent toxicity units (by extrapolation from 2,3,7,8-tetrachlorodibenzodioxin (TCDD))

Table 9.10   Concentrations of various polychlorinated dioxin and furane congeners in the Belaya River

Congeners (ET, µg l^{-1})	Upstream of Ufa	Downstream of Ufa
2,3,7,8-TCDD	1.8	0.7
Sum of other TCDD	26.2	7.8
Sum of other PnCDD	7.0	1.0
Sum of HxCDD	3.3	0.8
1,2,3,4,6,7,8-HpCDD	8.7	8.5
Sum of other HpCDD	1.4	21.6
OCDD	35.1	56.5
Total PCDF	4.2	3.1

ET   Equivalent toxicity units
TCDD   Tetrachlorodibenzodioxins
PnCDD   Pentachlorodibenzodioxins
HxCDD   Hexachlorodibenzodioxins

HpCDD   Heptachlorodibenzodioxins
OCDD   Octachlorodibenzodioxins
PCDF   Polychlorinated dibenzo-furanes

aldehydes, acids, etc.) were detected which had been formed as intermediate products during their transformation in the natural environment.

In some river sections, usually downstream of the site of industrial effluent discharge, a broader range of organic pollutants was found including alkanes, formaldehyde, acetaldehyde, methanol, volatile

Table 9.11   Drinking water pollution by polychlorinated dioxins and furane in the city of Ufa

	Concentration (ET µg l^{-1})	
Water intake	Before chlorination	After chlorination
Southern	0.90	0.99
Northern	0.33	1.65
Kovshovy	1.18	0.22
Demsky	3.54	8.85
Izyaksky	8.42	0.38

ET	Equivalent toxicity units	PCDF	Polychlorinated dibenzo-furanes
PCDD	Polychlorinated dibenzo-dioxins		

Table 9.12   Concentrations of various organic compounds in the River Tom (Kemerovo, August 1988)

Compound	Concentration (µg l^{-1})
Octane-2 (isomers)	112
2-Ethylhexene-1	12
Dimethylbenzene	5
Bis(2-chlorethyl ether)	65
Bis(2-chlorpropyl ether)	63
Acenaphtene	2.2
Fluorene	5
Dinitrobenzene	10

chlorinated hydrocarbons and ethers and nitro-derivatives of benzene and toluene. In the plume, their concentrations exceeded 10 µg l^{-1} and rapidly decreased downstream. Data obtained within Kemerovo city in 1988 are presented in Table 9.12 (Petrosyan, 1992).

The marked variability in the organic pollution concentration is possibly due to the organic contaminants' pathways to the river and their rate of transformation. Particularly typical is an increase in concentrations of some chemicals during snowmelt flooding and heavy rain and a decrease in concentrations in summertime.

9.8.3 Sheksna river (Upper Volga basin)

The Sheksna river near the city of Cherepovets with large-scale steel, chemical and other industries was studied in detail in 1989–90. Small tributaries of the Sheksna river (e.g. Koshta, Yagorba) received large amounts of various effluents and their mouth sections were heavily polluted. Typical of the region is a periodic occurrence in water samples of toxic pollutants such as PAHs and PCBs.

Effluents from the coke plant at the Cherepovets steel works were detected as the PAH source. In cases of insufficient treatment or accidental release, PAHs are discharged initially to the Koshta river and then into the Sheksna river, with subsequent deposition and accumulation in bottom sediments. The PCB pollution source was not identified, but bottom sediments in the Koshta and Yagorba mouth sections were heavily contaminated with PCB up to 8–12 µg g^{-1} dry weight. The PCB was identified primarily as a mixture of Arochlor 1254 and Arochlor 1260. Presence of PAHs and PCBs in bottom sediments was confirmed by an additional survey in 1992–94.

A detailed study of the distribution of PAHs and PCBs in water bodies has not been undertaken so far. Negative impacts of PAHs and PCBs on aquatic biota occur due to the transfer of pollutants through the food chain. Tumours and various lesions have been reported in the fish landed in the polluted area. A marked PCB concentration, up to 6 µg g^{-1} was reported in fish tissue, particularly in the liver and brain.

9.8.4   Irtysh river

Pollution of the Irtysh river in the Ust-Kamenogorsk region was studied in 1990. Industrial effluents and snowmelt flood water were the main sources of organic pollutants. The organic pollutants present were typical of water bodies located in urban areas and consisted namely of hydrocarbons of various composition and products of their transformation (e.g. alcohols, aldehydes, ketones, acids and phthalates). Concentrations of individual compounds were between 0.1–10 µg l^{-1}.

Polychlorinated biphenyls were also found in water samples. The PCB composition was identified as Pyralene 3010 and Arochlor 1254 and originated from effluents of an electric condenser works located upstream of Ust-Kamenogorsk. Although fast currents and a pebble/sandy river bed were not favourable for PCB accumulation, the compound was traced in bottom sediments downstream of the city.

9.8.5   The Northern Dvina river basin

An international scientific cruise to study pollution in the Arkhangelsk region was organised by the Centre of Independent Ecological Programmes in 1993. Of the extensive data collected during the expedition, only the results for PAH, PCDD and PCDF assays in the Northern Dvina river basin are presented (Assessment, 1994).

The researchers obtained the data on PAH pollution by high-performance liquid chromatography which allowed a reliable assay of chemicals such as benzo(a)pyrene, fluoranthene, dibenzo(a,h)anthracene, pyrene and indeno(1,2,3-c,d)pyrene. Fluoranthene was detected in 31 of the 33 samples taken, with a concentration ranging from 0.8–322 ng l^{-1} (typical range 20–45 ng l^{-1}). Pyrene was identified in 24 samples with concentrations varying from 1.5–1,670 ng l^{-1} (typical range 10–75 ng l^{-1}). Dibenzo(a,h)antracene and indeno(1,2,3-c,d)pyrene were identified in about half the samples and usually at low concentrations, although the first sample taken from the Onega river near the town of Kargopol showed concentrations exceeding 500 ng l^{-1}.

Concentrations of hazardous carcinogenic compounds such as benzo(a)pyrene were < 5 ng l^{-1} in most samples (which is the MAC for this congener) although the concentration of benzo(a)pyrene in the Mekhrenga river at 1 km upstream of the mouth was 98.7 ng l^{-1} and in the Onega river section near Kargopol it was 116 ng l^{-1}.

Dioxin and furane concentrations in the Northern Dvina river and its tributaries, the Vychegda, Sukhona, Yemets and Pinega were estimated by chromato-mass-spectrometry. Several pulp and paper mills located on the banks of the above rivers contributed significantly to pollution of the Northern Dvina by PCDD and PCDF.

Over 5 ng kg^{-1} equivalent-toxicity units (ET) were identified in bottom sediments in the Northern Dvina mouth near Arkhangelsk. Silt samples taken near the wastewater discharge site of the Solombal pulp and paper mill in Arkhangelsk contained 10.9 ng kg^{-1} ET. Sediment toxicity in the upstream section of the Vychegda river varied from 0.6–1.8 ng kg^{-1}. Silt samples taken near the Syktyvkar pulp and paper mill showed total concentrations of dioxins and furanes of 84 ng kg^{-1} ET. The high persistence of dioxins and furanes in water bodies should be noted. Bottom sediment toxicity in the Puksa river, a tributary of the Yemets, reached 1.8 ng kg^{-1} although the nearest pulp and paper mill had not been operating for a long time.

In the city of Novodvinsk, pulp and paper mill wastewaters contained 5.2 ng l^{-1} of dioxin and furane and tap water contained 3.5 ng l^{-1} ET. In view of the different congener composition in these samples, the PCDD and PCDF formation was assumed to occur during water chlorination. Dioxin and furane concentrations in the drinking water sources in Arkhangelsk were 5.2 ng l^{-1} and in tap water they were 10.1 ng l^{-1}, confirming the above assumption.

9.8.6   Aquatic ecosystems in the Republic of Buryatia

In 1992–94, concentrations of organic pollutants were studied in some rivers and lakes in the Republic of Buryatia, in drinking water in Ulan-Ude and in groundwater in the Muyysk region. The results showed that the major pollutants included aliphatic hydrocarbons, aliphatic carbonic acids, detergents, phthalates, phenols and some organochlorine compounds including PCB. The most typical pollutants are shown in Tables 9.13 to 9.15 (Dugarova *et al.*, in press).

Table 9.13 Concentrations of organic pollutants in rivers of the Buryatia Republic

Pollutant (µg l⁻¹)	Selenga[1]	Uda	Muya	Kindikan	Muyakan	Irokinda	Khamota	Ona
Aliphatic hydrocarbons	56.8–204.2	3.38	2,396	88.8	493.0	120.6	10.8	51.8
Aliphatic carbonic acids and their ethers	31.2–48.6	0.15	14.2	16.2	13.0	3.5	13.7	47.1
Polycyclic aromatic hydrocarbons	–	1.20	–	–	–	–	–	–
Phthalates	116.4–222.4	0.03	119.4	322.4	327.6	66.5	36.0	52.9
Phenols	3.8	0.12	–	5.6	–	–	–	–
Organochlorides	–	0.95	–	–	–	–	0.7	1.3

[1]   Samples were selected 8 km above the inflow   inflow in the Selenga River
of the Uda River and 12 km below the Uda

Table 9.14 Concentrations of organic pollutants in lakes of the Buryatia Republic

Pollutant (µg l⁻¹)	Arangatuy	Kolok	Kotokel	Orot
Aliphatic hydrocarbons	26.8	194.5	67.2	12.4
Aliphatic carbonic acids and their ethers	7.4	45.6	20.8	15.1
Phthalates	64.0	558.0	257.2	74.9
Phenols	1.0	–	–	–

Table 9.15 Concentrations of organic pollutants in groundwaters and drinking water of the Buryatia Republic

Pollutant (µg l⁻¹)	Groundwaters at Taksimo	Drinking water in Ulan-Ude
Aliphatic hydrocarbons	5,048	1.35
Aliphatic carbonic acids and their ethers	60.8	66
Polycyclic aromatic hydrocarbons	–	3.1
Phenols	13.2	–
Phthalates	1,411	100
Organochlorines	477.6	–
Other	1,320	166

9.8.7   The Klyazma river basin, Moscow region

In the winter of 1993–94, the surface water pollution in the Klyazma river basin in the Schelkovo region, located in the north-eastern part of the Moscow region, was assessed. In addition to the Klyazma river, the Shalovka, Ucha, Vorya and Shirenka rivers are polluted by effluents of the local industries (Brodsky *et al.*, 1995).

Analysis of 16 samples collected at different sites showed that specific compounds arising from local pollution sources, as well as other contaminants, were usually found in almost all samples. High concentrations of specific substances were not found, whereas the concentration of the usual organic contaminants varied widely and included:

- Phthalates (concentration range 4.2–163.7 µg l^{-1}, typical concentration 4.3–12.6 µg l^{-1});
- Alkylphenols (concentration range 2.9–41.4 µg l^{-1}, typical concentration 2.9–16.1 µg l^{-1});
- Total petroleum products (concentration range 106–26,690 µg l^{-1}, typical concentration 1,187–9,479 µg l^{-1});
- Complex ethers of aliphatic dicarbonate acids (concentration range 0.45–6.05 µg l^{-1}, typical concentration 1.7–5.5 µg l^{-1});
- Organochlorine compounds (concentration range 0.06–48.1 µg l^{-1}, typical concentration 1.9–10.1 µg l^{-1});
- Alkybenzene (concentration range 0.8–15.2 µg l^{-1}, typical concentration 1.1–11.7 µg l^{-1});
- Aliphatic carbonic acids (concentration range 72–1,526 µg l^{-1}, typical concentration from 152–989 µg l^{-1}).

Baseline organic pollutants of the Klyazma river basin were represented by specific 'urban' chemicals (e.g. petroleum products, organochlorine compounds) and products of biotreatment (e.g. fatty acids and their ethers, phthalates).

9.9   References

(All references are in Russian unless otherwise stated).

Analytical review of environmental background pollution by organochlorine compounds and polycyclic aromatic hydrocarbons in the territory of some East European countries, 1982–1989 1990 Co-ordinating Centre of CMEA States on the Problem 'Global Environmental Monitoring System', Gidrometeoizdat, Moscow.

Asin, V.I., Zatula, A.I., Kivshik, L.S., Kovtun, V.G., Pyatakova, A.M., Rykhtenko, L.I., Frantsuzova, A.S. and Yurchenko, A.I. 1986 Water pollution by herbicides from drainage water of rice paddies. *Water Resources*, **6**, 101–111.

Assessment of Dioxin Contamination and Revealing Dioxin Emission Sources in the Arkhangelsk Region 1993–1994 1994 Moscow, 173 pp. (In English).

Brodsky, E.S., Lukashenko, I.M., Kalinkevich, G.A., Balashova, S.P., Klyuev, N.A. and Peshkov, A.S. 1995 Composition of baseline organic pollutants in surface waters in the Schelkovo region of the Moscow. *Ecological Chemistry*, 4(3), 188–193.

Demchenko, A.S., Brazhnikova, L.V. and Tarasov, M.N. 1977 Transport of herbicides used in rice growing by drainage waters of the Proletarsk irrigation network. *Hydrochemical Transactions*, **5**, 66–71.

Dogadina, T.V., Kraynyukova, A.N. and Mironenko, V.I. 1976 On the impacts of some pesticides on algoflora in drainage canals. In: *Formation and Control of Surface Water Quality*. No. 3, Kiev, 28–30.

Dugarova, I.D., Doroshkevich, L.S., Funtov, A.V., Lebedev, A.T. and Petrosyan, V.S. (In press) Water ecosystems in the Republic of Buryatia and some data on their chemical pollution. *Water Resources*.

Dzhuraev, A.D. [Ed.] 1986 Elaboration of methods for calculating river transport of organic and biological compounds, micro-elements and pesticides from small-scale water catchment areas in different physico-geographical conditions. Report on R&D (Final)/SANII No. 79027458, Inv. No. B 935702, Tashkent, 1,994 pp.

Final Report on Results of Fulfilling the Dioxin Programme in the Republic of Bashkortostan 1994 Ufa, 53 pp.

Gevorkian, S.G. [Ed.] 1978 Studies of the actual level of residual pesticides in major rivers and water bodies in Trans-Caucasus. Report on R&D (Final)/VNIIGINTOKS on Topic No. 9, No. 79048401, Inv. No. B 774713, Yerevan, 59 pp.

Gevorkian, S.G., Pluziach, A.I. and Sakhalian, E.O. 1971 Residual pesticides. In: *Residual Pesticides and the Preventive Measures against their Contamination of Food Stuffs, Fodder and the Environment*. Proceedings of the II All-Union Conference, Tallinn, 244–248.

Gontovaya, N.A. and Akselrod, F.M. 1973 Residual pesticides in drinking water sources in Moldova. Abstracts of a Joint Conference of Hygienists, 121–122.

Gorbatenky, G.G. and Sinelnikova, A.A. 1987 Pesticide impacts on water quality and aquatic biota in water bodies of the Moldovian SSR. State-of-the-art and prospects for improving methodological guidelines of chemical and biological monitoring of terrestrial surface waters. Volume 3, Rostov-on-Don, 125–126.

Hydrochemical Institute 1983–1992 Yearbooks of Surface Water Quality in the USSR, 1982–1991. Obninsk, VNIIGMI-MCD, Rostov-on-Don.

Hydrochemical Institute 1993 Yearbooks of Surface Water Quality in the Russian Federation, 1992. Obninsk, VNIIGMI-MCD, Rostov-on-Don.

Hydrochemical Institute 1994 Yearbooks of Surface Water Quality in the Russian Federation, 1993. Obninsk, VNIIGMI-MCD, Rostov-on-Don.

Korotova, L.G., Tarasov, M.N. and Demchenko, A.S. 1989 Migration of hexachlorocyclohexane, metaphos and parathion in water catchment areas. *Hydrochemical Transactions*, **16**, 165 pp.

Krasilschikov, D.G. 1971 Residual pesticides. In: *Residual Pesticides and the Preventive Measures against their Contamination of Food Stuffs, Fodder and the Environment*. Proceedings of the II All-Union Conference, Tallinn, 299–301.

Likhovidova, T.P. 1986 Migration and transformations of thiocarbamates at a rice paddy irrigation network. Candidate of Science (Geogr.) Thesis/GHI, Rostov-on-Don, 23 pp.

Maslova, O.V., Shebunina, N.A. and Komarovskiy, F.J. 1990 Accumulation and distribution of resistant pesticides in ecosystems of the Southern Kily branch of the Danube and Dnieper-Bug gulf. *Hydrobiological Journal*, 26(4), 62–66.

Medovar, A.A., Ivanov, L.N., Pismennaya, M.V. and Man'ko, N.A. 1987 Issues of pesticide monitoring in Southern Ukraine ecosystems. State-of-the-art and prospects of improving methodological guidelines for chemical and biological monitoring of terrestrial surface waters. Volume 3, Rostov-on-Don, 9–10.

Motuzinsky, N.F. [Ed.] 1984 Elaboration of integrated studies and recommendations on rational use and protection of small rivers in major economic regions of the country. Report on R&D (Intermediate)/VNIIGINTOKS, Topic 0.85.01-03 D-4.D-6, Kiev, 62 pp.

Orlova, A.P. and Yaroshenko, L.V. 1976 Pesticide and mineral fertiliser concentrations in return water in an Uzbekistan irrigation network. Studies of pesticide degradation in water. Formation and Control of Surface Water Quality, Issue 1, Kiev, 106–109.

Pesticides in Surface and Ground Waters 1993 Issue 1, Information Review, Series 87. Monitoring of the State of Environment, Obninsk, 75 pp.

Petrosyan, V.S. 1992 Analysis of organic contaminants in natural and drinking waters. In: *Sixth Russian-Japanese Symposium on Analytical Chemistry*. Moscow and St Petersburg, 110–116. (In English).

Review of USSR Surface Water Quality Based on Observation Results from 1975–1981 1976–1982 Rostov-on-Don.

Sakhalian, E.O. [Ed.] 1983 Studies of toxic chemical sources, their distribution in water and bottom sediment in Lake Sevan and measures for their control. Report on R&D (Final)/VNIIGINTOKS Branch, No. 8002418B, Inv. No. 2840031959, Yerevan, 41 pp.

Sinelnikova, A.A. 1987 Insecticides concentrations and their transformation in surface waters. Candidate of Science (Chem.) Thesis/GHI, Rostov-on-Don, 24 pp.

Sinelnikova, A.A., Gorbatenky, G.G., Bogonina, Z.S. and Davydova, I.F. 1990 Pesticide pollution levels in major water bodies in the Moldovian SSR. *Experimental Water Toxicology*, 4, 4–9.

Toryanikova, R.V. and Karaseva, T.A. 1992 Pesticide concentration and behaviour in surface waters in Uzbekistan. *Transactions of the Central Asian Regional Research Institute*, 142, 80–100.

Toryanikova, R.V., Nishankhodzhaeva, S.A. and Beloborodova, N.F. 1984 Risks of pond pollution by residual pesticides. *Studies of Environmental Pollution*, 1, 19–25.

Vrochinsky, K.N. 1975 Main results and achievements in open water sources for sanitary protection against pesticides. Proceedings of the V All-Union Conference on Pesticides, Kiev, 28–29.

Vrochinsky, K.N. [Ed.] 1977 Study of pesticide contamination of a major USSR river (Southern Bug) and evaluation of hygienic recommendations and standards. Report on R&D (Intermediate)/ VNIIGINTOKS on Topic No. 21, No. 80069743, Inv. No. B 884528, Kiev, 55 pp.

Yearbook of Pesticide Monitoring in Environmental Bodies in the Russian Federation 1992 Book 2, Part 1, Obninsk, 289–353.

Yearbook of Residual Pesticide Content in Environmental Bodies in Some Regions of the Soviet Union 1989 Book 2, Part 3, Obninsk, 41–76.

Yearbook of Residual Pesticide Content in Environmental Bodies in the Soviet Union 1990 Book 2, Part 3, Obninsk, 176–225.

Yearbook of Residual Pesticide Content in Environmental Bodies in the Soviet Union 1991 Book 2, Part 1, Obninsk, 201–245.

Zatula, A.I., Kivshik, L.S. and Bushtez, S.P. 1987 Pollution of some water bodies in Kalmykia by herbicides used in rice growing. State-of-the-art and prospects of improving methodological guidelines of chemical and biological monitoring of terrestrial surface waters. Rostov-on-Don, Volume 2, 37 pp.

Zatylny, V.P. 1989 Distribution of organochlorine pesticides in the Krasnooskolsk reservoir. VINITI, Kiev, 19 pp.

Zholudeva, R. 1979 Transport of the herbicide 2,4-D amine salt by drainage water of irrigation networks. Problems of water protection and use. Minsk, 81–91.

Appendix 9

Table I   COD and BOD of surface waters in the European part of the former USSR

Table II COD and BOD of surface waters in the Asian part of the former USSR

Table III Petroleum products and surfactant concentrations in surface waters of the former USSR

Table IV Phenol concentrations in surface waters of the former USSR

Appendix 9 Table I COD and BOD of surface waters in the European part of the former USSR

Water body	Observation period	COD (mg l⁻¹)			BOD (mg l⁻¹)		
		n	Range	Mean	n	Range	Mean
River Neman, upstream	1975–1979	31	12–35	21	52	1.4–3.7	2.3
	1980–1984	52	21–45	32	59	1.5–2.8	2.2
	1985–1989	60	18–44	32	60	1.5–3.0	2.5
	1990–1991	24	15–58	29	24	1.1–3.1	2.3
River Neman, middle reach	1975–1979	58	14–34	24	53	1.1–5.4	2.7
	1980–1984	60	11–38	25	62	1.5–3.1	2.2
	1985–1989	59	24–62	40	59	3.1–7.0	4.8
	1990–1991	24	17–40	31	24	2.2–6.8	4.6
River Neman, mouth	1986–1989	47	45–82	60	47	3.5–12	6.5
	1990–1991	22	34–109	60	22	3.6–23	6.6
Kaunas Reservoir	1975–1979	21	12–40	23	19	0.5–3.5	1.5
	1980–1984	24	12–51	22	25	1.2–2.9	2.0
	1985–1989	26	17–36	26	26	1.4–4.4	2.9
	1990–1991	9	19–30	26	8	2.2–6.8	3.7
River Neris, middle reach	1975–1979	59	11–30	26	8	0.6–4.5	3.3
	1980–1984	56	15–39	21	53	1.8–4.0	2.8
	1985–1989	59	19–45	27	64	2.1–4.6	3.3
	1990–1991	24	16–33	30	60	1.5–6.3	3.3
River Neris, mouth	1975–1979	57	15–38	25	55	0–5.2	2.8
	1980–1984	56	15–50	28	60	1.9–4.7	3.3
	1985–1987	35	23–61	38	35	3.9–6.9	5.0
River Piarnu, middle reach	1978–1979	7	14–41	28	7	0–4.2	1.8
	1980–1984	20	18–83	37	20	0.6–5.3	1.7
	1985–1989	22	6–46	22	20	0–3.3	1.4
	1990–1991	9	27–51	37	9	0–2.2	1.3
River Piarnu, mouth	1976–1979	11	26–49	38	8	0.9–5.6	2.7
	1980–1984	20	20–76	46	20	0–7.0	3.0
	1985–1989	22	2–64	33	20	0–7.4	3.0
	1990–1991	9	23–77	45	9	0–2.1	1.3
River Western Dvina, upstream	1976–1979	11	13–35	27	26	1.0–5.0	2.0
	1980–1984	9	16–53	29	35	1.2–3.0	2.0
	1985–1989	27	16–54	35	35	1.1–2.6	1.8
	1990–1991	14	13–56	35	14	1.2–3.4	2.0
River Western Dvina, middle reach	1976–1979	11	18–57	35	40	1.2–2.7	1.8
	1980–1984	15	27–41	32	60	1.1–3.1	1.9
	1985–1989	44	16–44	31	58	0.6–2.8	1.6
	1990–1991	24	13–63	41	24	0–4.2	1.7
River Western Dvina, mouth	1975–1979	21	26–81	45	20	1.7–7.3	4.7
	1980–1984	59	22–67	38	60	0.9–4.4	2.6
	1985–1989	60	32–56	43	60	1.1–3.5	2.3
	1990–1991	23	26–56	40	23	1.1–3.9	2.3
Riga Reservoir	1976–1979	65	14–78	40	61	0.6–2.7	1.5
	1980–1984	59	21–54	35	57	0.7–2.6	1.6
	1985–1989	59	29–49	39	60	1.1–2.7	1.8
	1990–1991	33	27–60	40	23	1.1–2.5	1.7

Continued

Appendix 9 Table I Continued

Water body	Observation period	COD (mg l^{-1})			BOD (mg l^{-1})		
		n	Range	Mean	n	Range	Mean
River Venta,	1978–1979	16	15–47	27	14	1.0–8.0	2.1
upstream	1980–1984	49	17–44	30	49	1.1–6.2	2.7
	1985–1989	57	17–38	27	57	1.8–4.8	3.0
	1990–1991	31	17–32	22	21	0.8–3.9	2.3
River Venta,	1975–1979	37	23–44	31	33	1.4–6.0	3.9
middle reach	1980–1984	49	22–57	38	49	1.1–5.0	3.6
	1985–1989	58	24–50	36	58	2.4–5.9	3.0
	1990–1991	21	20–53	30	21	1.0–4.5	3.9
River Venta,	1975–1979	82	16–53	32	83	0.6–3.1	1.9
mouth	1980–1984	59	16–43	30	58	1.0–4.2	2.3
	1985–1989	64	19–39	39	64	0.7–3.4	1.9
	1990–1991	24	20–46	30	24	1.3–3.2	2.1
River Western Bug,	1979	5	22–343	133	5	1.8–5.5	3.9
middle reach	1980–1984	27	32–185	83	27	1.5–9.9	5.0
	1985–1989	53	24–94	61	54	3.7–14	8.0
	1990–1991	35	12–77	53	25	1.4–7.8	3.9
River Neva,	1975–1979	12	18–31	23	17	0–2.3	1.6
source	1980–1984	22	5–51	22	28	0.7–5.4	3.0
	1985–1989	27	8–34	22	26	0–3.3	1.5
	1990–1991	11	13–37	19	11	0.5–2.4	1.3
River Neva,	1975–1979	22	10–44	25	34	0.8–3.2	2.1
mouth	1980–1984	77	15–30	23	94	1.1–4.3	2.3
	1985–1989	28	10–30	19	30	0.5–3.4	1.5
	1990–1991	11	10–26	16	11	0.7–3.8	1.6
River Northern Dvina,	1975–1977	17	23–82	54	7	0.5–5.0	2.1
upstream	1985–1989	50	24–52	40	52	1.1–4.0	2.4
	1990–1991	43	18–62	40	42	1.7–4.8	3.3
River Northern Dvina,	1986–1989	38	25–52	38	39	1.3–3.1	2.4
middle reach	1990–1993	39	28–52	40	39	1.0–4.1	2.2
River Northern Dvina,	1976–1979	210	23–72	42	133	0.6–4.9	2.3
mouth	1980–1984	235	26–52	43	219	1.3–4.2	2.5
	1985–1989	194	25–56	38	193	1.5–3.7	2.4
	1990–1993	143	25–54	33	133	0.8–4.2	2.1
River Sukhona,	1981–1984	25	20–83	43	22	0.8–7.4	4.2
upstream	1987–1989	37	33–56	45	36	1.2–6.1	4.1
	1990–1993	52	25–58	44	49	1.2–6.6	3.5
River Sukhona,	1976–1979	18	28–166	66	18	1.0–23	5.4
mouth	1980–1984	30	10–99	43	32	2.1–5.7	3.8
	1985–1989	55	40–66	51	55	0.9–6.2	3.3
	1990–1993	51	32–68	49	51	1.0–4.8	2.4
River Vychegda,	1975–1979	24	6–53	25	–	–	–
upstream	1980–1984	34	16–51	34	7	1.0–8.0	4.1
	1985–1989	37	11–59	36	15	0.9–8.0	4.5
	1990–1993	28	6–51	28	13	2.2–7.5	5.6

Continued

Appendix 9 Table I Continued

Water body	Observation period	COD (mg l⁻¹)			BOD (mg l⁻¹)		
		n	Range	Mean	n	Range	Mean
River Vychegda,	1975–1979	100	16–66	38	30	0.6–10	2.6
mouth	1980–1984	142	10–51	30	95	1.0–9.6	3.9
	1985–1989	61	17–68	38	55	1.3–8.0	3.6
	1990–1993	49	17–54	33	38	1.6–4.9	2.6
River Pechenga,	1975–1979	34	7–20	16	34	0–1.8	0.8
middle reach	1980–1984	59	6–23	13	60	0–1.7	1.1
	1985–1989	61	4–23	13	61	0–1.7	0.9
	1990–1993	52	5–20	12	52	0–1.8	1.1
River Pechenga,	1975–1979	41	5–18	12	37	0–1.4	0.8
mouth	1980–1984	60	6–24	14	60	0–1.6	1.0
	1985–1989	64	5–29	12	64	0.5–1.6	1.0
	1990–1993	52	6–21	12	52	0–1.8	1.0
River Danube,	1979	2	139–167	153	6	0.5–5.0	3.0
mouth	1980–1984	1	–	33	1	–	4.9
	1985–1989	108	5–81	27	111	0.4–6.5	3.2
	1990–1991	41	7–38	22	45	0.7–6.7	3.2
River Prut,	1986–1989	16	5–28	12	16	0.8–6.5	3.8
upstream	1990–1991	7	9–30	16	7	2.1–4.9	2.7
River Prut,	1986–1989	53	3–36	19	52	2.3–8.6	4.5
middle reach	1990–1991	25	9–29	21	25	1.9–9.3	4.3
River Prut,	1976–1979	15	5–58	21	12	1.4–7.9	4.1
mouth	1980–1984	22	9–60	29	22	1.0–3.9	2.3
	1985–1989	30	13–80	26	30	0.7–14	3.5
	1990–1991	13	14–39	24	13	1.7–16	5.5
River Dniester,	1975–1979	12	5–113	22	8	0.6–2.5	1.9
upstream	1980–1984	27	2–88	17	24	0.9–4.5	2.4
	1986–1989	14	10–51	19	13	1.7–6.2	3.4
	1990–1991	10	7–23	12	10	2.2–3.9	2.6
River Dniester,	1975–1979	56	8–124	34	54	1.6–9.6	5.1
middle reach	1980–1984	60	19–74	40	60	1.3–5.8	3.5
	1986–1989	59	12–30	22	59	1.0–4.4	2.6
	1990–1991	20	14–35	22	20	0.6–11	2.7
River Dniester,	1975–1979	59	10–86	26	34	1.3–9.9	4.8
mouth	1980–1984	60	20–57	36	60	1.1–4.3	2.2
	1986–1989	62	13–31	20	62	1.2–4.9	2.6
	1990–1991	19	14–33	22	19	1.0–3.8	2.4
River Reut,	1975–1979	66	26–102	58	47	1.8–13	8.1
middle reach	1980–1984	61	38–91	61	60	4.5–13	8.6
	1986–1989	60	21–52	35	61	1.5–9.1	6.1
	1990–1991	21	12–65	35	21	1.5–15	5.8
River Reut,	1976–1979	17	19–283	61	13	1.1–7.0	3.9
mouth	1980–1984	33	34–85	56	33	1.7–7.0	4.0
	1986–1989	52	21–50	36	51	1.6–15	5.2
	1990–1991	20	28–48	36	20	1.5–9.9	5.6
River Southern Bug,	1986–1989	34	5–40	16	30	1.6–15	4.6
upstream	1990–1991	19	10–65	24	19	1.9–11	4.9

Continued

Appendix 9 Table I Continued

Water body	Observation period	COD (mg l⁻¹)			BOD (mg l⁻¹)		
		n	Range	Mean	n	Range	Mean
River Southern Bug, middle reach	1975–1979	13	6–60	27	8	0–4.5	2.0
	1980–1984	42	20–75	40	3	1.0–3.5	2.7
	1985–1989	44	6–33	19	16	1.3–8.0	4.0
	1990–1991	21	10–65	24	19	0.6–11	5.4
River Southern Bug, mouth	1975–1979	9	5–160	32	5	0–8.9	2.0
	1980–1984	25	4–97	33	18	1.3–13	4.3
	1985–1989	31	5–36	13	31	0–8.2	4.1
	1990–1991	9	8–33	18	9	2.3–11	6.5
River Dnieper, upstream	1975–1979	28	3–38	21	25	0.7–12	3.8
	1980–1984	60	19–37	27	61	2.0–5.2	3.4
	1985–1989	60	16–44	28	60	1.7–5.2	3.4
	1990–1993	23	11–36	23	26	1.4–8.5	3.3
River Dnieper, middle reach	1976–1979	19	13–144	40	15	0.6–10	3.7
	1980–1984	32	15–100	42	22	0.5–5.2	2.4
	1985–1989	29	5–81	30	23	0.7–5.8	2.1
	1990–1991	16	14–43	25	16	0.6–2.3	1.6
River Dnieper, mouth	1988–1989	13	10–72	35	13	0–6.7	2.2
	1990–1991	17	16–53	31	17	0.7–7.6	2.8
Kiev Reservoir	1980–1984	65	16–103	33	30	0–6.0	1.3
	1985–1989	23	6–61	29	15	0.7–5.3	1.7
	1990–1991	7	18–32	25	7	0.5–2.0	1.0
Kremenchug Reservoir	1985–1989	45	14–43	28	44	0–6.5	2.5
	1990–1991	40	20–45	32	37	1.6–9.0	5.0
Dnieprodzerzhinsk Reservoir	1986–1989	28	10–61	29	30	1.1–5.8	2.4
	1990–1991	15	11–57	31	16	1.8–3.6	2.4
Dnieper Reservoir	1977–1978	6	14–36	23	6	1.6–8.0	4.4
	1988–1989	16	28–48	33	16	2.8–4.6	3.8
	1990–1991	16	24–40	33	16	2.1–5.0	3.4
Kakhovka Reservoir	1975–1978	23	13–168	47	27	0–3.5	1.8
	1988–1989	28	16–63	38	28	0.7–7.6	3.6
	1990–1991	41	16–43	27	41	0.6–6.1	2.9
River Berezina, middle reach	1975–1979	24	11–44	28	52	1.3–4.8	2.6
	1980–1984	51	18–45	33	60	1.2–2.4	1.8
	1985–1989	62	16–49	32	61	1.6–2.8	1.9
	1990–1991	24	21–56	33	24	1.1–3.7	2.1
River Berezina, mouth	1980–1984	25	20–46	32	28	0.6–3.3	1.9
	1985–1989	35	20–44	30	36	0.6–2.9	1.9
	1990–1991	14	17–50	32	14	0.7–3.0	1.8
River Svisloch, middle reach	1975–1979	35	22–72	43	46	12–52	35
	1980–1984	55	18–43	30	58	5.6–25	13
	1985–1989	63	10–44	27	61	3.9–14	7.6
	1990–1991	24	12–43	26	24	3.0–18	9.1
River Svisloch, mouth	1975–1979	12	0–44	21	25	1.2–5.1	2.7
	1980–1984	33	14–36	25	35	1.3–3.8	2.5
	1985–1989	35	10–34	34	35	1.1–3.2	2.0

Continued

Appendix 9 Table I Continued

Water body	Observation period	COD (mg l⁻¹)			BOD (mg l⁻¹)		
		n	Range	Mean	n	Range	Mean
River Desna, middle reach	1975–1979	53	11–28	20	75	0.5–5.1	2.2
	1980–1984	63	7–37	18	65	0.7–3.5	1.8
	1985–1989	68	11–39	22	68	0.9–3.7	2.0
	1990–1991	55	12–49	26	55	1.0–4.8	2.6
River Desna, mouth	1987–1989	36	13–30	22	21	1.0–6.1	2.5
	1990–1991	21	8–37	23	20	2.4–5.2	3.0
River Don, upstream	1975–1979	36	5–31	15	36	1.2–5.4	2.9
	1980–1984	53	10–42	22	53	0.9–3.9	3.4
	1985–1989	40	10–27	18	40	0.7–4.7	2.3
	1990–1993	32	11–24	16	32	0.8–3.6	2.3
River Don, middle reach	1975–1979	22	12–59	23	13	0–3.1	< 0.5
	1980–1984	31	24–50	20	6	1.0–3.0	2.0
	1985–1989	36	13–32	20	32	2.0–4.1	3.3
	1990–1993	18	11–32	20	18	2.0–7.8	4.0
River Don, mouth	1975–1979	45	0–27	25	50	0–5.0	1.3
	1980–1984	5	28–72	47	66	1.2–5.7	3.4
	1985–1989	35	26–93	61	61	1.5–7.9	4.2
	1990–1993	44	12–108	72	67	1.9–5.5	3.8
Tsimlyansk Reservoir	1975–1979	51	15–46	28	43	0.6–8.3	4.0
	1980–1984	79	4–40	17	81	0–8.9	3.9
	1985–1989	86	11–29	19	86	1.1–7.5	3.6
	1990–1993	42	10–31	23	42	2.0–8.5	4.9
River Seversky Donets, upstream	1979	10	3–42	20	16	1.0–4.7	2.7
	1980–1984	64	10–46	27	64	0.9–4.9	2.7
	1985–1989	67	18–58	34	66	1.6–10	5.2
	1990–1993	53	17–60	32	53	1.6–11	4.9
River Seversky middle reach	1985–1989	158	60–164	100	159	2.1–6.9	4.0
	1990–1993	151	49–91	65	154	2.5–5.7	3.7
River Seversky Donets, mouth	1975–1979	42	18–128	77	38	0–7.4	3.9
	1980–1984	42	9–99	52	32	3.6–6.5	4.5
	1985–1989	30	13–45	25	30	1.6–9.3	4.8
	1990–1993	23	24–49	37	23	1.2–11	4.5
River Kuban, upstream	1975–1979	29	0–21	8	29	0–1.4	0.5
	1980–1984	10	8–54	31	11	0.8–2.6	1.4
	1985–1989	19	7–54	27	11	1.3–5.7	2.7
	1990–1992	9	7–23	14	9	1.4–5.8	3.7
River Kuban, middle reach	1975–1979	22	0–40	8	44	1.0–8.5	3.9
	1980–1984	26	13–50	26	64	1.4–9.2	4.4
	1985–1989	57	17–38	28	60	1.7–5.7	2.6
	1990–1993	48	12–37	23	48	1.4–7.0	4.0
River Kuban, mouth	1975–1979	129	0–33	16	101	0–4.5	1.4
	1980–1984	79	7–39	20	124	0.8–3.3	1.7
	1985–1989	131	10–45	25	121	1.1–3.6	2.0
	1990–1993	94	10–31	20	94	1.0–3.5	1.5
River Belaya, upstream	1980–1984	22	0–80	47	–	–	–
	1985–1989	29	6–44	25	29	1.1–3.7	2.3
	1990–1993	21	5–22	14	21	1.2–2.9	2.4

Continued

Appendix 9 Table I Continued

Water body	Observation period	COD (mg l⁻¹)			BOD (mg l⁻¹)		
		n	Range	Mean	n	Range	Mean
River Belaya, mouth	1981–1984	13	16–65	32	7	1.6–2.9	3.8
	1985–1989	20	11–48	26	18	1.2–3.7	2.6
	1990–1993	16	6–30	17	15	1.2–4.8	2.5
River Rioni, upstream	1978–1979	13	2–31	8	16	1.2–3.3	2.4
	1980–1984	17	2–29	11	23	2.2–3.2	2.7
	1989–1991	23	2–9	5	23	1.2–3.9	1.8
River Rioni, middle reach	1983–1984	15	6–13	8	15	2.0–2.6	2.3
	1985–1989	58	5–14	9	60	3.0–2.9	3.5
	1990–1991	21	2–15	7	21	1.9–3.0	2.6
River Rioni, mouth	1979	7	6–22	11	10	1.4–3.5	2.4
	1980–1984	53	2–49	16	58	2.4–3.0	2.8
	1985–1989	56	4–12	9	60	2.1–3.0	2.6
	1990–1991	21	3–16	7	21	2.1–8.6	2.9
River Inguri, middle reach	1978–1979	17	6–310	130	25	0–6.5	3.0
	1980–1984	48	7–210	71	44	3.1–9.5	6.6
	1985–1989	58	6–98	30	57	0.6–12	5.4
River Inguri, mouth	1982–1984	5	2–47	19	–	–	–
	1985–1989	20	3–50	19	3	0–0.5	0.4
River Terek, middle reach	1975–1979	28	0–80	40	25	0–7.4	3.5
	1980–1984	39	9–60	31	19	0.9–4.3	2.1
	1985–1989	53	10–31	20	53	1.3–4.4	2.7
	1990–1993	46	14–34	22	46	2.0–5.1	2.7
River Kura, upstream	1978–1979	14	0–24	10	–	–	–
	1980–1984	49	6–27	16	35	1.6–2.0	1.8
	1985–1989	142	13–22	7	141	0–3.3	1.6
River Kura, middle reach	1978–1979	15	0–43	11	33	1.7–5.6	3.4
	1980–1984	42	6–29	20	46	1.9–9.3	5.1
	1985–1989	54	6–25	16	58	0.9–10	4.6
River Kura, mouth	1978–1979	7	4–12	8	36	0–1.8	0.9
	1980–1984	43	6–16	8	49	0.5–2.4	0.6
	1985–1989	44	9–20	13	59	1.2–2.0	1.1
	1990–1991	24	10–25	17	24	0.5–5.2	1.0
River Araks, middle reach	1977–1979	12	6–31	19	–	–	–
	1980–1984	20	6–28	14	20	1.6–3.4	2.6
	1985–1989	30	14–41	27	30	1.0–4.5	2.7
	1990–1991	10	18–38	29	11	1.9–3.9	2.6
River Araks, mouth	1983–1984	11	8–16	12	11	1.1–5.1	2.6
	1985–1989	44	10–24	16	58	1.5–4.0	2.7
	1990–1991	20	10–26	17	21	1.0–5.0	2.6
River Razdan, middle reach	1980–1984	60	7–26	15	60	1.3–6.0	2.9
	1985–1989	54	17–39	28	52	1.2–3.6	2.5
	1990–1991	22	13–62	30	23	1.1–5.4	2.0
River Razdan, mouth	1980–1984	59	7–39	20	60	3.4–31	13
	1985–1989	35	14–58	35	51	3.6–36	15
	1990–1991	22	13–45	34	22	2.6–20	12

Continued

Appendix 9 Table I Continued

Water body	Observation period	COD (mg l⁻¹)			BOD (mg l⁻¹)		
		n	Range	Mean	n	Range	Mean
River Volga, upstream	1977–1979	10	16–43	30	6	1.7–2.2	2.0
	1980–1984	14	11–38	27	19	1.3–3.0	1.9
	1985	4	26–28	27	4	1.4–1.8	1.7
River Volga, mouth	1975–1979	28	10–116	32	42	1.2–4.8	2.7
	1980–1984	82	7–31	19	80	1.1–5.3	3.0
	1985–1989	212	8–26	17	199	1.3–5.0	2.9
	1990–1993	190	12–31	19	232	1.2–3.9	2.4
Rybinsk Reservoir, northern part	1975–1979	33	17–50	35	3	1.4–2.1	1.8
	1980–1984	39	27–52	38	39	1.9–5.0	3.5
	1985–1989	65	39–84	49	64	1.0–7.0	3.6
	1990–1993	69	24–58	42	69	1.4–4.5	2.9
Rybinsk Reservoir, central part	1979	6	28–38	34	2	0–1.4	0.9
	1980–1984	38	24–37	31	18	1.4–6.6	3.4
	1985–1989	56	21–43	32	53	0.9–4.6	2.5
	1990–1993	40	16–39	30	40	0.9–2.7	1.6
Gorky Reservoir	1975–1979	57	23–45	35	62	0.7–4.5	2.3
	1980–1984	59	20–58	35	56	0.9–5.6	3.0
	1985–1989	56	19–44	31	60	1.0–6.4	3.2
	1990–1993	48	20–43	30	48	1.0–6.8	2.8
Cheboksary Reservoir	1976–1979	16	16–29	22	23	0.5–7.5	1.9
	1980–1984	50	20–34	26	52	0.8–4.8	2.5
	1985–1989	59	16–30	23	60	1.0–4.1	2.4
	1990–1993	48	16–35	24	48	1.2–5.2	3.2
Kuybyshev Reservoir	1975–1979	16	12–35	26	29	1.4–8.6	3.4
	1980–1984	40	16–39	27	40	1.3–6.1	3.5
	1985–1989	53	22–44	32	51	1.4–4.1	2.5
	1990–1993	59	20–40	29	60	1.1–3.9	2.7
Saratov Reservoir	1975–1979	11	19–46	23	22	0.5–8.7	3.0
	1980–1984	30	14–39	24	30	1.0–7.9	2.6
	1985–1989	30	18–44	30	30	1.0–4.4	3.4
	1990–1993	24	16–40	27	24	1.0–3.4	2.2
Volgograd Reservoir	1975–1979	137	13–35	23	53	0.6–3.0	1.4
	1980–1984	93	15–32	23	92	0.8–3.3	1.9
	1985–1989	93	18–38	26	92	0.9–4.7	2.1
	1990–1993	73	16–35	24	73	1.1–3.0	2.1
River Oka, middle reach	1975–1979	24	11–30	17	38	1.5–4.5	2.8
	1980–1984	53	12–32	21	53	1.2–4.4	2.6
	1985–1989	45	13–29	20	42	1.6–7.6	4.3
	1990–1993	51	10–39	25	50	2.1–10	4.8
River Oka, mouth	1975–1979	26	14–40	25	42	0.9–4.5	2.3
	1980–1984	38	14–40	25	31	0.6–8.2	2.6
	1985–1989	45	15–34	25	48	1.1–6.3	3.7
	1990–1993	48	15–40	27	48	0.9–7.0	4.1
River Kama, upstream	1975–1979	–	–	–	36	1.0–3.5	2.2
	1980–1984	52	16–44	30	44	0.7–2.7	1.6
	1985–1989	80	14–35	24	44	0.6–3.8	1.9
	1990–1993	51	15–34	24	37	1.0–2.6	1.8

Continued

Appendix 9 Table I Continued

Water body	Observation period	COD (mg l⁻¹)			BOD (mg l⁻¹)		
		n	Range	Mean	n	Range	Mean
River Kama,	1985–1989	55	10–53	25	48	0.6–7.1	3.0
mouth	1990–1993	46	16–61	40	48	1.0–7.5	3.9
Votkinsk Reservoir	1975–1979	–	–	–	67	1.0–4.3	2.3
	1980–1984	63	18–90	39	80	0.5–3.8	1.8
	1985–1989	79	13–41	28	80	0.6–3.8	1.8
	1990–1993	53	21–50	34	63	1.0–3.1	2.0
River Vyatka,	1975–1979	62	12–40	25	44	0.7–4.9	2.7
mouth	1980–1984	66	9–40	24	45	0–4.5	2.1
	1985–1989	68	5–37	20	63	0–5.4	2.1
	1990–1993	52	7–32	18	53	0–1.8	0.8
River Ural,	1975–1979	34	9–31	18	37	0.7–3.6	2.1
upstream	1980–1984	50	11–33	20	57	0.9–3.4	1.9
	1985–1989	–	–	–	58	0.9–2.8	1.7
	1990–1993	28	10–39	22	40	0.8–2.4	1.5
River Ural,	1975–1979	46	21–51	32	84	1.4–4.0	2.6
middle reach	1980–1984	171	19–49	30	173	2.1–4.0	3.1
	1985–1989	179	19–43	28	180	2.1–4.0	3.0
	1990–1993	178	19–47	33	178	2.2–4.4	3.2
River Ural,	1975–1979	47	7–56	25	78	0.7–3.7	2.0
mouth	1980–1984	111	4–50	22	122	1.2–5.5	3.0
	1985–1989	185	9–21	15	210	1.1–2.9	2.0
	1990–1991	123	10–27	18	123	1.9–2.8	2.5

COD Chemical oxygen demand   n Number of samples
BOD Biochemical oxygen demand

Appendix 9 Table II COD and BOD of surface waters in the Asian part of the former USSR

Water body	Observation period	COD (mg l⁻¹)			BOD (mg l⁻¹)		
		n	Range	Mean	n	Range	Mean
River Amu Darya, upstream	1979	8	9–29	18	7	1.1–5.2	2.0
	1980–1984	59	10–22	16	50	0–2.6	1.3
	1985–1989	48	7–16	12	43	0–2.7	1.3
	1990–1992	28	4–24	12	27	0–4.2	1.2
River Amu Darya, middle reach	1987–1989	31	0–25	9	29	0.5–4.0	1.0
	1990–1991	22	0–22	12	21	0.5–3.7	1.1
River Amu Darya, mouth	1975–1979	53	0–25	7	52	0–4.0	1.9
	1980–1984	59	11–20	16	53	0–3.6	1.7
	1985–1986	24	9–26	15	22	0–5.5	1.0
	1990–1992	27	11–20	15	22	0–2.4	1.0
Kara-Kum Canal	1980–1984	58	0–28	14	62	0.7–2.9	1.5
	1985–1989	53	0–22	9	52	0.6–3.6	1.9
	1990–1991	23	0–23	11	23	0–3.0	1.0
River Zeravshan, upstream	1975–1979	20	0–18	5	10	0–4.2	2.8
	1980–1984	51	0–18	5	53	0.7–3.9	2.1
	1985–1989	32	0–13	6	30	1.0–5.4	2.7
	1990–1991	10	0–16	10	12	1.7–3.8	2.9
River Zeravshan, middle reach	1975–1979	38	0–25	8	39	0–3.4	1.6
	1980–1984	53	6–27	15	57	0–0.9	0.4
	1985–1989	60	3–24	10	61	0–1.0	0.6
	1990–1992	29	5–46	21	29	0–1.9	0.8
River Vakhsh, middle reach	1980–1984	34	0–25	14	31	0–5.6	1.8
	1985–1989	34	0–19	10	32	0–1.8	1.1
	1990–1991	10	5–26	11	12	0–1.2	0.4
River Vakhsh, mouth	1975–1979	24	0–25	10	44	0.5–3.8	2.1
	1980–1984	60	0–29	12	58	0–3.1	1.6
	1985–1989	59	0–17	6	60	0.6–2.4	1.4
	1990–1991	19	0–19	11	24	0.5–2.8	1.5
River Kafirnigan, middle reach	1975–1979	43	0–13	6	52	0.6–4.2	2.3
	1980–1984	58	0–17	7	60	0.6–2.7	1.6
	1985–1989	57	0–18	5	57	0.5–2.3	1.3
	1990–1991	15	0–13	8	18	0.5–2.6	1.6
River Kafirnigan, mouth	1975–1979	15	0–11	6	–	–	–
	1980–1984	50	0–32	12	30	0–3.6	1.6
	1985–1989	59	0–18	9	59	0.7–3.4	1.7
	1990–1991	19	0–25	11	24	0.8–2.3	1.1
River Syr Darya, upstream	1975–1979	38	0–19	7	52	0–2.2	1.0
	1980–1984	60	7–20	14	55	0–1.4	0.7
	1985–1989	60	6–17	10	57	0–1.7	1.0
	1990–1992	27	2–18	10	27	0–1.6	0.7
River Syr Darya, middle reach	1975–1979	19	0–47	21	–	–	–
	1980–1984	21	2–36	15	31	0.7–3.8	1.7
	1985–1989	94	9–45	20	99	1.8–3.6	3.5
	1990–1991	39	12–27	21	39	1.7–3.3	2.4

Continued

Appendix 9 Table II Continued

Water body	Observation period	COD (mg l⁻¹)			BOD (mg l⁻¹)		
		n	Range	Mean	n	Range	Mean
River Syr Darya,	1975–1979	16	0–40	18	–	–	–
mouth	1980–1984	14	4–49	20	35	0–5.2	2.0
	1985–1989	50	7–40	19	50	0.8–6.4	2.9
	1990–1991	23	15–28	22	23	2.0–3.8	2.0
River Naryn,	1978–1979	3	7–15	10	–	–	–
upstream	1980–1984	15	0–19	11	18	0.5–6.8	2.3
	1985–1989	32	2–19	10	35	0.5–3.8	1.8
	1990–1991	16	2–16	8	16	0–2.3	1.0
River Naryn,	1975–1979	25	0–46	14	29	0–13	1.7
mouth	1980–1984	33	3–22	12	33	0.6–1.8	1.1
	1985–1989	47	3–20	10	47	0–1.9	1.2
	1990–1991	8	4–28	15	8	0–2.6	1.1
River Chirchik,	1977–1979	32	3–11	6	28	0–5.2	1.6
middle reach	1980–1984	55	3–12	7	55	0–3.5	1.6
	1985–1989	60	4–15	9	59	0–2.1	0.9
	1990–1992	29	0–9	4	29	0–2.3	0.8
River Chirchik,	1985–1989	59	11–22	15	58	0.9–5.2	2.8
mouth	1990–1992	28	5–19	12	27	1.2–5.5	3.1
River Ob,	1975–1978	149	3–19	9	132	1.1–6.2	2.8
upstream	1980–1984	167	6–37	16	175	1.0–5.7	3.0
	1985–1989	–	–	–	359	0.8–4.6	2.6
River Ob,	1976–1979	24	3–26	13	12	0–5.2	1.9
middle reach	1980–1984	126	3–15	9	122	1.1–14	5.0
	1985–1989	56	4–14	8	159	1.0–8.6	4.0
	1990–1993	48	1–12	8	137	1.7–12	5.8
River Ob,	1976–1979	32	26–41	24	24	0.6–7.3	4.0
downstream	1980–1984	63	9–38	21	64	0.9–5.4	2.6
	1985–1989	67	10–36	20	67	0.9–3.8	1.9
	1990–1993	47	8–40	22	47	0–2.7	1.6
Novosibirsk	1975–1979	40	4–12	8	24	0.7–6.4	2.0
Reservoir	1980–1984	37	4–11	8	37	1.0–4.0	2.3
	1985–1989	15	4–10	7	19	0.7–7.5	2.4
	1990–1993	3	5–10	8	3	1.2–3.4	2.4
Ust-Kamenogorsk	1975–1979	23	6–24	11	7	0–1.1	0.6
Reservoir	1980–1984	36	8–19	12	27	0.8–3.2	1.7
	1985–1989	56	8–22	15	55	1.2–2.4	1.8
	1990–1991	35	6–35	19	35	0.8–5.0	2.2
River Irtysh,	1976–1979	36	0–12	7	36	1.0–6.1	3.4
upstream	1980–1984	62	6–34	19	64	1.0–4.2	2.2
	1985–1989	58	6–27	14	58	1.1–3.6	2.1
	1990–1991	23	5–52	23	23	0.9–5.3	2.7
River Irtysh,	1976–1979	52	7–20	13	40	0.6–3.5	2.2
middle reach	1980–1984	95	8–22	14	94	0.6–3.9	2.1
	1985–1989	90	8–24	15	90	0.5–5.1	2.4
	1990–1993	73	6–31	16	73	1.8–3.4	2.5

Continued

Appendix 9 Table II Continued

Water body	Observation period	COD (mg l⁻¹)			BOD (mg l⁻¹)		
		n	Range	Mean	n	Range	Mean
River Irtysh,	1975–1979	18	16–55	33	6	0.8–3.4	2.3
mouth	1980–1984	59	27–46	35	54	1.3–5.8	3.3
	1985–1989	76	14–76	38	56	1.0–10	3.2
	1990–1993	37	28–65	44	40	1.2–8.6	4.1
River Tobol,	1983–1984	19	19–40	29	39	1.0–3.0	2.0
middle reach	1985–1989	60	20–42	30	60	1.7–3.4	2.2
	1990–1991	24	24–60	37	24	1.4–3.0	2.1
River Tobol,	1975–1979	33	21–58	40	10	0.4–11	3.7
mouth	1980–1984	46	26–56	42	4	0.5–7.0	1.9
	1985–1989	60	21–73	45	28	0–7.5	4.3
	1990–1993	40	16–61	38	36	1.1–7.1	3.2
River Tom,	1975–1979	18	3–18	7	16	0.5–3.6	2.4
upstream	1980–1984	53	2–12	7	54	1.0–3.7	2.2
	1985–1989	58	3–15	7	61	1.0–3.7	2.0
	1990–1993	43	3–14	7	43	1.0–3.5	2.0
River Tom,	1977–1979	19	3–17	8	13	0–4.9	2.6
mouth	1980–1984	115	3–31	13	127	1.3–9.0	5.3
	1985–1989	301	4–29	12	201	0.8–4.8	2.6
	1990–1993	72	4–22	10	73	1.1–3.8	2.1
River Yenisey,	1978–1979	9	4–33	10	6	1.0–1.8	1.8
upstream	1980–1984	23	4–35	16	24	0.6–5.7	2.6
	1985–1989	37	7–28	15	37	0.9–3.0	1.6
	1990–1993	27	5–38	15	26	0–3.8	1.4
River Yenisey,	1985–1989	31	5–34	14	31	0.6–3.2	1.9
middle reach	1990–1993	27	7–34	15	27	0.9–3.1	1.9
River Yenisey,	1985–1989	28	6–77	20	28	0–7.7	1.5
mouth	1990–1993	21	4–39	18	22	0.5–3.7	1.4
Krasnoyarsk	1975–1979	70	4–13	7	69	0.5–2.8	1.6
Reservoit	1980–1984	90	4–18	10	90	0–3.3	1.6
	1985–1989	69	7–14	10	69	0–2.9	1.7
	1990–1993	54	6–14	9	54	0.6–2.9	1.8
River Lower Tunguska,	1977–1979	7	18–66	28	5	0–2.8	1.8
upstream	1980–1984	17	6–69	26	12	1.2–9.0	3.4
	1985–1989	15	9–133	36	17	0.5–10	3.5
	1990–1993	12	7–65	26	11	0.8–6.4	2.4
River Lower Tunguska,	1978–1979	8	26–49	34	8	0.7–3.0	1.8
mouth	1980–1984	25	0–83	37	23	0.7–3.8	2.0
	1985–1989	25	17–130	46	24	0.6–5.9	2.1
	1990–1992	14	22–68	42	14	0.6–2.9	1.4
Irkutsk Reservoir	1975–1979	64	3–8	5	70	0–2.9	1.4
	1980–1984	80	5–7	6	77	0–1.8	1.0
	1985–1989	52	5–10	7	52	0–2.0	1.0
	1990–1993	46	5–16	8	45	0–1.8	0.9
River Angara,	1985–1989	31	2–19	9	31	0–3.5	1.4
middle reach	1990–1993	32	2–15	8	32	0.8–2.6	1.6

Continued

Appendix 9 Table II Continued

Water body	Observation period	COD (mg l⁻¹)			BOD (mg l⁻¹)		
		n	Range	Mean	n	Range	Mean
River Angara, mouth	1976–1979	18	5–44	14	18	0.6–7.2	2.0
	1980–1984	23	10–49	24	23	0–5.8	2.4
	1985–1989	50	11–48	21	50	0.8–3.4	1.7
	1990–1993	21	10–44	17	21	0.2–3.1	1.8
River Vikhoreva, mouth	1975–1979	73	61–162	100	82	3.6–18	8.8
	1980–1984	47	73–176	123	60	4.6–13	8.2
	1985–1989	54	70–228	139	54	3.5–13	7.6
	1990–1993	51	27–124	73	51	1.7–13	5.7
River Selenga, middle reach	1975–1979	58	3–29	14	52	0.6–6.6	2.8
	1980–1984	58	5–26	13	61	0–3.4	1.4
	1985–1989	59	6–23	15	54	0.5–3.6	1.6
	1990–1993	47	8–26	16	48	0.6–2.0	1.2
River Selenga, mouth	1975–1979	59	5–31	17	55	0.5–5.9	3.1
	1980–1984	61	6–33	17	57	0.5–3.8	1.9
	1985–1989	64	7–31	16	59	0.6–4.0	2.0
	1990–1993	47	8–26	15	47	0.7–3.1	1.8
River Lena, upstream	1975–1979	19	0–47	16	21	0–3.1	1.2
	1980–1984	20	6–47	18	20	0–3.1	1.7
	1985–1989	18	7–48	19	18	0.7–2.2	1.4
	1990–1993	16	8–45	23	16	0.5–3.2	1.9
River Lena, middle reach	1975–1979	27	8–64	26	47	0–5.6	2.8
	1980–1984	78	20–67	42	79	0.6–6.0	2.8
	1985–1989	74	15–68	38	73	0.6–5.6	2.9
	1990–1993	33	7–38	27	33	1.3–4.6	3.0
River Lena, mouth	1980–1984	53	7–26	13	54	0–4.3	2.1
	1985–1989	62	6–26	14	68	0.8–2.9	1.7
	1990–1993	45	0–31	17	51	0.6–1.8	1.3
River Vilyuy, upstream	1975–1979	22	28–68	41	22	0–8.6	2.4
	1980–1984	36	28–64	44	27	0–5.5	2.0
	1985–1989	34	20–71	44	23	0–3.0	1.5
	1990–1993	28	18–101	36	28	0–3.6	1.8
River Vilyuy, mouth	1975–1979	23	13–67	41	8	0–4.9	2.1
	1980–1984	31	18–74	47	28	0–5.1	2.5
	1985–1989	35	27–63	44	26	0.6–3.9	2.3
	1990–1993	9	12–58	29	2	0–3.0	1.8
River Aldan, upstream	1975–1979	39	4–45	18	44	0.7–4.2	2.4
	1980–1984	69	3–50	19	71	0.8–3.5	2.0
	1985–1989	65	0–54	18	86	0.7–2.8	1.5
	1990–1993	56	2–46	17	55	0.9–3.0	1.9
River Aldan, mouth	1975–1979	17	6–60	34	11	0–2.2	1.1
	1980–1984	11	3–123	32	26	0–7.9	3.5
	1985–1989	31	0–82	26	29	1.5–8.7	3.9
	1990–1993	42	3–39	17	41	0.8–5.0	2.1
River Amur, upstream	1975–1979	22	10–65	28	19	0–5.9	2.0
	1980–1984	31	5–57	25	30	0–6.5	2.2
	1985–1989	30	2–53	20	29	0–4.7	1.9
	1990–1993	20	6–44	22	20	0.5–6.4	2.5

Continued

Appendix 9 Table II Continued

Water body	Observation period	COD (mg l⁻¹)			BOD (mg l⁻¹)		
		n	Range	Mean	n	Range	Mean
River Amur,	1975–1979	33	13–39	26	40	0.7–2.7	1.5
middle reach	1980–1984	36	8–39	24	36	1.1–2.6	1.9
	1985–1989	37	3–27	15	36	0.7–3.0	2.1
	1990–1993	36	9–26	16	36	0.6–3.0	1.5
River Amur,	1975–1979	39	12–37	35	54	0–2.4	1.5
mouth	1980–1984	57	9–40	21	57	0.6–3.4	1.7
	1985–1989	97	2–26	13	103	1.0–3.9	2.1
River Ussuri,	1975–1979	24	3–32	23	36	0.8–7.1	3.2
middle reach	1980–1984	39	8–49	26	63	0.5–6.5	2.6
	1985–1989	64	5–43	18	64	0–5.9	2.2
	1990–1993	47	8–30	19	48	0.5–3.8	1.7
River Ingoda,	1975–1979	47	12–127	49	36	1.6–11	5.6
middle reach	1980–1984	45	14–71	48	62	1.2–21	5.2
	1985–1989	52	16–74	38	60	1.0–21	4.8
	1990–1993	48	13–51	29	48	0–11	3.0

COD   Chemical oxygen demand   n   Number of samples
BOD   Biochemical oxygen demand

Appendix 9 Table III Petroleum products and surfactant concentrations in surface waters of the former USSR

Water body	Observation period	Petroleum products (mg l^{-1})			Surfactants (mg l^{-1})		
		n	Range	Mean[1]	n	Range	Mean
River Neman, upstream	1975–1979	54	0.03–1.2	0.41	54	0–0.14	0.04
	1980–1984	56	0–0.40	0.13	59	0–0.07	0.03
	1985–1989	60	0–0.30	0.14	60	0–0.7	0.03
	1990–1991	24	0–0.80	0.14	24	0–0.09	0.03
River Neman, middle reach	1975–1979	52	0.06–6.0	1.0	53	0.04–0.45	0.16
	1980–1984	58	0.05–0.46	0.24	60	0–0.08	0.03
	1985–1989	59	0.09–0.39	0.13	31	0–0.02	0.01
	1990–1991	24	0–0.02	< 0.02	9	0.01–0.02	0.02
River Neman, mouth	1986–1989	36	0–0.03	< 0.02	21	0–0.02	0.01
	1990–1991	21	0–0.02	< 0.02	14	0–0.05	0.02
River Neris, middle reach	1975–1979	48	0.04–1.9	0.98	51	0–0.27	0.07
	1980–1984	57	0.05–0.32	0.20	54	0–0.06	0.02
	1985–1989	60	0–0.20	0.12	45	0–0.02	0.01
	1990–1991	24	0–0.02	0.01	10	0–0.04	0.02
River Piarnu, mouth	1976–1979	10	0–0.78	0.20	12	0–0.13	0.03
	1980–1984	19	0–0.50	0.19	20	0–0.03	0.01
	1985–1989	22	0–0.92	0.18	22	0–0.19	0.03
	1990–1991	9	0–0.43	0.14	8	0–0.07	0.03
River Western Dvina, upstream	1976–1979	26	0–1.2	0.26	26	0–0.56	0.04
	1980–1984	34	0.06–0.77	0.37	35	0–0.07	0.03
	1985–1989	34	0.05–0.52	0.26	30	0–0.15	0.02
	1990–1991	14	0.03–0.70	0.13	14	0–0.08	0.03
River Western Dvina, mouth	1975–1979	13	0.04–0.15	0.12	18	0–0.11	0.05
	1980–1984	39	0.04–0.21	0.12	60	0.02–0.14	0.08
	1985–1989	58	0.04–0.10	0.07	47	0.03–0.08	0.05
	1990–1991	23	0–0.07	0.05	23	0.02–0.04	0.03
River Venta, upstream	1978–1979	13	0.05–0.41	0.20	6	0.02–0.06	0.04
	1980–1984	45	0.10–0.72	0.29	26	0–0.10	0.02
	1985–1989	53	0–0.44	0.12	26	0–0.02	0.01
	1990–1991	21	0	0	10	0–0.03	0.01
River Venta, middle reach	1975–1979	38	0.10–1.5	0.65	18	0.04–0.48	0.20
	1980–1984	48	0.10–1.0	0.39	26	0–0.16	0.03
	1985–1989	58	0–0.60	0.19	26	0–0.02	0.01
	1990–1991	21	0–0.04	< 0.02	10	0–0.04	0.02
River Venta, mouth	1975–1979	47	0.05–0.12	0.07	82	0–0.14	0.07
	1980–1984	29	0.03–0.10	0.06	59	0–0.12	0.06
	1985–1989	61	0.04–0.11	0.06	53	0.02–0.08	0.04
	1990–1991	24	0.03–0.06	0.03	11	0–0.05	0.02
River Western Bug, middle reach	1979	5	0–0.24	0.14	5	0.09–0.19	0.14
	1980–1984	27	0–1.1	0.08	27	0–0.43	0.19
	1985–1989	41	0–0.25	0.16	54	0.09–0.89	0.34
	1990–1991	–	–	–	25	0.01–0.15	0.06
River Neva, source	1975–1979	15	0–0.39	0.08	14	0–0.02	0.01
	1980–1984	24	0–1.7	0.17	29	0–0.06	0.02
	1985–1989	27	0–0.13	< 0.02	27	0–0.04	0.01
	1990–1991	11	0–0.04	0.02	11	0–0.04	0.01

Continued

Appendix 9 Table III Continued

Water body	Observation period	Petroleum products (mg l^{-1})			Surfactants (mg l^{-1})		
		n	Range	Mean[1]	n	Range	Mean
River Neva,	1975–1979	34	0–0.72	0.26	31	0–0.15	0.03
mouth	1980–1984	52	0–0.37	0.11	68	0–0.04	0.02
	1985–1989	29	0–0.09	< 0.02	30	0–0.04	0.01
	1990–1991	11	0–0.07	0.02	11	0–0.04	< 0.01
River Northern Dvina,	1975–1977	16	0–0.26	0.08	13	0–0.10	0.03
upstream	1985–1989	35	0–0.13	0.07 (0.03)	44	0–0.03	0.02
	1990–1993	18	0–0.20	0.04 (0.06)	20	0–0.07	0.02
River Northern Dvina,	1976–1979	205	0–0.56	0.17	143	0–0.04	0.01
mouth	1980–1984	220	0–0.31	0.09	27	0–0.05	0.01
	1985–1989	127	0–0.08	0.03 (0.03)	60	0–0.05	0.01
	1990–1993	48	0–0.08	0.04 (0.05)	30	0.03	0.01
River Sukhona,	1976–1979	14	0–2.5	0.31	18	0–0.05	0.02
mouth	1980–1984	30	0–1.5	0.23	29	0–0.05	0.02
	1985–1989	48	0–0.08	0.04 (0.02)	34	0–0.04	0.02
	1990–1993	50	0–0.08	0.02 (0.04)	23	0–0.06	0.03
River Vychegda,	1975–1979	17	0–0.56	0.08	20	0–0.13	0.04
upstream	1980–1984	33	0–0.10	0.09	34	0.01–0.07	0.02
	1985–1989	36	0.02–0.12	0.06 (0.05)	37	0–0.03	0.01
	1990–1993	28	0–0.14	0.04 (0.01)	8	0–0.04	0.01
River Vychegda,	1975–1979	57	0–1.2	0.33	61	0–0.15	0.03
mouth	1980–1984	132	0–0.17	0.08	151	0–0.09	0.05
	1985–1989	58	0.02–0.20	0.09 (0.03)	58	0–0.05	0.02
	1990–1993	49	0–0.10	0.04 (0.01)	25	0–0.03	0.01
River Pechenga,	1975–1979	35	0–1.2	0.25	17	0–0.20	0.04
middle reach	1980–1984	60	0–0.22	0.10	20	0–0.16	0.03
	1985–1989	58	0–0.07	0.04	25	0–0.26	0.02
	1990–1993	52	0	0	7	0–0.02	< 0.01
River Pechenga,	1975–1979	43	0–1.9	0.28	22	0–0.10	0.03
mouth	1980–1984	58	0–0.30	0.09	20	0–0.08	0.04
	1985–1989	67	0–1.4	0.16	29	0–0.03	0.01
	1990–1993	52	0–0.05	< 0.02	7	0–0.02	< 0.01
River Danube,	1979	17	0–0.60	0.22	14	0–0.11	0.05
mouth	1980–1984	67	0–0.81	0.19	64	0.01–0.08	0.04
	1985–1989	105	0–0.56	0.15	101	0–0.03	< 0.01
	1990–1991	43	0–0.55	0.20	43	0–0.10	0.03
River Prut,	1976–1979	14	0–0.59	0.11	12	0–0.04	0.01
mouth	1980–1984	22	0–0.76	0.21	22	0–0.28	0.04
	1985–1989	30	0–0.39	0.12 (0.01)	29	0–0.15	0.05
	1990–1991	13	0–0.24	0.06 (0.01)	13	0–0.08	0.02
River Dniester,	1975–1979	8	0–3.0	0.31	7	0–0.38	0.09
upstream	1980–1984	27	0–0.68	0.08	27	0–0.13	0.01
	1986–1989	15	0–0.41	0.06 (0.02)	15	0–0.17	0.04
	1990–1991	10	0.02–0.80	0.23 (0.20)	10	0–0.12	0.03

Continued

Appendix 9 Table III Continued

Water body	Observation period	Petroleum products (mg l⁻¹)			Surfactants (mg l⁻¹)		
		n	Range	Mean[1]	n	Range	Mean
River Dniester,	1975–1979	47	0–0.35	0.15	39	0–0.05	0.02
mouth	1980–1984	60	0–0.63	0.22 (0.01)	59	0–0.06	0.02
	1986–1989	61	0–0.25	0.12 (0)	49	0–0.08	0.03
	1990–1991	19	0–0.22	0.09 (0)	12	0–0.06	0.01
River Reut,	1975–1979	42	0–0.45	0.26	37	0–0.12	0.04
middle reach	1980–1984	61	0–0.46	0.17 (0)	61	0–0.07	0.03
	1986–1989	60	0–0.25	0.18 (0.01)	49	0–0.08	0.04
	1990–1991	21	0–0.20	0.05 (0.01)	13	0–0.04	0.01
River Reut,	1976–1979	18	0–2.0	0.25	17	0–0.05	0.02
mouth	1980–1984	33	0–0.6	0.23 (0)	33	0–0.06	0.03
	1986–1989	51	0–0.15	0.08 (0)	40	0–0.10	0.05
	1990–1991	20	0–0.20	0.04 (0)	12	0–0.05	0.02
River Southern Bug,	1986–1989	27	0–0.48	0.08 (0)	35	0–0.02	< 0.01
upstream	1990–1991	19	0–0.06	< 0.02 (0)	19	0	0
River Southern Bug,	1975–1979	13	0–1.9	0.20	18	0–0.78	0.14
middle reach	1980–1984	14	0–0.56	0.16 (0.35)	45	0.01–0.11	0.05
	1985–1989	24	0–1.4	0.06 (0)	43	0–0.03	0.01
	1990–1991	20	0–0.28	0.05 (0)	21	0.01–0.03	0.02
River Southern Bug,	1975–1979	11	0–0.86	0.19	14	0–0.13	0.06
mouth	1980–1984	24	0–1.1	0.14 (0.19)	27	0–0.16	0.05
	1985–1989	29	0	0	32	0–0.20	0.01
	1990–1991	8	0–0.08	0.02 (0)	9	0	0
River Dnieper,	1975–1979	18	0–0.59	0.32	23	0–0.26	0.08
upstream	1980–1984	40	0–0.28	0.16	59	0.02–0.08	0.05
	1985–1989	51	0–0.35	0.10	60	0–0.07	0.03
	1990–1993	22	0–1.8	0.10	26	0.02–0.11	0.06
River Dnieper,	1976–1979	29	0–2.3	0.51	32	0–0.20	0.13
middle reach	1980–1984	32	0–0.53	0.29 (0.29)	33	0–0.12	0.06
	1985–1989	29	0–0.33	0.07 (0.07)	30	0–0.20	0.04
	1990–1991	16	0–0.24	0.04	16	0–0.25	0.07
River Dnieper,	1988–1989	13	0–0.25	0.12 (0.01)	13	0–0.08	0.01
mouth	1990–1991	17	0–0.08	0.02 (0.03)	17	0–0.20	0.04
Kiev Reservoir	1980–1984	47	0–0.88	0.30 (0.21)	70	0–0.16	0.06
	1985–1989	17	0–0.38	0.05 (0.03)	22	0–0.13	0.03
	1990–1991	7	0–0.10	0.02	7	0–0.19	0.10
Dnieprodzerzhinsk	1986–1989	30	0–0.03	< 0.02 (0)	27	0–0.61	0.12
Reservoir	1990–1991	12	0–0.04	< 0.02 (0.01)	16	0–0.26	0.11
Kakhovka Reservoir	1975–1978	8	0.04–0.20	0.08	27	0–0.27	0.04
	1988–1989	28	0–0.26	0.10 (0.01)	28	0–0.07	< 0.01
	1990–1991	41	0–0.05	0.03 (0.01)	41	0–0.06	0.03
River Berezina,	1975–1979	52	0.03–1.0	0.32	51	0–0.12	0.05
midddle reach	1980–1984	59	0.04–1.1	0.46 (0.02)	59	0–0.07	0.02
	1985–1989	61	0–1.0	0.32 (0.04)	60	0–0.04	0.02
	1990–1991	24	0–0.46	0.08 (0.01)	24	0–0.14	0.03

Appendix 9 Table III Continued

Water body	Observation period	Petroleum products (mg l⁻¹)			Surfactants (mg l⁻¹)		
		n	Range	Mean[1]	n	Range	Mean
River Berezina,	1980–1984	28	0–0.88	0.26 (0)	28	0–0.14	0.04
mouth	1985–1989	34	0–0.40	0.19 (0.01)	33	0–0.12	0.05
	1990–1991	14	0–0.08	0.04 (0)	14	0–0.12	0.05
River Svisloch,	1975–1979	48	0.07–2.2	0.66	47	0.02–2.2	0.21
middle reach	1980–1984	60	0.03–1.1	0.38 (0.02)	60	0.02–0.37	0.17
	1985–1989	63	0.08–1.2	0.42 (0.08)	63	0.03–0.16	0.10
	1990–1991	24	0.06–0.89	0.28 (0.03)	24	0.02–0.15	0.09
River Svisloch,	1975–1979	24	0.03–0.66	0.24	26	0–0.21	0.05
mouth	1980–1984	34	0–0.80	0.43 (0.01)	35	0–0.06	0.02
	1985–1989	33	0.04–0.29	0.16 (0.01)	35	0–0.03	0.01
River Desna,	1975–1979	39	0–0.31	0.15	51	0–0.68	0.07
middle reach	1980–1984	57	0–0.13	0.06	64	0–0.11	0.04
	1985–1989	68	0–0.04	< 0.02	68	0–0.07	0.02
	1990–1991	53	0	0	55	0–0.02	< 0.01
River Don,	1975–1979	33	0.02–1.0	0.40	34	0–0.03	0.02
upstream	1980–1984	51	0–0.11	0.05	51	0–0.02	< 0.01
	1985–1989	40	0–0.29	0.07 (0.12)	40	0–0.02	< 0.01
	1990–1993	32	0.03–0.06	0.05 (0.06)	32	0–0.02	< 0.01
River Don,	1980–1984	21	0–1.7	0.62	30	0–0.12	0.03
middle reach	1985–1989	46	0–0.21	0.08	30	0–0.14	0.05
	1990–1993	12	0–0.26	0.06	18	0–0.05	0.02
River Don,	1975–1979	80	0–1.0	0.26	81	0–0.13	0.07
mouth	1980–1984	56	0.06–0.70	0.28	67	0–0.11	0.04
	1985–1989	60	0–0.45	0.18	61	0–0.20	0.05
	1990–1993	64	0–0.70	0.15	67	0–0.08	0.03
River Northern Donets,	1975–1979	13	0.02–0.22	0.08	13	0–0.06	0.03
upstream	1980–1984	58	0–0.17	0.06 (0.21)	63	0–0.11	0.06
	1985–1989	64	0.02–0.16	0.08 (0.32)	67	0.02–0.10	0.05
	1990–1993	51	0–0.53	0.13	53	0–0.05	0.02
River Northern Donets,	1975–1979	46	0.10–3.5	1.3	47	0.05–0.52	0.20
mouth	1980–1984	47	0.17–1.4	0.65	45	0–0.40	0.20
	1985–1989	28	0–1.2	0.28	30	0.05–0.10	0.07
	1990–1993	23	0–0.60	0.12	23	0.05–0.09	0.08
River Kuban,	1975–1979	33	0–0.62	0.13	37	0–0.16	0.04
upstream	1980–1984	22	0–1.6	0.31	28	0–0.10	< 0.01
	1985–1989	16	0.13–0.90	0.30	20	0–0.02	< 0.01
	1990–1992	6	0.20–1.1	0.42	9	0	0
River Kuban,	1975–1979	138	0–1.5	0.37	136	0–0.07	0.01
mouth	1980–1984	119	0.08–1.2	0.42	120	0–0.12	0.05
	1985–1989	118	0.05–0.44	0.20 (0.31)	109	0–0.10	0.04
	1990–1993	94	0.09–0.57	0.26 (0.29)	46	0–0.03	0.01
River Rioni,	1979	10	0.06–0.21	0.09	11	0–0.03	0.02
mouth	1980–1984	57	0.03–0.23	0.09	57	0–0.06	0.02
	1985–1989	60	0–0.57	0.18 (0.55)	59	0–0.13	0.06
	1990–1991	20	0.03–0.38	0.15	21	0–0.16	0.10

Continued

Appendix 9 Table III Continued

Water body	Observation period	Petroleum products (mg l⁻¹)			Surfactants (mg l⁻¹)		
		n	Range	Mean[1]	n	Range	Mean
River Terek,	1975–1979	28	0–3.8	0.94	27	0–0.60	0.04
middle reach	1980–1984	40	0.10–1.1	0.38	38	0–0.23	0.08
	1985–1989	51	0–1.0	0.34	28	0–0.09	0.02
	1990–1993	41	0.04–0.45	0.26	22	0–0.20	0.03
River Kura,	1978–1979	20	0–0.18	0.04	23	0–0.30	0.02
upstream	1980–1984	55	0–0.15	0.05	53	0–0.06	0.02
	1985–1989	57	0–0.17	0.06	56	0–0.06	0.03
River Kura,	1978–1979	–	–	–	38	0.02–0.06	0.07
mouth	1980–1984	42	0–0.16	0.06	44	0.02–0.07	0.04
	1985–1989	59	0–0.32	0.13	59	0–0.06	0.03
	1990–1991	24	0–0.48	0.20	29	0–0.10	0.03
River Araks,	1983–1984	7	0–0.10	0.06	11	0.03–0.07	0.04
mouth	1985–1989	58	0–0.18	0.05	58	0–0.06	0.04
	1990–1991	21	0–0.29	0.10	21	0–0.07	0.03
River Razdan,	1980–1984	60	0.05–0.85	0.30 (0.11)	60	0	0
mouth	1985–1989	53	0–0.70	0.20 (0.16)	53	0–0.18	0.04
	1990–1991	21	0–0.24	0.11 (0.11)	21	0–0.03	< 0.01
River Volga,	1978–1979	2	0.20–0.50	0.32	4	0–0.04	< 0.01
upstream	1980–1984	15	0–0.49	0.15	17	0–0.05	0.01
River Volga,	1975–1979	34	0–0.67	0.26	31	0–0.10	0.02
mouth	1980–1984	53	0–1.3	0.26	81	0–0.05	0.02
	1985–1989	193	0–1.2	0.26	165	0–0.04	0.02
	1990–1993	164	0–1.6	0.33	57	0–0.08	0.03
Rybinsk Reservoir,	1979	6	0.04–3.9	0.55	3	0–0.04	0.01
central part	1980–1984	38	0.08–0.92	0.30	18	0–0.50	0.02
	1985–1989	56	0.08–0.62	0.27	21	0–0.07	0.03
	1990–1993	40	0.10–0.53	0.27	26	0–0.06	0.02
Gorky Reservoir	1975–1979	62	0–1.1	0.54	61	0–0.20	0.08
	1980–1984	57	0.12–1.5	0.66	60	0–0.20	0.09
	1985–1989	60	0.10–1.4	0.52	60	0–0.16	0.07
	1990–1993	48	0.10–1.4	0.40	47	0–0.10	0.05
Cheboksary Reservoir	1976–1979	17	0–0.90	0.24	8	0–0.05	0.01
	1980–1984	57	0.15–1.8	0.57 (0.09)	51	0–0.15	0.08
	1985–1989	62	0–1.5	0.36 (0.69)	60	0–0.13	0.06
	1990–1993	48	0–0.53	0.17 (0.01)	48	0–0.12	0.06
Kuybyshev Reservoir	1975–1979	22	0–2.2	0.41	19	0–0.10	0.04
	1980–1984	40	0–0.30	0.12 (0.01)	29	0–0.13	0.04
	1985–1989	53	0–0.55	0.11 (0.02)	50	0–0.08	0.03
	1990–1993	59	0–0.66	0.18 (0.06)	58	0–0.07	0.03
Saratov Reservoir	1975–1979	21	0–0.22	0.06	17	0–0.06	0.03
	1980–1984	30	0–0.26	0.08	30	0–0.18	0.05
	1985–1989	30	0–0.22	0.07	30	0–0.07	0.03
	1990–1993	24	0–0.27	0.07	24	0–0.10	0.04
Volgograd Reservoir	1975–1979	143	0–0.16	0.08	148	0.02–0.13	0.07
	1980–1984	82	0–0.32	0.12	81	0.02–0.16	0.05
	1985–1989	89	0–0.17	0.04 (0.07)	80	0–0.04	0.02
	1990–1993	56	0–0.07	0.03 (0.04)	73	0–0.05	0.02

Continued

Appendix 9 Table III Continued

Water body	Observation period	Petroleum products (mg l⁻¹)			Surfactants (mg l⁻¹)		
		n	Range	Mean[1]	n	Range	Mean
River Oka,	1975–1979	48	0.20–0.73	0.53	42	0.03–0.16	0.10
middle reach	1980–1984	58	0.09–0.51	0.28	58	0.02–0.08	0.05
	1985–1989	46	0.09–0.27	0.18 (0.08)	47	0.03–0.09	0.06
	1990–1993	53	0.08–0.53	0.40 (0.08)	44	0.05–0.34	0.17
River Oka,	1975–1979	43	0–0.75	0.26	29	0–0.87	0.12
mouth	1980–1984	33	0.32–2.8	0.98 (0.07)	31	0–0.20	0.07
	1985–1989	49	0–0.55	0.23 (0.29)	48	0.01–0.21	0.07
	1990–1993	49	0–1.0	0.19 (0.01)	49	0.01–0.23	0.08
River Vyatka,	1975–1979	49	0–0.55	0.28	39	0–0.06	0.02
mouth	1980–1984	60	0–2.8	1.0 (0.06)	67	0–0.13	0.03
	1985–1989	68	0.20–1.	0.58 (0)	42	0–0.10	0.04
	1990–1993	52	0–1.0	0.40	29	0–0.05	0.02
River Ural,	1975–1979	83	0–0.56	0.22	48	0.02–0.30	0.10
middle reach	1980–1984	173	0.16–0.40	0.27 (0.43)	173	0.02–0.15	0.07
	1985–1989	180	0.05–0.24	0.12 (0.12)	180	0.02–0.14	0.06
	1990–1993	178	0–0.17	0.08 (0.03)	177	0–0.09	0.04
River Ural,	1975–1979	35	0.02–0.26	0.11	50	0–0.11	0.02
mouth	1980–1984	41	0–0.34	0.11	45	0–0.06	0.03
	1985–1989	103	0–0.18	0.04 (0.05)	63	0–0.05	0.03
	1990–1991	120	0.03–0.12	0.06 (0.01)	26	0.02–0.04	0.03
River Amu Darya,	1979	8	0–0.32	0.05	8	0.02–0.12	0.05
upstream	1980–1984	57	0–0.50	0.14 (0.07)	52	0–0.19	0.06
	1985–1989	48	0–0.62	0.12 (0.11)	48	0–0.03	< 0.01
	1990–1992	28	0–0.11	0.03 (0.05)	27	0–0.09	< 0.01
River Amu Darya,	1975–1979	32	0–8.0	2.2	56	0–0.12	0.05
mouth	1980–1984	52	0–0.37	0.09 (0.01)	52	0–0.14	0.03
	1985–1986	24	0–0.49	0.07 (0.05)	24	0–0.03	0.01
	1990–1992	27	0–0.11	0.04	27	0–0.03	< 0.01
River Kafirnigan,	1975–1979	3	0	0	–	–	–
mouth	1980–1984	38	0	0	46	0–0.02	< 0.01
	1985–1989	35	0	0	34	0–0.03	< 0.01
	1990–1991	17	0–0.08	0.02	14	0–0.06	0.02
River Syr Darya,	1980–1984	59	0–0.45	0.10 (0.02)	59	0–0.14	0.04
upstream	1985–1989	60	0–0.40	0–0.05 (0.05)	60	0–0.03	0.01
	1990–1992	27	0–0.23	0.05	27	0–0.02	< 0.01
River Syr Darya,	1975–1979	8	0–0.38	0.11	19	0–0.09	0.02
mouth	1980–1984	38	0–0.26	0.11 (0.22)	28	0–0.14	0.03
	1985–1989	54	0–1.0	0.23 (0.18)	29	0–0.13	0.03
	1990–1991	22	0–0.18	0.06 (0.01)	10	0–0.07	0.04
River Naryn,	1975–1979	8	0.22–1.5	0.64	21	0–0.18	0.02
mouth	1980–1984	28	0–1.5	0.38	33	0.01–0.06	0.02
	1985–1989	47	0–0.05	0.03 (0.01)	47	0.01–0.16	0.06
	1990–1991	7	0–0.08	0.02 (0.03)	8	0–0.08	0.06
River Ob,	1976–1979	23	0.05–6.2	1.2	25	0–0.70	0.13
middle reach	1980–1984	126	0.06–0.88	0.33	126	0–0.06	0.02
	1985–1989	57	0.07–0.81	0.25	50	0–0.03	0.01
	1990–1993	48	0.06–0.55	0.25	35	0–0.03	0.01

Continued

Appendix 9 Table III Continued

Water body	Observation period	Petroleum products (mg l⁻¹)			Surfactants (mg l⁻¹)		
		n	Range	Mean[1]	n	Range	Mean
River Ob,	1976–1979	28	0.11–3.4	0.40	28	0–0.46	0.14
mouth	1980–1984	63	0.10	0.31	65	0–0.06	0.02
	1985–1989	49	0.08–0.78	0.40	51	0–0.04	0.01
	1990–1993	46	0.23–2.6	0.91	28	0–0.06	0.02
River Irtysh,	1976–1979	10	0–0.24	0.09	34	0–0.03	< 0.01
upstream	1980–1984	52	0.04–0.32	0.14	30	0–0.07	0.01
	1985–1989	57	0.05–0.32	0.22 (0.02)	28	0–0.07	0.02
	1990–1991	23	0–0.72	0.18	11	0–0.20	0.04
River Irtysh,	1975–1979	41	0.07–0.42	0.24	28	0–0.24	0.09
mouth	1980–1984	60	0.08–0.36	0.19 (0.07)	56	0–0.20	0.08
	1985–1989	68	0.03–2.5	0.76 (0.21)	50	0–0.15	0.04
	1990–1993	45	0.29–2.1	0.90	31	0–0.06	0.01
River Tom,	1977–1979	19	0.10–3.4	0.42	12	0–0.15	0.04
mouth	1980–1984	122	0.10–0.93	0.39	111	0–0.12	0.03
	1985–1989	198	0.10–0.62	0.31	205	0–0.05	0.01
	1990–1993	73	0.12–0.66	0.34 (0.29)	30	0–0.06	0.01
River Yenisey,	1978–1979	7	0.03–1.2	0.57	7	0–0.04	0.01
upstream	1980–1984	24	0–2.6	0.94 (0.34)	24	0–0.13	0.03
	1985–1989	38	0–0.80	0.38 (0.23)	38	0–0.03	0.01
	1990–1993	27	0–1.0	0.48 (0.15)	27	0–0.06	0.01
River Yenisey,	1985–1989	31	0–1.2	0.37 (0.22)	31	0–0.05	0.01
middle reach	1990–1993	27	0–1.2	0.50 (0.12)	27	0–0.11	0.02
River Yenisey,	1985–1989	28	0–0.40	0.20	28	0–0.16	0.03
mouth	1990–1993	22	0–0.60	0.17	22	0–0.04	< 0.01
River Lower	1978–1979	8	0.40–2.2	1.1	7	0–0.02	< 0.01
Tunguska, mouth	1980–1984	25	0.12–1.6	0.60	25	0–0.19	0.05
	1985–1989	25	0–1.8	0.32	24	0–0.03	< 0.01
	1990–1992	14	0–0.50	0.29	14	0–0.05	0.02
Irkutsk Reservoir	1975–1979	55	0–0.13	0.07	15	0–0.02	< 0.01
	1980–1984	60	0–0.06	0.02	21	0–0.05	0.01
	1985–1989	51	0–0.11	0.03	17	0–0.04	< 0.01
	1990–1993	46	0–0.14	0.06	15	0	0
River Angara,	1976–1979	19	0–4.4	1.2	18	0–0.27	0.07
mouth	1980–1984	23	0–3.0	0.84	23	0–0.08	0.02
	1985–1989	49	0–1.0	0.50	50	0–0.06	0.01
	1990–1993	21	0–1.0	0.40	21	0–0.05	0.02
River Selenga,	1975–1979	55	0–0.68	0.23	55	0–0.04	0.02
mouth	1980–1984	68	0–0.38	0.13 (0.03)	65	0–0.06	0.02
	1985–1989	62	0–0.22	0.09 (0.03)	51	0–0.04	0.02
	1990–1993	47	0–0.16	0.04 (0.02)	46	0–0.04	0.01
River Lena,	1975–1979	19	0–0.12	0.04	10	0–0.03	0.01
upstream	1980–1984	20	0–0.17	0.02 (0)	20	0–0.07	0.02
	1985–1989	18	0–0.84	0.11 (0)	18	0–0.04	0.01
	1990–1993	15	0–0.30	0.12 (0.02)	16	0–0.04	< 0.01

Continued

Appendix 9 Table III Continued

Water body	Observation period	Petroleum products (mg l⁻¹)			Surfactants (mg l⁻¹)		
		n	Range	Mean[1]	n	Range	Mean
River Lena,	1980–1984	59	0–0.17	0.06 (0.03)	35	0–0.05	0.02
mouth	1985–1989	92	0–0.10	0.04 (0.02)	51	0–0.04	0.01
	1990–1993	45	0–0.12	0.04	28	0–0.03	0.01
River Vilyuy,	1975–1979	31	0–1.2	0.32	23	0–0.10	0.04
mouth	1980–1984	31	0–1.6	0.23	31	0–0.10	0.03
	1985–1989	35	0–0.30	0.11	35	0–0.04	0.03
	1990–1993	9	0–0.16	0.08	9	0–0.06	0.03
River Aldan,	1975–1979	18	0–4.8	0.77	15	0.02–0.17	0.06
mouth	1980–1984	29	0–1.0	0.32	30	0–0.29	0.05
	1985–1989	30	0–1.1	0.20	31	0–0.08	0.02
	1990–1993	42	0–0.33	0.11	42	0–0.04	0.02
River Amur,	1975–1979	41	0–0.20	0.08	43	0–0.13	0.04
middle reach	1980–1984	36	0–0.20	0.09 (0.02)	37	0–0.10	0.03
	1985–1989	35	0–0.20	0.08 (0.07)	37	0–0.22	0.07
	1990–1993	35	0–0.03	< 0.02 (0.01)	34	0–0.12	0.05
River Ingoda,	1975–1979	46	0.16–2.9	0.91	44	0.02–0.59	0.18
middle reach	1980–1984	63	0.13–0.82	0.41 (0.04)	63	0.02–0.35	0.10
	1985–1989	61	0.02–0.64	0.24 (0.02)	60	0–0.21	0.07
	1990–1993	48	0–0.24	0.10 (0.02)	48	0–0.12	0.04

n   Number of samples

[1]   Where known, the average hydrocarbon component (mg l⁻¹) is presented in brackets

Appendix 9 Table IV Phenol concentrations in surface waters of the former USSR

Water body	Observation period	Number of samples	Range (μg l^{-1})	Mean (μg l^{-1})
River Neman, upstream	1975–1979	42	0–18	2
	1980–1984	59	0–11	4
	1985–1989	41	0–2	1
River Neman, middle reach	1975–1979	54	0–5	3
	1980–1984	60	0–5	2
	1985–1989	59	0–3	1
	1990–1991	24	0–4	2
River Neman, mouth	1986–1989	47	0–3	2
	1990–1991	21	0–3	2
River Neris, mouth	1975–1979	52	0–6	3
	1980–1984	54	0–3	2
	1985–1987	35	0–3	2
River Western Dvina, upstream	1976–1979	20	0–12	3
	1980–1984	35	0–8	3
	1985–1989	28	0–15	2
	1990–1991	14	0–12	2
River Western Dvina, mouth	1975–1979	11	2–24	5
	1980–1984	30	0–28	8
	1985–1989	30	0–9	3
	1990–1991	12	0–5	2
River Venta, mouth	1975–1979	50	0–6	3
	1980–1984	29	0–21	5
	1985–1989	35	0–4	3
	1990–1991	12	0–5	4
River Western Bug, middle reach	1980–1984	27	0–17	4
	1985–1989	54	0–3	2
	1990–1991	13	0–38	6
River Northern Dvina, mouth	1978–1979	195	0–14	3
	1980–1984	221	0–5	< 1
	1985–1989	192	0–3	< 1
	1990–1993	76	0	0
River Sukhona, mouth	1976–1979	21	0–38	6
	1980–1984	30	0–37	2
	1985–1989	54	0–4	< 1
	1990–1993	50	0–6	1
River Vychegda, mouth	1975–1979	76	0–13	4
	1980–1984	127	0–14	7
	1985–1989	59	0–7	2
	1990–1993	37	0–10	5
River Pechenga, mouth	1975–1979	20	0–49	11
	1980–1984	20	0–19	5
	1985–1989	40	0–14	6
	1990–1993	52	0–6	3
River Danube, mouth	1985–1989	91	0–9	5
	1990–1991	45	4–8	6
River Prut, mouth	1976–1979	14	0–8	1
	1980–1984	22	0–40	3
	1985–1989	30	0–18	5
	1990–1991	13	0–11	2

Continued

Appendix 9 Table IV Continued

Water body	Observation period	Number of samples	Range (µg l^{-1})	Mean (µg l^{-1})
River Dniester,	1975–1979	48	0–5	1
mouth	1980–1984	59	0–6	2
	1986–1989	62	0–12	3
	1990–1991	19	0–24	2
River Reut,	1976–1979	19	0–5	1
mouth	1980–1984	33	0–8	2
	1986–1989	51	0–15	5
	1990–1991	20	0–9	2
River Southern Bug,	1978–1979	14	0–46	6
mouth	1980–1984	22	0–60	5
	1985–1989	32	0	0
	1990–1991	9	0	0
River Dnieper,	1980–1984	56	0–19	5
upstream	1985–1989	60	0–5	2
	1990–1993	25	0–15	2
River Dnieper,	1976–1979	23	0–25	4
middle reach	1980–1984	31	0–48	3
	1985–1989	29	0–15	4
	1990–1991	16	0–2	< 1
River Dnieper,	1988–1989	12	0–15	3
mouth	1990–1991	17	0–9	1
River Berezina,	1980–1984	30	0–18	4
mouth	1985–1989	24	0–2	1
River Don,	1975–1979	25	0–2	< 1
upstream	1980–1984	48	0	0
	1985–1989	36	0	0
River Don,	1975–1979	83	0–10	3
mouth	1980–1984	65	0–18	6
	1985–1989	63	0–11	4
	1990–1993	21	0–10	2
River Seversky Donets,	1980–1984	13	0	0
upstream	1985–1989	62	0	0
	1990–1993	31	0	0
River Seversky Donets,	1975–1979	48	0–17	5
mouth	1980–1984	46	0–11	5
	1985–1989	30	0–9	5
	1990–1993	23	0–5	2
River Kuban,	1975–1979	139	0–7	1
mouth	1980–1984	118	0–15	4
	1985–1989	119	0–10	3
	1990–1993	94	0–3	1
River Rioni,	1979	10	4–16	9
mouth	1980–1984	57	5–24	9
	1984–1989	58	3–11	7
	1990–1991	21	0–17	7
River Kura,	1978–1979	4	2–15	6
upstream	1980–1984	50	0–16	7
	1985–1989	54	3–13	7

Continued

Appendix 9 Table IV Continued

Water body	Observation period	Number of samples	Range ($\mu g\ l^{-1}$)	Mean ($\mu g\ l^{-1}$)
River Kura, middle reach	1978–1979	30	0–32	10
	1980–1984	47	2–28	17
	1985–1989	57	3–21	13
River Kura, mouth	1980–1984	39	0–11	5
	1985–1989	59	5–15	9
	1990–1991	24	3–15	8
River Araks, mouth	1983–1984	11	3–15	8
	1985–1989	58	4–18	9
	1990–1991	21	7–18	9
River Razdan, middle reach	1980–1984	59	0–10	5
	1985–1989	19	0–20	5
River Volga, mouth	1975–1979	41	0–9	5
	1980–1984	84	0–6	2
	1985–1989	214	0–10	2
	1990–1993	258	0–13	3
Kuybyshev Reservoir	1975–1979	23	0–45	4
	1980–1984	39	0–25	9
	1985–1989	53	0–15	6
	1990–1993	60	0–9	2
River Oka, mouth	1975–1979	43	0–18	4
	1980–1984	32	0–25	3
	1985–1989	48	0–6	1
	1990–1993	49	0–3	< 1
River Ural, mouth	1975–1979	36	0–10	1
	1980–1984	41	0–10	2
	1985–1989	70	0	0
	1990–1993	41	0–4	< 1
River Amu Darya, upstream	1979	7	16–68	40
	1980–1984	56	0–29	11
	1985–1989	48	0–7	3
	1990–1992	28	0–21	5
River Amu Darya, mouth	1975–1979	57	0–28	10
	1980–1984	57	0–31	12
	1985–1986	22	0–12	2
	1990–1992	27	0–15	5
River Syr Darya, upstream	1975–1979	49	0–33	11
	1980–1984	60	0–33	12
	1985–1989	60	0–7	3
	1990–1992	27	0–15	4
River Syr Darya, mouth	1975–1979	19	0–27	10
	1980–1984	18	0–23	6
	1985–1989	58	0–4	1
	1990–1991	22	0–6	3
River Naryn, mouth	1975–1979	15	0–3	< 1
	1980–1984	23	0–4	< 1
	1985–1989	37	0–3	1
	1990–1991	8	0–2	1

Continued

Appendix 9 Table IV Continued

Water body	Observation period	Number of samples	Range (µg l^{-1})	Mean (µg l^{-1})
River Ob, middle reach	1976–1979	26	0–40	9
	1980–1984	124	0–28	9
	1985–1989	55	0–14	7
	1990–1993	48	0–21	9
River Ob, mouth	1976–1979	28	0–8	1
	1980–1984	65	0–11	2
	1985–1989	67	0–7	2
	1990–1993	47	0–13	4
River Irtysh, upstream	1976–1979	32	0–3	2
	1980–1984	30	0	0
	1985–1989	24	0–6	< 1
	1990–1991	51	0–9	1
River Irtysh, mouth	1975–1979	41	0–20	5
	1980–1984	57	0–12	5
	1985–1989	66	0–16	5
	1990–1993	45	3–21	14
River Tom, mouth	1980–1984	126	0–34	12
	1985–1989	205	0–22	6
	1990–1993	70	0–19	4
River Yenisey, middle reach	1985–1989	30	0–13	3
	1990–1993	27	0–12	4
River Lower Tunguska, mouth	1978–1979	8	0–7	3
	1980–1984	25	0–25	8
	1985–1989	25	0–13	5
	1990–1992	14	3–20	12
Irkutsk Reservoir	1975–1979	56	0–5	2
	1980–1984	60	0–6	2
	1985–1989	52	0–4	< 1
	1990–1993	46	0	0
River Angara, mouth	1976–1979	15	0–15	11
	1980–1984	22	0–8	3
	1985–1989	49	0–25	4
	1990–1993	21	0–23	8
River Lena, upstream	1975–1979	11	0–4	< 1
	1980–1984	19	0–12	2
	1985–1989	18	0–15	2
	1990–1993	15	0–5	1
River Lena, mouth	1980–1984	61	0–4	1
	1985–1989	94	0–10	4
	1990–1993	47	0–13	5
River Amur, middle reach	1975–1979	17	0–14	3
	1980–1984	35	0–18	5
	1985–1989	37	0–9	3
	1990–1993	33	0–16	7
River Ingoda, middle reach	1975–1979	43	0–26	13
	1980–1984	62	0–18	7
	1985–1989	56	0–12	4
	1990–1993	48	0–12	4

Chapter 10[*]

MICROBIAL POLLUTION

Water may serve as a source of many infectious diseases including cholera, typhoid, paratyphoid, dysentery, brucellosis, leptospirosis, hepatitis and poliomyelitis. Infectious diseases related to water are widespread at present despite all measures taken to stem them. According to some researchers, 80 per cent of all diseases are related to the abuse of sanitary standards for water supply (Zakharchenko *et al.*, 1993). Thus, in 1993 there was an increase in the number of water-related intestinal infections (On the sanitary and epidemiological situation, 1993), with 17 such outbreaks, 10 of which were dysentery. The worst affected areas were in the North Osetia Republic (340 people were taken ill), and in the regions of Arkhangelsk (340 people), Krasnodar (181 people) and Krasnoyarsk (171 people). These outbreaks of intestinal infections were linked to significant microbial pollution of water supply sources. Another example of infection through water supply was an outbreak of gastrointestinal infections and typhoid in 1993 in the Rostov Region, where 300 people were taken ill in the town of Volgodonsk. In some regions of the former USSR, cholera *Vibrio* were registered annually in surface waters.

10.1 Microbiological indicators of water quality

In 1718, Peter I issued a law banning the citizens of St Petersburg from discharging sewage directly into rivers, canals and drains in order to protect water reservoirs of the city from faecal pollution (Abakumov, 1981). In 1873, an *ad hoc* commission of the Physicians' Society in the town of Kazan adopted a 'Programme for periodic surveys of water quality supplied to the town including microbiological components'. In 1887, the significance of microbiological analysis of water quality and the importance of systematic water quality monitoring in natural water bodies was recognised (Tcherbakov, 1877). At a similar time, practical recommendations regarding microbiological analysis of water were given in a series of lectures on hygiene (Dobroslavin, 1874).

At present, microbiological surface water monitoring is performed using a number of indicators which reflect the state of the water body

---

[*] *This chapter was prepared by V.A. Abakumov and Y.G. Talayeva*

and the epidemiological safety of water for different types of use and enable the assessment of trends in water quality (Abakumov, 1980; Talayeva and Zaydenov, 1986).

10.1.1 Organic matter pollution

In 1994, a century had passed since R. Koch suggested a microbial assessment of water quality using a microbial number (MN) index. This index is the number of microbial colonies grown in 1 ml of sampled water on a standard meat-peptone agar (MPA). Such a nutrient medium in a laboratory situation enables the selective growth of saprophytic bacteria. Therefore the index can be referred to as 'total number of saprophytic bacteria' or in short 'saprobic number'. Three synonymous terms are also frequently used for this index: 'microbial number', 'total number of MPA bacteria', and 'MPA bacteria number'. Microbial number is sometimes referred to as 'total microbe number' or 'number of microbes'. However, this is not correct because MN refers exclusively to the group of bacteria which grows on meat-peptone agar.

From an ecological aspect, MPA bacteria are micro-organisms which consume organic matter arising from dead organisms and animal excrement. From a physiological aspect, saprophytic bacteria are heterotrophs which require organic substances as a carbon source. Because MPA bacteria are indicators of organic water pollution, MPA micro-organisms can be referred to as saprobic micro-organisms, saprobes or saprobionts.

A good indicator of the ecological and microbiological state of water bodies is the ratio of the total number of bacteria on membrane filters to the number of MPA bacteria. Here MPA bacteria number means microbial number as suggested by Koch. The total number of bacteria on membrane filters from sampled water is, in reality, the total number of all microbes. It is determined by filtering a known volume of water through a standard membrane filter. The filter with the filtrate is dyed and dried and the microbes are counted on a part of the filter under a microscope. This so called 'method of direct counting' (i.e. without growing bacteria in colonies) considers not only MPA micro-organisms but also other types of bacteria including sulphur bacteria, nitrogen-fixing bacteria, nitrifying bacteria and various anaerobic bacteria; it therefore allows the determination of the total number. The classification of water quality by these microbiological indicators, which are used in the regular monitoring of water bodies in Russia (and in the former USSR), is given in Table 10.1.

10.1.2 Faecal contamination

As a preventive measure against water-related intestinal infections, the sanitary-epidemiological service controls water quality by monitoring coliforms and using a coli-index. The coli-index (CI) reflects the number

Table 10.1 Classification of water quality by microbiological indicators

Class of water	Quality of water	A Total number of bacteria (10^6 cells per ml)	B Saprophytic bacteria (10^3 cells per ml)	Ratio (A:B)
1	Very good	< 0.5	< 0.1	> 10^3
2	Good	0.6–1.0	0.1–5.0	10^2–10^3
3	Moderately polluted	1.1–3.0	5.1–10.0	> 10^2
4	Polluted	3.1–5.0	10.1–50.0	> 10^2
5	Poor	5.1–10.0	50.1–100.0	10^2
6	Very poor	> 10	> 100	< 10^2

Source: Abakumov, 1983

of coliform bacteria in 1 litre of water. Coliform bacteria belong to the family Enterobacteriaceae and ferment both glucose and lactose at 37 °C. Other synonyms for 'coliform bacteria' include 'lactose-positive colibacillus', 'coliforms', and 'bacteria of the coli-group'. The presence of coliform bacteria indicates a high probability of faecal contamination of water by human beings or warm-blooded animals, especially when they occur in purified mains water which could not have had contact with sewage waters. The coli-index is a relative unit. Besides being a 'microbial number', the coli-index is a standard index used by the State Sanitary and Epidemiology Service of Russia (Figure 10.1).

Within the coliform bacteria there is an even smaller indicative group of bacteria, *Escherichia coli*. *E. coli* has particular properties in addition to those which are similar to coliform bacteria, namely that they ferment lactose not only at 37 °C, but also at 44 °C, that they cannot use sodium citrate as a carbon source and that they produce indole in a peptone medium with tryptophan. In comparison with coliform bacteria in general, *E. coli* indicate definite faecal contamination of water. The index of faecal *Escherichia coli* (IFEC) is the concentration of faecal *E. coli* in 1 litre of water. The index is used for the description of the most polluted waters. For example, an IFEC of 100 means that in 1 litre water there are 100 cells of faecal *E. coli* (Table 10.2).

10.1.3 Viral contamination

Coliphage viruses, or to be more precise, bacteriophages of coliform bacteria, attack and lyse bacteria of the coliform group. If coliforms are considered as indirect indicators of the possible presence of pathogenic intestinal bacteria, coliphages are indirect indicators of the possible presence in water of pathogenic intestinal viruses (enteroviruses). In order to determine the presence of coliphages, water samples are placed on agar-covered Petri dishes, the entire surface of which have been

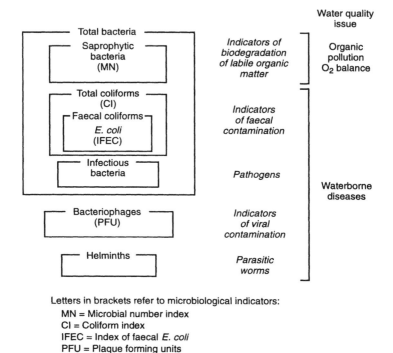

Letters in brackets refer to microbiological indicators:
MN = Microbial number index
CI = Coliform index
IFEC = Index of faecal *E. coli*
PFU = Plaque forming units

Figure 10.1   Microbial indicators used in water quality surveys

inoculated with a suspension of an appropriate coliform bacteria. Where coliforms are infected by phage and lysed, lighter-coloured zones (plaques) are seen on the bacterial "lawn" which develops after incubation. It is considered that 1 plaque corresponds to 1 phage unit. The coliphage content is calculated for 1 litre of water and expressed in phage units or in plaque-forming units (PFU). In cases of extreme pollution, water is checked for the presence of bacterial and viral pathogens and for pathogenic protozoa.

10.1.4 Water quality standards

The standards by which microbiological pollution of surface waters is measured are given in Table 10.3. For this purpose, surface waters are subdivided into two categories. The first category (I) contains water bodies used for drinking and domestic purposes. The second category (II) contains water bodies used for recreation and for those within town borders. Guidelines for the protection of surface waters from pollution ban the discharge of pathogen-containing wastewaters into water bodies. Wastewaters may only be discharged into water bodies after treatment and disinfection, and with a microbial concentration not exceeding a coli-index of 1,000 and a coliphage count of 1,000 phage units

Table 10.2 The detection frequency of *Campylobacter* and *Salmonella* from the River Moskva and its tributaries with different *Escherichia coli* counts

E. Coli count	Campylobacter			Salmonella		
	n	No. of positive identifications	Isolation (%)	n	No. of positive identifications	Isolation (%)
Temperature > 15 °C						
IFEC < 5,000	15	1	7	12	0	0
IFEC > 5,000	39	12	31	30	14	47
Total	**54**	**13**	**21**	**42**	**14**	**33**
Temperature < 15 °C						
Total	**74**	**58**	**78**	**43**	**30**	**70**
All temperatures						
Total IFEC < 5,000	21	4	19	12	0	0
Total IFEC > 5,000	107	67	63	72	11	15
Total 5,000 < IFEC < 10,000	24	6	25	24	0	0
Total 10,000 < IFEC < 50,000	38	23	61	32	8	25
TOTAL	**128**	**71**	**55.5**	**85**	**44**	**52**

n   Number of samples   Source: Dorodnikov, 1992
I C   Index of faecal *E. coli*

Table 10.3 Standards for assessing microbial water pollution in surface waters

Indicators	Category I[1]	Category II[2]
Pathogens	Should not contain pathogens	
Coliforms	< 1,000 cells per litre	< 5,000 cells per litre
Coliphages	< 100 phage-units per litre	
Helminths	Should not be detected in 1 litre	

[1] Water bodies used for drinking water supply   Source: Protection, 1988
[2] Water bodies used for recreation

(Protection, 1988). In accordance with accepted standards for drinking water, the coli-index may not exceed 3 and the total number of saprophytes should not exceed 100 cells per ml (Drinking water, 1982).

The epidemiological control of drinking water quality from centralised drinking water supply systems is performed by bacteriological laboratories located at drinking water supply stations. Water is analysed twice a day using two main microbiological indicators: total microbial number and the coli-index. In addition, two to three times a month, sanitary and epidemiological institutions analyse drinking water quality using the same microbiological indicators. When permissible levels of microbial pollution are exceeded, water analysis is undertaken for a second time.

In cases of hazardous health conditions, drinking water is analysed for the presence of both bacterial and viral pathogens. At present, routine analysis includes the determination of coliphages as indirect indicators of viral contamination of drinking water.

Control of drinking water quality from non-centralised water supplies (i.e. wells, springs, irrigation ditches) is performed by laboratories of sanitary and epidemiological institutions much less frequently, about once every three months. In cases of epidemiologically unfavourable situations, drinking water is analysed more frequently using a wider range of microbiological indicators.

10.2 Microbial pollution of surface waters

Microbial agents of intestinal infections usually enter water bodies through wastewaters of municipal sewage systems, rain water and live-stock effluents, especially when these wastewaters are not adequately treated and disinfected. Hospital wastewaters are also important poten-tial sources of contamination. Microbial contamination of rivers, especially those flowing through densely populated areas, is usually high and the presence of pathogenic microflora is frequently detected (See data of the Information Centre of the State Committee of Sanitary and Epidemiological Control of Russia, (Appendix 10 Table I)). Specific examples of microbial water pollution are presented below in more detail for various regions of the former USSR.

10.2.1 Moskva and Neva rivers basins

The microbiological state of the Moskva and Neva rivers are similar in many respects (Abakumov, 1993, 1994) and this is of special interest because the two largest cities, Moscow and Leningrad (St Petersburg), are situated on them.

In the River Moskva, prior to 1992, from its source to the town of Zvenigorod, the total number of bacteria did not exceed 10^6 cells per ml, the number of saprophytes did not exceed 10^3 cells per ml and their ratio was approximately 1,000. Within the city of Moscow, the number of saprophytes reached 9×10^4 cells per ml and the ratio decreased to 100. The highest number of bacteria was recorded downstream of Moscow (9×10^5 cells per ml and ratio 100). Downstream of the Besedy settle-ment, the total number of bacteria and the number of saprophytes began to decrease due to intensive self-purification processes. In the River Moskva and its tributaries, *Salmonella* and *Campylobacter* were wide-spread (Table 10.2).

In the 1980s water at the River Neva source complied, in 72 per cent of cases, with requirements for drinking water supply, but within the city of Leningrad, water quality deteriorated significantly. The total number of bacteria increased to 5×10^6 cells per ml. The number of saprophytes reached 10^5 cells per ml. Bacteria and fungal growth were also noted in the river. Pollution of the Neva was most evident downstream of the River Okhta inflow. Here the number of $E.$ $coli$ (IFEC) reached $7.0 \times 10^6 - 1.5 \times 10^7$ in various arms of the Nevka river delta. However, the River Chernaya was characterised by even higher levels of microbial pollution which, until relatively recently, received the communal wastewaters of all districts of Leningrad. The IFEC value varied in this river within 10^5–10^8 and the number of positive $Salmonella$ samples ranged from 58 per cent to 90 per cent. The most polluted areas in the Neva bay were situated close to Leningrad, as well as in the northern and central part of the region (Talayeva, 1988a).

10.2.2 Moldovian water bodies
In recent years microbial indicators of water quality in Moldova have remained relatively constant (Abakumov, 1993, 1994). In the River Prut the number of saprophytes varied from 0.5×10^3 to 10^4 cells per ml, the ratio varied from 400–1,000. In the River Dniester the number of saprophytic bacteria ranged from 10^3 to 3.8×10^5 cells per ml, with the highest numbers as well as the smallest ratio (< 100) usually recorded in the Dubossary reservoir. The number of faecal $E.$ $coli$ varied within 10^4–10^6 cells per ml. These indicators were closely correlated with the results of detection of $Salmonella$ ($r = 0.87$) which were found in 8–50 per cent of samples. $Salmonella$ counts and the indicators of their frequency of detection were found to be more variable compared with the coli-index, and they were more closely correlated with the sanitary and topographic characteristics of the sampling sites (Shlyakhov et $al.$, 1983). During in-depth studies of pathogen occurrence in the 1970s, $Salmonella$ occurred only at sampling sites located downstream of major sources of faecal contamination. Here, contamination by $Salmonella$ was recorded in 10–14 per cent of samples. $Leptospira$ occurred at most sampling sites and their occurrence in samples from areas of wastewater discharge was 20–44 per cent.

10.2.3 Angara river basin and the Irkutsk and Bratsk reservoirs
In 1993, in the basin of the River Angara, the Lake Baikal outlet, there was a significant decrease in water quality based on general microbiological indicators (Table 10.4) (Abakumov, 1994). Despite the general deterioration in the microbial state of the River Angara, the change in the coli-index was less marked in its tributaries and reservoirs. Beyond areas directly under the influence of wastewaters, the coli-index as a rule did not exceed 5×10^3. The discharge of wastewaters from the cities of Irkutsk and Angarsk increased the number of coliforms to 10^5 cells per litre.

At the end of the 1970s, intense bacteriological and virological research was carried out on the River Angara (Plyasunov et $al.$, 1979; Talayeva, 1979). This showed important trends in the distribution of pathogenic bacteria and viruses, as well as regular bacterial self-purification in the

Table 10.4 Total number of bacteria (A), number of saprophytic bacteria (B), and the ratio total bacteria:saprophytic bacteria (A:B) in the basin of the River Angara

Station	1992			1993		
	Total bacteria (A) (10^6 cells per ml)	Saprophytic bacteria (B) (10^3 cells per ml)	A:B	Total bacteria (A) (10^6 cells per ml)	Saprophytic bacteria (B) (10^3 cells per ml)	A:B
1 Irkutsk Reservoir						
1.1 Angara source, Baikal outlet	0.22–1.11	0.09–0.43	2,444–10,000	0.9–1.1	0.4–1.8	614–2,763
1.2 Patrony village	1.44–1.73	1.21–8.13	593–1,430	1.6–1.9	2.2–30.6	63–706
1.3 Irkutsk city water supply abstraction site	–	–	–	2.2–3.0	20.2–38.5	73–111
2 Angara						
2.1 Irkutsk, 6 km upstream of wastewater discharge from a municipal sewage treatment plant (right bank)	1.41–2.42	0.16–27.0	90–11,313	1.1–2.7	0.5–15.7	95–4,208
2.2 Irkutsk, 2 km downstream of wastewater discharge from a municipal sewage treatment plant (right bank)	1.2–2.24	0.82–44.75	87–2,122	1.1–4.0	2.7–84.0	48–775
2.3 Irkutsk, 2 km upstream of wastewater discharge from a municipal sewage treatment plant (left bank)	1.21–2.96	1.85–14.00	179–1,124	0.8–2.4	3.3–57.0	43–361
2.4 Irkutsk, 0.5 km downstream of wastewater discharge from a municipal sewage treatment plant (left bank)	1.73–5.64	0.4–47.90	97–5,110	1.1–2.8	2.8–73.9	31–718
2.5 5.5 km upstream of Angarsk town	1.76–3.99	4.6–44.40	165–394	1.1–2.7	14.6–51.2	48–105
2.6 Angarsk	2.25–3.8	3.45–45.7	68–693	1.4–3.4	24.9–56.3	40–66
2.7 0.9 km downstream of Angarsk town	2.17–4.32	2.80–26.90	161–886	1.3–2.1	7.8–35.8	46–164

Table 10.4 Continued

Station	1992			1993		
	Total bacteria (A) (10^6 cells per ml)	Saprophytic bacteria (B) (10^3 cells per ml)	A:B	Total bacteria (A) (10^6 cells per ml)	Saprophytic bacteria (B) (10^3 cells per ml)	A:B
3 Bratsk reservoir						
3.1 Usole-Sibirskoye	1.48–3.25	4.55–20.15	89–458	1.2–2.3	18.7–32.9	33–81
3.2 2 km downstream of Usole-Sibirskoye	1.68–5.68	6.6–54.25	105–350	1.3–2.3	29.1–49.4	42–67
3.3 0.5 km upstream of Svirsk town	1.21–2.92	1.45–40.38	55–924	1.3–2.9	23.4–61.5	39–86
3.4 0.5 km downstream of Svirsk town	2.37–6.32	1.35–94.5	43–387	1.9–4.6	24.8–89.6	47–99
4 River Irkut						
4.1 Water abstraction point at Shelekhov town	1.48–2.29	0.15–56.2	41–10,266	1.6–5.0	26.0–543.0	10–62
4.2 Irkutsk, 4 km downstream of the River Olkha mouth	1.5–2.76	0.35–71.35	39–4,286	2.6–6.7	59.8–1,549.5	10–43
4.3 Irkutsk, 0.5 km downstream of the discharge of waste waters from a furniture factory	2.44–4.39	0.85–99.1	44–2,871	2.2–5.9	48.7–864.0	10–46
5 River Olkha						
5.1 0.5 upstream of Shelekhov town	1.68–1.98	1.15–15.1	123–1,722	1.7–4.0	22.4–402.5	10–76
5.2 Shelekhov	2.0–5.49	6.55–88.5	62–382	2.5–4.7	57.8–454.5	10–44
5.3 1.8 km downstream of Shelekhov town	1.51–180	0.45–7.2	217–1,119	3.0–5.0	67.1–558.0	9–45

Source: Abakumov, 1994

river. In the River Angara within the borders of Irkutsk, 31 per cent of samples contained *Salmonella* with concentrations not exceeding 100 cells per ml and 16 per cent of samples during summer contained non-agglutinative *Vibrio*, NAG-*Vibrio*. In the river around the town of Usole-Sibirskoye, NAG-*Vibrio* occurred during summer in 30 per cent of samples and some *Vibrio*-El Tor (cholera *Vibrio* which cause epidemics) were also found. For 60 km, from Usole-Sibirskoye to the town of Svirsk, bacterial self-purification of water was less evident. Water abstracted near Svirsk contained *Salmonella*, including the *Salmonella* species responsible for typhoid. The frequency of detection of *Salmonella* in water was correlated with the coli-index. Thus, with a coli-index > 10^5, *Salmonella* occurred in 29 per cent of samples, whereas with a coli-index of 10^2–10^4 *Salmonella* occurred in 9 per cent of samples. When a coli-index < 10^4 was obtained, in 90 per cent of river samples, *Salmonella* did not occur. NAG-*Vibrio* occurred in 13 per cent of samples when the coli-index was 10–1,000.

Virological studies showed a low level of pollution in terms of intestinal viruses. Of 170 samples, only two samples from the water abstraction site at the town of Angarsk had the poliomyelitis virus Type I. In terms of the cytophatogenic dose (CPD), at 50 per cent degeneration, their index was 1.35 CPD_{50}. The number of phages of *E. coli* in those samples was 10^3 PFU per litre. The absence of intestinal viruses in samples of river water situated downstream of the discharge of wastewaters might be explained by their low concentrations which are beyond the limits of sensitivity of the concentration methods employed.

10.2.4 Water bodies of the Aral Sea basin
In the areas situated around the Aral Sea, especially in the Kzyl Orda region of Kazakhstan and Karakalpakistan in Uzbekistan, there is a high risk to the population of infectious intestinal waterborne diseases.

In the lower reaches of the River Syr Darya, near the town of Kzyl Orda, the total microbe number reached 1.7×10^7 cells per ml and the coli-index was 2.3×10^7 cells per litre; the frequency of detection of coliphages was approximately 100 per cent and that of enteroviruses exceeded 72 per cent. In the mid-1970s, the total microbial number and the coli-index varied within 1,800–175,000 cells per ml and 280–23,800 cells per ml respectively (Medical and ecological problems, 1993).

In the Kyzketken canal, the main water body in the town of Nukus, the number of samples with a coli-index exceeding the norm (> 10^4 per litre) was greater than 58 per cent, the number of samples which contained *Salmonella* and potentially pathogenic bacteria reached 13 per cent, and the number of samples which contained *Shigella* exceeded 3 per cent. Data concerning pathogenic enterobacteria in water bodies of Karakalpakistan are presented in Table 10.5.

Table 10.5   Percentage occurrence of pathogenic enterobacteria in samples from
surface waters of Karakalpakistan (Uzbekistan)

Area	1986	1987	1988	1989	1990
Nukus (town)	6.1	9.4	26.5	15.4	0.6
Beruni district	9.8	9.4	2.0	5.4	33.3
Muynak district	16.1	4.4	–	2.4	–

Source: Medical and ecological problems, 1993

10.3   Microbial pollution of drinking water

In the former USSR, 13.1 per cent of water samples from water supply
systems did not meet State standards. The greatest percentage of
samples which did not comply with standards was in Turkmenistan
(18.3 per cent), Kyrgyzstan (17.5 per cent), Kazakhstan, Azerbaijan and
Estonia. In 48 per cent of towns of the former USSR, drinking water
samples did not comply with standards. In 12 per cent of towns, the
number of samples which did not comply with coli-index standards
increased in spring and summer by up to 60–70 per cent or more. In
30 per cent of towns, the number of samples which did not comply with
standards was greater than 20 per cent. Surveys in the central and
southern climatic zones of the country showed that the level of bacterial
pollution of drinking water did not guarantee its safety in terms of potential viral contamination. Thus, in the drinking water of two Uzbekistan
towns which formally complied with standards according to bacteriological indicators, coliphages (which were used as indicators of the viral
contamination of water) were found in 73 per cent and 93 per cent of
samples and enteroviruses in 30 per cent and 80 per cent of samples
(Talayeva, 1988a).

   The quality of water in water supply systems depends on many
factors, including the level of pollution of the supply source, the sanitary
and technological state of the water treatment plant and the distribution
network (Simkalov, 1993). Data concerning the quality of water in the
distribution system by microbiological indicators in Russia are
presented in Appendix 10 Table II. The worst affected areas in terms of
tap water quality are in regions of Central Asia around the Aral Sea.
Thus, in the Kzyl Orda Region, 67 per cent of the water supply is from
centralised water mains, but the quality of the water is poor. Up to
60 per cent of water samples from municipal water supply systems and
approximately 38 per cent from rural water supply systems did not
comply with standards. In six settlements, where approximately
40 per cent of the region's population live, inhabitants still use highly
polluted waters from the Syr Darya river. The characteristics of tap

Table 10.6 Percentage of samples not complying with standards for coliforms in mains water in the Kzyl Orda region of the Syr Darya basin, Kazakhstan

Town	1985	1986	1987	1988	1989	1990
Yanykurgan	28.7	32.4	13.5	27.3	12.7	4.3
Dzhusaly	62.4	53.8	43.4	62.2	45.2	28.0
Terenosek	12.8	14.3	13.3	14.8	11.9	12.8
Dzhalagash	63.4	70.5	54.1	20.4	19.9	16.7
Kzyl Orda	22.3	16.5	14.8	13.4	11.3	15.4
Tasbuget	21.2	23.2	23.3	22.6	30.3	31.7
Chiili	10.3	24.8	21.8	20.5	34.8	14.7
Saksaulsk	40.0	50.0	40.0	28.6	41.7	35.7
Aralsk	25.2	24.4	8.5	16.0	11.4	5.0

Source: Medical and ecological problems, 1993

water quality in the Kzyl Orda region, based on bacteriological indicators, from 1985 to 1990 are presented in Table 10.6. Although there was a gradual improvement in water quality from 1985 to 1990, the quality of drinking water the population received remained poor. In 1990, 4–32 per cent of samples, depending on the location, did not comply with standards. In addition, pathogenic enterobacteria were constantly found in the tap water of the region.

There is a general shortage of drinking water in the Lower Syr Darya and Amu Darya basins. Up to 40 per cent of the local rural population use drinking water directly from wells, while in some settlements water is delivered in special tanks. In the Lower Amu Darya basin near the Aral Sea, there was some improvement in water quality from 1985 to 1990, but pathogenic bacteria remained present in water causing a potential epidemic threat to the population (Table 10.7). For many years, tap water in the town of Nukus had high levels of microbial pollution. However, in 1990, the quality of drinking water in Nukus, and in Karakalpakistan in general, gradually improved due to the sanitary and technical measures introduced.

Microbiological indicators characterising the quality of tap water in the Khorezm region in Uzbekistan from 1986 to 1990 are presented in Table 10.8. Data highlight the great heterogeneity of pollution in different towns and rural districts.

10.4 Seasonal dynamics of microbial water pollution

Pathogenic bacteria
The most common bacterial pathogens, *Salmonella* and *Shigella*, usually enter surface waters with untreated communal sewage, but they tend not to reproduce in water. Their presence in water downstream of

Table 10.7   Bacteriological indicators of drinking water quality in Karakalpakistan in the Lower Amu Darya basin, Uzbekistan, 1985–90

Town/region	1985	1986	1987	1988	1989	1990
Samples not complying with standards (%)						
Tap water						
Nukus town	29.0	16.5	5.6	6.4	22.9	0
Beruni district	18.3	0	0.7	1.0	0.7	0
Muynak district	1.6	9.2	8.2	7.1	11.0	0
Well water						
Nukus town	–	–	50.0	94.4	63.6	100
Beruni district	0	–	38.5	53.9	0	5.8
Muynak district	6.2	1.0	1.3	0	17.2	0
Detection of pathogenic bacteria (%)						
Tap water						
Nukus town	1.4	2.3	0	3.0	5.2	4.2
Beruni district	0	0	0	0	0	0
Muynak district	0	3.0	2.6	0.4	0	0
Well water						
Nukus town	–	–	40.0	11.1	0	0
Beruni district	–	0	1.1	0	0	0
Muynak distric	–	7.5	5.3	0	0	–

Source: Medical and ecological problems, 1993

wastewater discharges depends on the number of sick people and human carriers and also depends on the level of dilution of the wastewaters in the water body. The increased rate of sickness from intestinal infections in summer and autumn suggests an increase in the numbers of pathogenic bacteria entering the water during that time. However, processes of self-purification in summer are more intensive than in winter. These two factors balance the content of pathogenic bacteria in surface waters close to the wastewater discharge points over the year.

Intestinal viruses
In surface water bodies in spring and autumn, there is generally an increase in intestinal viruses due to two principal reasons. The increase in autumn is due to an increase in the infection rate of the population with intestinal viruses and, in spring and autumn, inoculation of the population (mainly children) with live poliomyelitis vaccines occurs.

Cholera Vibrio
The cholera pathogen can be dormant in open water bodies and reproduce actively in warm periods (at water temperatures exceeding 20–22 °C). In summer, the cholera *Vibrio* occurs not only in the warmer

Table 10.8 Microbiological indicators of tap water quality in the Khorezm Region, Uzbekistan, 1986–90

Town/district	Samples not complying with the coliform standard (%)					Detection of pathogenic bacteria (%)					Detection of coli-phages (%)				
	1986	1987	1988	1989	1990	1986	1987	1988	1989	1990	1986	1987	1988	1989	1990
Urgench	10.6	7.4	15.7	5.7	19.5	0	0	0	0	0.1	45.2	33.3	26.2	39.6	29
Khiva	26.5	59.3	46.2	25.1	30.3	0.3	0.09	0.2	0.4	0.5	0	0	0	0	0
Druzhba	3.8	12.7	1.2	4.4	9.8	0	0	0	0	0	2.5	1.3	0.3	4.6	7.2
Khazarasi	10.0	19.7	20.4	18.1	25.4	0	0	0	0	0	16.1	8.9	5.7	–	–
Bagat region	27.1	12.5	34.7	6.3	19.1	0	0	0	0	0	0	–	–	–	–
Yashnaryk region	12.6	14.4	12.4	7.7	16.4	0	0	0	1.5	0	22.2	25.6	32.7	7.6	–
Khanki region	24.6	40.7	53.8	21.5	55.8	0	0	0	0	0.2	–	7.9	10.8	0	0
Urgench region	7.3	30.5	27.1	38.9	30.1	0	5.7	9.0	0	0.7	3.9	0	–	0	0
Koshkunyr	17.0	24.0	21.1	10.3	15.2	0	0	0.8	0	0.2	4.2	1.3	15.0	0	0
Shavat	5.6	17.6	9.6	4.2	4.7	0	0	0	0	0	0	3.7	0	2.0	0
Gurlen	9.0	21.4	13.9	11.7	21.6	0	5.4	0.5	4.6	0.5	–	0	0	0	0
Yangibazar	–	61.1	0	28.1	52.1	0	0	0	0	0	–	0	0	0	0

Source: Medical and ecological problems, 1993

south, but also in central and even northern parts of the country where shallow waters characterised by slow water renewal and high organic levels are abundant. This leads to a wide dispersion of the cholera *Vibrio* in surface waters in summer and, consequently, to an increase in poten- tial epidemic outbreaks.

10.5   Assessment of the epidemic potential of polluted water
When assessing the epidemic potential of polluted water, it must be considered that diseases may be caused by the presence of relatively small numbers of pathogenic and potentially pathogenic intestinal bacteria. For example, there was a case recorded where 77 people were taken ill after drinking water in which the coli-index was 260–4,200 and the concentration of *Klebsiella* was 70–200 cells per litre. Seventy per cent of those taken ill were shown serologically to have had klebsiellas of serovariants k13 and k27. A relationship was established between the quality of the drinking water consumed and the outbreak of intestinal diseases among children (Artemov *et al.*, 1983a,b).

The presence of intestinal bacterial and viral pathogens in water is closely related to socio-hygienic factors. Violations in regulations for water supply and use and the inadequate state of communal sanitation facilities have an overwhelming effect on the occurrence and spread of intestinal infections. During an assessment of the epidemic situation undertaken with the aim of working out preventive and anti-epidemic measures, it became necessary to assess the sanitary state of different towns and administrative territories. Such methods were worked out and tested in areas with different climatic and sanitary conditions. This in turn allowed integral indicators to be substantiated, which character- ised separate components of sanitary conditions and their ranking in relation to the degree of potential epidemic outbreaks. For each sanitary factor, a complex of important and informative indicators has been defined and evaluation tables have been drawn up (Talayeva and Zaydenov, 1986).

10.6   Assessment of combined effects of microbial and chemical pollution on human health
The elaboration of criteria for the integrated assessment of microbial and chemical pollution effects on human health is at present the most urgent task (Abakumov, 1984; Meybeck *et al.*, 1989). In view of this, experimental research and monitoring were performed which identified certain trends and established a reduction in immunoresistance of laboratory animals to infectious diseases by various chemical compounds. Karimova (1985) showed that laboratory mice subjected for a month to water containing hexavalent chrome salts at concentrations of 0.1, 1.0 and 10.0 mg l^{-1} became increasingly sensitive to *Salmonella* species, and the chance of typhoid was increased by 1.2, 4 and 25 times

respectively (immunological tests and LD_{50}). Similar trends were observed for laboratory animals subjected to other chemical pollutants (Agapova, 1985; Talayeva, 1988b). In further works, the experimental research was supported by studies of the morbidity of the population under varying conditions of chemical and microbial water pollution (Kulaev, 1986). Thus, the results of scientific research may serve as a basis for establishing the criteria of epidemic safety and for setting permissible levels of microbial pollution according to different water uses, taking into consideration the existing background of chemical pollution.

In addition, the chemical pollution of surface waters may influence micro-organisms, promoting their transfer from a saprophytic state to a parasitic one, and inducing mutations and the appearance of new features (Butyanova and Nacheva, 1992). Chemical pollution may also produce a significant effect on the correlation of indicators and pathogenic micro-organisms, because chemical compounds may influence saprophytic and pathogenic micro-organisms by suppressing or stimulating their reproduction (Nemyrya, 1978; Talayeva et al., 1979). Consequently, under conditions of chemical pollution, microbial indicators cannot always serve as a reliable indicator of the safety or epidemic potential of water bodies. In such cases it may be necessary to use direct detection methods for pathogens in the water. This raises questions about the necessity of setting standards for factors which influence human health unfavourably because these factors may be of a different nature (chemical, biological or physical) and must take into account the combined effects of these factors on organisms in different climatic and geographical conditions.

10.7   Methods for the protection of water sources from microbial pollution

To prevent microbial pollution of water close to water abstraction points from surface reservoirs or underground aquifers, a sanitary protection area must be imposed which consists of three consecutive zones. In the first zone, at the site of water abstraction and mains construction, it is prohibited for people to live or to use water resources for any purposes. In the second zone, the borders of which are defined according to local conditions, the discharge of wastewaters and the use of water resources are only allowed with strict sanitary control. In the third zone, including the whole drainage basin of small and medium-sized rivers and individual parts of large river basins, there is systematic monitoring of the sanitary and epidemiological state of towns and cities and infectious morbidity. Unfortunately, these simple rules are frequently violated. Their strict observation would be a direct and effective way reducing microbial pollution of water sources. An additional effective rule is the strict enforcement of guidelines for wastewater discharge, and the obligatory purification and disinfection of these waters to a safe level, i.e.

with a coli-index not exceeding 1,000, with coliphages not exceeding 1,000 and with the obligatory absence of pathogenic microflora.

10.8 References

(All references are in Russian unless otherwise stated).

Abakumov, V.A. 1980 Bacteria as indicators of environmental conditions and features of their populations. In: *Ecological Monitoring Problems and Ecosystems Modelling*. Volume 3, Gidrometeoizdat, Leningrad, 24–51.

Abakumov, V.A. 1981 The history of water quality control based on hydrobiological indicators. In: *Fundamentals of Water Quality and Control Based on Hydrobiological Variables*. Proceedings of the All-Union Conference, Gidrometeoizdat, Leningrad, 46–74.

Abakumov, V.A. 1983 Hydrobiological analysis of surface waters and bottom sediment. In: *Manual on Methods of Hydrobiological Analysis of Surface Waters and Bottom Sediment*. Gidrometeoizdat, Leningrad, 7–21.

Abakumov, V.A. 1984 Hydrobiological control of freshwater. In: *Scientific Information Bulletin*. Transactions of the Inter-Agency Scientific and Technical Council on Environmental Protection, Problems and Rational Use of Natural Resources by the State Committee of Science and Technology, Volume 9, VINITI, Moscow, 47–64.

Abakumov, V.A. [Ed.] 1993 *Yearbook on the State of Surface Water Ecosystems in Russia and Neighbouring Countries*. VNIIGMI-MCD, Obninsk, 396 pp.

Abakumov, V.A. [Ed.] 1994 *Yearbook on the State of Surface Water Ecosystems in Russia and Neighbouring Countries*. VNIIGMI-MCD, Obninsk, 402 pp.

Agapova, T.M. 1985 Basis and approbation of the totality of quantitative criteria for the hygienic assessment of non-specific resistance of organisms subject to chemical pollution (experimental and nature research). Synopsis of a Thesis, USSR Academy of Medical Sciences, Moscow, 24 pp.

Artemova, T.Z., Kiseleva, B.S. and Gegechkori, M.I. 1983a Critical intestinal diseases caused by the use of water infected with *Klebsiella*. In: *Hygienic Research of Biological Pollution of the Environment*. Minzdrav, Moscow, 12–14.

Artemova, T.Z., Talayeva, Y.G. and Kiseleva, B.S. 1983b The relationship between the quantitative level of *Klebsiella* in drinking water and the rate of critical intestinal infections in an organised collective. In: *Hygienic Research of Biological Pollution of the Environment*. Minzdrav, Moscow, 14–15.

Butyanova, A.P. and Nacheva, E.B. 1992 Pollution of water bodies. In: Proceedings of the Congress of the All-Union Hydrobiological Society, Murmansk, 25–27.

Dobroslavin, A.P. 1874 Hygiene. A Course of Public Health Part II. St Petersburg.

Dorodnikov, A.I. 1992 Basis of sanitary-bacteriology indices of epidemic safety of surface waters in terms of Campylobacteria. Synopsis of Thesis, Russian Academy of Medical Sciences, Moscow, 24 pp.

Drinking Water 1982 GOST-2874-82, Goskomstandart, Moscow, 7 pp.

Indicators of Environmental Quality 1992 Statistical Report No. 18 on the Sanitary Condition of the Regions in 1991 in Russia for each Administrative Territory. State Committee on Sanitary and Epidemiological Surveillance, Moscow, 350 pp.

Karimova, R.I. 1985 Experimental basis for the assessment method of joint influences of chemical and bacterial water pollution on the immunoresistance of organisms. Synopsis of Thesis, IOKG, Moscow, 23 pp.

Kulaev, P.J. 1986 Joint impact of organochlorine pesticides and microbial water pollution on the morbidity rate of intestinal infections. Synopsis of Thesis, Minzdrav, Moscow, 24 pp.

Medical and Ecological Problems of the Aral Crisis 1993 VINITI, Moscow, 102 pp.

Meybeck, M., Chapman, D. and Helmer, R. [Eds] 1989 *Global Freshwater Quality: A First Assessment*. Blackwell Reference, Oxford, 306 pp. (In English).

Nemyrya, V.I. 1978 The influence of some chemical substances on *Salmonella* in water. In: *Actual Issues of Sanitary Microbiology*. Minzdrav, Moscow, 107–108.

On the Sanitary and Epidemiological Situation in the Country 1993 Monthly Information Bulletin of Gossanepidnadzor of the Russian Federation, Volume 7, 1–2.

Plyasunov, A.K., Bagdasaryan, G.A., Talayeva, Y.G., Grigorieva, L.V., Rabyshko, E.V., Bedareva, L.I. and Sukhocheva, I.F. 1979 *Microbiological Aspects of Water Quality Control of Regulated Water Bodies Subject to Chemical Pollution*. IOKG, Moscow, 200 pp.

Protection of the Environment 1988 The Scientific Information Bulletin (Moscow), **9**, 34–39.

Shlyakhov, E.N., Potapov, A.I. and Merenuk, G.V. 1983 *The influence of water bodies' biological pollution on the infection rate of the population. In: Hygienic Research of Biological Pollution of the Environment.* Minzdrav, Moscow, 219–220.

Simkalov, A.N. 1993 The state of water supply in the Russian Federation and the rate of infection of the population in relation to water factors in 1992. *Health of the Population and Environment Bulletin*, **8**, 1–16.

Talayeva, Y.G. [Ed.] 1979 Research on self-purification processes of water bodies in Siberia. NTIC, State Registration No. 80009643 B 82640, Moscow, 207 pp.

Talayeva, Y.G. [Ed.] 1988a *Hygienic Aspects of Biological Pollution of the Environment*. Minzdrav, Moscow, 147 pp.

Talayeva, Y.G. [Ed.] 1988b The influence of the joint impact of microbial and chemical (pesticides) factors of water pollution on the morbidity rate of the population through bacterial intestinal infections. All-Russian NTICS, State Registration No. 01.86.0014402, Moscow, 207 pp.

Talayeva, Y.G. and Zaydenov, A.M. [Eds] 1986 Guidelines on the epidemiological assessment of sanitary and hygienic conditions for intestinal infections prophylaxis. (Approved by Minzdrav, USSR, 6 June 1986, N28-6/20), Minzdrav, Moscow, 25 pp.

Talayeva, Y.G., Bogatyreva, M.D. and Yaroshevskiy, V.N. 1979 Research on the influence of surfactants on pathogenic enterobacteria in water. In: *Microbiological Aspects of Quality Control of Regulated Water Bodies*. IOKG, Moscow, 150–155.

Tcherbakov, A.J. 1877 Methods for Hygienic Research. Part I. Quantitative and Qualitative Analysis of Drinking Water. St Petersburg.

Zakharchenko, M.I., Goncharuk, E.I. and Koshelev, N.F. 1993 *Modern Eco-hygiene Problems*. Part 1, Naukova Dumka, Kiev, 69–70.

Appendix 10

Table I Percentage of samples not complying with microbiological standards in surface waters of the Russian Federation

Table II Microbiological quality of water in the distribution system in the Russian Federation

Appendix 10 Table I Percentage of samples not complying with microbiological standards in surface waters of the Russian Federation

| Region | Category I: Water bodies used for drinking water supply | | | | | | Category II: Water bodies used for recreation | |
| | Coliform bacteria (coli-index) | | Pathogen occurrence | | Agents of helminthiasis dangerous to humans | | Coliform bacteria (coli-index) | |
	1991	1992	1991	1992	1991	1992	1991	1992
Northern Region	82.0	–	4.3	–	0.2	–	97.2	–
Arkhangelsk region	78.0	66.2	5.1	4.0	0	–	90.6	95.9
Vologda region	82.8	80.7	6.6	3.7	0.6	0.4	97.8	90.8
Murmansk region	100.0	100.0	–	–	0	–	100.0	97.8
Republic of Karelia	83.6	79.6	1.5	1.9	0	–	97.6	100.0
Komi Republic	100.0	100.0	–	–	0	–	100.0	89.6
North-western Region	95.2	–	3.5	–	6.4	–	93.3	–
St Petersburg	99.3	33.0	3.2	5.4	3.0	–	100.0	96.5
Leningrad region	89.3	60.3	4.6	2.0	14.4	–	89.9	91.7
Novgorord region	100.0	88.2	–	–	0	–	100.0	79.3
Central Region	97.1	–	7.0	–	0.7	–	92.6	–
Pskov region	94.4	94.4	7.0	5.6	0	–	84.0	80.3
Bryansk region	100.0	100.0	–	–	–	–	99.2	96.9
Vladimir region	89.7	68.6	37.9	15.7	1.1	–	86.8	80.2
Ivanovo region	100.0	100.0	–	–	–	–	100.0	98.6
Tver region	94.8	100.0	1.6	–	–	–	98.6	83.1
Kaluga region	98.4	100.0	3.7	–	–	–	100.0	87.9
Kostroma region	100.0	97.5	–	2.5	–	–	100.0	97.2
Moscow city	–	–	0.65	–	–	–	98.6	87.1
Moscow region	100.0	95.7	–	–	–	99.0	100.0	–
Orel region	–	–	–	–	–	–	78.6	97.2
Ryazan region	64.9	89.3	9.5	33.3	–	41.9	75.0	95.0
Smolensk region	–	–	–	–	–	–	98.8	97.0
Tula region	–	–	–	–	–	–	70.7	47.5
Yaroslavl region	89.9	100.0	1.7	8.3	1.7	1.2	93.4	89.9

Continued

Appendix 10 Table I Continued

Region	Category I: Water bodies used for drinking water supply						Category II: Water bodies used for recreation	
	Coliform bacteria (coli-index)		Pathogen occurrence		Agents of helminthiasis dangerous to humans		Coliform bacteria (coli-index)	
	1991	1992	1991	1992	1991	1992	1991	1992
Volga-Vyatka Region	89.5	—	1.9	—	—	91.2	—	—
Kirov region	100.0	86.4	0	2.3	—	—	99.6	98.5
Nizhny Novgorod region	87.3	63.2	2.3	2.5	—	—	86.8	85.6
Mari Republic	100.0	100.0	0	—	—	—	73.9	89.3
Mordva Republic	—	—	0	—	—	—	96.4	95.6
Chuvash Republic	100.0	75.0	0	—	—	—	100.0	87.0
Central-Chernozem Region	—	—	—	—	—	—	68.7	—
Belgorod region	—	—	—	—	—	—	95.3	94.3
Voronezh region	—	—	—	—	—	—	44.0	54.0
Kursk region	—	—	—	—	—	—	100.0	99.2
Lipetsk region	—	—	—	—	—	—	89.8	83.7
Tambov region	—	—	—	—	—	—	99.4	100.0
Volga Region	92.3	—	1.7	—	—	—	73.9	—
Astrakhan region	100.0	100.0	—	—	—	—	100.0	96.3
Volgograd region	56.9	100.0	1.9	1.9	—	—	79.6	92.1
Penza region	100.0	100.0	—	—	—	—	74.0	96.0
Samara region	98.7	98.6	1.8	—	—	—	91.7	98.0
Saratov region	99.8	94.2	2.4	2.1	—	—	98.4	98.3
Ulyanovsk region	4.0	48.2	8.0	44.4	—	—	84.2	84.3
Republic of Kalmykia	—	—	—	—	—	—	—	—
Tatarstan Republic	100.0	100.0	—	—	—	4.76	66.0	91.1

Continued

Appendix 10 Table I Continued

Region	Category I: Water bodies used for drinking water supply						Category II: Water bodies used for recreation	
	Coliform bacteria (coli-index)		Pathogen occurrence		Agents of helminthiasis dangerous to humans		Coliform bacteria (coli-index)	
	1991	1992	1991	1992	1991	1992	1991	1992
North-Caucasus Region	81.4	–	0.4	–	0.5	–	87.4	–
Krasnodar region	100.0	97.6	–	2.4	0	–	97.4	69.4
Adigey Republic	100.0	–	–	–	0	–	100.0	100.0
Stavropol region	56.5	71.9	1.2	7.8	2.6	–	55.6	48.6
Karachay-Cherkes Republic	79.3	83.1	–	–	0	–	100.0	71.1
Rostov region	100.0	93.4	0.8	1.5	0	–	99.2	88.1
Dagestan Republic	27.8	100.0	–	–	0	–	26.4	100.0
Kabarda-Balkar Republic	100.0	100.0	–	–	0	–	87.7	100.0
North Osetia Republic	–	–	–	–	–	–	–	100.0
Checheno-Ingush Republic	100.0	100.0	–	–	0	–	100.0	100.0
Ural Region	88.6	–	6.1	–	2.0	–	90.0	–
Kurgan region	100.0	100.0	–	–	0	–	88.3	89.4
Orenburg region	100.0	100.0	–	–	0	–	100.0	86.8
Perm region	82.9	64.7	17.1	2.4	0	2.4	65.6	75.1
Sverdlovsk region	85.2	94.0	4.4	0.7	5.2	0.7	87.2	74.3
Chelyabinsk region	100.0	56.9	–	5.9	0	–	100.0	78.2
Republic of Bashkiria	95.5	98.2	4.5	3.7	0	–	99.7	93.0
Republic of Udmurtia	82.4	72.2	–	–	0	–	98.6	68.0
West-Siberian Region	95.5	–	2.0	–	–	–	93.9	–
Altai region	98.7	100.0	1.1	–	–	–	97.9	87.7
Gorny-Altai Republic	94.6	–	–	–	–	–	96.9	100.0
Kemerovo region	91.8	81.5	2.5	6.1	–	–	91.2	70.0
Novosibirsk region	91.3	84.7	13.5	3.9	–	–	91.8	77.3
Omsk region	100.0	100.0	1.4	1.6	–	16.0	78.4	100.0
Tomsk region	–	–	–	–	–	–	100.0	100.0
Tyumen region	100.0	39.4	–	4.2	–	–	100.0	22.2

Continued

Appendix 10 Table I Continued

Region	Category I: Water bodies used for drinking water supply						Category II: Water bodies used for recreation	
	Coliform bacteria (coli-index)		Pathogen occurrence		Agents of helminthiasis dangerous to humans		Coliform bacteria (coli-index)	
	1991	1992	1991	1992	1991	1992	1991	1992
East-Siberian Region	96.0	–	4.6	–	0.6	–	81.7	–
Krasnoyarsk region	96.4	98.7	4.5	0.7	0.9	–	90.0	98.5
Khakassia Republic	100.0	83.3	–	–	0	100.0	100.0	100.0
Irkutsk region	93.2	100.0	6.8	9.9	0	–	94.5	99.0
Chita region	100.0	100.0	–	–	0	–	55.6	100.0
Buryatia Republic	100.0	100.0	–	–	0	–	100.0	100.0
Tuva Republic	–	–	–	–	–	–	100.0	100.0
Far-East Region	54.0	–	0.1	–	1.7	–	85.7	–
Primorsk region	100.0	94.1	0	–	0	–	100.0	100.0
Khabarsovsk region	21.9	87.0	–	6.5	–	7.6	39.5	95.8
Yevrei Republic	100.0	100.0	–	–	–	–	–	100.0
Amur region	96.5	58.6	–	–	–	3.4	90.0	49.0
Kamchatka region	100.0	–	–	–	–	–	100.0	–
Magadan region	100.0	–	0	–	0	–	100.0	–
Sakhalin region	57.6	100.0	0	–	48.0	4.2	66.0	–
Yakutia (Sakha) Republic	38.6	87.3	0.2	–	0	26.1	100.0	97.11
Baltic Region	100.0	–	1.6	–	–	–	99.7	–
Kaliningrad region	100.0	100.0	1.6	1.8	–	–	99.7	–
Russia	85.4	88.0	3.1	2.3	1.2	3.14	88.3	86.4

Source: Indicators, 1992

Appendix 10 Table II Microbiological quality of water in the distribution system in the Russian Federation

Region/Republic	Samples not complying with the coliform standard (% of total number of samples)		Samples with ≥ 20 coliforms/litre (% of the number of samples not complying with standards)	
	1991	1992	1991	1992
Northern Region				
Arkhangelsk region	23.9	22.4	5.7	17.5
Vologda region	17.0	100.0	4.6	73.1
Murmansk region	11.9	1.3	4.1	0.5
Republic of Karelia	12.6	15.6	6.1	9.8
Komi Republic	4.8	4.0	4.3	2.0
North-western Region				
St Petersburg	0.3	0.7	0.1	0.1
Leningrad region	9.4	100.0	4.4	23.1
Novgorod region	18.4	16.1	4.9	7.5
Central Region				
Pskov region	13.5	13.2	4.4	4.4
Bryansk region	14.4	13.2	5.2	7.0
Vladimir region	13.4	12.1	4.6	5.5
Ivanovo region	19.8	19.1	4.6	9.1
Tver region	16.9	15.0	6.0	8.2
Kaluga region	29.4	25.1	5.4	14.1
Kostroma region	18.4	14.0	4.2	6.7
Moscow city	1.2	2.3	1.0	1.0
Moscow region	8.4	9.6	3.4	2.3
Orel region	14.2	15.1	5.3	7.4
Ryazan region	16.2	18.4	5.4	9.6
Smolensk region	24.0	18.1	5.4	7.5
Tula region	7.1	5.1	5.3	8.1
Yaroslavl region	20.2	17.3	2.7	8.2
Volga-Vyatka Region				
Kirov region	19.2	19.0	5.0	8.1
Nizhny Novgorod region	12.9	11.1	4.9	6.1
Mari Republic	19.0	20.2	2.9	5.3
Mordva Republic	15.9	14.1	3.7	3.5
Chuvash Republic	6.9	7.2	6.4	6.3
Central-Chernozem Region				
Belgorod region	11.4	8.6	3.4	7.6
Voronezh region	2.4	1.8	5.0	6.9
Kursk region	11.9	11.1	2.7	2.8
Lipetsk region	7.5	7.7	6.3	4.0
Tambov region	20.4	17.0	4.7	9.6
Volga Region				
Astrakhan region	21.7	23.6	5.7	13.3
Volgograd region	12.4	11.5	2.0	4.0
Pensa region	5.2	6.0	3.5	6.0
Samara region	8.2	8.6	4.1	4.0
Saratov region	16.6	16.3	1.8	10.0
Ulyanovsk region	13.6	13.0	4.1	3.7
Republic of Kalmykia	49.3	38.5	–	14.1
Tatarstan Republic	10.4	9.2	4.4	4.7

Continued

Appendix 10 Table II Continued

Region/Republic	Samples not complying with the coliform standard (% of total number of samples)		Samples with ≥ 20 coliforms/litre (% of the number of samples not complying with standards)	
	1991	1992	1991	1992
Northern Caucasus Region				
Krasnodar region	6.7	6.3	3.5	2.3
Adigey Republic	10.3	8.9	2.1	6.0
Stavropol region	13.9	–	5.5	
Karachay-Cherkes region	35.9	46.5	5.9	14.4
Rostov region	13.1	13.5	4.8	5.1
Dagestan Republic	21.0	15.8	5.3	6.5
Kabarda-Balkar Republic	17.0	19.2	3.7	5.4
North Osetia Republic	5.2	5.1	1.4	0.6
Checheno-Ingush Republic	19.2	13.9	3.7	5.4
Ural Region				
Kurgan region	13.5	12.5	4.7	3.8
Orenburg region	14.3	11.5	4.1	4.2
Perm region	10.5	9.0	5.1	5.1
Sverdlovsk region	9.7	7.1	5.4	3.1
Chelyabinsk region	13.5	–	4.2	–
Republic of Bashkiria	7.2	7.3	3.8	2.3
Republic of Udmurtia	7.2	7.4	4.3	3.3
West-Siberian Region				
Altai region	8.9	8.1	3.4	1.9
Gorny-Altai Republic	11.0	6.5	–	–
Kemerovo region	12.7	13.9	5.1	5.7
Novosibirsk region	13.4	9.7	4.9	4.7
Omsk region	17.2	23.0	7.4	11.1
Tomsk region	13.2	8.9	2.8	3.1
Tyumen region	12.3	11.4	4.3	5.4
East-Siberian Region				
Krasnoyarsk region	13.9	12.3	4.7	5.3
Khakassia Republic	6.9	10.4	7.6	4.2
Irkutsk region	5.0	33.9	3.7	–
Chita region	11.2	9.3	3.8	3.3
Buryatia Republic	13.0	10.7	4.2	4.2
Tuva Republic	16.0	10.5	2.7	7.2
Far-East Region				
Primorsk region	23.7	100.0	6.1	60.0
Khabarovsk region	21.2	17.2	4.7	9.1
Yevrei Republic	22.1	12.8	7.8	9.8
Amur region	21.7	17.2	6.0	7.6
Kamchatka region	12.8	9.9	5.4	4.4
Magadan region	11.3	5.6	3.3	1.3
Sakhalin region	16.5	16.2	3.9	7.2
Chukotka region	–	16.9	–	7.5
Yakutia (Sakha) Republic	12.9	8.4	5.6	3.2
Baltic Region				
Kaliningrad region	25.7	22.9	3.4	10.2
Total in Russia	**12.8**	**12.4**	**4.6**	**5.5**
Total in Russia, 1993		**11.2**		**4.8**

Source: Indicators, 1992

Chapter 11*

ACIDIFICATION

The influence of atmospheric pollution on the acidification of surface
waters became acute in the 1950s as a result of the acidification of precipitation in a number of industrialised countries. The dramatic impact this
has had on aquatic organisms is related to the fact that throughout their
evolution, pH has been one of the most stable environmental parameters
of water quality. The evolution of pH resulting from natural processes,
such as lake formation and its related biological successions, occurs on a
temporal scale that enables adaptation 'mechanisms' of biocoenoses to
develop fully. Since the rate of anthropogenic acidification is several times
faster than the adaptation capacity of biocoenoses acidification inevitably
results in the degradation of these aquatic communities.

The degradation of biocoenoses due to acidification of water bodies
commences with impacts on the least resistant organisms. Subsequently,
a 'chain reaction' occurs in the biocoenoses leading to degradation of
biological communities when established inter-population links are
destroyed. As a result of anthropogenic acidification, these changes in
aquatic community structure can be detected long before environmental
acidification exceeds the adaptation capacity of most species originally
present in the acidified water body. The most important economic impact
of the above processes is on fisheries, including sports and recreational
fisheries. Primarily, it reduces the stock of valuable commercial fish
species, such as salmon. A sharp reduction in species diversity and slower
metabolic processes indicate the profound irreversible impacts of anthropogenic acidification on water bodies.

Acidification of natural waters as a result of acid precipitation may
occur in areas with underlying acid rocks, and with soils of low buffering
capacity which are weakly saturated with bases (Glazovskaya, 1981,
1988). For anthropogenic acidification of surface waters, two factors are
generally necessary: the occurrence of acid precipitation and the natural
predisposition of the land to acidification. The predisposition of surface
waters to acidification is found in formerly glaciated regions, over
granite or other non-carbonated bedrock, in regions with quartz sands

---

*   *This chapter was prepared by G.M. Chernogaeva, V.A. Abakumov and
V.A. Kimstach*

and also in regions where soil is relatively old and considerably weathered and leached. When considering the natural predisposition of water bodies to acidification, it is also important to consider factors such as surface area, size of catchment and trophic status of the water body. Small rivers in wet regions rich in wetlands, or with the above-mentioned lithological properties, are mostly predisposed to acidification.

11.1   Acidity of atmospheric precipitation

Natural values of pH in precipitation are generally within the pH 5–6 range, although sometimes the values recorded have been lower at pH 4.6. The main pollutant causing an increase in the acidification of precipitation is sulphur dioxide (SO_2) emitted into the atmosphere in large quantities. In the process of acidification, SO_2 is converted into sulphuric acid. Equally important are nitrogen oxides (NO_x) which form nitric acid. At present, the typical acid precipitation composition is two thirds SO_2 and one third nitrogen oxide (NO). The greatest emissions of SO_2 are related to energy production (coal and fuel burning) and smelters, and those of NO are, as a rule, the result of motor exhaust emissions.

Monitoring of atmospheric deposition of acidifying compounds (solid and liquid) in the former Soviet Union did not show a direct correlation between concentrations of sulphates and pH values of atmospheric precipitation. This is due to the neutralisation of acids in precipitation in most industrial aerosols and, in the South, by the presence of natural dust.

Acidic precipitation has been noted mainly in regions adjacent to the western border. The absence of acidification over most of the territory is explained by the overall balance between anions and cations. The ion balance depends on the composition of anthropogenic emissions and the inputs into the atmosphere of calcium and magnesium carbonates from aeolian erosion. Thus, for processes of acidification, the main role is played not by the absolute concentration of sulphates but by the ionic balance, which should be preserved in those cases where the reduction in sulphur emissions is impossible.

A map showing the distribution of average annual pH values in precipitation over the European territory of the former USSR from 1956–61 is presented in Figure 11.1 (Drozdova et al., 1964). This map shows that in the 1950s and 1960s, acidification of precipitation on the above mentioned territory did not occur and there was a tendency for a pH increase from north to south which corresponded to trends in the natural qualities of surface waters.

A more detailed picture of spatial variation in pH from snow cover monitoring data during the period of maximum snow accumulation in 1988 is presented in Figure 11.2 (Vasilenko et al., 1989). Higher acidification (pH 4–5) occurs over very limited areas, often not marked on the map scale. These areas are situated along the western border of the former USSR in the regions of Kamenets-Podolsk, Rakhov, Khusta,

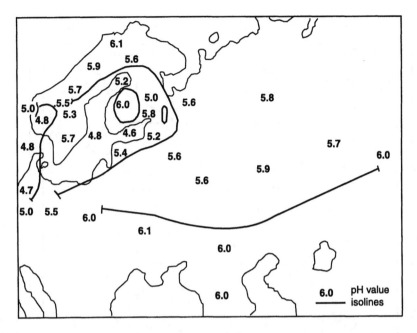

Figure 11.1 Average annual pH values in precipitation over the European part of the country, 1956–61 (After Drozdova *et al.*, 1964)

Velikiye Berjezny, Rava-Russkaya, Pruzhany and Liepaya. This acidification is caused by the influence of long-range transboundary air pollution. Regions with higher acidification are also found in the form of isolated spots in the industrial regions of Donbass, the Ukrainian industrial region, Lipetsk and the industrial region of the Urals. Similar low pH values are also found in some areas along the Ob gulf and in the village of Amderma; these are probably related to the influence of atmospheric transfer from the industrial regions of Vorkuta and Norilsk.

Areas with pH 5–6 occur over a large part of the north west of the country, in Karelia and other regions adjacent to the western border of the former USSR. These areas with pH 5–6 are typical of many industrial regions such as Donbass, Kuzbass, Krivbass, Moscow and Irkutsk. A wide band with weak acidity is noted from the south west to north east of the West-Siberian lowland and also the region of the Chukchi Peninsula which cannot be explained by industrial causes. These regions, especially in the West-Siberian lowland, are characterised by a low concentration of aerosols in the air due to the wetlands and floodplains in these territories. In these regions there is practically no industry, and Chukchi is surrounded by the Arctic Ocean (East Siberian and Bering Seas). Sulphate emissions from the North-American continent into Chukchi is not great but cannot be excluded. Another potential source is volcanic activity (e.g. Kamchatka). Weak acidification in these regions is

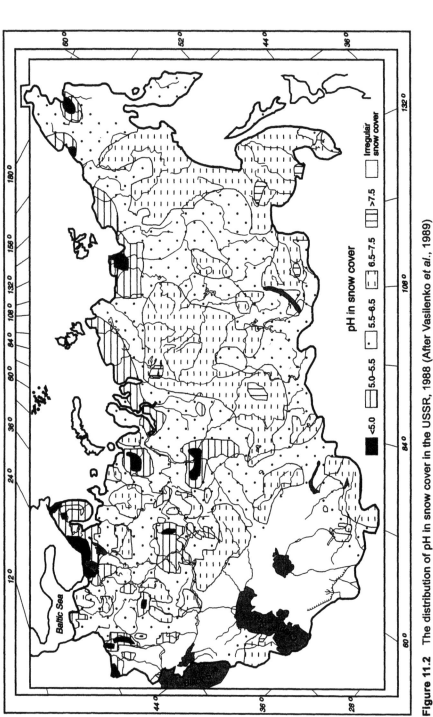

Figure 11.2 The distribution of pH in snow cover in the USSR, 1988 (After Vasilenko *et al.*, 1989)

related to an ion imbalance as a result of low concentrations of alkaline compounds in the clean air, as has been found in a number of relatively pristine regions around the world.

Almost over the entire country, with the exception of East Siberia, greater deposition of acidifying compounds occurs in cold periods with extreme values recorded in February–March. Low values of snow pH (4.2–5.0) are found in small areas, on the periphery of industrial regions with large emissions of sulphur. These areas include the north and east of the industrial regions of the Urals, around Leningrad, along the western border of the former USSR in Karelia, the Ukraine, Byelorussia, north of Kiev and the centre of the European territory. On the Kola Peninsula, values of pH 5.0 are recorded around non-ferrous metallurgical industries. In the Carpathians, snow cover acidification is caused by transboundary atmospheric transport and occurs over the entire mountain region.

11.2   Natural acidity of surface waters

At present, the pH in surface waters varies from 5.8 to 8.5. The lowest pH is characteristic of northern regions and the highest pH is characteristic of southern regions. The pH value for most river waters in winter is 6.8–7.4 and in summer it is 7.4–8.2. Rivers draining wetlands have higher concentrations of hydrogen ions and their pH often drops to 6.0. Southern rivers sometimes have pH values up to 8.5. Minimum pH values ≤ 5.0 are found in the small forest lakes of the Kola Peninsula.

The variety in chemical composition, including pH, in river and lake waters is due to the different environmental conditions of each catchment area. This variation in surface water composition in the former Soviet Union is related to general geographic trends described by Alekin in the 1950s (see Chapter 2). This is demonstrated by an increase in water mineralisation and a pH shift towards alkaline from north to south (Alekin, 1970).

Trends in pH and the influence of acidification on the biomass of benthic organisms and fish are shown in Tables 11.1 and 11.2 respectively (Kitaev, 1984). Trends in the quality of surface waters are related primarily to climatic factors which control the level of soil leaching, and to the presence of soluble salts in the soil. In addition to climatic factors, permafrost and the type of underlying rocks have to be considered (see Chapter 2). For the entire northern part of Siberia and for the vast territory of East Siberia, the presence of permafrost promotes homogeneity and low mineralisation of water in comparison with the European territory. In the northern part of the European territory and West Siberia, the occurrence of thick Quaternary deposits accounts for the trends in the qualitative composition of surface waters. These deposits determine, to a large extent, a decrease in mineralisation and pH compared with arid regions.

Table 11.1 Relationship between lake pH and benthic biomass and the percentage of lakes in five pH and biomass classes in taiga and mixed forest of the European territory

pH	Benthic biomass (g m^{-2})					Average biomass (g m^{-2})	No. of lakes
	< 1.25	1.25–2.5	2.5–5	5–10	> 10		
Taiga (% of lakes)							
< 5	57.2	42.8	–	–	–	1.33	7
5–6	53.3	21.4	13.1	10.3	1.9	2.44	107
6–7	26.2	27.6	19.8	7.7	8.7	3.38	505
7–8	26.3	21.1	25.6	18.2	8.8	4.28	308
> 8	8.9	11.1	22.2	31.2	26.5	7.44	45
Mixed forest (% of lakes)							
< 5	66.7	33.3	–	–	–	1.25	3
5–6	36.4	45.5	18.1	–	–	1.87	11
6–7	52.6	16.0	16.0	16.0	–	2.58	25
7–8	19.3	19.4	23.7	26.9	10.7	5.10	93
> 8	9.6	16.9	21.9	16.9	34.7	7.55	260

Source: Kitaev, 1984

Table 11.2 Relationship between lake pH and fish biomass and the percentage of lakes in five pH and biomass classes in taiga and mixed forest of the European territory

pH	Fish biomass (g m^{-3})					Average biomass (g m^{-3})	No. of lakes
	< 0.5	0.5–1	1–2	2–4	> 4		
Taiga (% of lakes)							
< 5	40.0	40.0	20.0	–	–	0.75	5
5–6	54.2	25.0	16.0	4.2	–	0.77	24
6–7	51.8	14.3	17.8	14.3	1.8	1.10	56
7–8	9.1	36.3	18.2	18.2	18.2	2.21	11
> 8	–	–	–	100.0	–	3.00	1
Mixed forest (% of lakes)							
< 5	–	–	–	–	–	–	–
5–6	11.1	66.7	11.1	11.1	–	1.04	9
6–7	–	14.3	42.8	28.6	14.3	2.46	7
7–8	5.6	11.1	16.7	55.5	11.1	3.02	18
> 8	–	33.4	–	11.1	55.5	4.60	9

Source: Kitaev, 1984

With a decrease in the size of drainage area, the influence of local conditions becomes greater. Lithology, hydrogeological conditions, relief, soils and vegetation define the variation in pH within natural vegetation zones. The influence of all these factors (climate, geology, size of lakes, soil-vegetation relationships, etc.) which produce a higher acidity of water has been observed in small Karelian forest lakes (lambas). This has led to numerous studies of natural and anthropogenic acidification of the aquatic environment and the influence of this process on the ecology of basins in the regions of Karelia and of regions similar in terms of physical and geographical conditions, such as the Kola Peninsula (Baranov, 1958; Kharkevich, 1960; Timboeva *et al.*, 1969; Meliantsev, 1978; Abakumov *et al.*, 1986a,b; Moiseenko and Yakovlev, 1990).

11.3   Lake water acidification

In the former USSR, there are over 2,850,000 lakes, 95 per cent of which are freshwater lakes. Almost all of the lake-water volume is concentrated in 15 lakes, each of them having a surface area of more than 1,000 km^2. In terms of chemical composition, almost all freshwater lakes are of the Ca^{2+}-HCO^{3-} type, with an almost neutral reaction (Nikanorov, 1989). The chemical composition of lake water is closely related to that of the rivers which feed the lakes. The influence of river water is greatest on small lakes with small drainage areas. In large lakes with large drainage areas, the influence of the hydrochemical composition of individual tributaries is reduced. The influence of atmospheric deposition directly on the lake surface is small. The smaller the drainage area, the greater the influence of local conditions (lithology, hydrology, relief, soil and vegetation) within natural vegetation zones and the greater the sensitivity of lake waters to acid precipitation. Thus, if there is a predisposition to acidification, acid precipitation exerts the greatest influence on small lakes with a surface area less than 1 km^2. Regions where a predisposition to acidification occurs include:

- Glaciated regions on granite or other siliceous bedrock, covered by alluviums and soils of similar lithology;
- Regions with a large proportion of quartz sand deposits;
- Regions with a thick layer of weathered and lixiviated soils.

The majority of lakes on the Kola Peninsula and in Karelia belong to the first type of region.

Monitoring of surface water quality has shown that the pH of lakes with a surface area > 1 km^2 is either close to a neutral pH or is in the pH range of 6.5–8.5, depending on natural conditions. In large lakes, pH tends to be ≥ 7.0 and the temporary variations in pH correlate more with the change in concentration of bicarbonate ions than with the concentration of sulphate ions (Miaemets *et al.*, 1980). In lakes such as Ilmen (Novgorod Region) and Beloye (Moscow Region), there is a tendency to be

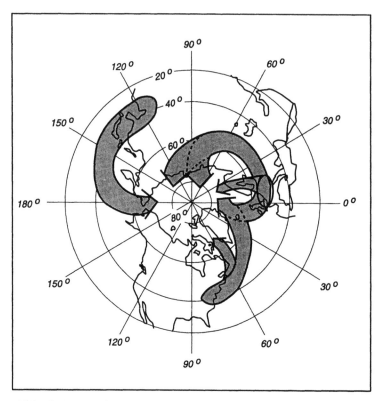

Figure 11.3 Pathways of atmospheric transfer of aerosols between moderate latitudes and the Arctic (After Nenonen, 1991)

slightly alkaline, and this is related to wastewater discharges containing pollutants (Pastukhova *et al.*, 1990).

11.4   Sources and effects of acidification on Karelia and Kola Peninsula lakes

The North West of the former Soviet Union, the Kola Peninsula and Karelia, is geochemically sensitive to acidity (EEA, 1994). Being on one of the pathways for long-range transboundary atmospheric transport of acidifying compounds from moderate latitudes to the Arctic (Figure 11.3), the North West is subject to increased fallout of sulphur and nitrogen compounds from the atmosphere. In addition, there are substantial local emission sources of acidifying compounds in the Murmansk Region and Karelia which contribute considerably to acidification of the environment in nearby territories of Russia, as well as in Finland and Norway. Two major smelters, 'Pechenganikel' and 'Severonikel' on the Kola Peninsula emit > 500,000 t a^{-1} of sulphur-containing gases into the atmosphere (Moiseenko, 1994). Atmospheric sulphur

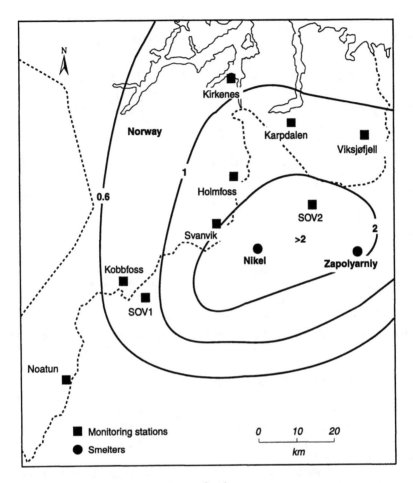

Figure 11.4 Total fallout of sulphur (g m^{-2} a^{-1}) in the border regions of Russia and Norway (After Acidification, 1991)

emissions from 'Pechenganikel', situated in the town of Nikel only 17 km from the Norwegian border, exceed the emissions of the whole of Norway by more than 3 times (SFT, 1994). The distribution of sulphur compound emissions near 'Pechenganikel' is shown in Figure 11.4. The sulphate ion concentration in lakes of the area corresponds to the distribution of atmospheric fallout (Figure 11.5).

It is important to note that, although the larger part of the Kola Peninsula (in terms of climatological and geological characteristics) is sensitive to acidification, 'Pechenganikel' is located in an area where the underlying rocks have a high buffer capacity (SFT, 1994). Therefore, in the Pechenga region, 81 per cent of investigated lakes had a pH value > 6.5 and only 6 per cent of lakes had pH < 6.0. Lakes with a high level of

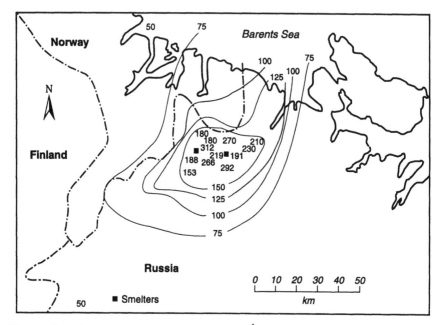

Figure 11.5   The distribution of sulphates (μeq l⁻¹) of non-marine origin in lakes in the border regions of Russia and Norway, 1989–90 (After Acidification, 1991)

acidity (pH < 5.0) are located mainly to the south of the smelter on other rock types. Unlike the Pechenga region, only 17 per cent of lakes investigated in the border region of Norway (Sør-Varanger County) had pH > 6.5, 47 per cent of lakes had pH < 6.0 and 10 per cent of lakes had pH < 5.0. This is a good example of the relationship between acidification and local conditions.

Atmospheric emissions of acidifying compounds from Karelia are roughly 4.5 times less than those from the Murmansk Region (Romanova, 1992). Although a number of gas emission sites from Karelia influence local acidification, its main source is due to long-range transboundary transport of pollution. This has been confirmed by the fact that, unlike the Kola Peninsula, the most acidified water bodies in Karelia are often located in regions which are far from local pollution sources (Figure 11.6).

The Karelian lamba lakes are acidified particularly strongly. These are typical forest lakes with a surface area < 1.0 km², with mineralisation as low as 15 mg l⁻¹ TDS and, as a rule, with acidic water of pH between 4.5 to 6.0. In a number of lakes, the lambas in particular, a reduction in pH (by 0.6–1.0) has occurred and the bicarbonate concentration has diminished in the past three decades (Abakumov *et al.*, 1986a,b, 1990b).

Alongside the pH changes of lake waters, specific aquatic biota composition and particularly features of lake zooplankton have also changed. There are between 7 and 42 species of zooplankton in the lakes

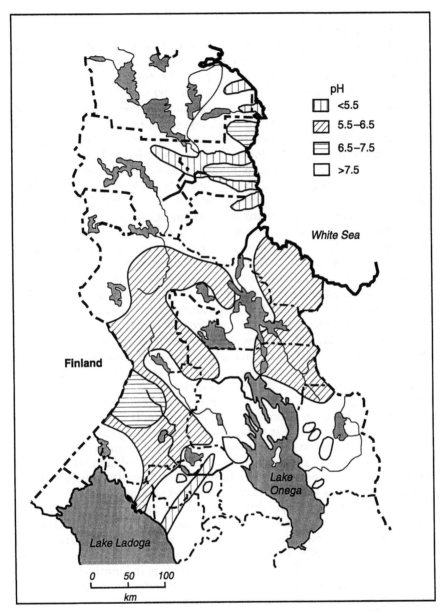

Figure 11.6   Acidity of surface waters in Karelia (After Livshits *et al.*, 1992)

investigated. Crustacean plankton dominate. More than half the
species inhabit all or most zoogeographical regions. The rest are typical
of northern lake fauna (typical of tundra and taiga zones):
namely *Holopedium gibberum, Bosmina obtusirostris, Daphnia cristata,*

Polyphemus pediculus, Bythotrephes cederstroemi, Eudiaptomus graciloides and *E. gracilis*.

Mean zooplankton counts vary from 4.9×10^3 to 21.1×10^3 individuals per m^3 and the average biomass varies from 0.11 to 0.38 g m^{-3}. In comparison with results obtained in 1965–68, the 1980 investigations revealed that the numbers of *Conochilus unicornis* and *Keratella* sp. had increased notably although, in general, zooplankton remained dominated by copepods and cladocerans. In some lakes, the number of species has been reduced by 11–17 per cent (Abakumov *et al.*, 1986a, 1987, 1991).

In most water bodies the dominant organisms have changed and the following species of crustacean plankton have decreased in number and biomass: *Daphnia hyalina galeata*, *Daphnia longispina*, *Bosmina longirostris*, *Eudiaptomus gracilis*, *Mesocyclops leuckarti* and *Heterocope appendiculata*. Many of these species prefer neutral or alkaline waters. Species associated with meso-acidic waters have become dominant, namely *Holopedium gibberum*, *Scapholeberis mucronata*, *Eurycercus lamellatus*, *Bosmina obtusirostris* var. *lacustris*, *Daphnia cristata* and *Polyphemus pediculus*. In some lakes there has been a rapid development of a few species of zooplankton tolerant of high acidity (pH 3.5–4.0), namely *Eucyclops serrulatus*, *Acantholeberis curvirostris*, *Scapholeberis mucronata* and *Graptoleberis testudinaria*.

The large cladoceran, *Limnosida frontosa*, a specimen typical of the aquatic fauna in the 1970s can no longer be found. Another large cladoceran, *Leptodora kindtii*, an important food source for plankton- and benthic-feeding fish, and in particular salmonids, has declined in number. Both species avoid high acidity and large amounts of humus substances.

In lakes with a high humus content, aluminium is bound in complexes and is less toxic than in clear water lakes. This probably explains the wide diversity of zooplankton in some lakes with coloured humic water (colour to 270°, pH around 5.5). Although the dominant and subdominant species in these lakes are acidophilic species, there are also a small number of species which are generally inhabitants of neutral and alkaline lakes.

Serious changes in zooplankton morphology have also occurred in the lakes investigated due to an increase in lake acidity. In some lakes the phenomenon of ecological and metabolic regression was discovered (Abakumov *et al.*, 1988, 1990a); for example, five species of zooplankton in a small acid lake in Karelia showed a decrease in length (Table 11.3). However, the dimensions of *Ceriodaphnia quadrangula*, an indicator organism of acidic lakes, increased slightly.

11.5 River water acidification

In general, the pH of river waters of the former Soviet Union is relatively neutral at present. The pH ranges between 6.0 and 8.1 in rivers of the Atlantic Ocean drainage basin, where calcium bicarbonate is dominant. In rivers of the Arctic Ocean drainage basin, pH ranges between 5.8 and

Table 11.3 Changes in individual sizes (mean ± S.E.) of crustacean plankton in Lake Lamba 3 (Karelia)

	Body length (µm)	
Species	1968	1983
Eudiaptomus graciloides	1,205 ± 15.0	1,050 ± 11.5
Cyclops scutifer	1,465 ± 14.0	1,375 ± 10.0
Holopedium gibberum	974 ± 26.0	925 ± 26.0
Diaphanosoma brachyurum	963 ± 26.0	850 ± 11.2
Daphnia longispina	1,463 ± 29.0	1,400 ± 14.0
Ceriodaphnia quadrangula	620 ± 15.5	700 ± 14.7

S.E. Standard error

7.6 with bicarbonate and calcium ions dominant. In rivers of the Pacific Ocean drainage basin, pH ranges between 6.1 and 7.5. The average annual pH in rivers of the southern slope of the European territory ranges between 6.9 and 8.4 and sodium and sulphate are the dominant ions. The same water type is found in rivers of the internal-drainage area of the Aral-Caspian region where pH ranges between 7.1 and 8.3.

In middle- and large-sized rivers, in spite of intra-annual variations, the average annual pH value remains close to neutral (6.5–8.5). At present the spatial distribution in pH reflects the general geographic trends in the quality of surface waters with a change to more alkali waters from humid regions to arid regions.

Rivers have also been affected by chemical inputs as a result of air pollution. Every year approximately 40×10^6 t of atmospheric sulphate is deposited on the territory of the former Soviet Union. A comparison of the spatial atmospheric load with the sulphate concentration in river waters, showed that the atmospheric component of sulphates in river water varies from 5 to 30 per cent on average, depending on the region (Table 11.4). In humid regions the sulphate river flux is double the sulphate concentration received from atmospheric fallout. This is especially true in forest areas because of the action of soil lixivation and flushing from forest litter. However, in arid regions the sulphate input from fallout is greater (Table 11.5).

Detailed calculations by Breslav *et al.*, (1985) and Chernogaeva (1986) have shown that, within the taiga zone, approximately 55.5 per cent of sulphates entered the river from soil flushing in spring, 18.5 per cent from groundwater discharge, 15 per cent from wet deposition and 11 per cent from snow melt. In the southern steppe, practically 90 per cent of sulphates entered rivers by groundwater discharge and 6 per cent by precipitation during the colder time of year. Thus, in southern Russia, the influence of precipitation on the sulphate concentration in river runoff was 2.5 times less than that in the north.

Table 11.4 Atmospheric component of the sulphate flux in selected river waters

River	Drainage area (km^2)	Water discharge (km^3 a^{-1})	Sulphate flux (10^6 t a^{-1})	Atmospheric component of sulphate (%)
Dnieper	482,000	45.3	2.4	29
Dniester	66,000	7.7	1.4	6
Western Dvina	84,700	20.2	0.6	16
Northern Dvina	350,000	119.0	6.7	5
Volga	1,380,000	260.2	19.9	14
Pechora	317,000	175.7	2.1	10
Yenisey	2,470,000	543.3	9.2	25
Amur	1,720,000	308.7	4.6	19

Source: Moshiashvili, 1992

Table 11.5 Sulphate fallout and outflow in the Ishim and Oka rivers

Station	Drainage area (km^2)	Fallout (t a^{-1}) Annual	Warm period	Cold period	Outflow (t a^{-1}) Annual	Warm period	Cold period
Oka River at Venderovo (high precipitation zone)	513	1,316	703	613	2,630	1,200	1,430
Ishim River at Udarnoe (arid zone)	202	688	275	413	137	19	118

Source: Chernogaeva et al., 1990a

Maximum values of sulphate runoff with snow-melt waters are typical of the southern regions of the forest zone (e.g. 0.47 t km^{-2} a^{-1} in the River Desna basin). This is related to a combination of factors such as high volume of snow, surface runoff and the maximum values of sulphate concentration in snow cover (Izrael et al., 1989; Vasilenko et al., 1989).

In regions formed of igneous rocks with low mineralisation (e.g. the Kola Peninsula and Karelia) and in regions of permafrost, the role of sulphates in atmospheric deposition is more important than in river fluxes (Table 11.6). Novikov (1990) noted that with the low and near-equal concentrations of chemical elements in precipitation, the difference between inflow and outflow in river waters was not so large. From the data of Rodin and Bazilevich (1965) and Dobrovolski (1973) it is possible to assume that sulphur retained in the drainage basin and not stored in the river bed enters the biological cycle. In the marsh ecotype of the Republic of Karelia and Arkhangelsk Region, an active process of sulphate reduction can be invoked during summer (Table 11.6).

Table 11.6 Average fluxes of major ions in the Republic of Karelia and
Arkhangelsk region

Region	Ca^{2+}	Mg^{2+}	Na^++K^+	HCO_3^-	SO_4^{2-}	Cl^-
Karelia						
Atmospheric input $(t\ km^{-2}\ a^{-1})$	0.24	0.09	0.89	1.52	1.88	0.92
River runoff $(t\ km^{-2}\ a^{-1})$	1.21	0.52	0.43	3.73	1.56	0.79
Arkhangelsk region						
Atmospheric input $(t\ km^{-2}\ a^{-1})$	0.23	0.09	0.98	1.81	1.65	1.48
River runoff $(t\ km^{-2}\ a^{-1})$	7.11	2.05	2.41	22.46	8.52	2.08

Source: Novikov, 1990

A comparison, based on 14 river basins in Estonia, of sulphate concentrations in surface snow-melt runoff from the watershed with the sulphate concentration in snow during the period of maximum snow cover showed that the ratio between them varied from 4.8 to 11.3 (Chernogaeva and Mohamedzhanova, 1990). Therefore the sulphate flushed during periods of floods and high water was not only due to the accumulation of sulphate in winter but also due to biological activity in the ecosystem.

The long-term monitoring of surface water acidification in the western regions of Russia, influenced by transboundary sulphur transport as well as local pollution sources, did not show a correlation between increased sulphate concentration in surface waters and pH (Chernogaeva *et al.*, 1990a,b).

11.6 Conclusion

To assess the influence of polluted precipitation on the acidity of surface waters, it is necessary to consider the ratio between natural and anthropogenic contributions and the acidic or alkaline components of surface waters. Important factors to consider are the natural acidity of soils, the presence of wetlands, the type of bedrock in the drainage area, the ratio between surface and groundwater components in river runoff and the presence of chemical substances in industrial atmospheric emissions which neutralise the acidity of precipitation.

The acidic precipitation which has been observed in some regions of the former USSR, with the presence of powerful sources of atmospheric SO_2 pollution, could lead to the acidification of some small rivers or lakes if overall conditions are appropriate.

The negative ecological consequences of atmospheric deposition of acidifying compounds can be observed in areas of low mineralisation and neutral or weakly acidic waters. These areas are mainly found in northern regions with podsol-marsh soils poor in bases and in acidic bedrocks which occur to the south in the European part of the country approximately to the latitude 55–58 °N and in Siberia to 50 °N. The acidification of natural waters in these areas will not occur in places with carbonate rock and limestone (e.g. some regions of Karelia, Byelorussia, the Urals, the Siberia-Lena-Aldan plateau, and areas along the Rivers Vilyuy and Olenek). In addition, the acidification of water bodies as a result of acid precipitation will not occur in territories south of 50–55 °N because of the high concentrations of alkaline elements in natural waters which neutralise the acidity.

11.7   References

(All references are in Russian unless otherwise stated).

Abakumov, V.A., Igolkina, E.D. and Svirskaya, N.L. 1986a Methods of prognosis of ecological consequences of water acidification. In: *Complex Methods of Environment Quality Control*. Moscow.

Abakumov, V.A., Kazakov, Y.E. and Svirskaya, N.L 1986b Hydrobiological consequences of anthropogenic lake acidification. Complex global monitoring of the biosphere state. Proceedings of the III International Symposium, Volume 3. Gidrometeoizdat, Leningrad, 221–226.

Abakumov, V.A., Igolkina, E.D. and Svirskaya, N.L. 1987 Ecological consequences of natural waters acidification due to acidic precipitation. Gidrometeoizdat, Moscow, 17 pp.

Abakumov, V.A., Svirskaya, N.L. and Igolkina, E.D. 1988 Structural changes in the zooplankton community under anthropogenic lake acidification. Ecological problems of the Baikal region. The III All-Union Scientific Conference Reports, Volume 3. Irkutsk University, Irkutsk, 3–33.

Abakumov, V.A., Igolkina, E.D. and Svirskaya, N.L. 1990a Trends in the ecological consequences of acidification of continental water bodies. In: *Assessment of Present and Future Environmental Conditions*. Transactions of the Institute of Applied Geophysics, Issue 76, Gidrometeoizdat, Moscow, 92–100.

Abakumov, V.A., Svirskaya, N.L. and Igolkina, E.D. 1990b Lake acidification and aquatic biocoenoses in Karelia. In: *Monitoring of Environmental Background Pollution*. Issue 6, Gidrometeoizdat, Leningrad, 179–185.

Abakumov, V.A., Svirskaya, N.L. and Igolkina, E.D. 1991 Zooplankton modification due to anthropogenic acidification. Poryarnaye Pravda, Murmansk, 153–157.

Acidification of surface waters, and nickel and copper in water and lake sediments in the Russian-Norwegian border areas 1991 Progress Report for 1989–1990, Apatity, Oslo, November 1991, 20 pp. (In English).

Alekin, O.A. 1970 The role of chemical composition of precipitation in the formation of river water composition. In: *Meteorology and Hydrology of Inland Waters and Oceanology*. Gidrometeoizdat, Leningrad, 140–149.

Baranov, I.V. 1958 Lake classification in the Karelian and Kola Peninsula limnological regions. Fisheries of Karelia, Issue 7, Petrozavodsk, 18–29.

Breslav, E.I., Vasilenko, V.N., Nazarov, I.M., Fridman, S.D. and Chernogaeva, G.M. 1985 The influence of snow cover in surface water sulphate pollution. *Meteorology and Hydrology*, **5**, 27–36.

Chernogaeva, G.M. 1986 The influence of anthropogenic sulphates on natural mineralisation of surface waters of the European territory of the USSR. In: *Monitoring of Environmental Background Pollution*. Issue 3, Gidrometeoizdat, Leningrad, 162–168.

Chernogaeva, G.M. and Mohamedzhanova, D.F. 1990 Sulphate budget in surface waters of Estonia. *Monitoring of Environmental Background Pollution*. Issue 6, Gidrometeoizdat, Leningrad, 186–193.

Chernogaeva, G.M., Petrukhin, V.A. and Gromov, S.A. 1990a Budget of pollutants in river basins of relatively pristine regions of the USSR. *Monitoring of Environmental Background Pollution*. Issue 7, Gidrometeoizdat, Leningrad, 171–175.

Chernogaeva, G.M., Fridman, S.D. and Vasilenko, V.N. 1990b Monitoring of natural water acidification in the western regions of the USSR. *Monitoring of Environmental Background Pollution*. Issue 6, Gidrometeoizdat, Leningrad, 175–178.

Dobrovolski, V.V. 1973 Methods and experiences in the assessment of chemical element transport in landscapes. In: *Topological Research Aspects in Substances Behaviour in Geosystems*. Irkutsk.

Drozdova, V.M., Petrenchuk, O.P., Seleznyova, E.S. and Svistov, P.F. 1964 The chemical composition of precipitation on the European territory of the USSR. Gidrometeoizdat, Leningrad, 209 pp.

EEA 1994 *European Rivers and Lakes. Assessment of their Environmental State*. European Environment Agency Environmental Monographs, Copenhagen, 122 pp. (In English).

Glazovskaya, M.A. 1981 Geochemistry of natural and anthropogenically impacted landscapes. In: *Pollutant Fluxes in Landscapes and the State of Ecosystems*. Nauka, Moscow, 7–41.

Glazovskaya, M.A. 1988 The geochemistry of natural and developed landscapes of the USSR. Vysshaja shkola, Moscow, 117 pp.

Izrael, Y.A., Nazarov, I.M., Pressman, A.Y., Rovinsky, F.Y. and Ryaboshapko, A.G. 1989 *Acid Rain*. Gidrometeoizdat, Leningrad, 269 pp.

Kharkevich, N.S. 1960 Data on small forest lakes (lambas) in Karelia. Petrozavodsk, 72–68.

Kitaev, S.P. 1984 Ecological base of lake bioproductivity in different natural vegetation zones. Nauka, Moscow, 207 pp.

Livshits, V.K., Lozovik, P.A., Filatov, N.N. and Sorokina, N.V. 1992 Aquatic ecological problems in Karelia. Ecological Transactions No. 1. Water basin state and atmospheric protection in Eastern Finland and the Republic of Karelia. Yoesuu, 34–41.

Meliantsev, V.G. 1978 Fishery results after fertilisation of small forest lakes of North West Russia. Inter-Institute collection of papers, Petrozavodsk, 7–16.

Miaemets, A.K., Bude, S.D. and Milman, I.S. 1980 The state of the Pskov-Chud lakes based on the results of a complex hydrological expedition. In: *Problems of USSR Large Lakes Research.* Gidrometeoizdat, Leningrad, 159–162.

Moiseenko, T.I. 1994 Acidification and critical loads in surface waters: Kola, Northern Russia. *Ambio,* **23**(7), 418–424. (In English).

Moiseenko, T.I. and Yakovlev, V.A. 1990 Anthropogenic transformation of aquatic ecosystems in the north of the Kola Peninsula. Nauka, Leningrad, 220 pp.

Moshiashvili, L.D. 1992 The influence of atmospheric sulphate pollution on the concentration of sulphur compounds in river water. In: *Geographic and Hydrologic Research.* Russian Academy of Science, Moscow, 139–145.

Nenonen, M. 1991 Report on acidification in Arctic countries. Anthropogenic acidification in a world of natural extremes. In: *The State of the Arctic Environment.* Arctic Centre, University of Lapland, Rovaniemi, 7–81. (In English).

Nikanorov, A.M. 1989 *Hydrochemistry.* Gidrometeoizdat, Leningrad, 351 pp.

Novikov, A.P. 1990 Geochemical features of surface waters with natural and anthropogenic anomalies and the chemical element balance in the taiga zone of the Russian plain. In: *Landscape and Geochemical Research of Anthropogenic Systems.* USSR Academy of Science, Moscow, 19–35.

Pastukhova, E.V., Golvina, M.V. and Karpova, N.N. 1990 The preliminary evaluation of acidifying compounds' influence on the surface waters of the European territory of the USSR. Scientific paper presented at the Sixth Task Force Meeting in Sweden, 25–26 October, 1990. Swedish Environment Protection Agency, Environmental Impact Assessment Department, 1–46. (In English).

Rodin, L.E. and Bazilevich, N.I. 1965 The dynamics of organic substances and the biological cycle of ash elements and nitrogen in the main vegetation types of the world. USSR Academy of Science, Moscow, 101 pp.

Romanova, N. 1992 The state of contamination of atmospheric air on the territory of the Republic of Karelia. Ecological Transactions No. 1. Water basin state and atmospheric protection in Eastern Finland and Karelia. Yoesuu, 46–47.

SFT 1994 Water pollution in the border areas of Norway and Russia. Summary Report 1989–1992, Oslo, 48 pp. (In English).

Timboeva, Z.P., Daukshta, A.S. and Tsukurs, T.M. 1969 On the productivity of some small acidotrophic lakes. The 15th Baltic Conference, Minsk, 51–64.

Vasilenko, V.N., Nazarov, I.M. and Fridman, S.D. 1989 Sulphur and nitrogen budget in the atmosphere of the USSR and their influence on forest ecosystems and soil. Monitoring problems and environmental protection. Proceedings of the 1st Soviet-Canadian Symposium, Tbilisi, USSR. Gidrometeoizdat, Leningrad, 191–199.

Chapter 12[*]

RIVER FLUXES OF DISSOLVED AND SUSPENDED SUBSTANCES

Gross river flux of dissolved and suspended substances into sea basins is defined as the amount of substances transported by the river to the land/sea boundary. Net river flux is determined as the amount of substances transported across a boundary (GESAMP, 1987). In defining gross and net river flux, it is important to determine the exact boundary between the river and the sea. This boundary is the area where there is no influence of sea water i.e. in a river cross section where flow direction is unilateral. Riverine material entering the zone of river and sea mixing undergoes quantitative and qualitative changes (e.g. sedimentation of suspended matter, aggregation and removal from solution of dissolved and colloidal substances, sorption and desorption). As a result, river material that reaches the open sea is significantly transformed (Gordeev, 1983). Therefore net river flux is the amount of substances which overcomes the geochemical and physical barrier of the river–sea zone and enters the estuary, shelf zone and continental slope, and finally reaches the deep sea.

The aim of this chapter is to assess gross river flux from the former Soviet Union without analysing the processes of transformation of river material in zones of mixed fresh and salt waters. According to the type of data available (in some cases only a few samples or even a single sample were taken over time, or sampling was performed unevenly over time), the simplest but relatively reliable algorithm to assess discharge was used as follows (Dolan et al., 1981; Temporary Guidelines, 1983):

$$L = \frac{1}{n}\sum_{j=1}^{n}C_j Q_j$$

where L = flux of substance, n = number of representative periods, C_j = average concentration in the representative period, and Q_j = water discharge in the representative period. The choice of representative period (month, quarter, season) was made taking into consideration seasonal variability of chemical composition of water and water discharge. As the

---

[*] *This chapter was prepared by V.V. Gordeev and V.V. Tsirkunov. Section 12.5 was prepared by V. Shlychkova, L. Korotova and N. Matveeva*

data were characterised by uneven intra- and inter-annual sample selection, different procedures of interpolation were applied.

The average long-term flux of substances has been taken as the arithmetic average for each year. The total flux of substances into sea basins was assessed using data for average long-term river water discharge and total water runoff into sea basins (see Chapter 1). The flux of substances into specific sea basins (e.g. Black and Azov Seas) has been computed on the basis of the individual river discharges (e.g. Southern Bug, Dnieper, Don) and on the basis of a coefficient accounting for the share of these rivers to the total water discharge into the sea basin. It was therefore assumed that undocumented rivers had concentrations similar to sampled ones.

12.1   Water discharge and flux of suspended matter

Estimates of water discharge and sediment load have been made in numerous publications (e.g. Lisitzina, 1978; Ivanov, 1985; Dzhaoshvili, 1986; Water Resources of the USSR, 1987; Tarasov et al., 1988). Data on rivers of the North Atlantic Ocean and especially of the Laptev and East Siberian Seas have been supplemented in recent years (Gordeev and Sidorov, 1993; Gordeev et al., 1996, 1997). The present chapter uses average long-term data on water and solid discharge over a period of 10 years, 1981–90 (Table 12.1). Where there was a lack of information from 1981–90, data prior to that period were used, especially for suspended matter discharge.

The former Soviet Union is divided into the basins of the Arctic, Pacific and Atlantic Oceans and endorheic regions (the Caspian Sea, Central Asia and Kazakhstan). Rivers of the Arctic Ocean discharge approximately 67 per cent of total water flow from the former USSR and only 32 per cent of suspended matter (Figure 12.1A). Specific discharge and total suspended solids (TSS) values (Table 12.1, Figures 12.1A,B) indicate that the river runoff decreases and the suspended load increases from the basins of the Arctic and Pacific to the Atlantic Oceans and endorheic basins.

In comparison with world river discharge values, rivers of the former Soviet Union account for 12 per cent of water discharge but only 2.3 per cent of sediment load. World water discharge has been estimated between 35,000 and 41,000 km^3 a^{-1} (e.g. Milliman, 1991), and solid discharge between 15×10^9 and 20×10^9 t a^{-1} (Milliman and Meade, 1983; Lvovich et al., 1989).

Over the past 30 years or more, river water discharge on the whole has not changed significantly under the influence of natural or anthropogenic factors (Figure 12.2). From 1955 to 1990, water discharge varied within 4–6 per cent of the average long-term values. However the greatest changes occurred in the endorheic basin of Kazakhstan and

Table 12.1 Average water discharge and discharge of suspended matter from rivers of the former USSR, 1981-90

River	Basin area (10³ km²)	Water discharge			Discharge of suspended matter			Source
		(km³ a⁻¹)	(m³ s⁻¹)	(l s⁻¹ km⁻²)	(10⁶ g a⁻¹)	(g m⁻³)	(t km⁻² a⁻¹)	
Arctic Ocean								
Barents and White Seas								
Onega	57	16.1	510	8.9	0.3	18	4.9	1, 2
Northern Dvina	357	106	3,360	9.4	3.8	35	10.6	
Mezen	78	27	860	11.0	0.9	32	11.1	
Pechora	322	140	4,450	13.7	13.5	80	32.4	
Whole basin	1,236	430	13,600	11.0	22.0	50	17.5	
Kara Sea								
Ob	2,990	404	12,800	4.3	16.5	38	6.4	1, 2
Yenisey	2,580	630	20,000	7.7	5.9	10	2.3	
Whole basin	6,248	1,344	42,600	6.8	33.2	22	5.0	
Laptev Sea								
Khatanga	364	85.3	2,700	7.4	1.7	20	4.6	2
Anabar	100	17.3	550	5.5	0.4	24	4.1	
Olenek	219	35.8	1,140	5.2	1.1	31	5.1	
Lena	2,486	525	16,650	6.7	17.6	34	7.1	
Omoloi	39	7	220	5.7	0.13	18	3.2	
Yana	238	34.3	1,090	4.6	3.5	103	14.8	
Whole basin	3,693	745	23,600	6.5	25.1	34	6.9	
East-Siberian Sea								
Indigirka	362	61	1,930	5.3	12.9	210	35.6	2
Alazeya	68	8.8	280	4.1	0.7	80	10.2	
Kolyma	660	132	4,190	6.3	16.1	120	24.3	
Whole basin	1,296	250	7,930	5.9	33.6	134	25.0	
Chukchi Sea								
Amguema	29.6	9.2	290	9.7	0.5	6.0	1.8	2
Whole basin	101	20.4	645	6.8	0.7	34	7.4	
Whole Arctic Ocean basin	**12,572**	**2,790**	**88,400**	**7.0**	**115**	**40**	**8.8**	

Continued

Table 12.1 Continued

River	Basin area (10³ km²)	Water discharge			Discharge of suspended matter			Source
		(km³ a⁻¹)	(m³ s⁻¹)	(l s⁻¹ km⁻²)	(10⁶ g a⁻¹)	(g m⁻³)	(t km⁻² a⁻¹)	
Pacific Ocean								
Bering Sea and the Pacific Ocean								
Anadyr	200	60	1,900	9.5	1.8	59	17	1, 3
Kamchatka	55.9	33.1	1,050	18.0	0.7	90	12	
Whole basin	570	202	6,400	11.2	7.9	56	13.9	
Sea of Okhotsk								
Amur	1,855	344	10,900	5.9	24.9	71	13.4	1, 3
Whole basin	1,695	564	17,900	10.5	35.1	70	12.5	
Sea of Japan								
Whole basin	124	37.2	1,180	9.8	0.96	26	10.0	4
Whole Pacific Ocean basin	**2,389**	**803**	**25,500**	**106**	**44**	**57.1**	**12.7**	
Atlantic Ocean								
Baltic Sea								
Neva	281	78.5	2,500	8.8	0.82	10	2.9	1, 3
Daugava	87.9	20.4	650	7.4	0.47	25	5.3	
Neman	98.2	19.7	620	6.4	0.66	39	6.7	
Whole basin	568	143	4,500	8.0	2.3	14.2	3.6	
Black and Azov Seas								
Dniester	72.1	10.7	340	4.7	0.41	100	6.1	1, 3
Southern Bug	63.7	3.4	110	1.7	0.5	21	19	1, 3
Dnieper	504	53.4	1,690	3.4	2.33	51	3.4	1, 3
Don	422	20.7	660	1.6	2.16	68	5.1	5
Kuban	57.9	13.4	420	9.1	1.3	136	22	5
Kodory	2.03	3.94	125	61.6	0.82	208	410	6
Inguri	0.89	1.58	50	56	0.13	82	146	
Chorokh	22.1	9.0	285	12.9	8.25	920	3,930	
Rioni	13.4	12.9	409	30.5	6.8	640	507	
Whole basin	1,367	137	430	5.5	27.8	230	20.9	1, 3
Whole Atlantic Ocean basin	**1,935**	**280**	**8,900**	**6.5**	**30.1**	**107**	**15.3**	

Continued

Table 12.1 Continued

River	Basin area (10³ km²)	Water discharge (km³ a⁻¹)	Water discharge (m³ s⁻¹)	Water discharge (l s⁻¹ km⁻²)	Discharge of suspended matter (10⁶ g a⁻¹)	Discharge of suspended matter (g m⁻³)	Discharge of suspended matter (t km⁻² a⁻¹)	Source
Endorheic basins								
Caspian Sea								
Volga	1,380	254	8,120	5.8	10.1	39	7.4	1, 5
Terek	43.2	8.4	266	6.2	16.3	1,940	380	7
Kura	188	15.9	504	2.7	36	2,000	–	8
Sulak	15.2	4.0	127	8.3	1.64	410	110	8
Samur	3.75	1.39	44	11.8	8.0	4,000	4,900	3
Ural	237	8.7	280	1.2	4.4	390	20	9
Whole basin	2,800	292	9,300	3.3	98	320	33	
Central Asia and Kazakhstan								
Syr Darya	219	2.16	68	0.3	5	760	23	10
Amu Darya	309	5.19	165	0.5	65	4,200	210	10
Whole basin	2,577	32.5	1,030	0.4	70	2,150	28	10
All Endorheic regions	5,300	325	10,300	1.9	168	500	31	
Whole Former USSR	22,196	4,198	133,000	6.1	357	84	15	

Sources:
1   Georgievsky, see Chapter 1
2   Gordeev et al., 1996
3   Lisitzina, 1978
4   Ignatova and Chudaeva, 1983
5   Bronfman, 1981
6   Dzhaoshvili, 1986
7   Alexeevsky, 1993
8   Mikhailova, 1993
9   Krasnozhon and Mazavina, 1988
10 Kuznetsov et al., 1987

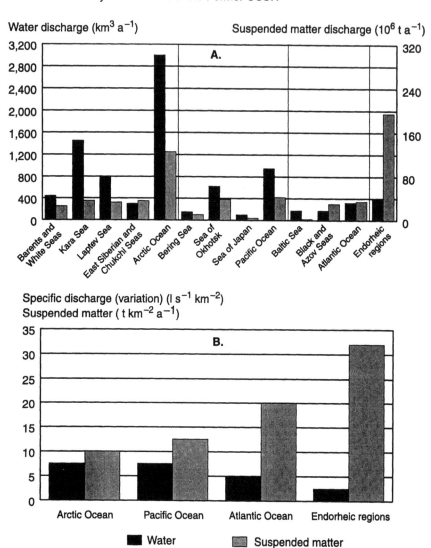

Figure 12.1 **A**. River discharge and suspended matter discharge; **B**. Specific discharges of water and suspended matter into various sea basins

Central Asia. By the mid-1980s the river discharge of the largest rivers of this region, the Syr Darya and Amu Darya, had decreased more than 10 times and in some years was completely absent (see Chapters 5 and 15). On the whole, during the above mentioned period, discharge in this basin was reduced by almost four times (Figure 12.2).

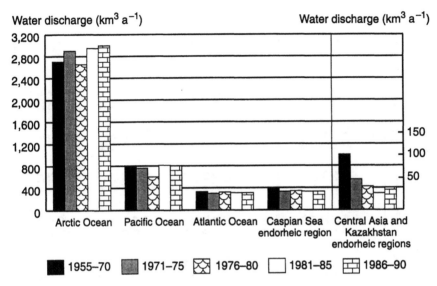

Figure 12.2 Variability in water discharge to various sea basins from 1955–90

12.2   Transport of major ions and silicon

12.2.1 Major ions

The first assessments of riverine ionic fluxes from the former Soviet Union commenced just after World War II and are considered as classical works conducted by Alekin and other researchers of the Hydrochemical Institute (Alekin, 1948, 1950, 1951; Eremenko *et al.*, 1953; Zenin and Protsenko, 1961; Alekin and Brazhnikova, 1964). Subsequently, there were other publications which gave assessments of ion fluxes both from the whole of the former USSR and from its regions (Voronkov, 1970; Kirsta, 1975; Peleshenko, 1975; Nikanorov and Tsirkunov, 1984, 1991; Koreneva and Konovalov, 1987; Konovalov *et al.*, 1991; Peleshenko and Khilchevsky, 1991; Tsirkunov *et al.*, 1992, etc.). Average data for individual ions and total dissolved solids (TDS) fluxes from the rivers of the former Soviet Union are presented in Table 12.2.

It is important to be able to assess error in estimates of the chemical fluxes of rivers. On average, errors made in calculating ion fluxes to sea basins are evaluated to ± 20–30 per cent. Assessments of major ions in river discharge are more exact compared with assessments of discharge of organic matter, nutrients and trace elements in the dissolved, and especially in the suspended, state. This is particularly true for TSS when the sampling frequency is low, i.e. less than monthly, and for trace elements when the analytical accuracy is low, and for both nutrients and TSS when seasonal variability is high.

Table 12.2 Average long-term annual flux of major ions of rivers in the former USSR

River (Station)	Period of observation	Water discharge (km³ a⁻¹)	Discharge of major ions (10⁶ t a⁻¹)							TDS specific discharge (t km⁻² a⁻¹)	Source
			Ca²⁺	Mg²⁺	Na⁺+K⁺	HCO₃⁻	SO₄²⁻	Cl⁻	TDS		
Arctic Ocean											
Barents and White Seas											
Onega (Porog)	1981–89	16.2	0.43	0.13	0.08	1.30	0.54	0.07	2.50	8.90	1
Northern Dvina (Ust-Pinega)	1981–90	103	2.83	0.68	0.82	8.03	3.65	0.56	16.7	48.0	1
Pechora (Ust-Tsilma)	1955–90	110	1.21	0.32	0.45	4.47	0.98	0.42	7.91	31.9	1
Mezen (Malonisogorskaya)	1963–89	20.1	0.29	0.08	0.13	1.24	0.21	0.07	2.05	36.3	2
Whole basin		431	8.20	2.08	2.55	25.9	9.27	1.93	50.2	40.6	
Kara Sea											
Ob (Salekhard)	1955–90	405	7.85	2.15	3.05	33.1	3.56	2.74	51.2	21.1	1
Yenisey (Igarka)	1952–90	588	9.72	2.24	4.95	35.5	6.86	5.73	64.8	26.5	1
Whole basin		1,344	23.8	5.93	10.8	92.6	14.1	11.4	157.0	25.1	
Laptev Sea											
Khatanga	1981–90	85.3	1.07	0.31	0.91	4.09	0.48	1.07	7.93	21.8	3
Anabar	1981–91	17.3	0.17	0.04	0.02	0.54	0.07	0.03	0.87	8.7	3
Olenek	1981–90	35.8	0.72	0.15	0.15	2.6	0.17	0.17	3.97	18.1	3
Lena	1981–90	525	8.42	2.32	6.14	27.3	6.46	8.98	59.6	23.4	3
Yana	1981–90	34.3	0.21	0.05	0.13	0.71	0.31	0.08	1.49	6.30	3
Whole basin		745	11.0	2.96	7.50	36.3	8.00	10.4	76.2	20.6	
East-Siberian and Chukchi Seas											
Indigirka	1981–90	61	0.70	0.14	0.06	1.73	0.83	0.11	3.57	9.9	3
Kolyma	1981–90	132	1.35	0.25	0.24	3.43	0.94	0.30	6.51	9.9	3
Amguema	1981–90	9.2	0.03	0	0.02	0.07	0.03	0.01	0.16	5.4	3
Whole basin		292.4	3.02	0.56	0.46	7.57	2.61	0.61	14.8	10.6	
Whole Arctic Ocean basin		**2,812**	**46**	**11.5**	**21.4**	**163**	**33.4**	**24.7**	**299**	**23.8**	
Pacific Ocean											
Bering Sea											
Anadyr (Snezhnoye)	1962–88	31.1	0.13	0.05	0.10	0.52	0.15	0.08	1.03	9.72	1
Kamchatka (Klyuchi)	1957–90	25.5	0.24	0.12	0.21	1.24	0.36	0.12	2.3	50.4	1
Whole basin		202	1.32	0.60	1.11	6.27	1.82	0.71	11.9	20.8	

Continued

Table 12.2 Continued

River (Station)	Period of observation	Water discharge (km³ a⁻¹)	Discharge of major ions (10⁶ t a⁻¹)							TDS specific discharge (t km⁻² a⁻¹)	Source
			Ca^{2+}	Mg^{2+}	$Na^{+}+K^{+}$	HCO_3^-	SO_4^{2-}	Cl^-	TDS		
Okhotsk and Japan Seas											
Amur-Komsomolsk	1981–90	317	2.83	0.72	1.57	9.24	1.98	1.24	17.1	9.9	2
Penzhyna (Kamenskoe)	1961–90	22.9	0.9	0.03	0.03	0.36	0.16	0.03	0.75	10.5	2
Whole basin, Sea of Okhotsk		564	4.85	1.25	2.66	15.9	3.56	2.11	29.6	17.5	
Whole basin, Sea of Japan		33.7	0.21	0.05	0.14	0.81	0.2	0.08	1.47	11.8	4
Whole Pacific Ocean basin		**800**	**7.82**	**2.21**	**4.61**	**27.3**	**6.45**	**3.5**	**51**	**21.3**	
Atlantic Ocean											
Baltic Sea											
Gauy (Valmiera)	1981–88	1.55	0.08	0.01	0.01	0.03	0.05	0.01	0.46	74.8	5
Venta (Kuldiga)	1981–88	2.58	0.18	0.04	0.03	0.63	0.11	0.04	1.03	124	5
Lielupe (Yelgava)	1981–88	2.14	0.21	0.05	0.04	0.55	0.26	0.06	1.19	99.2	5
Luga (Kingisep)	1986–91	4.34	0.14	0.06	0.03	0.53	0.08	0.05	0.88	68.8	2
Pregolia (Gvardeysk)	1981–91	3.00	0.25	0.02	0.04	0.80	0.17	0.10	1.38	101	2
Daugava (Daugavpils)	1981–88	14.2	0.50	0.10	0.12	1.88	0.34	0.14	3.09	47.8	5
Neman (Sovetsk)	1981–90	17.3	1.20	0.39	0.29	4.30	1.04	0.48	7.80	85.0	1
Neva (Novosaratovka)	1981–91	83.4	0.89	0.25	0.64	2.6	1.25	0.63	6.16	21.9	1
Whole basin		143	3.87	1.04	1.35	12.7	3.66	1.69	24.6	43.3	1
Black and Azov Seas											
Inguri (Divary)	1950–75	4.64	0.11	0.02	0.04	0.36	0.09	0.02	0.63	199	2
Bzyb (Dyirkhva)	1950–86	2.94	0.08	0.01	0.01	0.29	0.04	0.01	0.44	313	2
Rioni (Sakochakidze)	1981–90	15.4	0.54	0.18	0.11	1.98	0.35	0.07	3.31	249	2
Kodory (Varcha)	1966–90	4.53	0.08	0.01	0.02	0.27	0.054	0.01	0.45	225	2
Dniester (Bendery)	1981–90	8.54	0.59	0.22	0.54	1.94	0.99	0.54	4.81	72.8	1
Southern Bug (Alexandrovka)	1981–90	2.77	0.23	0.08	0.11	0.89	0.19	0.12	1.66	35.9	2
Kuban (Zaitsevo Koleno)	1979–85	4.47	0.25	0.05	0.21	0.70	0.47	0.13	1.79	39.1	2
Don (Aksai)	1981–90	19.3	1.60	0.70	2.82	3.87	4.14	2.83	15.6	37.1	2
Dnieper (Kakhovka reservoir)	1956–68	42.8	1.89	0.44	0.63	6.53	1.34	0.73	11.6	24.4	6
Whole basin		137	11.2	3.57	9.36	35.1	15.9	9.31	61.6	45.0	
Whole Atlantic Ocean basin		**280**	**12.1**	**3.65**	**8.16**	**38.6**	**15.4**	**8.47**	**86.2**	**44.5**	

Continued

Table 12.2 Continued

River (Station)	Period of observation	Water discharge (km³ a⁻¹)	Discharge of major ions (10⁶ t a⁻¹)							TDS specific discharge (t km⁻² a⁻¹)	Source
			Ca²⁺	Mg²⁺	Na⁺+K⁺	HCO₃⁻	SO₄²⁻	Cl⁻	TDS		
Endorheic basins											
Caspian Sea											
Volga (Verkhnelebyazhye)	1981–89	255	11.5	3.66	6.88	29.1	17.3	10.3	79.0	58.1	1
Terek (Kargalinskaya)	1981–90	7.23	1.49	0.14	0.30	1.36	0.91	0.34	3.63	97.1	2
Ural (Guriev)	1981–90	8.7	0.46	0.21	0.75	1.49	0.77	0.91	4.59	19.4	1
Kura (Salyany)	1981–90	12.2	0.83	0.41	3.10	2.46	3.37	1.35	10.4	55.4	2
Whole basin		293	26.3	4.6	11.4	35.8	23.2	13.3	101	36.4	
Kazakhstan and Central Asia											
Ily (Ushzharma)	1981–90	12.8	0.55	0.20	0.60	0.21	1.09	0.37	4.92	38.1	2
Turgai (Tusum)	1951–90	0.35	0.02	0.01	0.06	0.06	0.05	0.10	0.30	5.68	2
Syr Darya (Kazalinsk)	1981–90	2.38	0.31	0.17	0.52	0.42	1.84	0.36	3.59	16.4	2
Amu Darya (Kyzyl Djar)	1981–90	5.04	0.40	0.22	0.71	0.58	1.92	0.99	4.98	21.5	2
Chu (Ulanbel)	1956–75	0.65	0.06	0.03	0.10	0.22	0.20	0.06	0.67	9.90	2
Nura (Romanovskoye)	1950–75	0.53	0.03	0.15	0.08	0.10	0.09	0.09	0.41	9.09	2
Murgab (Takhta-Bazar)	1954–70	1.56	0.11	0.03	0.08	0.27	0.23	0.07	0.78	22.5	2
Tedjen (Puland-Khatun)	1951–70	1.04	0.08	0.06	0.03	0.23	0.42	0.29	1.35	19.1	2
Whole basin		32.5	2.07	1.16	2.90	2.78	7.77	3.10	22.6	9.04	
All Endorheic regions		325	28.3	5.7	14.2	38.5	30.7	16.3	123	23.2	
Whole territory of the USSR (exorheic regions)		3,892	66.8	17.5	34.7	228	56.6	37.4	436	25.8	
Whole territory of the USSR (exorheic and endorheic regions)		4,217	95.6	23.4	49.3	269	87.7	54.1	563	25.4	

TDS   Total dissolved solids

Sources:
1 Hydrochemical Institute, independent estimates by V. Koreneva and V. Tsirkunov, unpublished data
2 Tsirkunov, unpublished data
3 Gordeev et al., 1996
4 Ignatova and Chudaeva, 1983
5 Tsirkunov et al., 1992
6 Denisova, 1981

Total ionic discharge (10^6 t a^{-1})

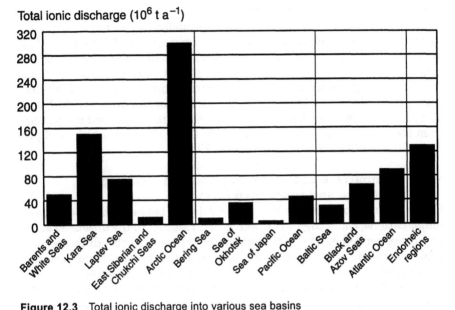

Figure 12.3 Total ionic discharge into various sea basins

As a rule, the most dominant ion discharged is bicarbonate. The exception to this is found in the Syr Darya, Amu Darya, Kura, Don and other rivers of arid regions where sulphates are the dominant anions discharged. In most rivers, calcium is the dominant cation discharged. River ranking for separate ions fluxes can be considerably different from the TDS flux. For example, the 10 rivers which have the greatest sulphate load are the Volga (17×10^6 t), Yenisey (6.9×10^6 t), Lena (6.5×10^6 t); Don (4.1×10^6 t), Northern Dvina (3.6×10^6 t), Ob (3.6×10^6 t), Kura (3.3×10^6 t), Amu Darya (1.9×10^6 t), Amur (1.9×10^6 t) and Syr Darya (1.8×10^6 t). The latter two rivers contained similar sulphate loads in the 1980s although water discharge of the Amur was more than 130 times greater than that of the Syr Darya.

The greatest TDS discharges are into the Kara Sea basin, because it has the highest river water discharge, and into the basin of the Arctic Ocean (Table 12.2 and Figure 12.3). Due to low mineralisation of river waters in the basin of the Pacific Ocean, ion discharge into this ocean is the lowest compared with the other ocean basins, although in terms of water discharge this basin has the second highest discharge (after the Arctic Ocean basin). Relatively high mineralisation of river waters in the basin of the Atlantic Ocean, and especially in endorheic regions, is the reason for high ion discharges in these basins. In the endorheic basins, high ion discharges, which have reached 130×10^6 t a^{-1}, have led to the accumulation of a significant amount of salts over the last 15–20 years.

Values of specific discharge (expressed in t km^{-2} a^{-1}) of TDS are given in Table 12.2. These values show the high variability of this important indicator of chemical weathering in the USSR, ranging from 2–3 t km^{-2} a^{-1} in both regions of ultra-fresh waters and in arid regions with very low surface runoff (< 1 l s^{-1} km^{-2}) up to > 200 t km^{-2} for mountain rivers in the Caucasus and parts of Central Asia. For most large river basins this value varies from 15 to 60 t km^{-2} a^{-1}. The average for the former USSR was 23 t km^{-2} a^{-1}. On the whole, the annual transport of dissolved substances from the former USSR was approximately 430 × 10^6 t a^{-1}, which is 12 per cent of the world discharge (Meybeck, 1979).

12.2.2 Comparison of Ca, Mg, Na, K and Si discharge in dissolved and suspended form

Under natural conditions, there is a correlation between the discharge of major cations, Ca^{2+}, Mg^{2+}, Na$^+$ and K$^+$ and silicon, in the dissolved state and in suspended matter, although they show considerable differences in their geochemical behaviour. For this reason, the two forms are not considered separately. Unfortunately, chemical analysis of river suspended matter has been much less frequent than the analysis of the dissolved forms. Therefore assessments of annual discharge of major cations and silicon (Table 12.3), based on the average suspended load and on the average chemical analysis of river suspensions, were sometimes based on only one or a few analyses. Subsequently, errors in such calculations may be much higher than errors in assessments of discharge of dissolved substances. However, variations in elements such as Si, Al, Fe, Ti in river suspension are not significant. The proportion of Si in suspended matter of various rivers globally differs by not more than 30–35 per cent (Martin and Meybeck, 1979).

Error calculations for the discharge of major elements in suspension were carried out for the basin of the White and Barents Seas. If it is assumed that the error made in calculating river suspended load is 50 per cent (Gordeev et al., 1996) and deviation from average concentration of silicon in suspension in the rivers Northern Dvina, Mezen and Pechora is 30 per cent (Morozov et al., 1974), the error in estimating particulate silicon discharge in this basin would be about 60 per cent. This error is probably close to the minimum when computing the discharge of most suspended chemical substances.

Even when considering the high errors in the assessment of discharge of suspended substances, in all basins the discharge of Ca, Mg, Na+K in the dissolved state was considerably greater than their discharge in suspension, except for mountainous Central Asia (Table 12.3 and Figure 12.4). For Si the situation was reversed. In endorheic regions the share of solid discharge for all elements analysed increased. This is due primarily to the high discharge of suspended matter (Figure 12.1). The share of discharge of particulate calcium increased greatly (from

Table 12.3 Discharge of major cations and silicon in dissolved and suspended state

River	Water discharge (km³ a⁻¹)	TSS (10³ t a⁻¹)	Ca (10³ t a⁻¹) Diss.	Susp.	Mg (10³ t a⁻¹) Diss.	Susp.	Na+K (10³ t a⁻¹) Diss.	Susp.	Si (10³ t a⁻¹) Diss.	Susp.	Source
Arctic Ocean											
Barents and White Seas											
Northern Dvina	106	3.8	2,800	87	680	64	800	200	217	930	1
Mezen	27	0.9	510	14	130	14	–	–	–	–	
Pechora	140	13.5	1,420	118	370	167	430	590	333	440	
Whole basin	372	22.0	7,090	270	1,800	302	2,210	1,000	1,010	5,470	
Dissolved discharge (as % of total)			96.3		86		68.8		15.6		
Kara Sea											
Ob	408	16.5	8,200	170	2,210	155	3,820	70	1,500	4,170	1
Yenisey	589	5.9	10,530	–	2,350	–	4,950	–	1,710	–	
Whole basin	1,345	33.2	23,800	350	5,930	310	10,800	1,440	450	8,400	
Dissolved discharge (as % of total)			98.5		95		88.2		35		
Laptev Sea											
Lena	525	17.6	820	760	2,320	210	6,140	550	1,030	5,160	2
Whole basin	745	25.1	11,000	1,100	2,960	300	7,500	780	1,310	7,350	
Dissolved discharge (as % of total)			91		91		90.6		15		
East-Siberian Sea											
Whole basin	250	33.6	2,280	1,450	460	400	440	1,040	370	9,850	2
Dissolved discharge (as % of total)			61		53.5		30		3.7		
Chukotsk Sea											
Whole basin	20.4	0.7	70	30	10	8	50	22	95	205	2
Dissolved discharge (as % of total)			70		56		69.5		32		
Whole Arctic Ocean basin	**2,741**	**115**	**44,800**	**3,200**	**11,200**	**1,320**	**20,900**	**4,280**	**7,340**	**31,300**	
Dissolved discharge (as % of total)			**93.3**		**89.5**		**83**		**19**		

Continued

Table 12.3 Continued

River	Water discharge (km³ a⁻¹)	TSS (10³ t a⁻¹)	Ca (10³ t a⁻¹)		Mg (10³ t a⁻¹)		Na+K (10³ t a⁻¹)		Si (10³ t a⁻¹)		Source
			Diss.	Susp.	Diss.	Susp.	Diss.	Susp.	Diss.	Susp.	
Pacific Ocean											
Bering Sea and Pacific Ocean											
Whole basin	147	7.9	960	70	440	30	810	280	1,400	1,800	2
Dissolved discharge (as % of total)			93.2		93.6		74		43.8		
Sea of Okhotsk											
Whole basin	723	35.1	6,220	300	1,600	130	3,500	530	1,780	8,530	2
Dissolved discharge (as % of total)			95.4		92.5		74		17.3		
Sea of Japan											
Whole basin	37.2	0.96	230	8	50	4	150	33	220	230	3
Dissolved discharge (as % of total)			96.6		93		82		49		
Whole Pacific Ocean basin	**760**	**44**	**7,410**	**380**	**2,090**	**160**	**4,370**	**1,540**	**3,400**	**10,560**	**2**
Dissolved discharge (as % of total)			**95**		**93**		**74**		**24.4**		
Atlantic Ocean											
Baltic Sea											
Whole basin	151	2.3	4,070	60	1,090	28	1,420	57	160	590	4
Dissolved discharge (as % of total)			98.5		97.5		96.1		21.3		
Black and Azov Seas											
Whole basin	131	27.8	81,600	850	2,600	280	6,820	880	530	7,500	2, 5
Dissolved discharge (as % of total)			90.6		90.3		88.6		6.6		
Whole Atlantic Ocean basin	**282**	**30.1**	**12,200**	**910**	**3,690**	**310**	**8,240**	**940**	**690**	**8,100**	
Dissolved discharge (as % of total)			**93**		**92.2**		**89.8**		**7.8**		

Continued

Table 12.3 Continued

River	Water discharge ($km^3\,a^{-1}$)	TSS ($10^3\,t\,a^{-1}$)	Ca ($10^3\,t\,a^{-1}$) Diss.	Susp.	Mg ($10^3\,t\,a^{-1}$) Diss.	Susp.	Na+K ($10^3\,t\,a^{-1}$) Diss.	Susp.	Si ($10^3\,t\,a^{-1}$) Diss.	Susp.	Source
Endorheic regions											
Caspian Sea											
Whole basin	305	98	27,300	5,500	4,770	1,000	11,900	2,900	850	25,400	2
Dissolved discharge (as % of total)			83.2		82.6		80.4		3.2		
Central Asia and Kazakhstan											6
Syr Darya	2.16	5	280	470	150	76	520	150	15	1,280	
Amu Darya	5.19	65	510	5,500	310	1,230	700	1,760	25	15,900	
Whole basin	32.5	70	2,070	5,970	1,160	1,300	2,900	1,900	200	1,720	
Dissolved discharge (as % of total)			33		47		60.4		1.1		
Whole basin of the endorheic regions	338	168	29,400	11,400	5,930	2,300	14,800	4,800	1,050	42,600	
Dissolved discharge (as % of total)			72		72		75.5		2.4		
Whole basin of the USSR territory	4,121	357	93,800	15,900	22,900	4,100	48,300	11,600	12,480	92,600	
Dissolved discharge (as % of total)			85.5		84.8		80.6		12		

TSS   Total suspended solids

Sources:
1   Morozov et al., 1974
2   Gordeev, 1983
3   Ignatova and Chudaeva, 1983
4   Emelyanov and Pustelnikov, 1975
5   Goliadze et al., 1975
6   Kuznetsov et al., 1987

Dissolved discharge
(% of total)

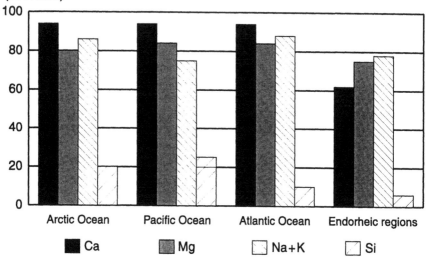

Figure 12.4   Proportion of dissolved discharge as a percentage of the total (dissolved and suspended) discharge to various sea basins for Ca, Mg, Na+K and Si

5–10 per cent in northern and central regions of the country up to 40 per cent for endorheic areas and to 60–90 per cent in the Rivers Amu Darya and Syr Darya). This was due, not only to high particulate discharge, but also to the high calcium carbonate content in suspension which is characteristic of regions with an arid climate. Natural factors (e.g. arid climate, low discharge of dissolved silicon, high discharge of suspended particles) also explain the significant role of solid discharge of silicon whereas the role of anthropogenic factors was minimal (Meybeck, 1981).

12.2.3 Changes in ion transport

Practically all types of human activity lead to an increase in salt concentrations in water bodies. The increase in concentrations of these ions, especially chlorides and sulphates, may be used as an indicator of the duration of anthropogenic impact. Estimates of changes in total ion transport and variations in chloride and sulphate transport for periods between 1950 and 1990 are given in Table 12.4 for the largest rivers exposed to salinisation. A comparison with the values of chloride and sulphate fluxes for both periods shows that the increase in ion flux caused by human activity has exceeded the natural flux in many river basins by 2–3 times.

The greatest increase in absolute values of sulphate flux was found in the basins of the Rivers Volga, Don, Kura and Neva; and for chloride flux in the basins of the Volga, Don, Amur, Kura and Neman (Table 12.4). The maximum relative increase in fluxes of these ions, reaching 200–300 per cent, were obtained for rivers in the basins of the Baltic (Luga, Neman, Neva, Daugava), Black and Azov (Don, Southern Bug, Dniester) and Caspian (Kura, Terek) Seas. The total increase in sulphate transport for all rivers listed in Table 12.4 (except rivers Syr Darya and Amu Darya) is 14.3×10^6 t, which is 71 per cent of the estimated background value of sulphate transport of these rivers in the 1950s. The greatest absolute increase in total ion fluxes occurred in the basins of the Volga (17.8×10^6 t), Don (3.2×10^6 t), Kura (2.9×10^6 t), Neman (2.4×10^6 t) and Neva (2.2×10^6 t). On the whole, for the rivers studied (except for the Amu Darya and Syr Darya) the increase in total ion flux was 34.7×10^6 t or 28.1 per cent of the level of total flux of these rivers in the 1950s.

A comparison of data relating to the increase in ion fluxes with official data of chloride and sulphate discharge from all types of wastewaters into river basins, as provided by Main Statistics (1990, 1991), showed that for most river basins the increase in sulphates and chlorides surpassed the evaluated anthropogenic inflows by 3–30 times or more. Rare cases where the increase in ion fluxes was less than the values obtained for wastewater discharge, can be explained by the fact that the latter take into consideration all discharges in the territory of the basin, including delta areas downstream of the last water quality monitoring station as well as direct discharges into estuarine regions and seas. For example, in the basin of the Kura river the values of Cl^- and SO_4^{2-} discharged with wastewaters were higher (in comparison with the increase in ion fluxes), because the data included 2 km^3 a^{-1} of irrigation drainage waters discharged directly into the Caspian Sea through the Main Shirvan and Mugan-Salyan drainage systems (with water mineralisation of 5 g l^{-1} and 16 g l^{-1} respectively) (Stepanov and Chembarisov, 1978; Main Statistics, 1990, 1991).

Such significant discrepancies between observed ion fluxes and wastewater discharge data for most river basins may be due to: i) an insufficient account of all types of wastewaters and their salt content, and ii) the high contribution of non-point sources of inflow of salts into water catchments. It is difficult to evaluate which of these has the most significant impact. For many river basins, non-point sources are probably the major sources of salts into water bodies.

The highest increase in specific ionic discharge has been noted for sulphates in the basins of the Kura, Neman, Pregolia and Dniester, and for chlorides in the basins of the Terek, Dniester, Pregolia and Don. Changes in specific sulphate discharge for large rivers of the eastern part of the European territory were comparable with data reported by Hem (1993) for the St Lawrence and Mississippi rivers. Thus, the increase in specific sulphate discharge from the beginning of this century in the St Lawrence river was 3.5–5.4 t km^{-2}, which corresponds to similar values for the Volga (4.4 t km^{-2}) and Don (5.4 t km^{-2}). The

Table 12.4   Trends in SO_4^{2-}, Cl^- and TDS loads (L) and their specific discharge (M) and their variations (ΔL and ΔM) in selected river catchments

River[1]	Period of observation	Water discharge (km³)	Measured ion discharge (10^3 t) SO_4^{2-} L	ΔL	ΔL(%)	Cl^- L	ΔL	ΔL(%)	TDS L	ΔL	ΔL(%)
Arctic Ocean											
White Sea basin											
Onega	1955–62	17.0	386	154	39.9	50.3	21.9	43.5	2,230	291	13
	1988–89	16.2	540			72.2			2,521		
Pacific Ocean											
Sea of Okhotsk basin											
Amur	1950–58	306	1,545	437	28.3	408	832	204	16,238	892	5.49
	1981–90	317	1,982			1,240			17,130		
Atlantic Ocean											
Baltic Sea basin											
Neva	1946–56	76.1	394	856	217	341	287	84.2	3,990	2,170	54.4
	1981–90	83.4	1,250			628			6,160		
Daugava	1947–60	15.0	112	229	204	24	116	483	2,030	1,060	52.2
	1981–89	14.2	341			141			3,090		
Lielupe	1947–60	2.3	169	91	53.8	11.7	45.4	388	791	399	50.4
	1981–88	2.14	260			57.1			1,190		
Neman	1955–62	18.0	303	740	244	133	352	265	5,360	2,430	45.3
	1981–90	17.1	1,043			485			7,790		
Pregolia	1956–62	3.45	81	85	105	38	59	155	1,177	199	16.9
	1981–91	3.0	166			97			1,376		
Luga	1956–65	2.83	17.5	65.7	375	11	41.3	375	468	416	88.9
	1986–91	4.34	83.2			52.3			884		
Black and Azov Sea basins											
Dniester	1955–62	7.93	493	500	101	212	323	152	3,218	1,587	49.4
	1981–90	8.54	993			535			4,805		
Southern Bug	1950–58	2.28	64	122	191	38	77	203	1,043	612	58.7
	1981–90	2.77	186			115			1,655		
Kuban	1957–61	6.06	343	125	36.4	77.7	50.3	64.7	1,710	83	4.85
	1979–85	4.47	468			128			1,793		
Don	1950–58	22.3	1,878	2,258	120	1,117	1,716	154	8,785	3,215	36.6
	1981–90	19.3	4,136			2,833			15,570		
Rioni	1963–70	13.3	282	68	24.1	46.2	22.2	48.1	2,860	450	15.7
	1981–90	15.4	350			68.4			3,310		
Endorheic regions											
Caspian Sea basin											
Volga	1950–58	241	11,352	5,948	52.4	5,476	4,849	88.6	61,240	17,800	29.1
	1981–89	255	17,300			10,325			79,000		
Terek	1950–58	7.30	655	258	39.4	135	208	154	268	95	35.4
	1981–90	7.23	913			343			363		
Kura	1950–58	15.5	1,271	2,098	165	921	428	46.5	7,508	2,878	38.3
	1981–90	12.2	3,369			1,349			10,386		
Kazakhstan and Central Asia											
Amu Darya	1955–62	38.5	5,143	–3,228	–62.8	3,456	–2,468	–71.4	19,688	–14,713	–74.7
	1981–90	5.04	1,915			988			4,975		
Syr Darya	1950–60	16.3	4,162	–2,324	–55.8	1,013	–655	–64.7	11,943	–8,353	–69.9
	1981–90	2.38	1,838			358			3,590		
Ily	1950–58	15.3	766	320	41.8	274	94	34.3	4,746	169	3.56
	1981–90	12.8	1,086			368			4,915		

TDS   Total dissolved solids

[1] Calculations are for the same catchment areas as in Table 12.2

Table 12.4 Continued

River[1]	Period of observation	Total effluents[2] (10³ t) SO$_4^{2-}$	Cl⁻	SO$_4^{2-}$ M	ΔM	Cl⁻ M	ΔM	TDS M	ΔM
Arctic Ocean									
White Sea basin									
Onega	1955–62	–	–	6.93	2.76	0.9	0.4	40.0	5.3
	1988–89			9.69		1.3		45.3	
Pacific Ocean									
Sea of Okhotsk basin									
Amur	1950–58	28.1	24.0	0.89	0.26	0.24	0.48	9.39	0.51
	1981–90			1.15		0.72		9.90	
Atlantic Ocean									
Baltic Sea basin									
Neva	1946–56	14.4	16.6	1.40	3.05	1.21	1.02	14.2	7.7
	1981–90			4.45		2.23		21.9	
Daugava	1947–60	20.3	93.2	1.74	3.55	0.38	1.81	31.4	16.4
	1981–89			5.29		2.19		47.8	
Lielupe	1947–60	–	–	14.1	7.6	0.98	3.78	65.9	33.3
	1981–88			21.7		4.76		99.2	
Neman	1955–62	230	777	3.3	8.1	1.45	3.83	58.4	26.5
	1981–90			11.4		5.28		84.9	
Pregolia	1956–62	–	–	5.96	6.24	2.79	4.34	86.5	14.5
	1981–91			12.2		7.13		101	
Luga	1956–65	–	–	1.37	5.13	0.86	3.23	36.6	32.5
	1986–91			6.5		4.09		69.1	
Black and Azov Sea basins									
Dniester	1955–62	75.6	90.2	7.46	7.54	3.21	4.88	48.7	24
	1981–90			15.0		8.09		72.7	
Southern Bug	1950–58	70.7	23.7	1.39	2.64	0.82	1.67	22.6	13.2
	1981–90			4.03		2.49		35.8	
Kuban	1957–61	53.2	29.8	7.47	2.73	1.69	1.1	37.3	1.8
	1979–85			10.2		2.79		39.1	
Don	1950–58	900	476	4.47	5.38	2.66	4.09	20.9	16.2
	1981–90			9.85		6.75		37.1	
Rioni	1963–70	–	–	21.2	5.1	3.47	1.67	215	34
	1981–90			26.3		5.14		249	
Endorheic regions									
Caspian Sea basin									
Volga	1950–58	3,160	6,423	8.35	4.35	4.03	3.56	45.0	13.1
	1981–89			12.7		7.59		58.1	
Terek	1950–58	21.6	44.0	17.5	6.9	3.61	5.56	7.17	2.54
	1981–90			24.4		9.17		9.71	
Kura	1950–58	4,101	2,917	6.78	11.2	4.9	2.29	40.0	15.4
	1981–90			18.0		7.19		55.4	
Kazakhstan and Central Asia									
Amu Darya	1955–62	1,285	1,553	22.3	–14.0	15.0	–10.7	85.2	63.7
	1981–90			8.29		4.28		21.5	
Syr Darya	1950–60	943	251	19.0	–10.6	4.63	–3.0	54.5	38.1
	1981–90			8.39		1.63		16.4	
Ily	1950–58	0.06	–	5.94	2.48	2.12	0.73	36.7	1.4
	1981–90			8.42		2.85		38.1	

[2] Indirect official estimates of point sources loads from Main Statistics, 1990, 1991

Source: Tsirkunov, unpublished data

specific sulphate discharge for the Mississippi is 7.7–9.6 t km^{-2} which is slightly less than values for the Don (9.9 t km^{-2}) and Volga (12.8 t km^{-2}). Long-term changes in concentrations of major ions in these rivers are similar. It has already been noted that the main increase in concentrations of major ions in USSR rivers was from the 1950s to the mid-to-end of the 1970s, while in the 1980s concentrations of major ions did not change significantly. The same situation applied to the large North American rivers. This supports Hem's conclusion (Hem, 1993) which was made for the St Lawrence, that additional inflow of salts to river catchments is well balanced by their increasing discharge into seas.

12.3   Fluxes of organic matter and nutrients

12.3.1 Organic matter

The first assessments of organic matter (OM) fluxes from the former USSR were made in the mid-1950s by Skopintsev and Krylova (1955) and Alekin and Brazhnikova (1964). Later, new studies were reviewed in a number of publications (Maltseva *et al.*, 1978; Romankevich and Artemyev, 1985; Tarasov *et al.*, 1985; Smirnov *et al.*, 1988). All data concerning organic matter fluxes are based on the results of the Goskomhydromet monitoring system in which, from 1976 to 1980, total organic matter in unfiltered samples was determined initially by the susceptibility of organic matter to oxidation by permanganate and later by bichromate (i.e. COD). To define organic matter concentrations, the values of permanganate and bichromate oxidisability were recalculated into total organic carbon (TOC) by multiplying by a corresponding conversion factor, and OM concentration was taken as 2 × TOC. Separate measurements of dissolved (DOC) and particulate organic carbon (POC) and the estimation of their fluxes were carried out for some rivers (Table 12.5). Data on average long-term fluxes of total OM measured during 1981–90 for 54 rivers of the former USSR are presented in Table 12.6 and Figure 12.5. It should be noted that the series of observations for OM fluxes is shorter and discontinuous compared with the observation periods for water discharge, and that they are of varying time periods for different rivers. The total error in the assessment of OM fluxes by some rivers is 20–50 per cent (Table 12.6). The error in OM fluxes for the entire sea basins may be higher. Comparing the fluxes of dissolved and suspended organic carbon showed that, except for some mountain rivers which are characterised by a very high load of suspended matter (e.g. the Chorokh), the flux of DOC always exceeded the flux of POC (Table 12.5).

In the 1980s, on average, approximately 70 × 10^6 t a^{-1} of OM were transported from the former USSR into sea basins, of which about 4.4 × 10^6 t a^{-1} were from the endorheic regions of Central Asia and the Caspian Sea basin. On the whole, the rivers of the former USSR

Table 12.5 Average concentrations and input of dissolved and particulate organic carbon to seas by various rivers

River	DOC (mg l⁻¹)	POC (mg l⁻¹)	POC (% from SM)	DOC+POC (mg l⁻¹)	$\frac{DOC}{DOC+POC}$ (%)	Discharge (10^6 t a^{-1})			Source
						DOC	POC	DOC+POC	
Northern Dvina	20.1	3.2	23.3	23.4	86	2.22	0.35	2.57	1
Pechora	12.7	0.3	16.0	13.0	98	1.64	0.04	1.68	1
Ob	9.1	0.9	2.0	10.0	91	2.78	0.27	3.05	2
Lena	6.6	1.1	3.8	7.7	86	4.56	0.74	5.30	3
Kuban	1.9	2.8	1.8	4.7	40	0.03	0.04	0.07	1
Don	4.2	1.7	3.8	5.9	71	0.41	0.05	0.46	1
Dnieper	4.9	0.9	16.4	5.8	84	0.26	0.05	0.31	1
Rioni	0.9	1.8	0.7	2.7	33	0.01	0.02	0.03	4
Chorokh	1.4	27.8	0.7	29.2	5	0.01	0.24	0.25	4

DOC   Dissolved organic carbon
POC   Particulate organic carbon
SM   Suspended matter

Sources:
1   Artemyev, 1993
2   Nesterova, 1960
3   Cauwet and Sidorov, 1994
4   Glagoleva, 1959

Table 12.6 Average long-term annual discharge of organic matter and nutrients into various sea basins

River	Organic matter $(10^3 \, t \, a^{-1})$	NO_3-N $(10^3 \, t \, a^{-1})$	PO_4-P $(10^3 \, t \, a^{-1})$
Arctic Ocean			
Barents and White Seas			
Onega	560	1.3	0.18
Northern Dvina	3,020	7.1	1.1
Pechora	2,330	5.7	4.5
Whole basin	9,590	27	9.6
Kara Sea			
Ob	6,600	42	28
Yenisey	11,900	53	6.3
Whole basin	24,700	118	55
Laptev Sea			
Lena	5,900	34	2.86
Yana	735	0.36	0.10
Whole basin	10,100	52.4	4.5
East-Siberian Sea			
Indigirka	457	3.6	0.76
Kolyma	980	5.1	1.22
Whole basin	2,150	13.0	3.0
Chukotsk Sea			
Whole basin	268	0.9	0.72
Whole Arctic Ocean basin	**46,900**	**210**	**72.7**
Pacific Ocean			
Bering Sea			
Whole basin	3,100	16.2	7.6
Sea of Okhotsk			
Amur	5,085	31.5	10.3
Whole basin	8,810	53.5	17.7
Sea of Japan			
Whole basin	790	3.37	0.83
Whole Pacific Ocean basin	**12,700**	**73.2**	**26.1**
Atlantic Ocean			
Baltic Sea			
Neva	1,280	23.0	0.94
Daugava	533	17.4	1.02
Neman	752	11.2	2.45
Whole basin	3,570	88.4	5.58
Black and Azov Seas			
Dniester	173	21.0	0.79
Dnieper	842	24.4	3.50
Don	392	18.6	1.54
Kuban	145	13.9	0.21
Whole basin	2,320	140	9.43
Whole Atlantic Ocean basin	**5,900**	**228**	**15.0**

Continued

Table 12.6   Continued

River	Organic matter (10^3 t a^{-1})	NO$_3$-N (10^3 t a^{-1})	PO$_4$-P (10^3 t a^{-1})
Endorheic regions			
Caspian Sea			
Volga	3,265	103	4.90
Kura	104	24.7	0.11
Terek	140	9.1	0.66
Whole basin	3,860	150	6.65
Central Asia and Kazakhstan			
Syr Darya	37.3	3.14	0.037
Amu Darya	60.7	3.16	0.055
Whole basin	506	42.6	0.89
All endorheic regions	**4,365**	**191**	**7.55**
Whole USSR territory	**69,900**	**704**	**121**

Source: Smirnov *et al.*, 1988; updated to 1990 by Smirnov for this publication

Organic matter discharge (10^6 t a^{-1})

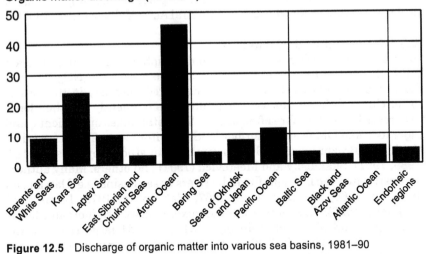

Figure 12.5   Discharge of organic matter into various sea basins, 1981–90

transported into the ocean about 5.5 per cent of the global flux of TOC estimated at 6×10^{14} g C a^{-1} (Meybeck, 1982).

Romankevich and Artemyev (1985) assessed the total flux of DOC and POC from the former USSR by multiplying the average DOC and POC concentration (expressed as a percentage of TSS) with water discharge and annual suspended load respectively. They obtained a value of 41×10^6 t a^{-1} of DOC and 18×10^6 t a^{-1} of POC.

Tarasov *et al.* (1991) analysed changes in OM transport by rivers under conditions of increasing anthropogenic impact during 1936–80. Generally, the flux of OM by rivers did not change significantly during this period. The wider use of fertilisers in agriculture caused an impact only in some regions. In the basins of the Arctic and Pacific Oceans, the fluxes of OM changed little with an overall tendency to decrease. In the basin of the Atlantic Ocean, the flux of OM increased mainly due to the impact of anthropogenic factors on small rivers.

Smirnov (personal communication) estimated the fluxes up to 1990 (Table 12.6) for practically the same 50 rivers as in the Tarasov *et al.* (1991) study mentioned above. During 1981–90, the tendency for a change in OM flux was minimal. An increase in OM fluxes occurred in the basin of the Atlantic Ocean and decreased in endorheic regions. In the latter case, this was due mainly to a decrease in OM transport by the Volga into the Caspian Sea. In general, changes in OM fluxes over the last 50 years did not exceed ± 10 per cent of the average long-term values and this is probably well within the error range.

In terms of the spatial distribution of OM fluxes (Figure 12.6), the highest values occurred in the north and north-west of the country declining to minimal values in the southern regions of the European part of the former USSR, in Kazakhstan and Central Asia (Maltseva *et al.*, 1987).

12.3.2 Nitrogen and phosphorus

The main inorganic forms of nitrogen and phosphorus, ammonia-nitrogen (NH_4-N), nitrite-nitrogen (NO_2-N), nitrate-nitrogen (NO_3-N) and phosphate-phosphorus (PO_4-P), were monitored regularly in the former Soviet Union. Other forms were monitored less frequently and their long-term records are therefore often absent. The total fluxes of nutrients, and major ions and OM from the former USSR territory were first assessed by Alekin and Brazhnikova (1964). They were later studied by Tarasov *et al.* (1988, 1991), Skakalsky and Meerovich (1991) and others. Many authors have estimated the nutrient fluxes by rivers or sea basins (e.g. Datsko and Guseinov, 1960; Almazov *et al.*, 1967; Maksimova, 1979; Laznik and Tsirkunov, 1990). The average annual fluxes of dissolved inorganic forms of nitrogen and phosphorus during 1981–90 for 54 large rivers of the former USSR are given in Table 12.6. In general, the error associated with nutrient load calculations is approximately 30–60 per cent for rivers, although for entire basins it may be even greater.

The data show that while approximately two thirds of major ions and OM enter the Arctic Ocean, nearly equal amounts of NO_3-N are transported by rivers to the Atlantic Ocean and Arctic Ocean and to the endorheic regions of the country. For inorganic phosphorus, the situation is similar to the OM distribution. This difference is probably related to the wider use of nitrogen fertilisers in the European part of the country and in Central Asia and Kazakhstan.

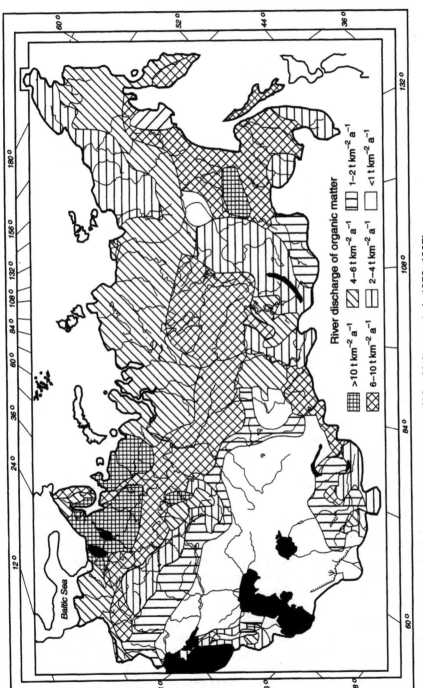

Figure 12.6   Specific river discharge of organic matter (After Maltseva *et al.*, 1978, 1987)

River discharge of organic matter

>10 t km^{-2} a^{-1} 4–6 t km^{-2} a^{-1} 1–2 t km^{-2} a^{-1}

6–10 t km^{-2} a^{-1} 2–4 t km^{-2} a^{-1} <1 t km^{-2} a^{-1}

Baltic Sea

Table 12.7 Input of different forms of dissolved nitrogen and phosphorus into the Laptev and East Siberian Seas

River	Input (10^3 t a^{-1})						
	NO$_3$-N	NH$_4$-N	DON	TDN	PO$_4$-P	DOP	TDP
Laptev Sea							
Khatanga	2.0	3.4	34.1	40.1	0.5	0.5	1.0
Anabar	0.5	0.7	4.3	5.5	0.1	0.1	0.2
Olenek	1.1	1.8	14.3	17.2	0.1	0.2	0.3
Lena	21.7	21.0	243.0	286.0	4.2	11.0	15.2
Yana	1.0	1.4	10.3	12.7	0.1	0.2	0.3
Rest of area (tundra)	1.4	2.4	12.8	16.6	0.1	0.2	0.3
Whole basin	28.3	30.7	319.0	378.0	5.1	12.2	17.3
East-Siberian Sea							
Indigirka	1.8	2.4	24.4	28.6	0.4	0.6	1.0
Kolyma	4.0	6.6	52.8	63.4	1.0	2.0	3.0
Rest of area (tundra)	1.7	2.9	15.4	20.0	0.1	0.2	0.3
Whole basin	7.5	11.9	92.6	112.0	1.5	2.8	4.3

DON   Dissolved organic nitrogen   TDP   Total dissolved phosphorus
DOP   Dissolved organic phosphorus
TDN   Total dissolved nitrogen   Source: Gordeev *et al.*, 1996

The structure of fluxes of inorganic nitrogen varies regionally. While in the Baltic, Black and Caspian Sea basins the NO$_3$-N fluxes prevail over the NH$_4$-N fluxes, the fluxes of NH$_4$-N into the seas of the Arctic and Pacific Oceans are higher (especially into the Kara Sea). The latter is explained by the presence of large wetlands in these basins and by the widespread occurrence of permafrost (Tarasov *et al.*, 1991; Skakalsky and Meerovich, 1991).

More detailed data on the fluxes of dissolved forms of nitrogen and phosphorus were reported by the Tiksi Branch of Goskomhydromet for the basins of the Laptev and East Siberian Seas (Table 12.7). This confirms the greater or equivalent fluxes of NH$_4^+$ compared with NO$_3^-$. An important feature is the prevalence of fluxes of dissolved organic forms of both nutrients over their inorganic forms. When environmental conditions (e.g. pH, oxygen saturation, temperature) in river waters change, there are transformations from one dissolved form to another and from the dissolved form into suspended forms and vice versa (the latter process is almost completely biological). The data show that it is not sufficient only to consider dissolved inorganic forms of nitrogen and phosphorus in order to assess the nutrient fluxes. Of the total global river flux of dissolved nitrogen (14×10^6 t N a^{-1}) estimated by Meybeck (1982), NO$_3$-N accounts for only 26 per cent (3.7×10^6 t a^{-1}); the remaining three quarters of the flux is organic nitrogen. A comparison of

NO_3-N flux from the former USSR territory into regional sea basins with global values showed that the proportion from the former USSR is about 14 per cent (including the anthropogenic influence).

Meybeck (1982) established a good correlation between concentrations of nitrogen and organic carbon in suspended matter for many rivers (C/N = 8.5 g g^{-1}). This allowed the estimation of the flux of particulate organic nitrogen (PON) by the flux of POC. It was found that the global flux of nitrogen in suspension (21×10^6 t a^{-1} N) significantly exceeded the total flux of dissolved nitrogen. Applying a similar relationship POC/PON = 8.5 to river flux from the former USSR, an approximate flux of PON of 2.1×10^6 t (combining organic N and adsorbed NH_4^+) was obtained, which is three times greater than the flux of NO_3-N.

The flux of dissolved organic phosphorus is about twice the flux of inorganic phosphorus. Globally, this ratio is 1.5:1.0. Data for the flux of suspended phosphorus from the former USSR are absent. For world rivers, the flux of total suspended phosphorus is significantly greater than dissolved phosphorus with suspended phosphorus accounting for 95 per cent of total phosphorus flux (Meybeck, 1982).

Data on nutrient transport before the 1980s showed that the NO_3-N flux increased by 80 per cent in the rivers studied, while in the whole drainage basins of the Baltic (USSR part only), Black, Azov and Caspian Seas it increased by 2–3 times (Tarasov et al., 1988, 1991; Skalalsky and Meerovich, 1991). In endorheic regions of Central Asia and Kazakhstan (Amu Darya and Syr Darya rivers), the NO_3-N and PO_4-P fluxes decreased because of the dramatic decrease in river water discharge. Observations from 1936 to 1990 on 54 rivers of the former USSR, although not always complete and considered as unreliable up to the 1980s, showed that on the whole, the trend of the early 1980s still remains. In the White and Barents Sea basins there was a reduction in NO_3-N flux with an increase in PO_4-P flux of 1.5 times. A similar situation was obtained in the basins of the East Siberian and Chukchi Seas, although over the last 20 years the PO_4-P flux remained at a stable but high level. Unfortunately, fragmentary data about the Laptev and Kara Seas do not allow for a complete picture of the whole Arctic basin.

In the Atlantic Ocean basin, there has been a general tendency for an increase in NO_3-N flux and a decrease in PO_4-P flux since the beginning of the 1990s. In the basins of the Black and Azov Seas, the NO_3-N flux increased significantly (up to 7–8 times). In the Baltic Sea basin, the NO_3-N flux decreased while the PO_4-P flux increased. In the Caspian Sea, a continuous reduction in fluxes of both nutrients occurred. In endorheic regions of Kazakhstan and Central Asia, in spite of sharp reductions in water discharge, the PO_4-P flux has not decreased, which may be interpreted as a constant increase in the anthropogenic load in the region.

12.4   Transport of metal trace elements

In river waters, most metal trace elements migrate in the form of suspended elements (Gordeev and Lisitzin 1978; Martin and Meybeck, 1979). Unfortunately, suspended forms of trace elements were rarely monitored in the former USSR (Konovalov, 1959; Konovalov *et al.*, 1968).

During sampling and analysis of dissolved trace elements in natural waters, there is a high risk of sample contamination at various steps of the procedure (see Chapter 8, Figure 8.1). It is therefore necessary to meet stringent guidelines and to use expensive, highly sensitive analytical methods. The analysis of natural waters for ultra-low concentrations of metals (about 10^{-7}–10^{-10} g l^{-1}) remains difficult. In many countries, including highly-developed ones, efforts for routine monitoring of most metals, particularly in the dissolved form and for unfiltered waters, have seldom led to positive results (GESAMP, 1987). Hence, accumulated and published data (mostly prior to 1985) of concentrations and river fluxes of trace elements in Russian and foreign literature are only of historic interest. It is therefore impossible to present a reliable assessment of the fluxes of micro-elements from the former USSR territory into sea basins.

Reliable data of trace metal concentrations in waters of major Arctic rivers, the Ob, Yenisey and Lena, have only been obtained during recent years (Cossa and Coquery, 1993; Martin *et al.*, 1993; Kravtsov *et al.*, 1994; Coquery *et al.*, 1995; Gordeev and Shevchenko, 1995; Rachold, 1995; Guieu *et al.* 1996). These show that downstream sections of the major Siberian rivers are actually characterised by very low concentrations of dissolved and particulate forms of Pb, Cd, Cu, Zn, Ni, As, Hg, etc. Dai and Martin (1995) showed that colloidal material (between 10^4 Dalton and 0.4 μm) made a significant contribution to the so-called 'dissolved' fraction of heavy metals in the Ob and Yenisey rivers (20–50 per cent Pb, 40–70 per cent Cd and Ni, up to 70–80 per cent Cu). The authors concluded that there was a fundamental role for colloidal fractions in determining the behaviour of trace metals in the river–sea mixing zone as well as in the control of their net discharge into the Kara Sea. The assessments of some dissolved and particulate trace metal inputs from the Ob, Yenisey and Lena rivers into the Kara and Laptev Seas and by all rivers from the Eurasian basin to the Arctic Ocean are presented in Tables 12.8 and 12.9. It is assumed that average concentrations of heavy metals in the three major Siberian rivers were representative for the natural river levels of the whole Eurasian basin. However, because data on seasonal changes in concentrations are absent, the assessments in Table 12.8 must be considered as preliminary.

Assessments of net river fluxes of dissolved metals from the Eurasian basin (Guieu *et al.*, 1996) and of fluxes from other sources to the Arctic Ocean including internal ones (inflows from other oceans, organic matter

Table 12.8 Riverine fluxes of dissolved and particulate heavy metals to the Arctic Ocean

River		Cu	Pb	Zn	Ni	Cd	As	Hg	Fe[1]
		Flux (t a^{-1})							
Ob[2]	Dissolved	850	6	160	530	0.3	–	0.2	12
	Particulate	840	260	1,700	630	3.3	–	0.8	940
Yenisey[2]	Dissolved	1,000	4	820	340	1.0	–	0.2	10
	Particulate	650	180	1,300	450	13.6	–	0.3	320
Lena[3]	Dissolved	300	9	180	160	2.8	80	0.4	12
	Particulate	490	400	2,500	550	4.6	–	4.0	590

[1] Values are 10^3 t a^{-1}
[2] Dissolved metals (except Hg) – Dai and Martin, 1995; Kravtsov et al., 1994; particulate metals (except Hg) – Gordeev et al., 1996
[3] Dissolved and particulate metals (except Hg) – Martin et al., 1993; dissolved and particulate Hg in all rivers – Cossa and Coquery, 1993; Coquery et al., 1995

Table 12.9 Gross river flux and net river flux of dissolved heavy metals to the Eurasian basin compared with other inputs to the Arctic Ocean

Metal	Gross river flux to Eurasian basin[1] (mol a^{-1})	Net river flux to Eurasian basin[2] (mol a^{-1})	Other fluxes to Arctic Ocean[3] (mol a^{-1})	Total fluxes to Arctic Ocean[2] (mol a^{-1})	Freshwater input (% of total)[2]
Cd	2.0×10^5	1.1×10^6	2.9×10^7	3.0×10^7	4
Cu	5.0×10^7	7.4×10^7	2.0×10^8	2.7×10^8	27
Ni	1.6×10^7	4.4×10^7	3.6×10^8	4.0×10^8	11
Zn	4.4×10^6	4.4×10^6	2.7×10^8	2.8×10^8	2

[1] Present study; concentrations of metals from Guieu et al., 1996
[2] Guieu et al., 1996
[3] Yeats and Westerlund, 1991

decomposition) are shown in Table 12.9 (Yeats and Westerlund, 1991). The data show that net river flux exceeds gross river flux 5–6 times for dissolved Cd, 2.5 times for dissolved Ni and 1.5 times for dissolved Cu, mainly due to desorption from particles in estuarine zones. Net and gross fluxes of dissolved Zn were equal. The data also show that the estuarine transformation of dissolved metal forms sometimes changes their gross river flux significantly, and the freshwater input of dissolved trace metals into the Eurasian sector of the Arctic Ocean appears to be rather small in comparison with all other sources (Cd 4 per cent, Cu 27 per cent, Ni 11 per cent, Zn 2 per cent).

12.5   Transport of some organic micropollutants

An estimation of the fluxes of total hydrocarbons, phenols, detergents, and total organochlorine pesticides (DDT and its metabolites, DDE and DDD, and HCH – the sum of α- and γ- isomers) into seas was made for 32 rivers of 11 sea basins during 1981 to 1990. In order to determine trends, comparisons were made for two five-year periods (1981–85 and 1986–90), using river water discharge and average concentrations of total hydrocarbons, phenols and detergents, and using average concentrations of organochlorine pesticides for periods of low water flow and periods of flood and high water (Table 12.10). As a rule, during 1981–85 more oil products, detergents and organochlorine pesticides, but fewer phenols, were transported than in 1986–90. In the case of the River Dnieper, in the absence of detailed data for each hydrological phase during the 1986–90 period, annual values of water discharge and concentrations of all substances were considered.

Of the total flux of HCH isomers, the proportion of α-HCH was 35–75 per cent and the proportion of γ-HCH was 25–65 per cent. Of the total flux of DDT products, the greatest proportion was DDT itself (50–100 per cent) whereas the proportion of its metabolites, DDE and DDD, constituted 0–50 per cent. A reduction in fluxes of organochlorine pesticides in most rivers during the second five-year period (1986–90) occurred as a result of a sharp reduction in use of HCH by 1990 and the complete cessation in use of DDT in river basins in the 1970s.

Generally, in the European part of the former USSR, especially in the river basins of the Baltic, Azov, Black and Caspian Seas, there was a significant increase in the fluxes of substances related to increasing anthropogenic load. A similar anthropogenic load affects the rivers of the semi-arid zones of Central Asia and Kazakhstan, although this is not adequately reflected in changes in the fluxes of substances due to the reduction (sometimes drastic) in water flow. The transport of substances into basins of the Arctic and Pacific Oceans did not vary significantly. For rivers such as the Lena, the flux of organic pollutants can be considered as characteristic of the natural background level of these products in such semi-arctic and arctic environments.

12.6   Conclusions

The former Soviet Union covered approximately one sixth of the globe and had a high level of industry and agriculture. Nearly all climatic zones occurred within the territory with the exception of the wet tropical zone. It is clear from the data presented that the level of anthropogenic impact varied in different regions.

A general evaluation of river discharges of water, sediment, salts and nutrients from the former Soviet Union in comparison with global values is presented in Table 12.11 for exorheic basins only (excluding the

Table 12.10 Average input of organic micropollutants into seas by various rivers

River	Sampling site	Period of observation	Average water discharge (km³ a⁻¹)	Petroleum products (10³ t a⁻¹)	Phenols (10³ t a⁻¹)	Detergents (10³ t a⁻¹)	ΣDDT (t a⁻¹)	ΣHCH (t a⁻¹)
Arctic Ocean								
White and Barents Seas basins								
Onega	Porog	1981–85	16.1	2.32	0.019	0.115	0.145	0.268
		1986–90	16.3	1.21	0.063	0.260	0.012	0.163
Northern Dvina	Ust-Pinega	1981–85	100	–	–	–	0.082	1.94
		1986–90	107	2.53	0.052	2.06	0.957	0.656
Mezen	Malonisogorskaya	1981–85	20.2	2.94	0.030	0.195	0.059	0.322
		1986–90	19.0	0.967	0.008	0.116	0	0.266
Kola	Kola	1981–85	1,595	0.242	0.005	0.130	0.031	0.142
		1986–90	1.45	0.022	0.019	0.017	0.062	0.025
Pechora	Naryan Mar	1981–85	143	77.1	0.154	1.94	0.955	5.83
		1986–90	138	6.11	0.195	2.06	1.83	4.83
Kara Sea basin								
Ob	Salekhard	1981–85	410	139	1.14	6.24	46.9	21.9
		1986–90	405	159	0.925	4.46	20.5	26.7
Yenisey	Igarka	1981–85	568	424	4.75	20.2	0	37.6
		1986–90	609	234	3.09	5.60	0.012	25.6
Laptev Sea basin								
Olenek	Tyumety	1981–85	30.0	1.27	0.093	0.460	–	–
		1986–90	39.2	2.97	0.117	0.337	0.109	0.070
Lena	Stolb	1981–85	547	41.8	0.784	9.88	15.2[1]	2.49[1]
		1986–90	546	51.6	1.58	4.48	0.758[2]	0.499[2]
East Siberian Sea basin								
Indigirka	Chokurdakh	1981–85	57	4.19	0.152	1.58	4.28[3]	0.320[3]
		1986–87	19.6	2.23	0.14	0.878	–	–
Kolyma	Kolymskoe	1981–85	9.6	5.19	0.166	1.81	–	–
		1986–88	58.2	10.1	0.302	1.25	0	0.098

Continued

Table 12.10 Continued

River	Sampling site	Period of observation	Average water discharge (km³ a⁻¹)	Petroleum products (10³ t a⁻¹)	Phenols (10³ t a⁻¹)	Detergents (10³ t a⁻¹)	ΣDDT (t a⁻¹)	ΣHCH (t a⁻¹)
Pacific Ocean								
Bering Sea basin								
Kamchatka	Klyuchi	1981–85	25.2	0	0.130	0.459	1.38	0.882
		1986–90	24.4	0.299	0.089	0.294	0.154	0.622
Sea of Okhotsk basin								
Amur	Bogorodskoye	1981–85	394	16.2	1.85	12.3	29.9	11.5
		1986–90	342	15.6	0.846	13.7	6.81[4]	4.77[4]
Penzhyna	Kamenskoe	1981–85	20.4	–	–	–	0.445	0.500
		1986–90	20.6	0.027	0.097	0.266	0.195	0.189
Atlantic Ocean								
Baltic Sea basin								
Neva	Leningrad	1981–85	83.4	14.5	0.143	1.28	6.44	2.30
		1986–90	83.4	0.77	0.170	0.500	3.66	0.693
Luga	Kingisep	1981–85	3.52	1.09	–	0.090	0.241	0.156
		1986–90	4.34	0.74	–	0.020	0.120	0.017
Narva	Narva	1981–85	12.8	1.30	0.024	0.190	0.080	0.140
		1986–90	16.5	1.14	0.040	0.314	0.016	0.053
Western Dvina	Daugavpils	1981–85	13.0	1.40	0.024	0.700	0.129	0.131
		1986–90	18.1	0.94	0.041	0.660	0.029	0.116
Naman	Smalininkai	1981–85	16.8	3.97	0.031	0.420	0.987	0.579
		1986–90	18.0	1.35	0.029	0.220	0	0.112
Pregolia	Gvardeysk	1981–85	3.2	0.55	0.009	0.086	0	0
		1986–90	3.06	0.17	0.005	0.041	0.015[5]	0.014[5]
Black and Azov Seas basins								
Danube	Kiliya	1981–85	122	28.9	No data	4.24	26.4	4.29
		1986–89	92.4	12.2	0.573	1.96	5.77	0.459
Dniester	Bendery	1981–85	9.94	1.78	0.011	0.200	1.21	0.287
		1986–90	6.96	0.890	0.023	0.210	0.507	0.255

Continued

Table 12.10 Continued

River	Sampling site	Period of observation	Average water discharge (km³ a⁻¹)	Petroleum products (10³ t a⁻¹)	Phenols (10³ t a⁻¹)	Detergents (10³ t a⁻¹)	ΣDDT (t a⁻¹)	ΣHCH (t a⁻¹)
Black and Azov Seas basins continued								
Southern Bug	Alexandrovka	1981–85	3.46	0.227	0.004	0.099	0.432	0.063
		1986–90	2.36	0.036	0	0.017	0.0004	0.005
Dnieper	New Kakhovka	1981–85	42.8	5.57	0.180	0.364	5.67	0.865
		1986–90	38.8	4.69	0.089	0.931	0.301	0.244
Sochi	Sochi	1981–85	0.53	0.012	0	0.011	0.037	0.033
		1986–90	0.52	1.49	0.0004	0.005	0.0002	0.007
Inguri	Darcheli	1981–85	0.74	0.061	0.009	0.014	0.055	0.018
		1986–90	1.42	0.084	0.012	0.041	0.173	0.028
Rioni	Sakochakidze	1981–85	13.2	1.36	0.139	0.058	0.352	0.299
		1986–90	17.5	1.97	0.106	0.862	2.41	0.408
Don	Rostov-on-Don	1981–84	22.2	13.4	0.076	0.857	0.268	0.375
		1986–90	19.2	4.78	0.080	0.903	0.272	0.401
Kuban	Tichovsky	1981–85	9.40	3.24	0.062	0.367	0.259	0.046
		1986–90	11.7	2.71	0.050	0.506	0.059	0.057
Caspian Sea basin								
Kura	Salyany	1981–85	13.6	0.900	0.079	0.510	1.52	0.320
		1986–90	12.4	0.762	0.081	0.400	0.021	0.060
Volga	Upper Lebyazhye	1981–85	248	89.8	0.570	3.21	6.72	5.10
		1986–90	264	95.2	0.762	3.10	8.01	1.01
Ural	Guriev	1981–85	7.66	0.628	0.011	0.281	0.499	0.214
		1986–90	9.86	0.661	0.007	0.267	0.190	0.178

ΣDDT Total dichlorodiphenyltrichloroethane
ΣHCH Total hexachlorocyclohexane

1 At Kusur, 1983–1984
2 At Kusur
3 At Vorontsovo
4 At Komsomolsk
5 At Kaliningrad

Table 12.11 Comparison of gross river input of water, sediment, salts and nutrients from the former USSR with the global river input[1]

	Former USSR	World	Former USSR (% of world total)
Area (10^6 km^2)	22.4	148.9	15.0
Exorheic water catchment area (10^6 km^2)	16.90	99.9	16.9
Water			
(10^3 km^3 a^{-1})	3.87	35.0[2]	11.0
(l s^{-1} km^{-2})	7.3	11.1	65.8
Sediment			
(10^6 t a^{-1})	189	15,000[2]	1.3
(t km^{-2} a^{-1})	11.2	150	7.5
Dissolved salts			
(10^6 t a^{-1})	0.43	3.8[3]	11.3
(t km^{-2} a^{-1})	26.0	38.0	67.9
TOC			
(10^6 t a^{-1})	33	395[4]	8.3
(t km^{-2} a^{-1})	1.95	3.95	49.3
NO$_3$-N			
(10^3 t a^{-1})	0.51	7.0[4]	7.3
(t km^{-2} a^{-1})	0.03	0.07	43
PO$_4$-P			
(10^3 t a^{-1})	0.11	1.0[4]	11.0
(t km^{-2} a^{-1})	0.007	0.010	70
Si			
(10^6 t a^{-1})	11.4	388[3]	2.9
(t km^{-2} a^{-1})	0.65	3.9	16.7

TOC Total organic carbon

[1] Discharges of water, sediment, salts and nutrients in the former USSR relate to the exorheic water catchment area. This is the reason for discrepancies in the percentage evaluations with the text

[2] Milliman (1991)
[3] Meybeck (1979)
[4] Meybeck (1982)

endorheic basins of the Caspian Sea, Central Asia and Kazakhstan and the other world endorheic basins).

Rivers of the former USSR delivered approximately one third less water and one order of magnitude less sediment per square kilometre compared with the global average. The input of dissolved salts (ionic input) and PO$_4$-P was proportional to water discharge.

The fluxes of organic matter, NO$_3$-N and especially silicates were not proportional to, and were lower than, water discharge. These relationships between water and dissolved salts, organic matter and silica fluxes were due to natural reasons, but for NO$_3$-N and PO$_4$-P anthropogenic impact was important.

Summarising the available data on chemical composition of river waters in the former USSR (i.e. major ions, organic matter, nutrients, trace elements) it was shown that, in general, pristine rivers are mostly found in the Arctic basin. Rivers of the Atlantic basin and the endorheic basins, in particular, are subject to anthropogenic influence. The Amu Darya, Syr Darya, Dnieper, Daugava, Neman and some other rivers are amongst the most polluted rivers, whereas when considering the concentrations of heavy metals, the three great Siberian rivers, Lena, Ob and Yenisey, are amongst the most pristine of the major rivers of the world.

12.7 References

(All references are in Russian unless otherwise stated).

Alekin, O.A. 1948 Hydrochemical classification of the USSR rivers. Proceedings of the State Hydrological Institute, 4(58), 32–69.

Alekin, O.A. 1950 Hydrochemical types of USSR rivers. Proceedings of the State Hydrological Institute, 25(79), 5–21.

Alekin, O.A. 1951 Ionic flux and average composition of river water for the USSR territory. Proceedings of the State Hydrological Institute, 33(87), 43–63.

Alekin, O.A. and Brazhnikova, L.V. 1964 *Flux of Dissolved Substances from the Territory of the USSR*. Nauka, Moscow, 144 pp.

Alexeevsky, N.I. 1993 Hydrological regime of the Terek river and its delta. In: *The Caspian Sea. Hydrology of the River Mouths of Terek and Sulak*. Nauka, Moscow, 15–32.

Almazov, A.N., Denisova, A.I., Maistrenko, Y.G. and Nakshina, E.P. 1967 *Hydrochemistry of the Dnieper River, Its Reservoirs and Tributaries*. Naukova Dumka, Kiev, 316 pp.

Artemyev, V.E. 1993 *Geochemistry of Organic Matter in the River-Sea System*. Nauka, Moscow, 190 pp.

Bronfman, A.M. 1981 Formation and anthropogenic changes in suspended matter transport to the Azov Sea. In: *Geographic Aspects of Study of the Azov Basin Hydrology and Hydrochemistry*. Civil Defence Publication, 43–57.

Cauwet, G. and Sidorov, I.S. 1994 Biogeochemistry of the Lena river: organic carbon and nutrient distribution. *Marine Chemistry*, 53, 211–228.

Coquery, M., Cossa, D. and Martin, J.-M. 1995 The distribution of dissolved and particulate mercury in three Siberian estuaries and adjacent arctic coastal waters. In: *Mercury as a Global Pollutant*. Kluwer Academic Publications, Dordrecht, Boston, London, 653–664. (In English).

Cossa, D. and Coquery, M. 1993 Mercury in the Lena delta and Laptev Sea. The Arctic Estuaries and Adjacent Seas: Bio-geochemical Processes and Interaction with Global Change. Third International Symposium, Svetlogorsk, Russia 19–25 April 1993. Abstracts Kaliningrad, p. 11–12. (In English).

Dai, M.H. and Martin, J.-M. 1995 First data on the trace metal level and behaviour in two major Arctic river/estuarine systems (Ob and Yenisey) and the adjacent Kara Sea (Russia). *Earth Planet. Sci. Lett.*, **131**, 127–141. (In English).

Datsko, V.G. and Guseinov, M.M. 1960 Transport of nutrients and organic matter by the River Don into the Azov Sea after regulation of its discharge. *Hydrochemical Materials*, **30**, 96–105.

Denisova, A.I. 1981 Trends in formation and methods of forecast of the hydrochemical regime of lowland river reservoirs (Dnieper cascade as an example). Doctoral thesis, Rostov-on-Don, 32 pp.

Dolan, D.M., Yui, A.K. and Geist, R.D. 1981 Evaluation of river load estimation methods of total phosphorus. *Journal of Great Lakes Research*, **7**, 207–214. (In English).

Dzhaoshvili, S. 1986 *River Sediments and Formation of Beaches on the Georgian Black Sea Shore*. Sabchota Sakartvelo, Tbilisi, 156 pp.

Emelyanov, E.M. and Pusternikov, O.S. 1975 Chemical composition of river and marine suspended matter of the Baltic Sea. *Geochemistry*, **6**, 918–932.

Eremenko, V.Y., Zenin, A.A. and Konovalov, G.S. 1953 Transport of dissolved matter by the River Kuban and its hydrochemical regime. *Hydrochemical Materials*, **23**, 30–53.

GESAMP 1987 Land/Sea Boundary Flux of Contaminants: Contribution from Rivers. GESAMP Reports and Studies No. 32, Joint Group of Experts on the Scientific Aspects of Marine Pollution, United Nations Educational, Scientific and Cultural Organization, Paris, 172 pp. (In English).

Glagoleva, M.A. 1959 Migration forms of elements in river waters. In: *Knowledge of Diagenesis of Sediments*. USSR Academy of Science, Moscow, 5–28.

Goliadze, N.S., Supatashvili, G.D. and Chikhradge, G.A. 1975 On the macrochemical composition of suspended matter of the rivers of Georgia. *Proceedings of the GSSR Academy of Science*, **79**(3), 649–652.

Gordeev, V.V. 1983 *River Input to the Ocean and Features of Its Geochemistry*. Nauka, Moscow, 160 pp.

Gordeev, V.V. and Lisitzin, A.P. 1978 Average chemical composition of river suspended matter and flux of river sediments to the oceans. *Report of the USSR Academy of Science*, **238**(1), 225–228.

Gordeev, V.V. and Shevchenko, V.P. 1995 Chemical composition of suspended sediments of the lower course of the Lena river and its mixing zone. In: *Russian-German Co-operation: Laptev Sea System*. Reports on Polar Research Volume 176, 154–169. Alfred Wegener Institut für Polar- und Meerforschung, Bremerhaven, FRG. (In English).

Gordeev, V.V. and Sidorov, I.S. 1993 Concentrations of major elements and their outflow into the Laptev Sea by the Lena river. *Marine Chemistry*, **43**(1–4), 33–46. (In English).

Gordeev, V.V., Martin, J.-M., Sidorov, I.S. and Sidorova, M.V. 1996 A reassessment of the Eurasian river input of water, sediments, major elements and nutrients to the Arctic Ocean. *American Journal of Science*, **296**, 664–691. (In English).

Gordeev, V.V., Paucot, H., Wollast, R. and Aibulatov, N.A. 1997 Grain size distribution of elements in suspended and bottom sediments of the Ob and Yenisey estuaries. *Estuarine and Coastal Shelf Science.* (Submitted).

Guieu, C., Huang, W.W., Martin, J.-M. and Yoon, Y.Y. 1996 Outflow of trace metals into the Laptev Sea by the Lena river. *Marine Chemistry,* **53**, 255–268. (In English).

Hem, J.D. 1993 Factors affecting stream water quality, and water-quality trends in four drainage basins in the conterminous United States, 1905–1990. In: *US Geological Survey, National Water Summary, 1990–1991.* Hydrologic events and stream water quality. US Geological Survey Water Supply Paper 2400, 67–92. (In English).

Ignatova, V.F. and Chudaeva, V.A. 1983 *River Solid Discharge and Shelf Sediments of the Japan Sea. (The Experience of a Comparative Study).* Vladivostok, 120 pp.

Ivanov, V.V. 1985 Continental runoff to the Arctic Ocean. In: *Atlas of the Arctic.* Goskomgidromet, Moscow, 92–93.

Kirsta, B.T. 1975 *Water Mineralisation, Chemical Fluxes of Rivers of Turkmenistan and Methods for Their Evaluation.* Ashkhabad, Ylym, 172 pp.

Konovalov, G.S. 1959 Transport of micro-elements in major USSR rivers. *Reports of the USSR Academy of Science,* **129**, 912–915.

Konovalov, G.S., Ivanova, A.A. and Kolesnikova, T.K. 1968 Dispersed and rare elements dissolved in water and contained in suspended matter of the main USSR rivers. In: *Geochemistry of Sediment Rocks and Ores.* Nauka, Moscow, 72–87.

Konovalov, G.S., Koreneva, V.I., Korenev, A.P. and Manikhin, V.I. 1991 Variations in fluxes of inorganic components from catchment areas into seas under anthropogenic impact and trends in its formation in mountain regions. In: Proceedings of the V All-Union Congress. Water Quality and the Scientific Basis for its Protection. Gidrometeoizdat, Leningrad, Volume 5, 189–196.

Koreneva, V.I. and Konovalov, G.S. 1987 Changes in trace elements fluxes from river basins into seas. *Hydrochemical Materials,* **100**, 159–170.

Krasnozhon, G.F. and Mazavina, S.S. 1988 Hydrochemical regime of the mouth of the River Ural. In: *Complex Study of the North Caspian Sea.* Nauka, Moscow, 5–40.

Kravtsov, V.A., Gordeev, V.V. and Pashkina, V.I. 1994 Dissolved forms of heavy metals in Kara Sea waters. *Oceanology,* **34**(5), 673–680.

Kuznetsov, N.T., Kliukanova, I.A. and Sanin, S.A. 1987 *Physico-Geographical Principles of Formation of Suspended Matter Composition in Rivers and Irrigation Systems.* Nauka, Moscow, 152 pp.

Laznik, M.M. and Tsirkunov, V.V. 1990 Changes in nutrient substances regimes and their transport by the Latvian SSR rivers. *Hydrochemical Materials,* **108**, 45–64.

Lisitzina, K.N. 1978 Sediment transport into oceans. In: *Sediment Transport, Its Study and Geographical Distribution.* Gidrometeoizdat, Leningrad, 78–92.

Lvovich, M.I., Bratseva, N.L., Karasik, G.Y., Medvedeva, G.P. and Meleshko, A.V. 1989 Map of modern land erosion. *Proceedings of the USSR Academy of Science Geography Series,* **3**, 17–30.

Main Statistics on Water Use in the USSR in 1989 1990 The Central Scientific Research Institute of Complex Use of Water Resources, Minsk, 45 pp.

Main Statistics on Water Use in the USSR in 1990 1991 The Central Scientific Research Institute of Complex Use of Water Resources, Minsk, 51 pp.

Maksimova, M.P. 1979 Criteria of anthropogenic eutrophication of river runoff and calculation of the anthropogenic component of river fluxes. *Water Resources,* **1**, 36–40.

Maltseva, A.V., Tarasov, M.N. and Smirnov, M.P. 1978 Estimates of the average long-term fluxes of organic substances into the seas and oceans from the territory of the USSR. *Proceedings of GOIN,* **142**, 116–121.

Maltseva, A.V., Tarasov, M.N. and Smirnov, M.P. 1987 The transport of organic substances from Soviet territories. *Hydrochemical Materials,* **102**, 119–218.

Martin, J.-M. and Meybeck, M. 1979 Elemental mass-balance of material carried by major world rivers. *Marine Chemistry,* **7**(2), 173–206. (In English).

Martin, J.-M., Guan, D.M., Elbaz-Poulichet, F., Thomas, A.J. and Gordeev, V.V. 1993 Preliminary assessment of the distribution of some trace elements (As, Cd, Cu, Fe, Ni, Pb and Zn) in a pristine aquatic environment: the Lena river estuary (Russia). *Marine Chemistry,* **43**, 185–199. (In English).

Meybeck, M. 1979 Concentration des eaux fluviales en elements majeurs et apports en solution aux oceans. *Rev. Geogr. Phys. Geol. Dyn.,* **21**, 215–246. (In French).

Meybeck, M. 1981 Pathways of major elements from land to the ocean through rivers. In: *River Inputs to Ocean Systems.* United Nations Environment Programme, Nairobi/United Nations Educational, Scientific and Cultural Organization, Paris, 18–30. (In English).

Meybeck, M. 1982 Carbon, nitrogen and phosphorus transport by world rivers. *American Journal of Science,* **282**, 401–450. (In English).

Mikhailova, M.V. 1993 Hydrological regime of the Sulak river and its delta. In: *The Caspian Sea. Hydrology of the River Mouths of Terek and Sulak.* Nauka, Moscow, 15–32.

Milliman, J.D. 1991 Flux and fate of fluvial sediment and water in coastal seas. In: *Ocean Margin Processes in Global Change.* John Wiley and Sons, Chichester, 69–90. (In English).

Milliman, J.D. and Meade, R.H. 1983 World-wide delivery of rivers sediments to the ocean. *Journal of Geology,* **91**, 1–21. (In English).

Morozov, N.P., Baturin, G.N., Gordeev, V.V. and Gurvich, E.G. 1974 On the composition of suspended and bottom sediments of the river mouth region of Northern Dvina, Mezen, Pechora and Ob. *Hydrochemical Materials*, **60**, 60–73.

Nesterova, T.D. 1960 Migration forms of elements in the Ob river. *Geochemistry*, **4**, 355–362.

Nikanorov, A.M. and Tsirkunov, V.V. 1984 Study of the hydrochemical regime and its long term variations in the case of some rivers in USSR. In: *Hydrochemical Balances of Freshwater Systems*. Int. Ass. Hydrol. Sci. Publ., Volume 150, 288–293. (In English).

Nikanorov, A.M. and Tsirkunov, V.V. 1991 Hydrochemical regime of USSR rivers. Analysis of long-term data. In: Proceedings of the V All-Union Congress (20–24 October 1986). Volume 5. Water Quality and the Scientific Basis for its Protection. Gidrometeoizdat, Leningrad, 336–344.

Peleshenko, V.I. 1975 *Assessment of the Inter-relationships of Chemical Composition of Different Natural Water Types (Example of lowland regions of the Ukraine)*. Visha Skola, Kiev, 168 pp.

Peleshenko, V.I. and Khilchevsky, V.K. 1991 Assessment of the anthropogenic impact on chemical composition of river waters in the Ukraine. In: Proceedings of the V All-Union Congress. Volume 5. Water Quality and the Scientific Basis for its Protection. Gidrometeoizdat, Leningrad, 225–230.

Rachold, V. 1995 Geochemistry of the Lena river suspended load and sediments. Preliminary results of the expedition in July/August 1994. In: *Russian-German Co-operation: Laptev Sea System*. Reports on Polar Research Volume 176, 272–279. Alfred Wegener Institut for Polar- und Meerforschung, Bremerhaven, FRG. (In English).

Romankevich, E.A. and Artemyev, V.E. 1985 Inputs of organic carbon into seas and oceans bordering the territory of the Soviet Union. In: *Transport of Carbon and Minerals in Major World Rivers*. Part 3. Mitt. Geol. Paleont. Inst., Univ. Hamburg, Heft 58, Hamburg, 459–470. (In English).

Skakalsky, B.G. and Meerovich, L.N. 1991 Present characteristics of nutrient fluxes from river watersheds of the European part of the USSR. Proceedings of the V All-Union Congress. Gidrometeoizdat, Leningrad. Volume 5, 174–184.

Skopintsev, B.A. and Krylova, L.P. 1955 Transport of organic substances by major rivers of the Soviet Union. *Reports of the USSR Academy of Science*, **105**(4), 770–773.

Smirnov, M.P., Tarasov, M.N., Maltseva, A.V., Kriuchkov, I.A. and Laki, G.I. 1988 River fluxes of organic substances from the USSR territory and their variations in time (1936–1980). *Hydrochemical Materials*, **103**, 67–83.

Stepanov, I.N. and Chembarisov, E.I. 1978 The influence of irrigation on river water mineralisation. Nauka, Moscow, 119 pp.

Tarasov, M.N., Smirnov, M.P. and Maltseva, A.V. 1985 Spatial and temporal structure and migration of organic matter in river water of the

USSR. In: *Geochemistry of Natural Water*. Proceedings of the II International Symposium, Rostov-on-Don. Gidrometeoizdat, Leningrad, 350–357.

Tarasov, M.N., Smirnov, M.P., Kriuchkov, I.A. and Laki, G.I. 1988 River fluxes of nutrients from the USSR territory and their variations in time (1936–1980). *Hydrochemical Materials*, **103**, 49–66.

Tarasov, M.N., Smirnov, M.P., Demchenko, A.S., Kriuchkov, I.A., Beschetnova, E.I., Korotova, L.G. and Melnikova, V.A. 1991 Transport of organic matter, nutrients and pesticides from the USSR territory under anthropogenic influence. In: Proceedings of the V All-Union Congress. Volume 5. Water Quality and the Scientific Basis for its Protection. Gidrometeoizdat, Leningrad, 169–174.

Temporary Guidelines for Calculation of Transport of Organic Compounds, Nutrients, Pesticides and Trace Metals by River Flow 1983. Gidrometeoizdat, Leningrad, Moscow, 3–22.

Tsirkunov, V.V., Nikanorov, A.M., Laznik, M.M. and Dongwei, Z. 1992 Analysis of long-term and seasonal river water quality changes in Latvia. *Water Research*, **26**(9), 1203–1216. (In English).

Voronkov, P.P. 1970 *Hydrochemistry of Local Runoff in the European Territory of the USSR*. Gidrometeoizdat, Leningrad, 188 pp.

Water Resources of the USSR and their Use 1987. State Water Register. Gidrometeoizdat, Leningrad, 45–46.

Yeats, P.A. and Westerlund, S. 1991 Trace metal distributions at an Arctic Ocean ice island. *Marine Chemistry*, **33**(2), 261–277. (In English).

Zenin, A.A. and Protsenko, A.V. 1961 Flux of dissolved substances by the Volga river into the Caspian Sea. *Hydrochemical Materials*, **34**, 60–64.

Chapter 13[*]

HYDROBIOLOGICAL ASSESSMENT

Hydrobiological monitoring is of significant importance for the assessment of both water quality and the ecological condition of water bodies. Monitoring systems based only on contaminant concentrations in the aquatic environment do not provide an adequate estimate of water quality or ecosystem integrity, because isolated analyses of certain chemical substances, without taking into consideration the whole ecological situation, do not reflect the complexity of aquatic system inter-relations.

13.1 Classification of the environmental state of water bodies and criteria applied

In Russia, hydrobiological assessment of water quality has long-standing traditions dating back to the middle of the last century, when a special Commission was established by the Kazan Naturalists Society to carry out a biological study of local lakes (Proceedings of the Naturalists Society, 1870). In 1974, the USSR Hydrobiological Service of Surface Water Monitoring was set up. This service used a set of methods which were consistently applied for the systematic study of rivers, lakes and reservoirs (Abakumov, 1983). Monitoring data were published annually (Abakumov, 1976–1983, 1990–1991, 1992–1994; Abakumov and Svirskaya, 1984–1991).

For the assessment of aquatic ecosystems, the following classes have been defined (Abakumov, 1991):

Class I   *Ecological integrity* reflecting pristine environments;

Class II   *Ecological stress* caused by moderate anthropogenic impact which stimulates the increase in species diversity and the intensity of biocoenosis metabolism;

Class III *Signs of ecological regression* which is an intermediate stage between ecological stress and ecological regression with some of the ecosystem components having certain regressive tendencies;

Class IV   *Ecological regression* is characterised by a decrease in species diversity, a decrease in spatial and temporal heterogeneity, an

*   *This chapter was prepared by V.A. Abakumov*

increase in entropy, simplified inter-species and trophic relations and a significant increase in intensity of biocoenosis metabolism caused by relatively strong anthropogenic impact;

Class V *Metabolic regression* is determined by very strong anthropogenic impact which results in a decrease in organic matter production and degradation processes and leads to a complete degradation of the biocoenosis.

The above ecological stages have a more integral biological meaning than the traditional saprobic categories. The saprobic system tends to reflect the ability of organisms to live in an aquatic environment with a certain pollution level, particularly organic pollution, and aquatic saprobity reflects the combined pollution of a water body. The Saprobic Index is calculated using the Pantle and Buck (1955) method modified by Sládecek (1973) and is based on two factors, the saprobic characteristics of indicator species (the characteristics of which are reviewed by Sládecek) and the relative abundance of indicator species. The Saprobic Index range is ≤ 1–4. Low indices correspond to good aquatic environmental conditions and water quality, whereas high indices (up to 4) correspond to poor conditions. The Saprobic Index is defined depending on the state of the phytoplankton, zooplankton and periphyton population. Generally phytoplankton and zooplankton are the indicators used in large water bodies and periphyton in small water bodes.

Different saprobic zones reflecting various intensities of water pollution can be defined. Saprobic zones were first proposed by Kolkwitz and Marsson (1908, 1909). Saprobic zones correlate with water quality and the Saprobic Index as follows:

Saprobic zone	Water quality	Saprobic Index
Xenosaprobic	Very clean	≤ 0.5
Oligosaprobic	Clean	0.5–1.5
β-mesosaprobic	Moderately polluted	1.5–2.5
α-mesosaprobic	Polluted	2.5–3.5
Polysaprobic	Very polluted	3.5–4.0

The Biotic Index devised by Woodiwiss (1981) is the most suitable method for formalising the assessment of benthic invertebrate condition. The Biotic Index is based on the presence or absence of key indicator organisms and is defined according to a special matrix where the most common indicator organisms of water quality are listed along the left vertical axis (indicators of clean water at the top of the axis with polluted water indicators at the bottom). Five stages of taxonomic diversity are listed along the upper horizontal line (the taxa which should be taken into account in the calculation were also defined by Woodiwiss). The

Table 13.1   Classification of ecological conditions of surface waters used in the former USSR

Ecological condition		Saprobic Index	Biotic Index
Class I	Ecological integrity	1.0–1.5	8–10
Class II	Ecological stress	1.5–2.0	6–8
Class III	Signs of ecological regression	2.0–2.5	4–6
Class IV	Ecological regression	2.5–3.0	2–4
Class V	Metabolic regression	3.0–4.0	0–2

benthic invertebrate condition can then be evaluated on a scale from 0 (total biocoenosis degradation) to 10 (most diverse community) according to the matrix. It should be noted that the Saprobic Index scale and the Biotic Index scale are in opposite directions. Therefore, the worst ecological conditions correspond to the highest Saprobic Index and the lowest Biotic Index (Table 13.1).

The Oligochaeta Index (Goodnight and Whitley, 1961) can also be used as an additional index to assess the condition of benthic invertebrate communities. It is defined as the percentage ratio of oligochaetes in the benthic population to the total benthic invertebrate population. The benthic invertebrate population is considered to be in good condition if oligochaetes do not exceed 60 per cent of the total population. When oligochaetes constitute 60–80 per cent of the zoobenthos population, conditions are poor and when the ratio exceeds 80 per cent, conditions are extremely poor.

13.2   Ecological state of water bodies in various hydrographic regions

The general degradation of freshwater ecosystems in the former USSR is the result of a constant search for short-term economic profit at the expense of long-term ecological damage that has lasted for decades. At the end of the 1980s and based on ecological surveys of freshwater ecosystems, it was concluded that if this negative trend continued, in the near future pristine freshwater would become the main limiting factor for human development in some regions (Abakumov and Sustchenya, 1991). Extreme conditions have already occurred in the region of the Aral Sea. However, critical ecological situations are also found in other regions.

In each hydrographic region there are large numbers of aquatic ecosystems which are not only in a state of ecological regression but also in a state of metabolic regression (Class IV–V). This is especially true of small rivers, many of which are of vital importance, despite their length, because of the high population densities in their basins.

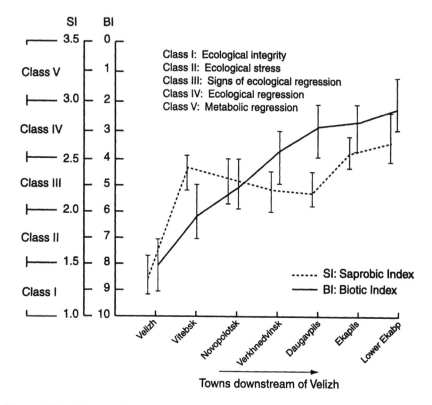

Figure 13.1   Changes in phytoplankton Saprobic Index and Biotic Index in some reaches of the Western Dvina river in the late 1980s

13.2.1 The Baltic hydrographic region

Western Dvina (Daugava) basin

In the Valdai highlands, the Western Dvina is still ecologically pristine (Class I). However, downstream, around the town of Velizh, anthropogenic stress occurs (Class II). The Biotic Index is usually between 7 and 9 (Figure 13.1).

In Byelorussia, the ecological state of the Western Dvina deteriorated under the impact of high discharges of industrial, domestic and agricultural wastewaters. Downstream of the town of Vitebsk, there are signs of ecological degradation which are most evident near the cities of Polotsk, Novopolotsk, Disna and Verkhnedvinsk (Class III).

In Latvia, most of the aquatic ecosystem of the Western Dvina (named Daugava in Latvia) is in a state of anthropogenic ecological stress (Class II). Around the cities of Daugavpils, Ekapils and Riga, evidence of ecological regression can be found (Class IV). The Saprobic Index reaches 2.7 and the Biotic Index is usually between 1 and 5.

Tributaries of the Western Dvina undergo rather different anthropogenic loads and differ greatly in their ecological state. Thus, the Rivers Ulla, Obol, Disna, Ushacha and Birveta are in a state of ecological stress (Class II). The values of the Biotic Index vary from 5 to 9. The Saprobic Index, based on algae and zooplankton, is within the β-mesosaprobic zone. The River Polota downstream of the town of Polotsk has been in a state of ecological regression for many years (Class IV). Ecological regression is also characteristic of the Rivers Feimanka (near Preyli city) and Laukyasa (Class IV).

The River Lielupe basin
The River Lielupe is characterised by ecological stress (Class II). However, downstream of Elgava city there is a significant deterioration in water quality which has led to increasing eutrophication. The Saprobic Index sometimes approaches 4.0 and the Biotic Index varies from 2 to 6.

Evidence of deterioration appeared in this basin at the beginning of the 1990s. The River Kulpe is the most polluted river in the basin. Near the city of Shaulyay the river ecosystem is in a state of metabolic regression (Class V). There is a sharp deterioration in water quality of the River Mushu (Musu) due to the confluence of the River Kulpe. The oxygen level decreases to 0.4 mg l^{-1} and the Biotic Index decreases to 0–2. In the lower reaches of the river, self-purification processes are well developed and the value of the Biotic Index increases to 6. On the River Sidabra, fish kills due to anoxia are frequent; downstream of the town of Ionishkis conditions of total anoxia have been periodically recorded.

The Venta river basin
The upper reaches of the River Venta are in a state of ecological stress with signs of ecological regression (Class II–III). Downstream of Kurshenay city, the river is in a state of ecological regression (Class IV) for most of its reach. This ecosystem state is especially typical downstream of the city of Mazhekyay. The Saprobic Index reaches 2.5–2.7 and the Biotic Index around the towns of Kurshenay, Mazhekyay, Piltene and Ventspils is between 1 and 7, with an average of 4.

In the Venta basin, the River Tsietsera is the most polluted river. In the early 1990s there was an increase in ammonium and nitrite which led to a general deterioration in the ecological state of the river.

The Neman basin
In its upper reaches the Neman river is in a state of ecological stress, with some evidence of ecological regression (Class II–III). Maximum values of the Saprobic Index do not exceed 2.0. Downstream, a decrease in benthic species diversity occurs. The Biotic Index near the city of Stolbtsy varies from 6 to 9, and from 3 to 6 near the town of Grodno. In the

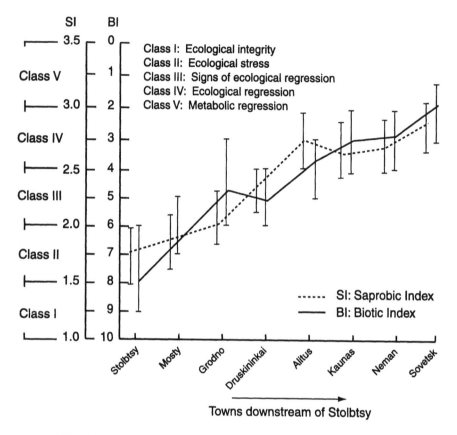

Figure 13.2   Changes in periphyton Saprobic Index and Biotic Index in some reaches of the River Neman in the late 1980s

upper reaches of the river, Trichoptera and Ephemeroptera (which indicate clean water) are numerous.

The middle and lower reaches of the Neman river are in a state of ecological regression with some evidence of metabolic regression (Class IV–V). The Saprobic Index, based on phytoplankton, around the town of Druskininkai does not exceed 2.3, but on the part of the river from the town of Alitus to the town of Sovetsk it reaches 3.0 and in some years 3.25. The Saprobic Index, based on zooplankton, near the cities of Neman and Sovetsk exceeds 2.5. The Oligochaeta Index reaches 97–100 and the Biotic Index is not greater than 4 (Figure 13.2).

The River Lideya near the town of Lida is in a state of ecological stress (Class II). The Saprobic Index based on phytoplankton and zooplankton is within the β-mesosaprobic zone. Benthic biocoenoses show signs of ecological regression (Class III). The Biotic Index is between 2 and 6. A similar ecological status has been noted for the Kotra river near the towns of Grodno and Skidel, with a Biotic Index between 2 and 3.

Some reaches of the Rivers Yozhka and Issa are in a state of ecological stress (Class II) whereas others are still ecologically pristine (Class I). Around the town of Grodno, in terms of the benthic biocoenosis, signs of ecological regression are observed (Class III). The Biotic Index based on benthic communities varies from 2 to 7.

The Western Bug basin

The Western Bug, upstream of the town of Busk, is mostly in a state of ecological stress and only in some reaches is it ecologically pristine (Class I–II). Downstream of the town of Busk, the ecological state of the river deteriorates sharply following confluence with the polluted waters of the River Plotva which carries the wastewaters of the town of Lvov. The dissolved oxygen concentration decreased to 5.16 mg l^{-1} and the average annual concentration of BOD_5 increased to 7 mg l^{-1}. Around the towns of Kamenka-Bugskaya and Sokal, the river is in a state of ecological regression (Class IV). The Saprobic Index reaches the upper border of the β-mesosaprobic zone and the Biotic Index frequently registers 2–3.

The River Plotva is the most polluted tributary of the Western Bug (Class IV). The Saprobic Index varies from 2.52 (downstream of Lvov) to 2.78 (at Busk). Around Lvov and Busk, the river shows metabolic regression with complete anoxia throughout the year (Class V). The average annual concentration of BOD_5 is 32.2 mg l^{-1} and the maximum level is 116 mg l^{-1}.

The ecosystem of the River Mukhovets, in the Kobrin-Brest area, is in a state of ecological stress (Class II). The biocoenosis has a high species diversity. The Saprobic Index is not greater than 1.88 and the Biotic Index is frequently between 5 and 6. Downstream of Brest city, signs of ecological regression are registered (Class III). There is a significant decrease in the diversity and in the number of plankton. The Saprobic Index increases to 2.0 but remains within the β-mesosaprobic zone. In some years, the benthic biocoenoses correspond to a state of ecological regression (Class IV).

The Rivers Ryta and Lesnaya have been ecologically pristine for many years and only the less resistant species in these ecosystems show signs of anthropogenic ecological stress (Class I–II).

13.2.2 The Barents hydrographic region

Lakes of the Kola Peninsula

The largest lake on the Kola Peninsula is Lake Imandra which has been subjected to significant anthropogenic loads. During half a century, the Monche bay of this lake has received wastewaters from the 'Severonikel' smelter, heavily polluted with numerous substances particularly heavy metals. The Belaya bay receives pollutants from the 'Apatite' plant, including oil products, phenols and metals. As a result of these inputs, the ecosystem is generally in a state of ecological regression and in highly polluted areas the ecosystem is in a state of metabolic regression (Class IV–V).

On the Kola Peninsula, the largest anthropogenic stress, according to hydrobiological indicators, is on the River Nuduay where wastewaters are discharged from the town of Monchegorsk. During the last seven years, the ecosystem of the river has been in a state of ecological and metabolic regression (Class IV–V) recording a Biotic Index of 1. The ecosystem of Lake Monche in the region of Monchegorsk is in a similar state (Class IV–V).

For many years, the river Kolos-Yoki has been grossly polluted, especially in its lower reaches. The river is in a state of ecological regression and in some parts, metabolic regression (Class IV–V). Benthic fauna is extremely poor with a Biotic Index not greater than 1–2.

The ecosystem of the River Pechenga, upstream of the inflow of the River Luottn-Yoki is in a state of ecological stress whereas downstream of the Nama-Yoki inflow it is in a pristine state (Class I–II). However, at the river mouth, signs of ecological regression are recorded (Class III) (Figure 13.3). In the River Patso-Yoki (Patsyoki) during the past few years, there has been a decrease in the number of zooplankton species (Table 13.2).

The Rivers Onega and Northern Dvina basins
The River Onega, and Lake Lacha out of which it flows, are ecologically pristine (Class I) until the inflow of the River Voloksha. In the area of the Voloksha confluence, the quality of water deteriorates sharply and signs of ecological regression are recorded (Class III). As a result of intensive self-purification processes, there is an improvement in water quality and signs of ecological regression disappear. For the remainder of the river until its discharge into the White Sea, the river is in a state of ecological stress (Class II).

The Northern Dvina river is in a state of ecological stress almost along the whole length of its flow (Class II). Downstream of Arkhangelsk and Kotlas, signs of ecological regression are noted (Class III). The Saprobic Index for zooplankton, as a rule, does not exceed 2.0 but downstream of Arkhangelsk and Kotlas it reaches 2.67.

The Rivers Sukhona, Vychegda and Vologda are in a state of ecological stress (Class II). Signs of ecological regression are noted around the towns of Sokol and Velikiye Ustyug on the Sukhona river and around the towns of Koryazhma, Solvychegodsk, Syktyvkar and Kotlas on the Vychegda river (Class III).

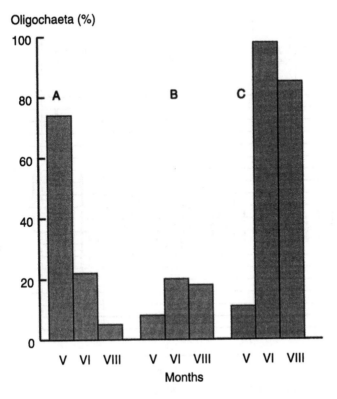

A : 0.1 km upstream of the inflow of the Luottn-Yoki
B : 0.5 km downstream of the inflow of the Nama-Yoki
C : river mouth

Figure 13.3   Dynamics of the relative number of Oligochaeta in the River Pechenga in the late 1980s

Table 13.2   Dynamics of zooplankton species diversity in the River Patso-Yoki (Patsyoki), Kola Peninsula

Year	Total number of species	Number of species		
		Rotifera	Cladocera	Copepoda
1990	39	15	14	10
1991	34	14	12	8
1992	34	11	13	10
1993	27	10	10	7

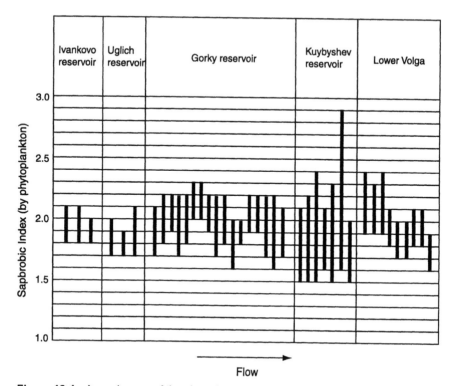

Figure 13.4   Annual range of the phytoplankton Saprobic Index on the Lower Volga and its reservoirs at various monitoring stations on the reservoirs in the late 1980s

13.2.3 The Caspian hydrographic region

The River Volga

Brief data on the ecological condition of five stretches of the River Volga: Ivankovo, Uglich, Gorky and Kuybyshev reservoirs and the Lower Volga are presented in this section. Annual variations of the Saprobic Index for phytoplankton and zooplankton and the Biotic Index for each of the five sections are presented in Figures 13.4, 13.5 and 13.6 respectively. The different stages of ecological condition for the five sections of the river, according to the Saprobic Index for phytoplankton and the Biotic Index for benthic invertebrates, are also presented in Table 13.3.

The Ivankovo reservoir is one the main sources of water supply for Moscow. It is in a state of ecological stress. The benthic biocoenosis is characterised by signs of ecological regression (Class II–III).

The ecological state of the Uglich reservoir is similar to the Ivankovo reservoir, being in a state of anthropogenic stress with obvious signs of ecological regression in the benthic biocoenosis (Class II–III). The Biotic Index is generally 2–3 with a maximum value of 5 being recorded.

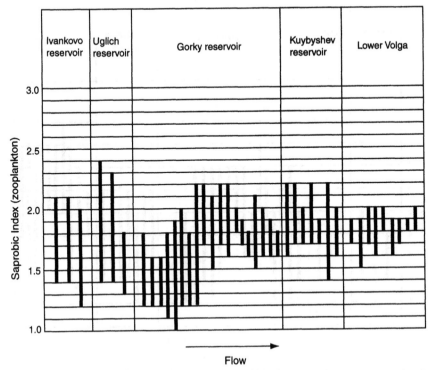

Figure 13.5   Annual range of the zooplankton Saprobic Index on the Lower Volga and its reservoirs at various monitoring stations on the reservoirs in the late 1980s

In the Gorky reservoir, some improvement in the ecological state was recorded in the 1990s. On the whole, the reservoir may be defined as being in a state of ecological stress (Class II). Benthic biocoenoses show less signs of ecological regression than in the Ivankovo and Uglich reservoirs. However, the ecological state of some parts of the reservoir differ noticeably, with certain parts showing stronger signs of eutrophication than others (Table 13.4). The Biotic Index, as a rule, does not exceed 6 and is usually within the range 3–6. However, at times the index falls to 1–2 and shows a dramatic degradation in the biocoenosis structure.

The Kuybyshev reservoir is mainly in a state of ecological stress with some signs of ecological regression (Class II–III). In particular, benthic biocoenoses show signs of ecological regression. The ecological situation is especially unfavourable near the towns of Zelenodolsk, Ulyanovsk, Tolyatti, Naberezhnye Chelny and Nizhnekamsk.

The Lower Volga is in a state of ecological stress (Class II). The only exception is the area of the Astrakhan nature reserve, but even here fish kills have been registered as the result of toxic effects. Organochlorine pesticides have also been found in the tissue of dead fish. In the main

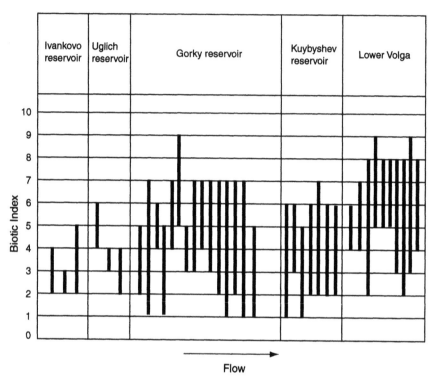

Figure 13.6   Annual range of the Biotic Index on the Lower Volga and its reservoirs at various monitoring stations on the reservoirs in the late 1980s

river, the Saprobic Index based on phytoplankton is 1.6–2.4, and 1.5–2.0 based on zooplankton indicators. The Biotic Index is generally 5–7 and in the Astrakhan reserve it is 7–9. The benthic invertebrates in the Lower Volga show great diversity.

Thus, in general, the most typical condition on the River Volga is ecological stress, followed by signs of ecological regression first showing at the Gorky reservoir (Class II–III). The benthic invertebrates (and bottom sediments) are in a worse ecological condition than the plankton. Signs of ecological regression, which are gradually changing to ecological regression, characterise the state of the benthic invertebrates (Class III–IV). Metabolic regression is occasionally observed at monitoring sites when an unfavourable combination of several factors occurs (Class V). Relatively constant metabolic regression is noted only in the Middle Volga region near large cities and industrial centres, in short river sections < 1–3 km long (Class V).

Table 13.3 Ecological state of five sections of the River Volga based on mean annual phytoplankton Saprobic Index and Woodiwiss Biotic Index, late 1980s

Section	Number of monitoring stations	Annual variations in index value	Number of stations within a given ecological condition				
			I	II	III	IV	V
Ecological state based on the Saprobic Index							
Ivankovo Reservoir	3	1.8–2.1	–	3	–	–	–
Uglich Reservoir	3	1.7–2.1	–	3	–	–	–
Gorky Reservoir	17	1.6–2.3	–	9	8	–	–
Kuybyshev Reservoir	7	1.6–2.9	–	6	1	–	–
Lower Volga	9	1.6–2.4	–	3	6	–	–
Ecological state based on the Biotic Index							
Ivankovo Reservoir	3	2–5	–	–	–	3	–
Uglich Reservoir	3	2–6	–	–	1	2	–
Gorky Reservoir	15	1–9	–	1	7	7	–
Kuybyshev Reservoir	7	1–7	–	–	3	4	–
Lower Volga	10	2–9	–	6	4	–	–

SI	Saprobic Index	III	Signs of ecological regression
BI	Woodiwiss Biotic Index	IV	Ecological regression
I	Ecological integrity	V	Metabolic regression
II	Ecological stress		

Table 13.4 The difference between sections of the Gorky Reservoir in the late 1980s according to quantitative plankton indicators

Quantitative indicators	Reservoir section	
	Rybinsk town to Yaroslavl city	Kostroma city to Gorky hydropower station
Phytoplankton		
Average number (10^3 cells per ml)	4	20
Maximum number (10^3 cells per ml)	18	80
Average biomass (mg l^{-1})	3	10
Maximum biomass (mg l^{-1})	34	38
Zooplankton		
Average number (10^3 individuals per m^3)	15	40
Maximum number (10^3 individuals per m^3)	25	300
Average biomass (g m^{-3})	0.3	7
Maximum biomass (g m^{-3})	0.6	35

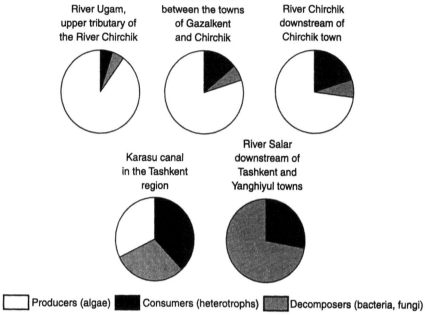

Figure 13.7   Changes in proportions of consumers, decomposers and producers in periphyton from ecologically pristine zones (River Ugam) to zones of heavy pollution (River Salar) in the River Chirchik basin (Uzbekistan), late 1980s

13.2.4 The Central Asian hydrographic region

Alpine and sub-alpine streams
Most mountain streams are in a pristine ecological state (Class I). A specific feature of these waters is a high species diversity (Abakumov and Talskikh, 1985; Talskikh, 1987). Functional structures of periphyton communities at different levels of pollution typical for this region are shown in Figure 13.7. The proportion of different ecological states for various water bodies of Uzbekistan is shown in Figure 13.8.

The Syr Darya basin
Along the entire length of the Syr Darya river, signs of ecological regression are evident (Class III). Highly mineralised drainage waters significantly increase the mineralisation of river water from 470–990 mg l^{-1} TDS during flood periods, to 995–2,035 mg l^{-1} TDS during low water flows (see Chapter 5). This causes the intensive development of sub-saline and halophilic algae.

From the upper reaches of the Syr Darya to the Kairakum reservoir, the Saprobic Index based on periphyton varies from 2.13 to 2.27 and the Biotic Index is generally 4. Downstream of the Kairakum reservoir,

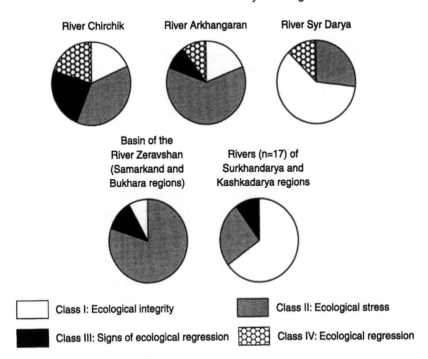

Class I: Ecological integrity   Class II: Ecological stress

Class III: Signs of ecological regression Class IV: Ecological regression

Figure 13.8 Proportion of ecological states for various water bodies in Uzbekistan, late 1980s

there is an improvement in the ecological state. The Saprobic Index does not exceed 2.0 and the Biotic Index increases to 5–6. Downstream of the Farkhade reservoir, the ecological state of the river deteriorates again. In the region of the inflow of the River Keles, the Saprobic Index reaches 2.3 and the Biotic Index occasionally decreases to 2–3. From the Chardarin reservoir to Kazalinsk, the Saprobic Index is generally 1.7–2.0 based on phytoplankton and 1.4–2.1 based on zooplankton (Abakumov and Talskikh, 1987; Talskikh, 1989).

In the upper reaches of Akhangaran river, upstream of the town of Angren, the river is ecologically pristine (Class I). The composition of biocoenoses is determined by cold water species. The Saprobic Index is 1.0–1.44 and the Biotic Index is 8–9. In the middle reach, which is also the longest reach, up to the Tyuyabuguz reservoir, the river is in a state of ecological stress. Downstream of the Tyuyabuguz reservoir, signs of ecological regression can be found and the Biotic Index decreases to 4 (Class II–III) (Figure 13.8).

The River Chirchik is one of the most polluted tributaries of the Syr Darya (Figures 13.8 and 13.9). Only in the upper reaches, upstream of the town of Gazalkent, is the Chirchik ecologically pristine (Class I). The Saprobic Index based on periphyton is 1.3–1.6 and the Biotic Index

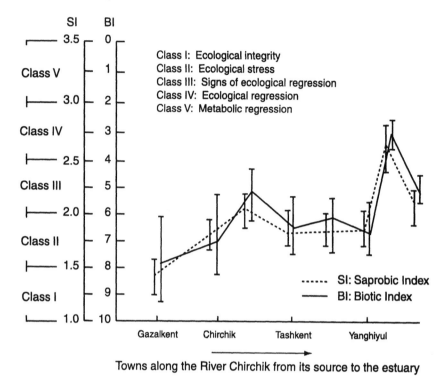

Figure 13.9 Changes in periphyton Saprobic Index and Biotic Index in some reaches of the River Chirchik in the late 19980s

is 6–9 (Figure 13.9). From the town of Gazalkent to the town of Chirchik, the river is in a state of anthropogenic ecological stress (Class II). Downstream of Chirchik, signs of ecological regression are noted (Class III): the Saprobic Index increases to 1.9–2.3 and the Biotic Index decreases to 4–6. In the middle reaches, the River Chirchik forms two branches which rejoin again further downstream. The left branch is characterised by weak signs of ecological regression and the Biotic Index increases to 5–7. The right branch is known as the Kalgan-Chirchik and is in a worse ecological state. The most significant ecological damage is caused by wastewaters from the biochemical plant at Yanghiyul and by the industrial and municipal wastewaters of Tashkent which enter the Kalgan-Chirchik via the River Salar. The Saprobic Index increases considerably to 2.5–2.9 and the Biotic Index decreases to 2–3. Here the biocoenoses are between a state of ecological regression and metabolic regression (Class IV–V). The lower reaches of the river, after the confluence of the two branches, are characterised by signs of ecological regression (Class III). Usually, the Saprobic Index is 2.0–2.3 and the Biotic Index is 4–5.

Table 13.5   Quantitative characteristics of phytoplankton pigment composition in the Kapchagai Reservoir (River Ily, Central Asia), late 1980s

Study period	Chlorophyll concentration (μg l^{-1})				Ratio a:c	Pigment index
	a	b	c	Total		
Spring	1.45	–	0.19	1.55	62.70	2.86
Summer	0.57	0.04	0.05	0.67	4.71	2.97
Autumn	0.62	0.78	0.63	2.03	3.64	2.36
Average	0.88	0.41	0.25	1.42	23.86	2.73

The basin of the River Ily

In the reach from the Kapchagai reservoir to the mouth, the River Ily is characterised by signs of ecological regression (Class III). The Saprobic Index, based on phytoplankton and zooplankton, varies from 1.2–2.0. The Kapchagai reservoir is in a state of ecological stress (Class II). Analysis of the chlorophyll level classifies the reservoir as oligo-mesotrophic (see Chapter 6), with seasonal changes in the taxonomic groups (Table 13.5), and a satisfactory ecological state. The pigment level is relatively consistent across the entire reservoir except for areas of river inflow where maximum concentrations occur. The Saprobic Index is generally 1.8–2.0 and rarely exceeds 2.3. For zooplankton, the dominant species are oligo-saprobic forms and the Saprobic Index is generally 1.1–1.9.

The large Alma-Ata lake, from which the city of Alma-Ata is supplied with fresh drinking water, and the upper reaches of the Great Alma-Atinka river, are in a pristine ecological state (Class I). In the river, at Alma-Ata city and further downstream, signs of ecological regression are recorded (Class III). The Saprobic Index is 0.5–1.0. Downstream of Alma-Ata the Saprobic Index increases to 2.3. The benthos in the upper reaches of the river is characterised by a large diversity of species. The Biotic Index does not decrease below 5. The poorest benthic invertebrate areas are near Alma-Ata city. Downstream of the city and to the river mouth, benthic invertebrate diversity increases again and the Biotic Index varies from 2–7.

The River Nura and the Samarkand reservoir

Along the length of the River Nura except in its upper reaches, the river is characterised by signs of ecological regression (Class III). Phytoplankton diversity is high and there are approximately 100 species present. The Saprobic Index, based on phytoplankton of the River Nura, is not lower than 1.5 and in regions of intensive phytoplankton development, it increases to 3.0–3.5. Here the number and biomass of zooplankton are highest and the Saprobic Index by zooplankton varies from 1.2–2.2.

In terms of benthic invertebrates, the Nura river may be divided into three consecutive reaches. In the upper reach, upstream of the reservoir, there are no wastewater discharges and the Biotic Index is the highest, at 6–7. In the middle reach of the river, the discharge of wastewaters from the town of Temirtau leads to polluted waters and poor biocoenoses. In the lower reaches of the river, where partial self-purification occurs, the Biotic Index, although low, sometimes increases to 5–7, indicating that there is a partial restoration of biocoenoses in the absence of large local sources of pollution.

In the Samarkand reservoir, phytoplankton diversity is high and approximately 130 species occur. From spring to autumn there is a noticeable increase in the Saprobic Index of phytoplankton from 1.0–2.6 in the spring, to 2.2–3.8 in the autumn. The Saprobic Index of zooplankton also changes from 1.4–1.7 in the spring to 1.2–1.9 in the autumn. The Saprobic Index based on periphyton varies during the year from 1.8–2.2. The Biotic Index is generally 4 although at monitoring stations close to wastewater discharge sites this score falls to 0–1.

The River Irtysh basin (within Kazakhstan)
Although the River Irtysh is a tributary of the River Ob and belongs to the Kara Sea hydrographic region, within Kazakhstan it has ecological aspects which are mostly characteristic of rivers of the Central Asian hydrographic region. In the 1990s, the ecological state of the Irtysh improved following a general decrease in the discharge of pollutants to the river. In the reaches around the town of Ust-Kamenogorsk, the Biotic Index varies in the range 1–8. The Saprobic Index, based on phyto-plankton and zooplankton, is generally 1.4–1.7 and in more polluted parts it increases to 2.0. There is a tendency for an increase in species diversity downstream except in areas affected by wastewater discharges. In particular, the composition of benthic invertebrates within Ust-Kamenogorsk city borders changes dramatically and becomes relatively poor.

The upper (mountain) reaches of Bukhtarma reservoir are in a state of ecological stress (Class II). In the lower reaches of the reservoir there are signs of ecological regression (Class III). The Saprobic Index based on phytoplankton is 1.4–2.2, and based on zooplankton is 1.3–1.7. The Biotic Index decreases to 1–2 in the lake (lower) part of the reservoir. The Ust-Kamenogorsk reservoir differs from the Bukhtarma reservoir by having a lower species diversity. The Saprobic Index based on phytoplankton and zooplankton is 1.5–2.4 and the Biotic Index is generally 1–4.

The River Bukhtarma, upstream of Zyryanovsk town, is ecologically pristine (Class I). Downstream of the town, the ecological state deterio-rates and species diversity declines. The average value of the Saprobic

Index from upstream to downstream increases from 1.2 to 1.9 and the average value of the Biotic Index decreases from 6 to 3.

In the River Ishim in Kazakhstan, signs of ecological regression are recorded (Class III). The characteristic values of the Saprobic Index based on phytoplankton are 1.7–2.4 and based on zooplankton they are 1.2–2.2, with higher values recorded in the region of Tselinograd city.

13.2.5 The Pacific Ocean hydrographic region

The basin of the River Amur
The ecological status of the Amur river is mainly influenced by wastewaters from industrial enterprises of the towns Blagoveschensk, Khabarovsk, Amursk, Komsomolsk-on-Amur and Nikolaevsk. Upstream of these towns, the ecosystem of the Amur is in a state of ecological stress (Class II) and is characterised by a high species diversity, a low Saprobic Index (1.5–2.0), a low Oligochaeta Index (up to 40 per cent) and a relatively high Biotic Index (5–7). Downstream of these towns, signs of ecological regression are obvious (Class III). Thus, in the river reach around the town of Khabarovsk, the Saprobic Index is 2.14 whereas downstream of the town, it increases to 2.5. Downstream of Amursk town, the Saprobic Index is approximately 2.5. The greatest anthropogenic impact originates from the towns of Komsomolsk-on-Amur and Nikolaevsk. Downstream of these towns the Saprobic Index reaches 2.5 (Figure 13.10).

The right bank tributary of the River Amur, the River Argun, is in a state of ecological stress (Class II) whereas in parts of the River Shilka, the left bank tributary of the Amur, signs of ecological regression are found (Class III). Downstream of Sretensk town, the river is in a state of ecological regression (Class IV). In winter the dissolved oxygen concentration decreases to 2.58 mg l^{-1}. The ecosystems of both of the tributaries of the River Shilka, the Onona and Ingoda, are also in a state of ecological regression (Class IV). The River Ingoda, at the town of Chita, has a Biotic Index which decreases to 1 and an Oligochaeta Index which reaches 70 per cent. Deficits in dissolved oxygen have been recorded in the downstream reaches of this river every year since 1987. In more recent years, in February and March, the oxygen concentration decreased to 1.73 mg l^{-1} which has resulted in occasional fish kills. The ecosystem of the River Chita at Chita town is in a state of metabolic regression (Class V).

The River Zeya, in the region of the Zeya nature reserve and its adjacent areas, is in an ecologically pristine state (Class I). At the town of Svobodny, the river is in a state of ecological stress (Class II) and at the town of Blagoveschensk, it is in a state of ecological regression (Class IV). The Zeya reservoir is in a state of ecological stress, but near the town

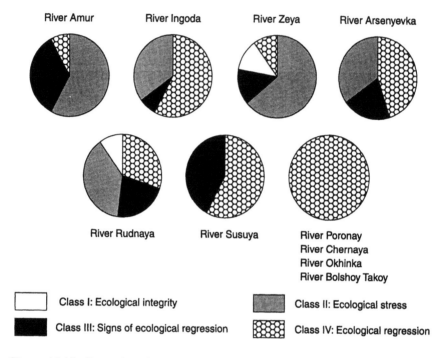

Figure 13.10 Proportion of ecological states for 10 rivers in the Pacific Ocean hydrographic region in the late 1980s

of Zeya the aquatic ecosystem is characterised by signs of ecological regression (Class II–III).

Most of the River Ussuri is in an ecologically pristine state (Class I). It is an important river for fisheries because it provides over-wintering pools for many valuable types of fish and numerous spawning grounds for the salmon *Oncorhynchus keta*. The River Khor is also in a pristine state (Class I) and it is an important tributary of the River Ussuri in terms of fisheries. Unfortunately, the pristine state is not typical of all rivers in the Ussuri basin. On the River Dachnaya, a deficit in dissolved oxygen concentrations has been recorded. The average annual oxygen concentration does not exceed 2.94 mg l^{-1} and in winter it falls to 2 mg l^{-1}. A similar oxygen deficit is also noted near the town of Arsenevsk downstream of which the Biotic Index decreases to 2 and the Saprobic Index is 3.0.

Rivers of the Primorsk Region

For many years, the River Rudnaya has been in a state of ecological regression (Class IV). In 1992 and 1993, there was a further deterioration in the ecological state of the river and in many areas the ecosystem was in a state of metabolic regression (Class V). Downstream of the town of Dalnegorsk, the zooplankton is dominated by pollution-tolerant species and the Saprobic Index varies from 3.6–4.1. The Rivers Spasovka and Kuleshovka suffer from anthropogenic stress (Class II), especially at the town of Spassk-Dalny where the Saprobic Index is 2.78–3.3.

The River Knevichanka is in a state of metabolic regression (Class V) being affected by wastewaters from industries in the town of Artem. The Saprobic Index based on zooplankton is 2.59 and based on periphyton is 2.9. The Rivers Razdolnaya, Komarovka, Rakovka near Ussuriysk town are in a similarly poor state. The Biotic Index in the River Razdolnaya falls to 4 and in its tributaries, the Rivers Komarovka and Rakovka, it falls to 2. In these rivers, the Oligochaeta Index increases to 50–65 per cent, the Saprobic Index based on periphyton is 2.7 and based on zooplankton ranges from 2.9 at Razdolnaya, to 3.4 at Rakovka and 3.55 at Komarovka. However, these rivers have extended reaches where the Saprobic Index based on zooplankton and periphyton do not exceed 1.6 and where the Biotic Index is 6–8 reflecting relatively good environmental conditions.

Rivers of Sakhalin Island
More than 60 per cent of water bodies examined on Sakhalin Island suffer from strong anthropogenic impact. Rivers such as Okhinka, Tym, Chernaya, Ay, Bolshoy Takoy, Susuya, Krasnoselskaya, Bolshaya Alexandrovka, Malaya Alexandrovka and Kozulinka are in a state of ecological regression (Class IV). For many years the River Okhinka at the town of Okha has been the most polluted river in the northern part of Sakhalin. This river has been increasingly polluted by organic substances over recent years. The River Tym is also in a state of ecological regression (Class IV). Downstream of Tymovskoye village, the Oligochaeta Index is 80 per cent and the Biotic Index does not exceed 2.

In 1992–93, there was some improvement in the ecological state of the Rivers Ekhabi, Erry, Kadylanyi, Piltun, Val, Dagi and Poronay. The latter is still in a state of ecological regression downstream of the town of Poronaysk (Class IV). Significant damage to the ecosystem of the River Poronay can be attributed to the polluted waters of a tributary, the River Chernaya, which is characterised in its lower reaches by metabolic regression (Class V).

In 1991–94, the ecological state of the River Lyutogi improved significantly. Along the river course, from the upper reaches to the mouth, the river is ecologically pristine (Class I). The river fauna is characterised by a great diversity of species which are characteristic of clean waters.

Most of the River Susuya is in a state of ecological regression (Class IV). The structure of its benthic communities is extremely simple. The Biotic Index is 1–2 and the Saprobic Index, based on periphyton, reaches the upper limit of the β-mesosaprobic zone at 2.46. Near Yuzhno-Sakhalinsk town, there are significant anthropogenic impacts on the River Krasnoselskaya where the Saprobic Index based on periphyton is 2.7.

Industrial wastewaters from the town of Alexandrovsk-Sakhalinsky also have a negative impact on the ecosystem of the Rivers Bolshaya Alexandrovka, Malaya Alexandrovka and Kozulinka which are in a state of ecological regression (Class IV).

13.3   Permissible ecological state of water bodies

The problem of management and improvement of the ecological state of water bodies can be solved using different approaches depending on the economic, social, aesthetic and scientific importance of the water body. Based on this concept, three categories of water bodies may be defined:

1.   Biosphere reserves or unique water bodies;
2.   Water bodies affected by moderate anthropogenic load;
3.   Water bodies significantly affected by anthropogenic load with already largely transformed ecosystems.

Each category has limits defining the state of the ecosystem. For example, for the third category of waters, although subjected to anthropogenic impact, an ecological state of metabolic regression (Class V) is not permissible. For the second category, a state of anthropogenic ecological regression (Class IV) is not permissible whereas for the first category of waters no anthropogenic ecological impacts are permissible (Izrael *et al.*, 1981).

Along with the general criteria of permissible ecological states for each water body, additional criteria may be introduced which take into account the individual requirements of specific water bodies. For example, in rivers where sturgeon or salmon occur, an additional criterion for the water body is the preservation of these valuable fish stocks. A determining factor must include a social optimum which takes into account the social consequences of environmental quality deterioration. By their nature, these consequences cannot be given a reliable monetary estimate (genetic effects of pollution, aesthetic value, etc.). Generally, the social optimum has a higher level of costs in comparison with the level reached at the economic optimum. The difference between both levels of costs may be great but to achieve the quality demands of the environment dictated by long-term social purposes, additional costs may be incurred (Abakumov, 1992).

13.4   References

(All references are in Russian unless otherwise stated).

Abakumov, V.A. [Ed.] 1976–1983 *Review of the Hydrobiological State of the USSR Surface Waters Based on 1975–1982 Monitoring Results.* VNIIGMI-MCD, Obninsk.

Abakumov, V.A. [Ed.] 1983 *Guidelines for the Hydrobiological Analysis of Surface Waters and Bottom Sediments.* Gidrometeoizdat, Leningrad, 240 pp.

Abakumov, V.A. [Ed.] 1990–1991 *Yearbook of the USSR Surface Water Ecosystem State According to Hydrobiological Indices in 1989–1990.* VNIIGMI-MCD, Obninsk.

Abakumov, V.A. [Ed.] 1991 *Ecological Modifications and Ecological Norms Setting Criteria.* International Symposium Publications, Gidrometeoizdat, Leningrad, 334 pp.

Abakumov, V.A. [Ed.] 1992 *Guidelines on the Hydrobiological Monitoring of Freshwater Ecosystems.* Gidrometeoizdat, St Petersburg, 318 pp.

Abakumov, V.A. [Ed.] 1992–1994 *Russia and its Neighbouring Countries' Surface Water Ecosystem State (According to Hydrobiological Indices) in 1991–1993.* VNIIGMI-MCD, Obninsk.

Abakumov, V.A. and Sustchenya, L.M. 1991 Hydrobiological monitoring of freshwater ecosystems and ways for its improvement. In: *Ecological Modifications and Ecological Norms Setting Criteria.* International Symposium Publications, Gidrometeoizdat, Leningrad, 384 pp.

Abakumov, V.A. and Svirskaya, N.L. 1984–1991 Data of background monitoring at hydrobiological monitoring stations. In: *Review of the Background State of the Environment in the USSR in 1983–1990.* Gidrometeoizdat, Moscow.

Abakumov, V.A. and Talskikh, V.N. 1985 Trends in periphyton community changes following environmental pollution. In: *Problems of Ecological Monitoring and Modelling.* Volume 8, Gidrometeoizdat, Leningrad, 44–59.

Abakumov, V.A. and Talskikh, V.N. 1987 Temporary structure of periphyton populations in pristine ecosystems. In: *Problems of Monitoring in Pristine Natural Environments.* Volume 5, Gidrometeoizdat, Leningrad, 97–107.

Goodnight, C.J. and Whitley, L.S. 1961 Oligochaetes as indicators of pollution. In: Proceedings of the 15th Industrial Waste Conference. Purdue University Eng. Ext. Ser., 106(45), 139–142. (In English).

Izrael, Y.A., Gasilina, N.K. and Abakumov, V.A. 1981 The Hydrobiological Surface Water Quality Monitoring and Control Service. In: *Scientific Principles of Hydrobiological Water Quality Control.* Second British-Soviet Seminar Publications, Gidrometeoizdat, Moscow, 7–15.

Kolkwitz, R. and Marsson, M. 1908 Öekologie der pflanzlichen Saprobien (Ecology of the plant saprobien). *Berichte der deutschen botanischen Gesellschaft Bd,* **26**(a), 505–519. (In German).

Kolkwitz, R. and Marsson, M. 1909 Öekologie der tierischen Saprobien (Ecology of the animal saprobian). *Internat. Revue der Hydrobiologie and Hydrographie Bd,* **2**, 126–152. (In German).

Pantle, R. and Buck, H. 1955 Die biologische Überwachung der Gewässer und die Darstellung der Ergebnisse (Biological monitoring of water bodies and the presentation of results). *Gas - und Wasserfach*, **96**(8), 1–604. (In German).

Proceedings of the Naturalists Society 1870 Proceedings of the Naturalists Society meetings of the Kazan University, First year, Kazan, 24–28.

Sládecek, V. 1973 System of water quality from a biological point of view. *Arch. Hydrobiol. Ergeb. Limnol.*, **7**, 218 pp. (In English).

Talskikh, V.N. 1987 Uzbekistan periphyton studies at different altitudes, flows and pollution levels. Thesis for Candidate of Science in Biology, Moscow State University, Moscow, 24 pp.

Talskikh, V.N. 1989 Periphyton populations state evaluation according to biotic indices. In: *Natural Waters Bioindexing and Biotesting Methods*. Volume 2, Gidrometeoizdat, Leningrad, 51–59.

Woodiwiss, F.S. 1981 Joint British-Soviet biological studies in Nottingham in 1977. In: *Scientific Principles of Hydrobiological Water Quality Control*. Second British-Soviet Seminar Publications, Gidrometeoizdat, Leningrad, 117–189.

Chapter 14*

THE LOWER DON BASIN

The River Don is one of the largest rivers in the European part of Russia (see General Appendix IV). It is 1,967 km long and the basin area is 422,500 km^2. The Don and its tributaries flow through 17 regions of Russia and the Ukraine and discharge into the Taganrog Bay of the Azov Sea. Many cities (Kharkov, Rostov-on-Don, Voronezh, Belgorod, Lugansk, etc.) are situated in its basin, as well as some of the largest industrial enterprises of the Donbass industrial region (Eastern Ukraine) and Southern Russia. The basin is densely populated (50–100 inhabitants per km^2) and is used intensively for agricultural production.

14.1 Principal physico-geographic characteristics

14.1.1 Climate, soil and vegetation
Due to its location, the Lower Don basin is under the influence of different air masses, in particular a cold Arctic air mass, a maritime air mass of the Atlantic Ocean, dry air from Kazakhstan and subtropical air from the Mediterranean basin.

The average annual air temperature fluctuates evenly over the basin from 4 °C in the north to 10 °C in the south. Seasonal differences occur between the north and south of the basin. Spring commences in the south in the first 10 days of March, whereas in the north, spring commences at the beginning of April. Summer is hot and dry and lasts for 100 days in the north to 140 days in the south. The first frosts occur over most of the basin in the first weeks of October. The average annual amount of precipitation increases in the Don region from south-east to north-west from 350 to 500 mm a^{-1}. Maximum precipitation occurs in summer and minimum precipitation (20–30 mm per month) occurs in February to March for the entire basin. The extent and depth of snow cover and the onset and duration of the melt period may change significantly every

---

*   This chapter was prepared by V.V. Tsirkunov, I.K. Akuz and A.A. Zenin. Section 14.5 was prepared by E.N. Bakaeva, V.A. Bryzgalo, and A.V. Zhulidov; Section 14.6 was prepared based on information provided by A.D. Semenov

year. The longest period with snow cover is 120–140 days and the thickest snowpack is 20–35 cm in the north of the basin.

There are two main types of vegetation cover in the Don region, forest and steppe. There are numerous forests in the north, mainly in watersheds, river valleys and ravines. In the south, there are no watershed forests and steppes are now replaced by fields. The extreme south-east, in the upper reaches of the River Sal, is mainly pasture land.

There are three main soil zones in the Don basin:

- Forest steppe with podsol, leached and typical chernozem (upper part of the basin);
- Steppe with typical and southern chernozem (middle and lower basin);
- Dry steppe with dark chestnut and chestnut soils (Volgograd and East Rostov regions).

The agricultural area covers 60 to 80 per cent, and in some places > 80 per cent, of the territory. The main crops are wheat and rye, together with corn, barley, sorghum and millet. Rice is cultivated in areas of irrigated farming. In addition, sugar beet, sunflower and mustard are grown and large areas are covered by grapevines and orchards.

14.1.2 Hydrology

The source of the Don is in the Central Russian highland. The upper reaches flow through a narrow valley which widens downstream, reaching 6–7 km in width and in some places 30 km, in middle reaches. The width of the river bed usually does not exceed 350 m. Downstream of Rostov-on-Don is the swampy Don delta which has an area of 340 km^2.

Near the town of Volgodonsk there is the Tsimlyansk reservoir (Figure 14.1). This reservoir lies on a north-east to south-west axis. It is over 240 km long, has an area of 2,702 km^2 and a volume of 23.8 km^3 when the water level is 36 m. Water level fluctuations may reach 2.5 m. Twenty-four small rivers feed the reservoir but their contribution to the freshwater input of the reservoir does not exceed 6–7 per cent.

Rivers of the basin are typical lowland rivers. Most of them are characterised by wide valleys, the presence of terraces, and a topographic asymmetry with steep right and gentle left slopes. The largest tributaries in the middle reaches of the Don are the rivers Tikhaya Sosna, Bityug, Khoper, Medveditsa, Ilovlya. The Lower Don receives the rivers Chir, Aksai Esaulovsky, Aksai Kurmoyarsky, Tsimla, Seversky Donets, Sal and Manych (Figure 14.1). The Seversky Donets is the largest tributary of the Don. It flows into the Don from the right bank, has a length of 1,053 km, and a watershed area of 90,900 km^2.

The water regime in the Don basin is recharged mainly from winter snow melt, groundwater and precipitation. Data concerning average long-term absolute and specific water discharge and distribution of water flow in seasons are presented in Table 14.1. The specific water

Figure 14.1   The basin of the River Don

Table 14.1 Hydrological characteristics of various rivers in the Don basin

Site[1] River (sampling station)	Distance of station from mouth (km)	Catchment area (km²)	Min. average monthly discharge (m³ s⁻¹)	Max. average monthly discharge (m³ s⁻¹)	Average long-term discharge (m³ s⁻¹)	Average long-term specific discharge (l s⁻¹ km⁻²)	Seasonal distribution of water discharge (%)		
							Spring	Summer/autumn	Winter
1 Don (Kazanskaya)	955	102,000	72	2,750	320	3.14	76.1	15.0	8.9
2 Don (Khovansky)	804	169,000	55	6,700	498	2.97	79.6	13.0	7.4
3 Don (Volgodonsk)	333	255,000	147	1,260	437	1.94	–	–	–
4 Don (Razdorskaya)	151	378,000	249	9,020	683	1.81	79.0	13.8	7.2
5 Don (Rostov)	44	421,000	–	–	–	1.62	–	–	–
6 Don (Azov)	0	422,500	–	–	–	1.62	–	–	–
7 Khoper (Dundukovsky)	45	60,600	32	933	126	2.20	79.5	15.1	5.4
8 Medveditsa (mouth)	66	33,700	12	664	63.2	1.87	78.4	16.3	5.3
9 Ilovlya (Ilovlinskaya)	22	9,110	1	82	7.4	0.85	77.7	14.3	8.0
10 Chir (mouth)	54	8,470	1.6	90	11.9	1.41	73.8	11.3	14.9
11 Aksai Esaulovsky (mouth)	36	2,110	0.3	26.4	1.6	0.76	79.4	8.9	11.7
12 Aksai Kurmoyarsky (mouth)	17	1,810	–	–	1.62	0.87	91.8	5.6	2.6
13 Sal (mouth)	205	19,500	2.6	70.2	9.91	0.51	74.6	17.6	7.8
14 Manych (mouth)	0	39,200	79	134	5.1	1.30	71.7	22.5	5.8
15 Egorlyk (New Egorlyk)	36	146,000	10.1	57.3	35.4	2.40	–	–	–
16 Middle Egorlyk (Shablievskoye)	19	2,170	0.1	1.8	0.34	0.16	61.3	26.7	12.0
17 Seversky Donets (Belaya Kalitva)	119	80,900	18	631	145	1.82	62.2	19.8	18.0
18 Bystraya (Apanaskin)	57	3,730	0.6	59.3	4.63	1.24	–	–	–
19 Kundryuchya (mouth)	22	2,150	1	67	4.8	1.66	66.4	20.0	13.6
20 Tuzlov (Nesvetai)	70	1,910	0	39.1	2.23	1.15	77.0	14.0	9.0

[1] Numbers refer to the sites shown in Figure 14.1

discharge is generally low (often $< 1 \, l \, s^{-1} \, km^{-2}$), which is typical of rivers in the steppe zone.

The average long-term value of water discharge of the Don is $28.7 \, km^3$. However, in recent decades, this value has been significantly reduced. The average discharge for the period 1949–90 was $20.5 \, km^3$, whereas for the period 1970–90 it was $19.1 \, km^3$. This reduction is mainly related to the loss of water during irrigation, an increase in evaporation and the reduction in discharge from small rivers. The reduction in inflow of freshwater into the Azov Sea from the two largest rivers of its basin, the Don and the Kuban, led to a considerable increase in the salinisation of the sea. The change in environmental conditions affected the biota significantly which was one of the principal reasons for a sharp reduction in the productivity of the sea.

The seasonal pattern of water discharge of the rivers of the Don region is generally characterised by high spring flood and low summer–autumn and winter flow (Table 14.1). The proportion of spring flow of various rivers in the basin is 60–90 per cent of the total annual volume of water flow. Therefore, small streams sometimes only flow in spring. Typical hydrographs showing discharge for the Rivers Chir and Middle Egorlyk are shown in Figure 14.2.

The water regime of many rivers in the Don basin has been altered by anthropogenic activity. Dams have been built on many rivers including Seversky Donets, Khoper and Medveditsa. The water regimes of the Sal and Manych rivers have been completely modified. To irrigate lands in these basins, water is transferred from the River Kuban and the Tsimlyansk reservoir. Many streams are transformed into a cascade of ponds for water supply and local irrigation. After the creation of the Tsimlyansk reservoir and the construction of the Konstantinovsk, Nikolaevsk and Kotchetovsk water management facilities, the water regime of the Lower Don changed significantly.

14.2   Use of water resources and problems of water quality management

14.2.1 Water use

The water resources of the Don (especially its lower basin) and the River Seversky Donets, where many large cities and industrial enterprises of the Donbass and Rostov regions are located, are used intensively. The construction of large water management systems (WMS) in the Don basin started at the end of the 1940s. In 1948, the Kuban-Egorlyk canal was built to irrigate dry land by flooding and to dilute with freshwater the highly mineralised waters of the River Manych, a tributary of the Don river. The largest inter-basin, multipurpose WMS was created in 1952 based on the Tsimlyansk reservoir and was designated for irrigation, navigation, fisheries and hydropower supply. River transport on

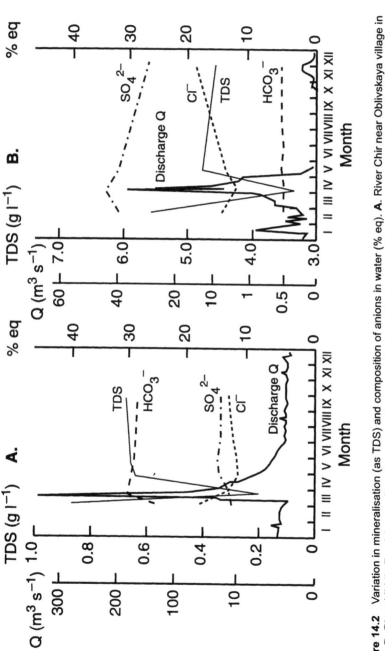

Figure 14.2 Variation in mineralisation (as TDS) and composition of anions in water (% eq). **A.** River Chir near Oblivskaya village in 1951; **B.** River Middle Egorlyk near Shablievskoye village in 1956 (After USSR Surface Water Resources, 1973)

the Don increased greatly after the construction of the Volga-Don canal (see Figure 14.1) and the Tsimlyansk reservoir and with the improvement of navigation conditions on the Lower Don.

In 1958, the construction of the Seversky Donets-Donbass canal (135 km long) was finished together with the associated Raigorodsk and Krasnooskolsk reservoirs. This inter-basin WMS was designated mainly for the water supply of towns and settlements in Donbass, but also for irrigation and power production.

The extreme regulation of discharge has a negative impact on the state of water bodies in the basin. In the Rostov Region alone, there are 16 reservoirs with a volume of more than $10 \times 10^6 \, \text{m}^3$ and 45 reservoirs with a volume of more than $1 \times 10^6 \, \text{m}^3$. An especially unfavourable effect has been created by the construction of closed earth dams on medium and small rivers which were built without adequate planning and were not provided with engineering installations. There are also more than 5,000 small ponds in the region which add to the already high rates of evaporation. Thus, in the basin of the River Sal, 558 ponds and reservoirs have been built, 447 of them without engineering design and with primitive water catchment installations. Loss of water by evaporation reaches 35–40 per cent of total volume.

The present water management scheme is extremely complicated and includes practically all types of water use. Every year, over 60 per cent of river discharge is used, of which 48 per cent is taken by industry (mainly by heat and power plants), 28 per cent by irrigation, 11.5 per cent for fish farming and 6.5 per cent by municipalities. Approximately 18 per cent is lost during transportation (data from the Don Basin Water Management Association (DBWMA)). The annual distribution of water consumption is uneven, with up to 60–65 per cent of the total water volume consumed during the low-water period from June to August and partially in September.

The total volume of water used in the Don basin in 1989 was $17.1 \, \text{km}^3$, including $14.2 \, \text{km}^3$ from surface waters (Main Statistics, 1990). Within the Rostov Region, water abstraction from surface and groundwaters increased between 1989 and 1992 from 5.73 to $6.88 \, \text{km}^3 \, \text{a}^{-1}$ (average $6.22 \, \text{km}^3$) with only 5 per cent coming from groundwater. At present, the Rostov Region, in terms of its total water resources, is among the regions characterised by a water deficit for industrial and drinking purposes. This deficit has restricted the establishment in this region of water-consuming industries and has created problems in supplying drinking water to the population.

Even though the recent industrial decline has tended to reduce water abstraction, the balance between water supply and demand for a number of rivers in the Don basin shows a deficit, especially in semi-dry (75 per cent of demand can be satisfied) and low water (95 per cent of demand can be satisfied) years. Free water resources in the basins of the Sal,

Manych, Big and Middle Egorlyk have been exhausted. In the Rivers Tuzlov and Kundryuchya, amongst others, water resources comprise mainly wastewaters from industrial enterprises and mines and irrigation returns.

14.2.2 Wastewater discharge

The negative impact on the water quality of the Lower Don is produced by the discharge of polluted wastewaters. On the whole, the total annual volume of wastewater discharge in the Don basin from municipalities, irrigation returns and mining waters was 10.65 km^3 in 1989, of which 7.64 km^3 met standards without treatment, 1.02 km^3 met standards after treatment and 1.58 km^3 did not meet standards (Main Statistics, 1990). Approximately 1 km^3 of wastewater was discharged into aquifers and landscape depressions. In the Rostov Region alone, according to data by the DBWMA, 222 enterprises discharge wastewaters in the Don basin, with over 24 per cent of the total volume of these wastewaters not meeting wastewater treatment standards.

The main sources of pollution are municipal waste, the chemical industry, coal industry and irrigation returns (especially from rice paddies). The largest point source of pollution is the Temernik river inflow in Rostov-on-Don into which up to 20,000 m^3 d^{-1} of untreated industrial and domestic wastewaters are discharged. This is one of the main factors leading to water quality deterioration near the mouth of the Don, which in turn has led to cases of intestinal diseases, the closing of beaches, etc.

Official data relating to the discharge of contaminants to the Don basin are presented in Table 14.2. However, for reasons discussed in Chapter 5, these data are significantly underestimated and imprecise, which can be highlighted by comparing data from the Rostov Region in 1990 and 1993. Although large amounts of contaminants are accumulated, transformed and transported from upstream to the lower reaches of the river, the role of these processes must not be exaggerated. For example, industrial wastewaters entering the Seversky Donets near the town of Rubezhansk (Ukraine) are not completely mixed with river waters up to the Taganrog Bay. This is significant because the discharge of wastewaters from the chemical plant 'Krasitel' in Rubezhansk makes up one quarter of the water discharges of the Seversky Donets (N.N. Trunov, personal communication).

The Lower Don is characterised by occasional, short-term, sharp peaks of pollutant concentrations caused by emergency discharges of considerable quantities of wastewaters. In some cases, it is only possible to detect these pollution peaks towards their tail end. However, more frequently, the peak discharges are reported by fishermen and fisheries workers who have sighted dying fish, or crayfish crawling onto the land.

Another negative influence on the water resources of the region results from the violation of water protection regulations, such as by the

Table 14.2   Pollutants discharged annually into rivers of the Don basin

Pollutants	Whole Don basin 1990[1]	Don basin – Rostov region 1990[2]	1993[3]
Suspended substances (10^3 t)	107.7	27.9	33.1
Dry residue (10^3 t)	2,925	559	1,433
Sulphates (10^3 t)	1,194	191	582
Chlorides (10^3 t)	452	114	267
Total phosphorus (t)	4,995	1,118	1,025
NH_4-N (t)	(5,120)	4,652	2,330
NO_3-N (t)	(1,989)	719	12,200
Hydrogen sulphide (t)	(33.2)	27.6	39.2
Iron (t)	1,362	782	240
Petroleum products (t)	4,610	3,150	1,168
$BOD_{full.}$ (10^3 t)	57.6	28.9	11.6
Detergents (t)	(651)	360	111
Fat/oil (t)	(9,086)	8,368	3,216
Copper (t)	48.1	25.8	4.7
Nickel (t)	16.2	19.9	15.2
Total chromium (t)	53.43	50.1	20.3
Zinc (t)	72.8	25.8	5.05
Manganese (t)	5.27	1.44	0.054
Arsenic (t)	–	0.88	0.44
Lead (t)	2.4	2.37	2.44
Aluminium (t)	(152)	4.98	59.9

BOD   Biochemical oxygen demand
[1]   Main Statistics, 1990. Values in brackets are 1991 data from the Don Basin Water Management Association
[2]   Ecological Review, 1993
[3]   Data from the Don Basin Water Management Association

ploughing of land up to the depth of the water table. This has led to small and medium rivers in the Don basin becoming silted, littered and their water resources becoming exhausted.

Following the breakdown of the former Soviet Union, the problem of water quality management in the Don basin has become a serious international problem, because the most intensive pollution of the Don basin occurs in the Donbass, Ukraine, from where it is transported by the Seversky Donets into Russia.

14.3   Surface water quality monitoring

Rivers in the Don basin are amongst the most studied water bodies in the former USSR. The earliest information about the chemical composition of river waters in the Don basin was recorded in 1915–18 in works by

Kashinsky, Chirvinsky and Raznitsyn amongst others. Regular studies of the chemical composition of surface water of the Don basin started in 1936 when Hydrological Yearbooks began publishing the results of hydrochemical observations. Data on the chemical composition of surface water in this territory were presented in hundreds of publications. At the same time, however, there is a lack of up-to-date general reports which give a comprehensive analysis of all the existing information on water quality, including accounts of the sources of pollution, and land and water use.

Ambient monitoring of surface water quality in the Don basin was carried out in the 1980s at 120 monitoring stations with more than 200 observation sites, of which 20 stations with 38 observation sites were situated on the River Don and 10 stations with 20 observation sites on the Seversky Donets. Every year at these sites, more than 1,000 water samples were taken and analysed according to the Guidelines of the USSR State Committee on Hydrometeorology and Environmental Control described in Chapter 4. The results of sample analyses are presented in Table 14.3. At many stations, the data series for water chemical composition covers 25–30 years or more. Hydrobiological observations were also regularly carried out. In particular, in the Lower Don basin there were 25 ambient monitoring stations at six tributaries, two River Don arms and two reservoirs. However, it should be noted that not all pollutants with a negative impact on the Don ecosystem are regularly monitored and very few observations are made for toxic compounds in sediments and biota.

14.4   Surface water quality

14.4.1 Natural water quality in the Don basin

The chemical composition of surface waters in the Don basin is characterised by great variety, which is related to differences in the geophysical and geological conditions prevalent where the water is formed. The natural mineralisation, expressed as total dissolved solids (TDS) of river waters in the Don basin varies from 0.1 to 7 g l^{-1} TDS, and the predominant anions vary from bicarbonates in the north-west to chlorides and sulphates in the south-east. Specific hydrochemical features can be clearly observed in local runoff. A series of hydrochemical maps showing the main phases of an average hydrograph have been produced based on data for major ions and total water mineralisation from the mid-1930s to the end of the 1960s on 50 small river catchments (USSR Surface Water Resources, 1973). An example of these maps for the period of winter low flow when groundwater is the main source of water in small rivers is presented in Figure 14.3. Under such conditions, the variation in composition and mineralisation of the water is greatest.

Table 14.3 Statistical distribution of water quality variables at key stations on rivers of the Don basin, 1980–93

River stations[1]		Ca^{2+}	Mg^{2+}	HCO_3^-	SO_4^{2-}	Cl^-	TDS	Si	Fe	NO_3-N	NH_4-N
1. Don	X_5	38.4	9.48	148	40.6	19.5	314	1.49	0.00	0.01	0.07
(Kazanskaya)	X_{50}	64.5	19.3	218	73.0	37.4	479	4.15	0.12	0.30	0.30
	X_{95}	96.6	34.4	376	148	63.1	704	11.0	2.09	0.84	1.26
2. Don	X_5	41.8	9.61	145	45.4	18.8	308	1.50	0.01	0.00	0.11
(Khovansky)	X_{50}	69.9	20.2	246	82.2	45.0	498	4.50	0.14	0.20	0.36
	X_{95}	98.5	33.9	340	179	73.6	727	9.87	1.58	0.72	1.14
3. Don	X_5	31.4	13.0	127	46.8	30.5	284	1.39	0.00	0.00	0.00
(Volgodonsk)	X_{50}	46.2	21.6	182	73.3	45.4	392	3.60	0.05	0.17	0.12
	X_{95}	56.7	43.1	247	138	71.2	545	8.74	0.45	0.56	0.39
4. Don	X_5	50.6	18.2	146	107	86.1	542	1.51	0.04	0.08	0.03
(Razdorskaya)	X_{50}	76.0	26.5	208	171	143	755	4.55	0.20	0.39	0.07
	X_{95}	109	46.4	299	231	216	1,100	9.37	1.94	1.85	0.40
5. Don	X_5	49.8	20.0	134	92.6	70.2	445	0.48	0.00	0.01	0.00
(Rostov)	X_{50}	84.0	37.0	194	197	141	759	3.40	0.25	0.20	0.06
	X_{95}	117	58.2	268	292	238	1,010	15.3	1.00	1.88	0.80
6. Don	X_5	49.8	22.8	145	84.8	68.2	475	0.60	0.00	0.01	0.00
(Azov)	X_{50}	79.8	38.6	190	195	160	774	3.40	0.23	0.15	0.06
	X_{95}	117	56.9	252	303	254	1,110	18.0	1.36	1.10	0.35
7. Khoper	X_5	39.8	6.88	124	52.0	24.0	289	2.05	0.00	0.00	0.02
(Dundukovsky)	X_{50}	75.9	22.0	224	98.4	51.5	549	4.80	0.13	0.16	0.21
	X_{95}	110	32.8	398	195	99.1	858	13.8	2.18	0.70	1.02
8. Medveditsa	X_5	40.1	9.5	100	56.2	31.4	277	1.42	0.01	0.00	0.02
(mouth)	X_{50}	76.6	21.3	226	100	80.1	590	5.20	0.11	0.12	0.21
	X_{95}	106	40.5	350	235	121	818	14.2	1.11	0.72	1.06
9. Ilovlya	X_5	48.7	6.89	147	62.3	51.6	407	0.49	0.00	0.00	0.01
(Ilovlya)	X_{50}	89.4	22.9	266	116	119	696	5.40	0.08	0.10	0.19
	X_{95}	129	44.0	379	242	197	993	13.4	0.56	0.43	0.87
10. Chir	X_5	32.5	10.4	129	38.6	20.7	274	1.02	0.00	0.00	0.06
(mouth)	X_{50}	53.5	18.5	190	78.3	44.3	439	3.55	0.05	0.27	0.23
	X_{95}	73.9	49.0	318	183	84.4	712	11.5	0.50	0.76	1.24
11. Aksai	X_5	39.7	12.2	131	43.7	36.3	298	0.80	0.00	0.00	0.05
Esaulovsky	X_{50}	49.1	22.4	184	82.6	51.8	443	2.35	0.05	0.13	0.23
mouth)	X_{95}	67.7	55.7	264	194	147	949	9.82	0.49	0.55	1.24
12. Aksai	X_5	31.7	13.2	120	35.5	34.2	294	0.85	0.00	0.00	0.00
Kurmoyarsky	X_{50}	52.5	22.3	191	85.4	52.3	442	2.90	0.04	0.15	0.23
(mouth)	X_{95}	79.2	76.5	262	441	385	1,560	12.4	0.34	0.55	0.94
13. Sal	X_5	73.6	30.0	130	211	153	863	1.23	0.13	0.07	0.04
(mouth)	X_{50}	115	53.4	247	375	320	1,370	4.10	0.32	0.68	0.10
	X_{95}	209	105	450	775	788	2,910	11.3	1.15	7.66	0.42
14. Manych	X_5	89.6	56.6	148	26.7	23.5	915	1.35	0.09	0.12	0.04
(mouth)	X_{50}	119	93.4	195	750	337	1,870	3.00	0.25	0.42	0.12
	X_{95}	164	132	314	906	576	2,480	8.32	1.25	2.69	0.45
15. Egorlyk	X_5	91.9	54.0	145	506	166	1,210	1.50	0.06	0.09	0.04
(New Egorlyk)	X_{50}	149	120	208	1,050	322	2,300	5.80	0.40	0.60	0.10
	X_{95}	217	216	332	1,910	487	3,710	13.3	3.26	4.53	0.70
16. Middle Egorlyk	X_5	114	120	154	676	296	2,560	1.70	0.62	0.11	0.07
(Shablievskoye)	X_{50}	171	199	333	1,860	523	3,950	6.55	0.36	0.52	0.13
	X_{95}	217	280	551	2,720	756	5,410	14.9	2.47	2.96	1.55

Continued

Table 14.3   Continued

River stations[1]		Ca^{2+}	Mg^{2+}	HCO_3^-	SO_4^{2-}	Cl^-	TDS	Si	Fe	NO_3-N	NH_4-N
17. Donets	X_5	96.9	21.6	193	161	193	840	1.50	0.08	0.12	0.02
(Belaya Kalitva)	X_{50}	144	37.9	269	320	298	1,300	5.75	0.30	0.76	0.12
	X_{95}	184	56.9	412	392	404	1,610	13.6	1.73	4.01	1.00
18. Bystraya	X_5	47.6	11.0	104	65.1	35.8	309	1.80	0.09	0.04	0.03
(Apanaskin)	X_{50}	102	43.6	241	305	221	1,150	6.15	0.26	0.50	0.09
	X_{95}	146	77.3	379	515	356	1,600	13.5	1.27	3.74	0.61
19. Kundryuchya	X_5	74.3	25.0	203	299	101	938	0.79	0.04	0.14	0.00
(mouth)	X_{50}	148	59.6	276	624	241	1,700	4.00	0.28	0.48	0.07
	X_{95}	223	105	442	1,030	397	2,400	14.0	1.85	5.78	0.22
20. Tuzlov	X_5	113	46.5	188	370	132	1,230	1.60	0.06	0.09	0.00
(Nesvetai)	X_{50}	184	96.2	297	1,020	244	2,230	4.50	0.24	0.45	0.01
	X_{95}	249	146	442	1,580	462	3,340	10.2	1.34	4.73	0.53

X_5, X_{50}, X_{95} Percentile of statistical distribution   [1]   Numbers refer to sites on Figure 14.4
TDS   Total dissolved solids

There are three types of water catchments in the Don basin. The first type includes those where the soil layer and underlying rocks have been well flushed removing easily-dissolved salts, chlorides and sulphates. Streams formed on this type of catchment have relatively poorly mineralised water (< 1 g l^{-1} TDS) during the whole year and belong to the Ca^{2+}–HCO_3^- group of waters. Catchments of the first type include small rivers of the Upper Don and most small rivers of the Middle Don, as well as catchments of the Rivers Chir and Aksenets in the Lower Don. The changes in mineralisation and relative composition of anions over a year, as well as a hydrograph of water discharge for rivers of this type, are presented in Figure 14.2A. Catchments of the second type are those where the soil layer and underlying rocks contain significant amounts of chlorides and sulphates. This results in river water with either a chloride or sulphate character such as the Lower Don tributaries of the Egorlyk, Middle Egorlyk, Tashla, Tuzlov and Great Nesvetai. These river waters, during the whole year, are characterised by high mineralisation (TDS up to 6 g l^{-1}) and a Na^+–SO_4^{2-} or Ca^{2+}–SO_4^{2-} composition (Figure 14.2A). The third type is intermediate and includes catchments where the easily-dissolved salts are relatively well flushed from the soil layer, while underlying rocks contain them in considerable quantities (e.g. the Rivers Buzuluk, Aksai Esaulovsky, Aksai Kurmoyarsky, Karpovka).

14.4.2 Characteristics of present day water quality

Information on the average concentration and range of water quality variables in the Don basin, based on data collected by the State monitoring network between 1980–93, is presented in Tables 14.3 and 14.4a,

Table 14.3 Continued

River stations[1]		Concentration (mg l^{-1})						Concentration (µg l^{-1})		
		PO$_4$-P	O$_2$	BOD$_5$	Petrol. P.	Phenols	Dets	Cu	Zn	DDT[2]
1. Don	X$_5$	0.019	5.19	1.24	0.00	0.000	0.00	0.00	0.00	0.054
(Kazanskaya)	X$_{50}$	0.144	10.5	3.00	0.09	0.000	0.04	4.00	1.00	0.083
	X$_{95}$	0.498	16.3	5.98	1.42	0.004	0.09	10.0	13.8	0.244
2. Don	X$_5$	0.022	6.70	1.10	0.00	0.000	0.00	0.00	0.00	0.000
(Khovansky)	X$_{50}$	0.136	11.6	3.22	0.10	0.001	0.03	3.00	1.00	0.055
	X$_{95}$	0.466	17.4	7.46	0.99	0.005	0.09	13.0	10.1	0.331
3. Don	X$_5$	0.010	5.13	0.93	0.00	0.000	0.00	0.00	0.00	0.083
(Volgodonsk)	X$_{50}$	0.075	9.83	2.71	0.06	0.001	0.02	2.00	1.00	0.162
	X$_{95}$	0.170	15.7	7.57	1.29	0.004	0.08	12.0	41.8	0.650
4. Don	X$_5$	0.024	8.20	1.85	0.02	0.001	0.05	4.00	1.00	0.000
(Razdorskaya)	X$_{50}$	0.060	12.9	4.20	0.12	0.003	0.07	7.50	10.0	0.012
	X$_{95}$	0.153	19.9	9.70	0.87	0.010	0.16	13.5	29.1	0.046
5. Don	X$_5$	0.004	6.95	1.43	0.00	0.000	0.00	0.00	0.00	0.032
(Rostov)	X$_{50}$	0.083	10.9	3.65	0.13	0.003	0.04	7.5	3.00	0.058
	X$_{95}$	0.210	14.9	7.79	0.92	0.015	0.12	14.0	18.0	0.740
6. Don	X$_5$	0.016	6.38	1.36	0.00	0.000	0.00	0.00	0.00	0.022
(Azov)	X$_{50}$	0.017	9.58	3.34	0.10	0.003	0.03	3.00	2.00	0.033
	X$_{95}$	0.259	14.1	6.76	0.68	0.017	0.11	10.0	19.0	0.200
7. Khoper	X$_5$	0.072	5.79	1.00	0.00	0.000	0.00	0.00	0.00	0.072
(Dundukovsky)	X$_{50}$	0.250	10.5	2.78	0.10	0.000	0.02	2.00	1.00	0.103
	X$_{95}$	0.701	15.1	6.51	0.78	0.005	0.11	11.0	10.0	0.447
8. Medveditsa	X$_5$	0.031	7.98	1.00	0.00	0.000	0.00	0.00	0.00	0.030
(mouth)	X$_{50}$	0.130	11.1	2.43	0.10	0.000	0.02	2.00	2.00	0.063
	X$_{95}$	0.380	15.1	5.04	0.61	0.004	0.07	11.0	13.0	0.275
9. Ilovlya	X$_5$	0.053	6.13	1.00	0.00	0.000	0.00	0.00	0.00	0.062
(Ilovlya)	X$_{50}$	0.179	9.50	2.40	0.07	0.000	0.02	3.00	2.00	0.125
	X$_{95}$	0.770	12.9	6.21	0.70	0.005	0.12	13.0	15.0	0.600
10. Chir	X$_5$	0.004	7.14	1.42	0.00	0.000	0.00	0.00	0.00	0.105
(mouth)	X$_{50}$	0.145	9.89	4.43	0.07	0.001	0.03	4.00	0.00	0.192
	X$_{95}$	0.473	13.5	10.4	0.32	0.007	0.11	11.5	23.8	0.355
11. Aksai	X$_5$	0.015	6.22	1.53	0.00	0.000	0.00	0.00	0.00	0.068
Esaulovsky	X$_{50}$	0.076	10.3	4.74	0.08	0.001	0.03	3.0	0.00	0.140
(mouth)	X$_{95}$	0.177	15.6	8.21	0.30	0.003	0.14	12.5	15.0	0.301
12. Aksai	X$_5$	0.012	6.39	1.39	0.00	0.000	0.00	0.00	0.00	0.206
(Kuroyarsky)	X$_{50}$	0.071	10.2	4.84	0.06	0.001	0.02	3.00	0.00	0.330
mouth	X$_{95}$	0.241	16.3	8.98	0.30	0.004	0.11	10.0	20.2	0.526
13. Sal	X$_5$	0.025	6.93	1.20	0.00	0.002	0.05	4.00	3.00	0.010
(mouth)	X$_{50}$	0.080	12.4	3.41	0.14	0.005	0.08	8.00	11.0	0.265
	X$_{95}$	0.183	19.1	9.32	0.81	0.020	0.18	14.0	24.8	0.633
14. Manych	X$_5$	0.019	7.44	1.52	0.03	0.012	0.05	4.00	0.50	0.000
(mouth)	X$_{50}$	0.080	11.9	3.20	0.16	0.003	0.07	7.00	9.00	0.016
	X$_{95}$	0.211	19.0	7.53	0.79	0.007	0.11	13.0	29.5	0.033
15. Egorlyk	X$_5$	0.012	3.16	1.21	0.01	0.002	0.05	4.85	4.00	0.000
(New Egorlyk)	X$_{50}$	0.060	11.2	3.20	0.01	0.001	0.05	4.85	4.00	0.015
	X$_{95}$	0.231	17.3	6.70	0.81	0.008	0.16	15.1	48.1	0.031
16. Middle Egorlyk	X$_5$	0.045	3.52	1.09	0.02	0.001	0.05	4.00	5.00	0.000
(Shablievskoye)	X$_{50}$	0.120	10.1	4.20	0.20	0.003	0.08	9.00	16.0	0.000
	X$_{95}$	0.935	17.9	15.5	1.40	0.010	0.14	14.0	54.0	0.000

Continued

Table 14.3   Continued

River stations[1]		Concentration (mg l^{-1})						Concentration (μg l^{-1})		
		PO$_4$-P	O$_2$	BOD$_5$	Petrol. P.	Phenols	Dets	Cu	Zn	DDT[2]
17. Donets	X$_5$	0.038	7.45	1.96	0.02	0.000	0.05	0.00	5.0	0.000
(Belaya Kalitva)	X$_{50}$	0.105	11.0	4.20	0.24	0.003	0.08	9.00	18.0	0.000
	X$_{95}$	1.010	16.6	8.20	1.14	0.015	0.28	19.7	75.8	0.030
18. Bystraya	X$_5$	0.027	8.60	2.20	0.01	0.002	0.04	1.90	2.00	0.000
(Apanaskin)	X$_{50}$	0.069	11.8	4.75	0.19	0.005	0.07	8.00	12.0	0.000
	X$_{95}$	0.204	17.5	9.70	1.94	0.012	0.18	15.1	37.7	0.030
19. Kundryuchya	X$_5$	0.003	8.25	1.22	0.00	0.014	0.05	1.70	1.05	0.000
(mouth)	X$_{50}$	0.072	10.2	2.50	0.14	0.005	0.08	8.00	10.0	0.184
	X$_{95}$	0.256	13.5	6.85	1.77	0.019	0.14	15.7	24.0	0.205
20. Tuzlov	X$_5$	0.016	8.5	1.9	0.00	0.001	0.05	5.00	3.00	0.000
(Nesvetai)	X$_{50}$	0.084	13.9	4.9	0.10	0.004	0.07	8.00	10.0	0.000
	X$_{95}$	0.350	21.2	10.5	1.14	0.008	0.10	16.1	35.7	0.019

X$_5$, X$_{50}$, X$_{95}$ Percentile of statistical distribution
BOD   Biological oxygen demand
Petrol. P.   Petroleum products
Dets   Detergents

DDT   Dichlorodiphenyltrichloroethane
[1]   Numbers refer to sites on Figure 14.4
[2]   For DDT statistical distribution = X$_{90}$, X$_{95}$ and X$_{max}$

b and c. Figure 14.4 is based on the data in Table 14.3 and shows the range of variation of concentrations and the maximum allowable concentrations (MAC) for sulphates, ammonium-nitrogen, easily oxidisable organic matter (as biochemical oxygen demand, BOD$_5$) and petroleum hydrocarbons for some sites on the Lower and Middle Don. A comparison of average values with median values shows that, in most cases, the distribution of concentrations is characterised by positive asymmetry, especially for heavy metals and pesticides. The distribution of concentrations is closer to normal for oxygen, BOD$_5$ and detergents.

A comparison of concentrations of water quality variables for the Don tributaries and for water bodies of the Seversky Donets basin (Tables 14.4a,b,c) shows that the latter is more contaminated, in particular with oil products, phenols, detergents, nutrients and most metals. However, the tributaries of the Don are characterised by higher pesticide concentrations than the rivers of the Seversky Donets basin due to greater agricultural production and irrigation, especially in the lower reaches of the Don.

Major ions
After the construction of the Tsimlyansk reservoir, the amplitude of inter-annual fluctuations of major ion concentrations in the Lower Don was reduced and their temporal variability changed (Figure 14.5). In more recent decades, maximum concentrations have been observed in spring with minimum concentrations in summer and early autumn when river waters are released from the reservoir which, during this

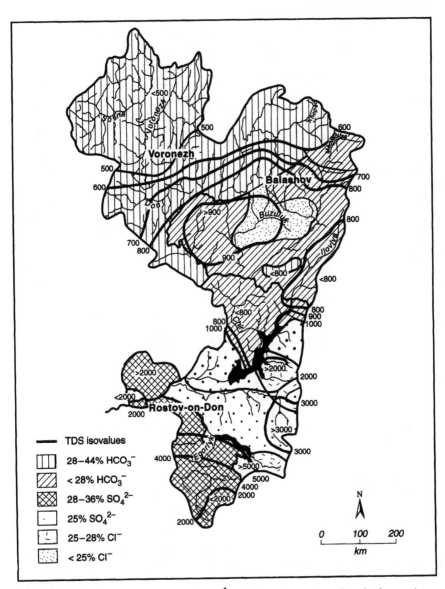

Figure 14.3   Mineralisation or TDS (mg l^{-1}) and composition of anions in river water during winter low water flow

season, still contains poorly mineralised snow melt water (Figure 14.5b). During the period of algal blooms, a pH increase of 8.5–9.25 in the reservoir water has been noted due to CO_2 consumption, and significant $CaCO_3$ precipitation has also been observed, resulting from a supersaturation of up to 730–1,045 per cent (Tsimlyansk, 1977). As a result of

Table 14.4a Statistical distribution of water quality variables in tributaries of the Don (excluding Seversky Donets), 1980–90

Indicator	Concentration (mg l⁻¹)[1]					
	X_5	X_{50}	X_{mean}	X_{95}	X_{99}	n
SO_4^{2-}	27.0	71.2	327	1,652	7,377	4,306
Cl^-	10.7	34.5	238	560	7,220	4,324
TDS	260	551	1,241	3,681	24,896	3,998
Dissolved oxygen	4.96	9.06	9.27	14.5	19.0	5,644
BOD_5	0.844	2.37	2.88	6.92	11.1	4,500
NH_4-N	0.032	0.249	0.456	1.748	3.810	4,783
NO_3-N	0.024	0.620	1.106	3.981	6.419	4,397
NO_2-N	0.000	0.021	0.052	0.256	0.498	4,812
Total Fe	0.000	0.142	0.248	0.882	2.291	4,781
Phenols	0.000	0.000	0.001	0.007	0.012	3,924
Petroleum products	0.000	0.051	0.119	0.515	1.631	4,442
Detergents	0.000	0.033	0.044	0.146	0.268	4,581
Cu	0.000	0.001	0.003	0.013	0.021	4,637
Zn	0.000	0.000	0.004	0.021	0.047	4,616
Ni	0.000	0.000	0.001	0.010	0.023	3,079
Cr	0.000	0.000	0.000	0.000	0.006	2,405
Pb	0.000	0.000	0.002	0.012	0.040	693
Hg	0.000	0.000	0.000	0.000	0.000	6
Cd	0.000	0.000	0.000	0.000	0.000	60
As	0.000	0.009	0.011	0.043	0.103	98
DDT (µg l⁻¹)	0.000	0.000	0.020	0.099	0.369	3,455
HCH (µg l⁻¹)	0.000	0.000	0.013	0.051	0.110	3,466
Formaldehyde	–	–	–	–	–	0

X_5, X_{50}, X_{95}, X_{99} represent the percentiles of statistical distribution
n   Number of samples
TDS   Total dissolved solids
BOD   Biochemical oxygen demand

DDT   Dichlorodiphenyltrichloroethane
HCH   Hexachlorocyclohexane
[1]   Concentrations in mg l⁻¹ unless otherwise indicated

this process, the concentrations of HCO_3^- and Ca^{2+} downstream at the Volgodonsk site are lower than in the upstream reaches (Table 14.3).

Processes of surface water salinisation in the Don basin are intensive (see Chapter 5). Concentrations of chlorides, sulphates and total dissolved solids often exceed MAC (Tables 14.3 and 14.4) and may be explained by both natural and anthropogenic factors.

The analysis of long-term trends of major ion concentrations has been carried out using non-parametric methods. For example, at the Aksai monitoring station on the Lower Don the median concentrations at the end of the 1980s to the beginning of the 1990s exceeded those of the late 1940s to early 1950s for Mg^{2+}, $Na^+ + K^+$, SO_4^{2-}, Cl^- and TDS by

Table 14.4b Statistical distribution of water quality variables in rivers of the Seversky Donets basin, 1980–90

Indicator	Concentration (mg l^{-1})[1]					n
	X_5	X_{50}	X_{mean}	X_{95}	X_{99}	
SO_4^{2-}	30.6	159	270	892	1,724	5,199
Cl^-	19.5	230	219	573	902	8,117
TDS	413	936	1,089	2,397	3,726	4,949
Dissolved oxygen	4.66	9.02	8.90	13.2	16.6	12,197
BOD_5	1.09	3.43	3.78	7.79	11.4	6,511
NH_4-N	0.000	0.634	1.030	3.403	6.469	8,740
NO_3-N	0.052	0.943	1.761	6.465	10.85	5,114
NO_2-N	0.000	0.036	0.086	0.407	0.940	6,561
Total Fe	0.000	0.204	0.315	1.039	2.231	5,819
Phenols	0.000	0.002	0.004	0.018	0.029	6,671
Petroleum products	0.000	0.205	0.348	1.432	3.127	6,322
Detergents	0.000	0.056	0.065	0.183	0.412	6,387
Cu	0.000	0.006	0.008	0.024	0.067	6,084
Zn	0.000	0.008	0.014	0.056	0.241	5,922
Ni	0.000	0.006	0.008	0.042	0.088	2,120
Cr	0.000	0.001	0.003	0.018	0.061	4,214
Pb	0.000	0.000	0.005	0.048	0.094	612
Hg	0.000	0.000	0.000	0.003	0.009	336
Cd	0.000	0.000	0.017	0.122	0.184	31
As	0.000	0.000	0.000	0.000	0.000	62
DDT ($\mu g\ l^{-1}$)	0.000	0.000	0.007	0.036	0.146	3,376
HCH ($\mu g\ l^{-1}$)	0.000	0.000	0.009	0.042	0.112	3,376
Formaldehyde	0.000	0.020	0.069	0.381	0.941	1,971

X_5, X_{50}, X_{95}, X_{99} represent the percentiles of statistical distribution
n   Number of samples
TDS   Total dissolved solids
BOD   Biochemical oxygen demand

DDT   Dichlorodiphenyltrichloroethane
HCH   Hexachlorocyclohexane
[1]   Concentrations in mg l^{-1} unless otherwise indicated

125 per cent, 184 per cent, 116 per cent, 190 per cent and 63 per cent respectively. For sulphates and chlorides (Figure 14.6) the increase in concentrations over time was variable, being most rapid from the late 1950s to the mid-1970s which, on the whole, corresponds to the period of faster growth in irrigated land area and the intensification of the use of fertilisers in agriculture. The increase in the concentrations of chlorides, sulphates, the sum of sodium and potassium and TDS was also notable for most tributaries of the Lower Don in the 1980s (Figures 14.6 B and C). Furthermore, the annual increase in water mineralisation reached 10–20 mg l^{-1} on the Rivers Sal, Egorlyk, Kundryuchya and Tuzlov.

Table 14.4c Statistical distribution of water quality variables in rivers of the Don basin as a whole, 1980–90

| Indicator | Concentration (mg l^{-1})[1] | | | | | |
	X_5	X_{50}	X_{mean}	X_{95}	X_{99}	n
SO_4^{2-}	29.2	93.8	246	939	1,990	13,200
Cl^-	14.2	96.4	192	557	1,044	16,124
TDS	309	619	987	2,427	5,388	12,547
Dissolved oxygen	5.15	9.42	9.48	14.3	18.4	28,128
BOD_5	1.00	3.04	3.50	7.78	11.5	17,806
NH_4-N	0.000	0.303	0.699	2.943	5.584	18,226
NO_3-N	0.019	0.631	1.266	4.714	8.078	13,204
NO_2-N	0.000	0.027	0.065	0.337	0.655	16,306
Total Fe	0.000	0.163	0.279	0.998	2.263	13,824
Phenols	0.000	0.001	0.003	0.014	0.026	16,478
Petroleum products	0.000	0.096	0.226	0.932	2.15	17,008
Detergents	0.000	0.043	0.052	0.153	0.302	15,921
Cu	0.000	0.004	0.005	0.019	0.036	15,919
Zn	0.000	0.004	0.008	0.032	0.080	15,665
Ni	0.000	0.000	0.004	0.021	0.059	6,187
Cr	0.000	0.000	0.002	0.013	0.033	7,235
Pb	0.000	0.000	0.003	0.020	0.084	1,729
Hg	0.000	0.000	0.000	0.003	0.011	507
Cd	0.000	0.000	0.003	0.005	0.117	166
As	0.000	0.000	0.004	0.017	0.063	320
DDT (µg l^{-1})	0.000	0.000	0.018	0.093	0.380	10,207
HCH (µg l^{-1})	0.000	0.000	0.009	0.044	0.110	10,215
Formaldehyde	0.000	0.020	0.069	0.381	0.941	1,971

X_5, X_{50}, X_{95}, X_{99} represents the percentiles of statistical distribution
n   Number of samples
TDS   Total dissolved solids
BOD   Biochemical oxygen demand

DDT   Dichlorodiphenyltrichloroethane
HCH   Hexachlorocyclohexane
[1]   Concentrations in mg l^{-1} unless otherwise indicated

Organic pollution and nutrients

Dissolved oxygen concentrations are generally satisfactory in the Don basin. On the whole, not more than 6–7 per cent of samples have an oxygen concentration < 6 mg l^{-1}. The median oxygen concentration decreases in the lower reaches of the river from 12.9 mg l^{-1} in Razdorskaya to 9.6 mg l^{-1} in Azov which is due to inputs of untreated or inadequately treated wastewaters.

The content of easily oxidisable organic matter (BOD_5) generally varies between 1 and 10 mg l^{-1} (Tables 14.3 and 14.4) and median concentrations for most monitoring stations are close to the MAC (Figure 14.4). Chemical oxygen demand (COD) values are much higher

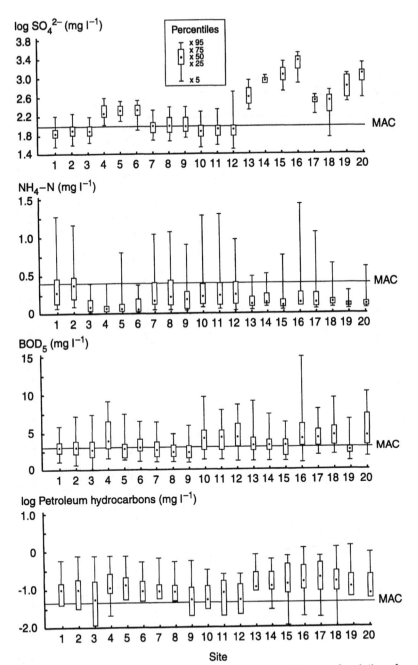

Figure 14.4 Box-wisker plots characterising the median and range of variation of concentrations of SO_4^{2-}, NH_4-N, BOD_5 and petroleum hydrocarbons, at various sites in the Don basin. For site location see Figure 14.1 and Table 14.1. MAC = Maximum allowable concentration

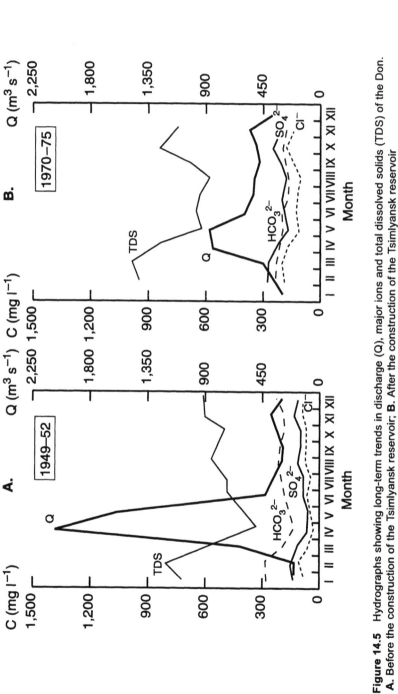

Figure 14.5 Hydrographs showing long-term trends in discharge (Q), major ions and total dissolved solids (TDS) of the Don. **A.** Before the construction of the Tsimlyansk reservoir; **B.** After the construction of the Tsimlyansk reservoir

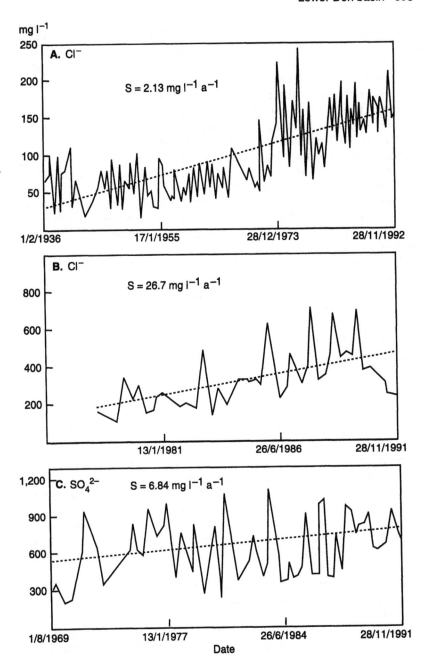

Figure 14.6 Long-term changes in quarterly-adjusted concentrations of chlorides.
A. Don river (Razdorskaya); **B.** Sal river (mouth); **C.** Quarterly-adjusted concentrations
of sulphates in the Kundryuchya river (mouth). S (slope) = Rate of annual increase

at 30–50 mg l^{-1} for the Lower Don and reach 100 mg l^{-1} for some observation sites on the Seversky Donets. There has been a trend for BOD$_5$ and COD to increase on the Lower Don (Razdorskaya, Aksai, Rostov and Azov) over the 10–15 years prior to 1995 (Figure 14.7A).

Of the nutrients, nitrite exceeds the MAC most often. Its median concentrations for most samples are close to the MAC. In the Don basin as a whole, NO$_2$$^-$ concentrations exceeded 1 MAC in 57.2 per cent of samples, 10 MAC in 6.62 per cent of samples, and 50 MAC in 0.34 per cent of samples. The corresponding values for NH$_4$$^+$ were 42.7 per cent, 1.96 per cent and 0.02 per cent. Of 13,204 NO$_3$$^-$ samples taken between 1980 and 1990, approximately 0.5 per cent only exceeded 1 MAC (9 mg l^{-1} N), and they were mostly from the Seversky Donets basin. Maximum concentrations and variability of phosphates occurred in the Seversky Donets, of silica in the mouth of the Don (towns of Rostov and Azov) and of iron in the highly mineralised tributaries of the Manych river.

The spatial distribution of nutrients shows a reduction in phosphates and ammonium and an increase in iron along the course of the Don. Nitrates decrease from Kazanskaya to Volgodonsk, increase at Razdorskaya and then again decrease significantly to Azov city. Analysis of long-term trends of nutrient concentrations showed that for most observation sites of the Lower Don an overall increase in phosphate (Figure 14.7B) and nitrite concentrations has occurred. Data for fluxes of substances to the Azov Sea from the Don river are presented in Chapter 12.

14.5   Characteristics of aquatic biota

The aquatic biota of the Lower Don is, to a great extent, defined by the processes that take place in the catchment under the influence of economic activities. The main factors which have influenced the aquatic biota of the region include changes in the hydrological and hydrochemical regime of the Lower Don as a result of the Tsimlyansk reservoir construction in 1952, as well as industrial, municipal and agricultural effluents. Thus anthropogenic impacts on the Lower Don over the last 30 years have resulted in changes in the dominant species, the disappearance of some species and an increase in the amplitude of biomass and population fluctuations.

The most significant changes have occurred in the Manych, Aksai, Kundryuchya, Tuzlov, Sal and Kagalnik rivers, in the Lower Don and in the Tsimlyansk and Veseloye reservoirs. Benthic biota are mostly affected in the River Don (area of Volgodonsk town) and in the Seversky Donets river (area of Lisichansk and Kamensk-Shakhtinsky towns).

Phytoplankton

The Lower Don phytoplankton is characterised by great species diversity, including almost 650 species. This diversity is related to the impact

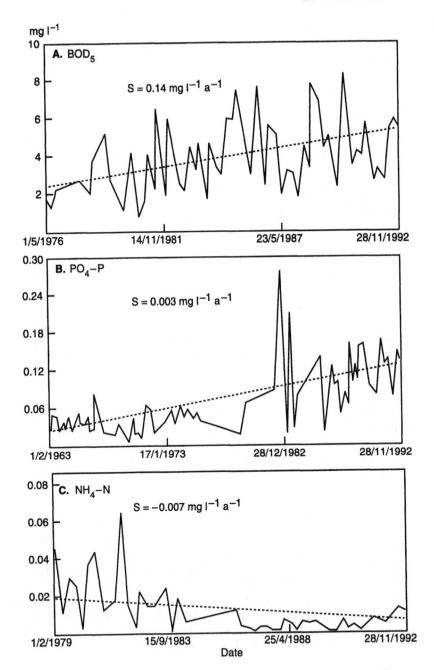

Figure 14.7   Long-term changes in quarterly-adjusted concentrations for: **A.** BOD5 (the Don at Rostov); **B.** PO4-P (the Don at Aksai); **C.** NH4-N (the Don at Aksai). S (slope) = Rate of annual increase

of the Tsimlyansk reservoir and the Don tributaries. Chlorophytes are dominant in terms of numbers of species, whereas cyanophytes and bacillariophytes have a dominant functional role in the ecosystem. The most common pollution indicators are β-mesosaprobes.

Inorganic nitrogen and phosphorus pollution has created favourable conditions for eutrophication in the reservoirs (Tsimlyansk in particular) and in the rivers (Seversky Donets in particular). In the Lower Don basin, particular features which have been observed include a decrease in species diversity (especially bacillariophytes), a predominance of cyanophytes (not only in summer and autumn, but also in spring), a reduction in the diversity and number of chlorophytes in summer and an increase in macrophyte biomass on the right bank of the Don where significant organic pollution occurs.

Zooplankton
The Lower Don zooplankton comprises approximately 60 species and includes the rotifers, cladocerans and copepods. The Tsimlyansk reservoir is the main source of zooplankton. On the whole rotifers and copepods dominate zooplankton in the Lower Don basin (with a significant proportion of calanoid copepods); *Daphnia longispina* and *D. hyalina* have disappeared and there is a wide amplitude in the fluctuation of zooplankton populations. On the whole, the Don right bank catchment has the greatest influence on the composition of fauna.

Periphyton
A regular feature of the succession of periphyton communities in the Lower Don is a decrease in species composition and an increase in the frequency of α- and β-mesosaprobic types. The least species diversity occurs in the Rostov Region where *Navicula* and *Nitzschia* algae are the dominant species.

Benthic invertebrates
The Lower Don benthic invertebrates comprise 40 species from the Nematoda, Polychaeta, Oligochaeta, Hirudinea, Crustacea, Insecta (Diptera: Chironomidae; Trichoptera) and Mollusca. Crustaceans and chironomid larvae dominate. The benthic invertebrates are characterised by a decrease in species diversity with a tendency for the relative population of oligochaetes to increase in the Manych, Aksai, Kundryuchya, Tuzlov and Sal rivers and the Veseloye reservoir. Reports of almost total benthic invertebrate annihilation have been made on the Seversky Donets river because of wastewater discharges.

14.6 The impact of pollution on the fisheries resources
One result of chronic and occasional emergency discharges to the Don has been the sharp deterioration in the state of biological resources of the

Lower Don and the Azov basin as a whole. This deterioration is reflected mainly by a sharp reduction in the fisheries resources with catches declining by 20–30 times or more for different species. Moreover, the number of valuable fish species in the Don basin has shown the greatest decline and these species include the white sturgeon (beluga, *Huso huso*), the Russian sturgeon (*Acipenser guldenstadti tanaicus*), the stellate sturgeon (*Acipenser stellatus*), the sterlet (*Acipenser ruthenus*), bream (*Abramis brama*), vimba (*Vimba vimba*) and pike-perch (*Stizostedion lucioperca*). Cases of mass fish kills have also been reported.

The Azov Fisheries Research Institute (AFRI) reported, as a result of river pollution, the loss of 'homing' of the most valuable sturgeon which resulted in more than three quarters of sturgeons not returning to the Don to spawn. Most fish, including the sturgeon, showed resorbtion of fish eggs, with the regeneration of the latter into fatty tissue. In addition, the number of fish with different pathological conditions increased. A sharp deterioration in the crayfish population also occurred with many highly-productive regions having lost their fishery value.

Because the data obtained by the monitoring network have generally not been sufficient to explain these deteriorations, the AFRI undertook a special survey to assess the level of micropollutants in water and sediment in the Lower Don and in the organs and tissues of aquatic biota. The research was carried out mainly on the River Don delta and Taganrog Bay. The confluence of the Rivers Seversky Donets and Sal were also studied. The main results of this study showed that various polluting substances were found along the Lower Don. The highest pollution levels were usually found in the lower reaches of the river, downstream of Rostov town. Amongst the pollutants which most often and significantly exceed MACs are petroleum products (concentrations exceed 1 MAC in 64.3 per cent of samples, 10 MAC in 11.1 per cent of samples and 50 MAC in 0.63 per cent of samples) and phenols (exceeding 1 MAC in 47.1 per cent of samples, 10 MAC in 7.7 per cent of samples and 50 MAC in 0.09 per cent of samples). High concentrations of petroleum products in the lower reaches of the Don are related to intensive water transport traffic and to their discharge from urbanised areas.

Petroleum products
At almost all sampling sites, the concentrations of petroleum products exceed the MAC. Data on volatile and non-volatile hydrocarbons and resins show that pollution of the Lower Don occurs due to local sources and long-range transport. From Rostov-on-Don to the river mouth, petroleum products from local sources dominate.

In the hydrocarbon analyses, the C_8–C_{23} alkanes and products of their degradation were identified. Many of them were present in water samples taken upstream and downstream of Rostov. This proves that both local and remote sources contribute to the amount of hydrocarbons in the Lower Don. The same trend occurs in the concentration and composition of the most dangerous group of hydrocarbons, the polycyclic aromatic hydrocarbons (PAH). Their total concentration ranged from 0.9 to 6.2 µg l^{-1} and they were composed of naphthalene (+methylnaphthalene), fluorene (+acenaphtene), phenanthrene, anthracene, dimethylbenzanthracene, pyrene, chrysene, benzo(a)pyrene, triphenylene, benzo(g)perylene, indenopyrene and benzo(b)fluoranthene.

In sediments, the total concentration of petroleum products varied widely from 0.06 to 13 mg g^{-1}. Sediment samples with increased concentrations are more clearly associated with pollution sources than water samples. An example of this is in the River Temernik, downstream of discharges from Azov and Rostov municipal wastewater treatment plants. In sediments, the proportion of resins and asphaltenes increases which reflects the increased role of chronic pollution in their composition. The composition of PAH compounds in sediments is the same as in water except for the presence of an additional compound, methylcholanthrene.

Pesticides

The intensive use of pesticides in water catchments, together with their improper storage and misuse have resulted in the frequent occurrence of these compounds in water. In 10–15 per cent of samples collected by monitoring networks, pesticides were detected and their concentrations were, at times, considerable. Thus in the Don basin during 1980–90, the concentration of DDT in more than 100 samples exceeded 0.38 µg l^{-1} (Table 14.3).

The distribution of pesticides in the Lower Don is similar to that of oil pollution. Upstream of Rostov, pesticide concentrations were lower (14–300 ng l^{-1}) than downstream of Rostov to the Taganrog Bay (40–480 ng l^{-1}). In terms of pesticide composition, isomers of HCH, metabolites of DDT and heptachlor were usually present. In samples taken upstream of Rostov, treflan and 1,2,4,5-tetrachlorobenzene were often found, whereas downstream saturn (ordram) was usually found. In sediment, DDT and its metabolites and isomers of HCH were most frequently found. The total concentration of pesticides was between 0.7 and 29 µg kg^{-1}.

Impacts of oil and pesticides on biota

High concentrations of pesticides in both water and sediment have been noted in the mouth of the Sal and in those parts where there is also a high concentration of petroleum products (e.g. the mouth of the River Temernik, municipal wastewater discharges and the arms of the Dry Kalancha and Wet Kalancha). Levels of oil and pesticides in the Lower Don are dangerous for the biota, especially for fish and crayfish. Coefficients of accumulation in these organisms are rather high which cause, at these levels of pollution, serious disruptions to vital fish functions, including reproductive processes. Analysis showed that in muscles and especially liver, and in fish eggs and brain tissue, large quantities of persistent pesticides had accumulated. For example, the concentration of organochlorine pesticides in pike-perch was 14–30 µg kg^{-1} in raw muscle tissue, 230–1,300 µg kg^{-1} in the liver, 850–2,900 µg kg^{-1} in brain tissue and 150–680 µg kg^{-1} in fish eggs and milt. Because the MAC for organochlorine pesticides in fish as a food product is 200 µg kg^{-1}, the muscle tissue of the pike-perch analysed was suitable for human consumption, but the internal organs could not be consumed as food.

Heavy metals
In the Don, heavy metal concentrations in water often exceed the MAC levels, especially for Cu where concentrations were higher than MAC values in more than 67 per cent of samples. However, the MAC value defined for Cu in the former USSR (1 µg l^{-1}) was not ecologically sound. Because water hardness in the Don is relatively high, the Cu concentration in the water at < 5 µg l^{-1} posed no apparent danger to the aquatic ecosystem.

The distribution of heavy metals in the water and sediment of the Lower Don in general resembled the distribution of petroleum products and pesticides, where downstream of Rostov pollution levels were higher than those upstream of Rostov. A significant increase in MAC was noted mainly for Cu with concentrations up to 10 µg l^{-1}, for Hg up to 0.25 µg l^{-1}, for Fe up to 800 µg l^{-1}, for Mn up to 110 µg l^{-1} and for Zn up to 25 µg l^{-1}. Concentrations of other metals such as Cr, Cd, Pb, Co, Ni, V, Mo and As in the Lower Don were lower than the MAC. In the sediment, metals were found in the following concentrations: Fe at 3.4–12 mg g^{-1}, Cr at 10–80 µg g^{-1}, Zn at 7–70 µg g^{-1}, Cu at 5–30 µg g^{-1}, Pb at 3–25 µg g^{-1}, Cd at 0.1–2.2 µg g^{-1} and Hg at 0.02–0.13 µg g^{-1}. All of these levels are very low and close to natural levels.

The levels of accumulation of heavy metals in fish muscle tissue and organs did not reach the MAC recommended by the Health Authorities. For example, Cu in muscles was < 0.3 µg g^{-1}, in liver < 7.4 µg g^{-1}, in gonads < 0.6 µg g^{-1} and in the brain < 1.6 µg g^{-1} with the MAC for Cu set at 10 µg g^{-1}. The concentration of Zn in the brain was < 15 µg g^{-1}, in the liver < 17 µg g^{-1}, with the MAC of Zn set at 40 µg g^{-1}. Mercury in the liver was < 0.15 µg g^{-1} with the MAC of Hg set at 0.4 µg g^{-1}. Nevertheless, these levels influence the physiological state of fish, in particular the reproductive capacity of male fish decreases.

Detergents
The concentration of synthetic surfactants in the water of the Lower Don is in the range 0.02–0.09 mg l^{-1}. There were no peak pollution events of detergents and the concentration of phenols was relatively low.

In samples of Don water taken downstream of the confluence with the River Temernik, there occurred, in addition to the above listed hydrocarbons, C_{11}–C_{18} organic acids, phthalic acid esters (dimethyl-, diethyl-, dibuthyl-, dihexyl-, dioctylphtalate), formaldehyde, acetaldehyde, hydroxymethyl benzene and other compounds.

Future research on the Lower Don aims to determine the concentration and stability of these compounds in water and sediment samples and to establish the main sources and pathways to water bodies.

14.7   Microbiological regime

The Rostov Science and Research Institute of Microbiology and Parasitology performed a five-year assessment (1988–92) of the sanitary and bacteriological state of the Tsimlyansk reservoir and the Lower Don. Twenty biotopes were studied to monitor the distribution and numbers of faecal indicator micro-organisms (coliforms, faecal coliforms and enterococci), faecally transmitted pathogenic bacteria belonging to the *Salmonella* genus and other potentially pathogenic bacteria associated with sewage (*Klebsiella*, *Proteus* spp. pseudomonads, acinetobacteria).

Analysis showed high microbial pollution along the length of the water body from the lower reaches of the Tsimlyansk reservoir to its mouth. In the lower reaches of the Tsimlyansk reservoir the mean coliform content was up to 29,600 microbes per litre, faecal coliforms 4,800 per litre, enterococci 2,500 per litre, *Klebsiella* 179,400 per litre, *Proteus* spp. 2,100 per litre, pseudomonads 445 per litre, acinetobacteria 758,400 per litre, and *Salmonella* 1.2 per litre.

There is a clear tendency for an increase in microbial pollution along the river flow. A sharp increase in the microbial pollution level is observed near settlements (Romanovskaya, Semikarokorsk and Bagaevskaya), at the points of confluence with the Lower Don tributaries (Seversky Donets, Sal, Temernik and Aksai) and downstream of wastewater discharge points from the towns of Volgodonsk, Rostov and Azov. Maximum microbial pollution was noted downstream of the Temernik inflow, with a mean coliform index of 25.59×10^6 per litre, faecal coliforms 4.7×10^6 per litre, enterococci 81,000 per litre, *Klebsiella* 4.61×10^6 per litre, *Proteus* spp. 14,740 per litre, pseudomonads 45,260 per litre, acinetobacteria 61×10^6 per litre and *Salmonella* 5.9 per litre.

It should be noted that several large intakes of municipal water supply systems for the towns of Shakhty, Novocherkassk, Rostov and Azov are located on the Lower Don. Water quality in the location of water intakes does not meet standards for water supply sources for all microbiological indicators. A high count of potentially pathogenic micro-organisms and the presence of *Salmonella* are evidence of an extremely unfavourable sanitary regime in the Lower Don.

14.8   Groundwater issues within the Rostov Region

14.8.1 Resources and formation of groundwater

The Rostov Region is situated in an arid zone, which limits the capacity for restoring natural groundwater resources. The average long-term groundwater discharge is $0.3 \, 1 \, s^{-1} \, km^{-2}$ which does not usually exceed 10 per cent of river discharge. The proportion of fresh and sub-saline groundwater is < 10 per cent of the total discharge (Akuz, 1979).

The occurrence in the Region of four artesian basins, the Volga-Khoper, Donets-Don, Azov-Kuban and Ergeni Artesian Basin (Figure 14.8), does not add significantly to available groundwater, because the recharge of these aquifers is only made at the periphery of the sedimentary platform structures where precipitation is sparse. These factors afford very little protection against pollution and exhaustion of water supplies.

Almost everywhere surface quaternary sediments form a weakly penetrating loam-silt layer approximately 80 m deep, which limits the atmospheric recharge of groundwater significantly. The exception to this is the 'open' part of the Donets coal basin and flooded terraces of large rivers, where the aquifer receives direct atmospheric input through a well penetrating unsaturated zone together with lateral inputs from river valley slopes (Lipatskova, 1965).

In the Rostov Region, the zone of impeded water exchange (chloride mineralised groundwater), related to poor drainage, is often reached by wells at depths of only 40–50 m (Ergeni and Azov-Kuban artesian basins).

The combination of geological, hydrogeological, hydrogeochemical and other natural features limits the possibility for the formation of significant groundwater resources suitable for drinking. Forecasts of such water resources, according to the latest assessment (Dubrovin, 1994), do not exceed $2.7 \times 10^6 \, m^3 \, d^{-1}$ (Table 14.5). Of this, the quantity of drinking water (with mineralisation up to $1 \, g \, l^{-1}$ TDS) does not exceed 65–70 per cent (i.e. approximately $1.57 \times 10^6 \, m^3 \, d^{-1}$). Most of it (about 80 per cent) is situated in the northern part of the region, in the Donets-Don and Volga-Khoper artesian basins. Thus, this water is at a significant distance from the most densely populated and industrially developed southern part of the Region, which suffers from an increasing deficit in drinking water supplies.

Exploration of fresh groundwater resources highlighted the limited possibility of locating significant water supply areas. Out of 37 areas studied in detail, only three sites were found which could reach a capacity of more than $80,000 \, m^3 \, d^{-1}$. Most areas have a capacity of $15,000–20,000 \, m^3 \, d^{-1}$. At many sites (especially those of the infiltration type), the actual capacity does not reach the predicted capacity as a result of changes in hydrological and other conditions (e.g. reduction in

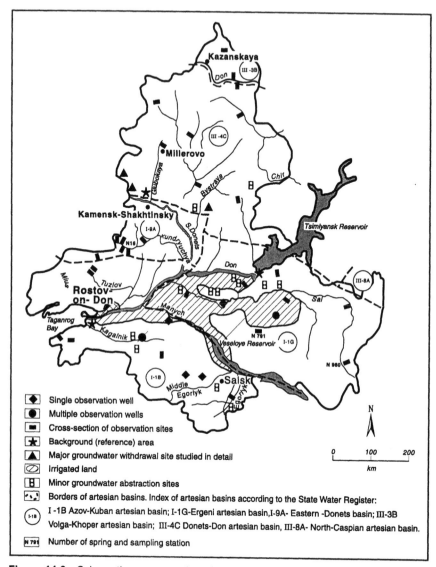

Figure 14.8   Schematic representation of the regional network of State groundwater monitoring in the Rostov Region

river flow and regulation of river discharge, pollution of surface and groundwater in the area of water extraction).

14.8.2 Groundwater monitoring system

In the early 1990s, the groundwater monitoring network in the Rostov Region included 205 wells. Wells are mainly grouped (usually 2–4) into

Table 14.5 Estimated and exploitable groundwater resources (with mineralisation up to 1.5 g l⁻¹ TDS), plus total abstraction and water use in the Rostov Region (as of 1 January 1994)

Hydrogeological regions	Groundwater resources		Abstraction (10^6 m³ a⁻¹)		Use (10^6 m³ a⁻¹)				
	Estimated total (10^3 m³ d⁻¹)	Potentially usable (10^3 m³ d⁻¹)	Total	Including mining waters	DDWS	IWS	AWS	IS	Total
Volga-Khoper artesian basin	264.0	3.4	3.55	–	1.46	0.07	2.02	–	3.55
Donets-Don artesian basin	11,876.0	504.6	180.31	80.78	63.36	4.87	31.31	–	99.53
Azov Kuban artesian basin	352.0	206.3	176.08	–	51.25	26.46	49.45	0.24	127.40
Ergeni artesian basin	468.0	93.5	49.27	–	5.81	3.60	39.84	0.02	49.27
Eastern Donets region	294.0	50.4	73.24	45.50	12.63	10.01	9.10	–	31.74
Total	2,655.9	858.2	441.80	126.28	134.51	45.01	131.72	0.26	311.48
Per river basin									
River Don	1,310.6	140.7	156.28	80.78					
River Seversky Donets	1,122.0	461.5	158.01	45.50					
River Western Manych	100.0	139.5	32.92	–					
Small rivers of Taganrog Bay	20.0	1,675	48.68	–					

DDWS   Domestic and drinking water supply
IWS   Industrial water supply
AWS   Agricultural water supply (excluding irrigation)
IS   Irrigation supply

Source: Adapted from Dubrovin, 1994

cross-sections to observe groundwaters near draining streams and rivers and into clusters or group piezometers to observe inter-layer aquifers. The observation network embraces the upper hydrodynamic zone of all the main artesian and river basins of the second and sometimes third orders (see Chapter 7) close to drinking water resources (Figure 14.8).

Thus, the regional (reference) observation network is designed to study the natural (background) groundwater regime as well as water quality with the aim of assessing trends and distribution patterns over a vast territory. This network is connected to a similar observation network in adjacent regions. For most stations, observations have been made for over 15 years.

The regional observation network includes both representative hydro-geological stations as well as sites of economic importance. These include sites with significant groundwater abstraction. Other sites include areas where, under the impact of economic activity, intensive pollution of groundwater may occur. Large irrigation systems and other areas with intensive agricultural technology (e.g. Novo-Alexandrovsk area of the Azov irrigation system), regions with a developed mining industry (e.g. Kundryuchya basin), industrial centres and cities may lead to pollution of water supplies, an increase in groundwater levels and land subsidence (e.g. Kamensk-Shakhtinsky, Volgodonsk, Rostov-on-Don).

The principles and guidelines for monitoring groundwater are described in Chapter 4.

14.8.3 Quality of groundwater

Long-term observations in the groundwater network show that, within the Rostov Region, the natural groundwater regime, on the whole, has been modified by economic activity. Specific features of these changes for the main artesian basins have been described by Topalov (1992) and are summarised here.

Groundwater of the Volga-Khoper and Donets-Don artesian basin is characterised by an increase in mineralisation by 2–4 times (from 70–80 mg l^{-1} to 270–280 mg l^{-1} TDS) during the seasonal rise in the water level. Watershed spaces are characterised by Na^+-Cl^--SO_4^{2-} waters with increased Ca^{2+} concentration and river terraces are charac-terised by Cl^--HCO_3^- water with approximately equal concentrations of Na^+ and Ca^{2+}. Inter-layer water is characterised by more stable miner-alisation and a higher concentration of Cl^- (150–300 mg l^{-1}) and lower HCO_3^- concentration (up to 70–100 mg l^{-1}). The concentration of SO_4^{2-} sometimes reaches 200 mg l^{-1} or more. The latter is connected with the presence of sulphides in water-containing rocks (up to 13 per cent of pyrite) (Topalov, 1992).

In the Eastern Donets hydrogeological region, the observation period at many sites is too short to allow the determination of patterns and trends. However, there are some sites where long-term observations were made and several trends have been observed. From 1970 to 1990, mineralisation almost doubled (Figure 14.9) with a simultaneous increase in SO_4^{2-}, Cl^- and NO_3^- concentrations. Maximum concentrations of nitrates occurred after the spring and autumn periods following the use of agricultural fertilisers.

In the Azov-Kuban artesian basin, the seasonal dynamics of water quality have not been studied, although observations allow several generalisations to be made about the increase in mineralisation and nitrates. In towns and villages, a rapid increase in groundwater level has occurred. For example, in the settlement of Tselina, an increase of 1 m a^{-1} occurred, which increased by 7 m over six years. Similar phenomena were observed elsewhere including Rostov-on-Don, Zernograd and Bataisk. The rise in levels is generally due to the widespread use of water for irrigation from water pipelines feeding from artesian aquifers or from rivers with a simultaneous absence of drainage systems. The cessation of groundwater use from wells, which were formerly the only source of water supply, has contributed to the rise in groundwater in rural settlements. As a result of the increase in water input to groundwaters, there has been a sharp rise in their level and water-logging of built-up areas has occurred.

In the Ergeni artesian basin during the last 20–25 years, there has been a slight tendency for an increase in groundwater mineralisation, which is probably related to the general rise in groundwater level. The NH_4^+, NO_3^- and NO_2^- concentrations in water have increased around agricultural lands. Such changes have occurred both in regions close to irrigated areas and far from them (Figure 14.9).

The greatest influence on the regime and quality of groundwater has been made by large irrigation systems, agricultural fertilisers, regulation of river discharge by reservoir construction, industrial wastewaters discharging into aquifers and changes in the natural balance of groundwaters following construction and development. On the whole, the change in the groundwater quality for the whole region shows that, even at considerable distances from direct point sources of pollution, there has been a regional deterioration in water quality. The main reasons for this are, most probably, diffuse sources of pollution from agricultural practices. The most evident illustration of such an influence is the change in concentration of nitrates, the concentration of which in groundwater has exceeded the MAC in many regions (Figure 14.9).

In addition to the abnormally high rise in the groundwater level in developed areas, there is a significant decrease in groundwater level where there is intensive abstraction of groundwater. This in turn has led to a reduction in the capacity of water abstraction and a negative impact on landscape, soil and other conditions. An example of such consequences is in the Seversky Donets valley where there has been significant abstraction of groundwater to provide drinking water for a

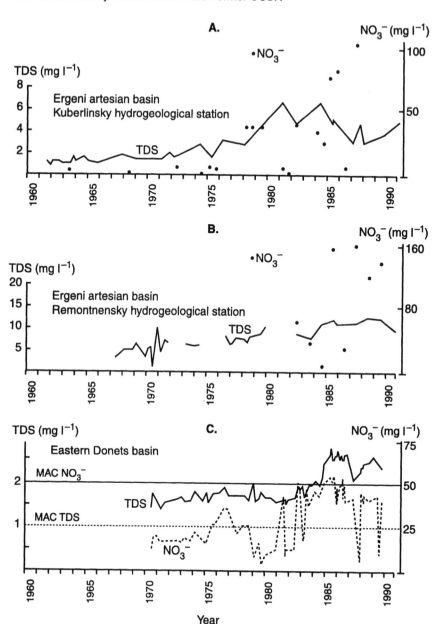

Figure 14.9 Long-term changes in mineralisation (expressed as TDS) and nitrate concentration in groundwater. **A** and **B**. Ergeni artesian basin; **C**. Eastern Donets basin (After Topalov, 1992)

number of towns, villages and coal-mining enterprises of the Rostov Region and Ukraine. In the Seversky Donets valley, from the mouth of the River Derkul to the River Belaya Kalitva, approximately 160,000 m^3 d^{-1} of groundwater are being abstracted for economic and drinking water supply (Dubrovin, 1994).

As a result of the regulation of the Seversky Donets discharge by water reservoirs, flooding of the flood plains has become rare. This has led to a loss of the alluvial layer as a regulator of water capacity for the largest aquifer in the Rostov Region, the Bolshoy-Sykhodolsk aquifer. The alluvial layer was a source for the annual recharge of groundwater resources of the aquifer located below the alluvium and extensively used during inter-flood periods. At the same time, the proportion of heavily polluted Seversky Donets waters in groundwaters abstracted from wells has increased (Privalenko et al., 1987).

Around the towns of Kamensk-Shakhtinsky and Belaya Kalitva, a reduction in groundwater extraction has been caused by the movement of polluted groundwater towards the site of water withdrawal due to infiltration from settling ponds, sludge dumps, industrial waste storage sites and, in some cases, lack of communal drainage systems (Moshkin, 1983; Topalov, 1992). In addition, around Kamensk-Shakhtinsky, the flood plain of the Seversky Donets has been intensively polluted by atmospheric dust fallout (on average 0.7 t km^{-2} a^{-1}) containing heavy metals (Hg, As, Sb, Cu, Cr and to a lesser extent, V and Ag).

14.8.4 Pollution of groundwater in a pilot irrigation area

A study of the influence of agriculture and irrigation on the pollution of groundwater within a pilot irrigation area (56 km^2) was carried out by the Regional State Geological Committee 'Yuzhgeologia' during 1981–90 near the village of Novo-Alexandrovka (Kagalnik basin, Azov-Kuban artesian basin). According to Velshanskaya (1990), based on official and (in her opinion) underestimated data, the application of nitrogen fertilisers and pesticides to arable land in the area was on average 21–42 t km^{-2} a^{-1}. Low levels of technology in applying agrochemicals, disregard for transportation and storage regulations (e.g. unused agrochemicals were left on fields in damaged containers or unpacked), and excessive use of aerial spraying has led to environmental pollution.

Nitrogen fertilisers rank first in terms of volume used and danger of polluting the environment. Excessive use has led to the accumulation of nitrogen compounds, nitrates in particular, in soil, agricultural products and groundwaters. The maximum concentrations of nitrates have been found in the upper layers of soil. Thus, in the top metre layer of soil in summer, the nitrate concentration is usually 100–250 mg kg^{-1}. About 10–20 per cent of nitrogen fertilisers, mainly nitrates, reach groundwater by infiltration following irrigation and precipitation. As a result, the concentration of nitrates in groundwater of irrigated areas in the growing season has exceeded the MAC by 2–5 times. Concentrations of nitrites in groundwaters of irrigated areas usually exceed those of dry land and usually vary from 0.001 to 0.7 mg l^{-1}; concentrations of ammonium have reached 1.3 mg l^{-1}.

The area studied is also characterised by seasonal dynamics in the mineralisation of groundwater with lower TDS in spring (3.3–4 g l^{-1}) whereas in autumn an increase in TDS occurs up to 4.5 g l^{-1}. Overall, the long-term tendency is for a gradual increase in mineralisation. Sulphate concentrations vary from 140 to 450 mg l^{-1} and there is a tendency for their accumulation in groundwater. This is connected with their inflow from irrigation waters which contain sulphates in relatively high concentrations. The chloride concentration in groundwater does not exceed 1.1–1.6 times the MAC and varies in different seasons. Boron, present in irrigation waters as well as in agricultural fertilisers, varies in groundwater from 1.8 to 4 times the MAC.

Extremely high concentrations of nitrogen compounds in groundwater have been recorded around human settlements. In particular, in wells used for domestic and drinking purposes, the nitrate concentration may reach 6–10 times the MAC and nitrite 12–20 times the MAC (1–1.6 mg l^{-1}). Such high concentrations are caused, in addition to the use of fertilisers, by the influence of local, domestic and livestock farming wastes.

In irrigated areas, groundwater pollution with pesticides also occurs. In the area studied alone, 35–40 pesticides are used. The presence of the most persistent pesticides, HCH (0.01–0.17 mg kg^{-1}), simazine (0.08 mg kg^{-1}), 2,4-D (0.01–0.06 mg kg^{-1}) and others, has been detected in the unsaturated zone. In each sample, between three and five types of pesticides have been detected including isomers of HCH, DDT, ramrod, treflan and dieldrin. The concentration and occurrence of pesticides increased significantly in the autumn as a result of the migration of pesticides from the unsaturated zone into groundwaters.

14.9   Conclusion

Within the Lower Don basin, the state of natural water is determined by the influence of industry and agriculture on water abstraction and use against a background of limited natural freshwater resources in this region. These factors have led to a fragile ecological situation, primarily due to the increasing pollution of surface and groundwaters in the region. An assessment of the major factors affecting the Lower Don is presented in Table 14.6.

Priority objectives for water resources management of the Don basin include:

- Stabilising water use and the sanitary situation in terms of pollution of surface and groundwater supply sources;

Table 14.6   An assessment of major problems in the Lower Don associated with water pollution

	Silting	Habitat alteration	Salinisation	Nitrate pollution	Eutrophication
Rivers and reservoirs	+ to ++	+ to ++	+ to ++	~	+
Groundwaters			+	+ to ++	
Biota, especially fish	+ to ++	+ to ++	+	~	+

	Organic matter pollution	Microbiological pollution	Heavy metals	Pesticides	Petroleum products
Rivers and reservoirs	+ to ++	+ to ++	+	+	+ to ++
Groundwaters	~ to + (u/d)	+	+ (u/d)	+ to ++	u/d
Biota, especially fish	+		+ to ++	+ to ++	+ to ++

~ No problem: concentrations considerably lower than MAC, or close to the natural level
+ Certain problem: use limitation
++ Major problem: heavy pollution
u/d Insufficient data

- Restoration and maintenance of a satisfactory ecological state of rivers, including small rivers and other water bodies, and their protection from pollution and over-abstraction;
- Serious economic incentives for encouraging water conservation activities and implementing ecologically sound and clean technologies;
- Co-ordination of industrial and drinking water supplies of large cities in the south of the Rostov Region and Donbass for joint use of surface and groundwater sources;
- Creation of modern comprehensive systems for monitoring natural waters in the Don basin, capable of producing reliable information necessary for implementing appropriate management techniques.

14.10 References

(All references are in Russian unless otherwise stated).

Akuz, I.K. 1979 The problems of use and the study of exploitation resources of ground waters of the Rostov Region. In: *Geological Structure and Prospecting of Rostov Region Minerals.* State University Publication, Rostov-on-Don, 91–97.

Dubrovin, O.A. 1994 Implementation of the state groundwater monitoring in the Rostov Region in 1993–1995 (Yearbook 1993). Geological Report of GGP 'Yuzhgeologia' Regional State Geological Committee, Roskomnedra, Rostov-on-Don.

Ecological Review 1993 Ecological Review N3, Rostov Regional Committee for Environmental Protection, Rostov-on-Don, 41 pp.

Lipatskova, E.N. 1965 Rostov Region Hydrogeological Map, Scale 1:500,000. Geological Report (Volga-Don Geological Department, Min. of Geol., RF), Rostov-on-Don.

Main Statistics on Water Use in the USSR in 1989 1990 The Central Scientific Research Institute of Complex Use of Water Resources, Minsk, 45 pp.

Moshkin, V.M. 1983 Impact assessment of industrial activity in the flood plain of the Seversky Donets. In: *The Lower Don Geology and Minerals*. Rostov-on-Don State University, Rostov-on-Don, 130–133.

Privalenko, V.V., Gorelov, G.M. and Moshkin, V.M. 1987 Geochemical assessment of atmospheric precipitation at the test and industrial site in the flood plain of the Seversky Donets. In: *Rostov Region Geological Structure and Minerals*. Rostov-on-Don State University Publication, Rostov-on-Don, 84–89.

Topalov, G.M. 1992 A study of the groundwater regime of the Rostov Region. Reports for the periods 1981–1985 and 1986–1990. Geological Report of the Regional State Geological Committee GGP, Roskomnedra, Rostov-on-Don.

Tsimlyansk 1977 *Tsimlyansk Watershed and Manych Water Reservoirs*. Gidrometeoizdat, Leningrad, 204 pp.

USSR Surface Water Resources 1973 Northern Caucasus. In: *Chemical Composition of Waters*. Volume 8. Gidrometeoizdat, Leningrad, 294–334.

Velshanskaya, L.A. 1990 Study of the regime of groundwater pollution by mineral fertilisers at a typical section of the Azov irrigation system. Engineering-Geological Report for the Period 1988–1990. Regional State Geological Committee, Ministry of Geology, USSR, State Register. N8-8-2/48, Rostov-on-Don, 141 pp.

Chapter 15[*]

THE AMU DARYA

The basin of the Amu Darya is situated between lat 34°36'N–43°45'N and long 58°15'E–75°07'E. Its length from north to south is 1,230 km and from west to east 1,470 km (Figure 15.1 and General Appendix IV). Its water catchment area, downstream of the confluence of the River Shera-bad, is 226,800 km^2. Sixty per cent of the catchment area is in the former Soviet Union and approximately 40 per cent in Afghanistan (USSR Surface Waters Resources, 1971). The Amu Darya is formed by the Rivers Pyandzh and Vakhsh and receives important tributaries in the first 180 km. The Amu Darya is the largest river in Central Asia and its natural discharge to the Aral Sea was, until 1961, relatively stable and depended mainly on snow and ice melt. In the 1950s, the actual discharge of the Amu Darya to the Aral (including losses due to evapo-transpiration in the delta flood plain) was 39.5 km^3 a^{-1} (Blinov, 1956). From 1961, a rapid growth in irrevocable water abstraction and regula-tion began which led to a sharp reduction in river discharge to the Aral Sea and at the end of the 1980s (data from 1987–89), the Amu Darya discharge varied between 1.1 and 16 km^3 a^{-1} (Izrael and Anokhin, 1991).

15.1 River regime

A specific feature of the Amu Darya basin is its clear division into an upper basin where the discharge is formed under the influence of predominately natural factors, and a lower part from which water is abstracted (Figure 15.1). The surface water resources of the Amu Darya, in a year of average water discharge measured at Termez sampling station, are 63.6 km^3 a^{-1} comprising discharges from the Pyandzh (51 per cent), Vakhsh (29.8 per cent), Kafirnigan (8.2 per cent), Surkhan-darya (5.3 per cent) and Kunduzdarya (5.4 per cent) (Rubinova, 1985).

In the upper basin (Figure 15.1, Zone I), river recharge is mainly from melting ice and snow (USSR Surface Waters Resources, 1971), with approximately 70 per cent of discharge forming in summer, up to 20 per cent in autumn and winter and 10 per cent in spring. Significant groundwater discharge occurs in almost all rivers, which constitutes

---

[*]   *This chapter was prepared by Y.A. Fedorov, R.A. Kulmatov and F.E. Rubinova*

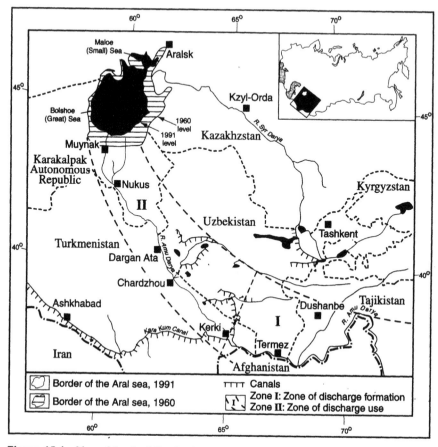

Figure 15.1   Map of the Aral Sea basin

25–30 per cent of the annual discharge. This discharge formation ceases when the river leaves the mountains.

In the lower basin (Figure 15.1, Zone II), the Amu Darya receives water from its tributaries only during high-water years, and a quantitative assessment can only be approximate because of the lack of appropriate data. The surface water resources, estimated at 76 km^3 a^{-1}, are assessed therefore by the inflow of water from Zone I, which forms up to 90 per cent of the inflow until Kerki city sampling station. When the Amu Darya reaches the lowlands, its waters are abstracted through irrigation canals such as the well-known Kara-Kum canal (Tsytsenko *et al.*, 1991). Due to a wide canal network (Figure 15.1) which runs through the lower flat basin, the water is widely distributed and either evaporates or infiltrates the river bed.

15.2   The Aral Sea

The Aral Sea is one of the largest inland water bodies in the world. The water and salt balance of the sea, its hydrological regime and biological productivity depend, to a great extent, on the discharge of the Rivers Syr Darya and Amu Darya. The Aral Sea is divided into the Bolshoe (Great) and Maloe (Small) Seas (Figure 15.1).

The unique original features of the Aral Sea prior to 1960 were (Blinov, 1956):

- An exceptionally beautiful blue colour of the water and an amazing transparency;
- The non-cyclical character of surface currents;
- Relatively low water salinity (9.6–10.3 g l^{-1});
- Over-saturation by calcium carbonate at 10 g l^{-1} salinity and as a result, carbonate salts sedimentation; and
- Increased sulphate ion concentration compared with seas and oceans.

Natural evolution of the Aral Sea was characterised by many long-term fluctuations in its level. Bottom sediment studies show that, in some geological periods, the Aral Sea level was as low as it is at present. At such stages, salinity increased by up to 30 g l^{-1} or more, which led to the sedimentation of salt minerals such as gypsum and mirabilite. Moreover, although it sounds paradoxical, negative anthropogenic influences on the Aral Sea ecosystem were already noted at the time of the Mongol invasion (beginning of the 13th century) and the Khorezm conquest by Timur (14th century) (Glazovsky, 1990). Studies of bottom sediment of that period showed increased water salinity and the presence of gypsum.

By the beginning of 1990, when the Amu Darya and Syr Darya runoff into the Aral Sea had practically disappeared, the area of the Bolshoe and Maloe Aral Sea decreased sharply. The shore line location in the early 1990s is shown in Figure 15.1. Mean salinity increased to 30 parts per thousand (ppt) and sedimentation of sulphate salts began. The Bolshoe Sea level dropped to 38.6 m above sea level (asl). The Maloe Sea level is 39.5 m asl. Because of this, the Maloe Sea water began to flow into the Bolshoe Sea through the Berg Strait. In comparison with 1780, the Bolshoe Sea level has fallen by 14.4 m (Izrael and Anokhin, 1991). The dramatic reduction in the sea's level has caused changes in its morphology. At present, 45 per cent of the Aral Sea has dried out and the surface area has decreased to 37,487 km^2. The Aral Sea volume has been reduced by almost 70 per cent and its mean depth by 43 per cent. The Syr Darya river has changed its course and now discharges directly into the Berg Strait (Tsytsarin and Bortnik, 1991).

This decrease in the water level of the Aral Sea has led to an enormous ecological catastrophe. A summary of the principal changes which have occurred include:

- Climate change in the Aral Sea region;
- A change in morphology of the Aral Sea and an increase in the area of dried out lake bed;
- An increase in the frequency of sand and salt storms;
- Water and salt balance disturbance;
- Extinction of certain species and drastic loss of fisheries;
- Degradation of the unique Aral Sea ecosystem.

15.3   Water use

Irrigation in some regions of the Amu Darya basin has occurred since ancient times. Currently, irrigation is the main consumer (> 90 per cent) of water resources followed by hydropower then industry, municipal use and fisheries (Rubinova, 1987). Irrigation occurs from April to September and in winter, saline lands are flushed out. For industrial and municipal needs, water is abstracted from rivers during the whole year.

Anthropogenic loads on the Amu Darya basin have increased together with the increase in irrigated lands. From 1950 to 1985, the irrigated area in the basin increased four times and water intake from rivers increased 3.2 times. In 1990, the total annual water intake for irrigation was 67.06 km^3 and for other water-consuming industries it was 5.0 km^3.

In the lower reaches of the Amu Darya, which occupy a wide delta region reaching from the Tyamyun gorge to the Aral Sea, water is lost by abstraction (irrigated farming) and by natural causes (evapotranspiration and infiltration). The changes in irrigated areas and principal crops grown from 1950 to 1988 are presented in Table 15.1. The data show that for irrigated land, both in the lower reaches in total and for the Amu Darya delta (downstream of the town of Nukus), changes in the crop structure included a decrease in some grain crops planted and the simultaneous increase in more water-consuming agricultural crops such as rice and fodder.

Between 1950 and 1988, the non-returnable water consumption volume for irrigated lands increased from 4.0 to $11.6 \text{ km}^3 \text{ a}^{-1}$ in the Amu Darya lower reaches in total and in its delta from 1.3 to $4.2 \text{ km}^3 \text{ a}^{-1}$. This corresponds approximately to the distribution of irrigated lands there. The increase in the volume of non-returnable water consumption is related primarily to the increase in areas irrigated (Table 15.1).

The total loss of discharge in the Lower Amu Darya, including both anthropogenic and natural losses during 1953–88, did not vary greatly (Table 15.2). However, the nature of these losses underwent significant changes. Until 1960, natural losses prevailed over anthropogenic losses and were approximately 60 per cent (Table 15.2). Subsequently the situation changed sharply; anthropogenic losses began to dominate and by 1981–85 their volume was on average $19 \text{ km}^3 \text{ a}^{-1}$. The increase in natural losses in 1986–88 was insignificant and was related mainly to the increase in water discharge and, probably, to an underestimation of

Table 15.1 Changes in irrigated area and their crop structure in the lower reaches of the Amu Darya and its delta

Year	Irrigated area (10³ ha)		Flooded area (10³ ha)		Crops grown in flooded areas (%)										
					Grain				Industrial crops				Animal fodder		
					Total		Rice		Total		Cotton		Total		
	Lower reaches	Delta	Lower reaches	Delta	Lower reaches	Delta	Lower reaches	Delta	Lower reaches	Delta	Lower reaches	Delta	Lower reaches	Delta	
1950	523	190	392	139	15	20	3	5	60	54	59	54	25	26	
1960	590	165	472	124	5	21	1	2	64	50	63	50	31	29	
1970	572	185	486	145	9	27	7	14	64	46	64	46	27	27	
1980	809	292	728	233	17	33	12	26	50	31	50	31	33	36	
1988	1,150	425	1,046	345	16	29	13	27	45	20	44	20	39	51	

Source: Tsytsenko *et al.*, 1991

Table 15.2   Total loss of discharge in the lower reaches of the Amu Darya due to anthropogenic and natural causes

Period	Anthropogenic		Natural	
	($km^3 a^{-1}$)	(%)	($km^3 a^{-1}$)	(%)
1953–55	8.4	37	14.3	63
1956–60	9.4	40	14.3	60
1961–65	11.0	62	6.6	38
1966–70	14.4	75	4.7	25
1971–75	17.6	80	4.3	20
1976–80	21.2	93	1.7	7
1981–85	19.3	96	0.8	4
1986–88	17.4	90	1.9	10

Source: Tsytsenko *et al.*, 1991

the volume of water extraction (Tsytsenko *et al.*, 1991). It should also be noted that, in the lower reaches of the Amu Darya, anthropogenic losses are irrevocable.

15.4   Sources of pollution

The main source of surface water pollution is irrigated farming. The long-term monoculture of cotton demanded the application of large amounts of mineral fertilisers and pesticides. Irrigated lands in the Amu Darya basin received, and are still receiving, on average 300–350 kg ha^{-1} of mineral fertilisers (the average figure in the former Soviet Union is 30 kg ha^{-1}). From the mid-1960s to 1988, there was a continuous increase in the volume of mineral fertilisers used, but this declined slightly in 1989 (Figure 15.2). The amount of pesticides (insecticides, herbicides, defoliants, etc.) used in the Amu Darya basin was on average 20–30 kg ha^{-1} which is 20–25 times greater than the average in the former Soviet Union (Shodiemetov, 1993).

Thus the main water pollution source in the Amu Darya is the return drainage waters from irrigated lands, as well as industrial and municipal wastewaters. Inclusion of saline soils into irrigated farming and imperfect traditional methods of irrigation led to the formation, in great quantities, of recurrent salty irrigation returns. In 1981–85, the average total volume of drainage runoff in the whole Amu Darya basin was 17.9 km^3 a^{-1}, and of this 10.3 km^3 a^{-1} returned to the Amu Darya. The total dissolved solids (TDS) of these waters varied from 0.7 to 8.0 g l^{-1} in different parts of the basin. Moreover, they were enriched with residues of mineral fertilisers, pesticides and other contaminants (Shodiemetov, 1993).

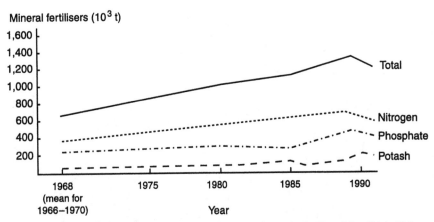

Figure 15.2   Temporal trends in the use of mineral fertilisers in the Republic of Uzbekistan

The anthropogenic impact on the Aral Sea, which has been sharply deteriorating since the 1960s, led to the appearance of another source of pollution, i.e. the sea itself. Additional amounts of salt, of natural origin from the dry sea bed, enter the atmosphere, and some of this also reaches the Amu Darya.

15.5   Characteristics of pollution and the influence of different factors on water quality

15.5.1 Mineralisation and major ions

The hydrochemical regime of rivers in the Amu Darya basin is character-ised by great spatial and temporal variability (Chembarisov, 1988). The Central Asian Regional Scientific Research Hydrometeorological Insti-tute (CARSRHI) carried out a regional study of the Amu Darya ionic discharge. Five regions were defined according to lithological composi-tion and stream runoff (Figure 15.3).

Within each region, mineralisation of tributaries increases down-stream when ice and snow melt contributions (as measured by specific runoff in $l\,s^{-1}\,km^{-2}$ or mm a^{-1}) decreases. In parallel, the area containing easily weathered sedimentary rocks increases and so does the proportion of rainfall and groundwater inputs in the river regime. The highest mineralisation of river waters is in the River Kzylsu (Figure 15.3), which is formed by easily eroded porous Quaternary rocks (Chembarisov, 1988). Mineralisation is not high for most parts where the water flows over weathering-resistant, intrusive rocks (Figure 15.3).

The hydrochemical regime of the Amu Darya is, to a great extent, determined by the water regime. The overall trend in TDS decreases with discharge. For waters of similar discharge, TDS values are higher

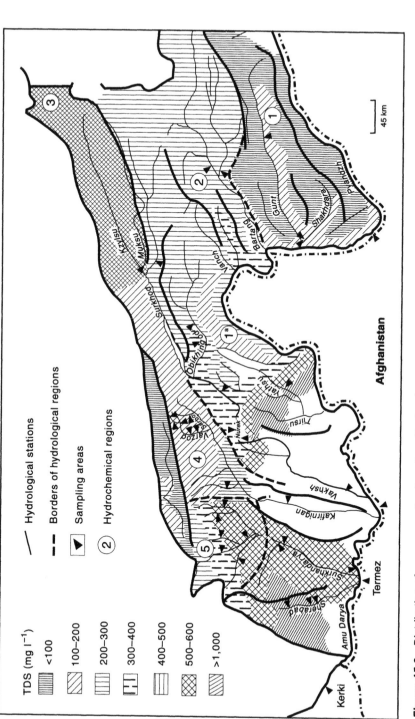

Figure 15.3 Distribution of mean annual water mineralisation (expressed as TDS) in the upper basin of the Amu Darya

during flood rise (April–June) than at the receding stage and during low water. Levels of TDS are also higher for dry years compared with wet years. Near the town of Termez, TDS values did not differ significantly from those of the Rivers Vakhsh and Pyandzh and, in 1976–90, practically did not increase. The inter-annual variation corresponds to the natural regime, where values of TDS exceeding maximum allowable concentrations (MAC) occur in some cases and for limited periods of time (Figure 15.4A).

Total dissolved solids increase in time and along the length of the river in proportion to the increase in discharge of return waters occurring downstream (Figure 15.4B,C). However, since the mid-1970s, TDS values have stabilised, and fluctuations from year to year are determined only by changes in water discharge. The seasonal variations of mineralisation, in general, are maintained up to the lower reaches of the Amu Darya (Figure 15.4). However, the number of cases and the length of periods exceeding the MAC increase along the river flow. Thus, at the most downstream observation site (Ilchik village)in the middle reaches, cases of mineralisation exceeding the MAC have been found frequently during the year. In the lower reaches of the river, from February/March to November, mineralisation has constantly exceeded the MAC. The increase in mineralisation over time and along the length of the river is accompanied by a change in the ion composition with an increase in Mg^{2+}, $Na^{+}+K^{+}$, SO_4^{2-} and Cl^{-} concentrations.

In Zone I of the Amu Darya basin, concentrations of Cl^{-}, $Na^{+}+K^{+}$, SO_4^{2-} and Mg^{2+} ions do not exceed MAC (Figure 15.5). In the Rivers Pyandzh, Vakhsh and Kafirnigan variations in major ion concentrations are generally similar, the two main exceptions being the HCO_3^{-} concentration in the Kafirnigan and the Ca^{2+} concentration in the Pyandzh. Concentrations exceeding the MAC for some major ions occur from the lower reaches of Zone I (upstream of Kerki), to Zone II of water use downstream.

In Zone II of the basin, values close to or exceeding the MAC for Mg^{2+}, $Na^{+}+K^{+}$, SO_4^{2-} are recorded near the towns of Dargan Ata and Nukus. An excess in MAC of Mg^{2+} occurs at TDS values > 800–850 mg l^{-1} (Figure 15.6). The concentration of $Na^{+}+K^{+}$ exceeds the MAC at TDS concentrations of 1,000 mg l^{-1} and SO_4^{2-} where it is 1,250 mg l^{-1}. The concentration of Cl^{-} reaches the MAC value near Dargan Ata when TDS values are 1,350 mg l^{-1} (Figure 15.6).

Considering the difference in the dynamics of mineralisation in the zone of discharge formation (Zone I) and the zone of water use (Zone II), concentrations of Mg^{2+}, $Na^{+}+K^{+}$ and SO_4^{2-} exceed MACs during low water flows upstream of Kerki sampling station, whereas in the middle and lower reaches of the Amu Darya concentrations exceed MACs almost over the entire year.

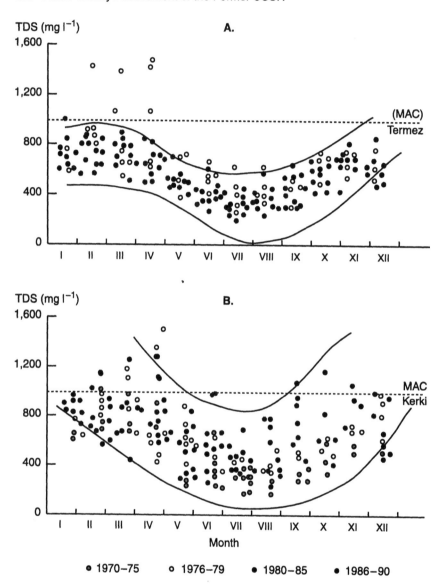

Figure 15.4   Inter-annual variation in TDS of the Amu Darya at various observation sites. **A**. Termez; **B**. Kerki. MAC = Maximum allowable concentration

The mean SO_4^{2-} concentration and mean TDS at the Amu Darya's most downstream observation site (Temirbay) over the period 1911–60 were 105 and 471 mg l^{-1} respectively and 619 and 1,640 mg l^{-1} respectively in 1981–85 (Hydrometeorology and Hydrochemistry, 1990). Thus, on average, the sulphate concentration in the Amu Darya increased six

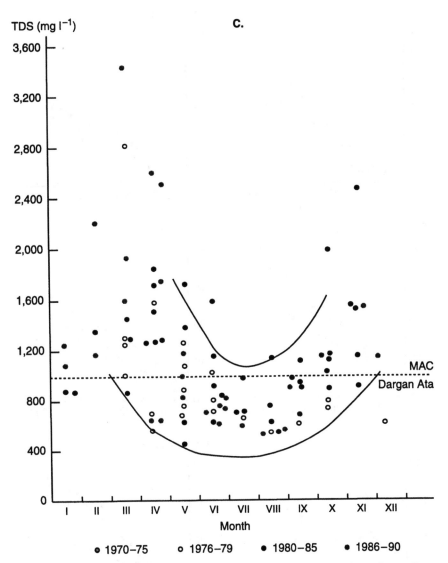

Figure 15.4C   Inter-annual variation in TDS in the Amu Darya at various observation sites at Dargan Ata. MAC = Maximum allowable concentration

times and TDS 3.5 times in just over 20 years. The absolute and relative increase in sulphate concentration was followed by a similar increase in sodium. The transformation of calcium–bicarbonate waters into sodium–sulphate waters also occurred. The increase in TDS and changes in composition of major ions are primarily connected with the almost complete regulation and use of the Amu Darya water resources and with

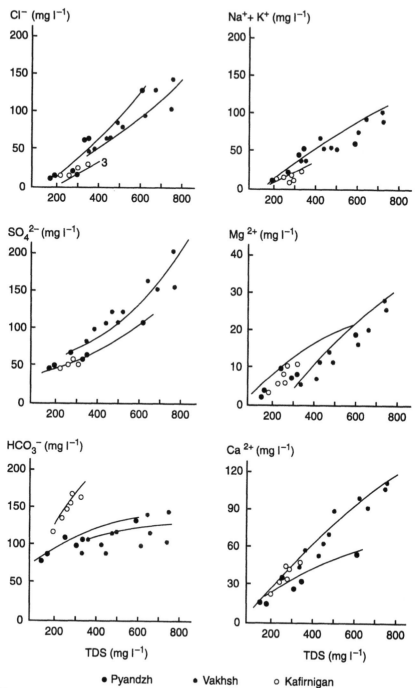

Figure 15.5 Correlation of long-term mean concentrations of major ions with TDS in three tributaries (Pyandzh, Vakhsh and Kafirnigan)

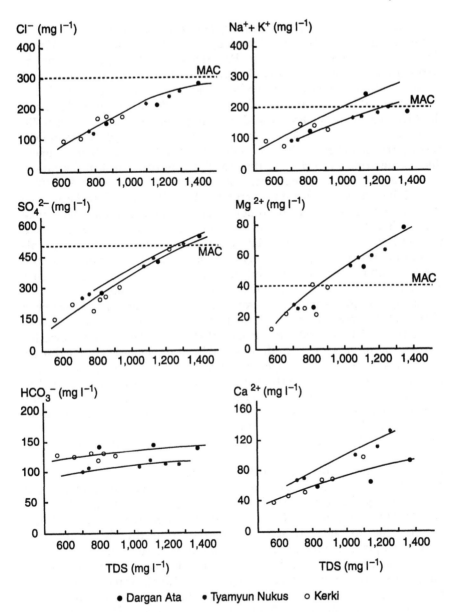

Figure 15.6   Correlation of long-term mean concentrations of major ions in the Amu Darya with TDS at various observation sites (Kerki, Dargan Ata and Tyamyun Nukus)

an increase in the proportion of strongly mineralised waste and drainage water discharged (Hydrometeorology and Hydrochemistry, 1990). However, for some authors (e.g. Fedorov et $al.$, 1993), an additional factor accounts for the profound changes in the chemical content and mineralisation of the Amu Darya lower reaches, namely the Aral Sea.

The annual average amount of salt that was deposited on the banks of the Aral Sea as the sea level was decreasing was 4.85×10^6 t in 1961–70, 2.84×10^6 t in 1971–80, and 4.82×10^6 t in 1981–85 (Hydrometeorology and Hydrochemistry, 1990). Therefore, the total amount of salt deposited on the dried out bed, from 1961 to 1985 was 101×10^6 t. If the rate of drainage of the sea did not decrease in 1985–90, another 24.1×10^6 t should be added to this total amount for 1961–90. Therefore, over 125.1×10^6 t of salt have been deposited on the drained sea bed since 1960, when the sea level began to drop. Sand and salt-marshes have formed on the drained sea bed fed by rising capillaries of highly mineralised groundwaters, separated bays and seepage lakes drying up at the surface. Salts, together with sand, are carried away by aeolian transport and can be moved up to 400–450 km by dust storms. Salt dust transfer has a negative influence on the Amu Darya delta, where up to 60 per cent of the total atmospheric flux of dust and sand is directed (Kes, 1983). From the mid-1970s, satellite images of the Aral region show powerful dust storms originating from the drained littoral regions of the north-east and east Aral Sea banks (Grigoriev, 1985). This has led to increased mass transport of salts and to anthropogenic impacts in the Amu Darya delta. According to Fedorov (1992), who used the isotope composition of sulphur from the Aral Sea as a 'tracer', the SO_4^{2-} concentration originating from Aral aeolian deposits increased in the water of a regulated section of the Amu Darya (Takhiatysh dam) by 30 per cent from 1970 to 1990.

15.5.2 Nutrients and suspended matter

The seasonal average for the concentrations of ammonium, nitrate and nitrite nitrogen in rivers of the basin varies from 0.01 to 0.36, 0.25 to 4.77 and 0.003 to 0.073 mg l^{-1} N respectively. The highest nitrogen contamination, due to the industrial discharge from the fertiliser plant 'Navoiazot', is in the lower reaches of the River Zeravshan.

The nutrient concentration in the Amu Darya is characterised by instability (Khomenko and Emelyanova, 1991). Concentrations of nitrogen and phosphorus may differ tens or hundreds of times during the year. Maximum values of different forms of nitrogen are 2.2 mg l^{-1} N for ammonium, 16 mg l^{-1} N for nitrate and 0.76 mg l^{-1} N for nitrite. Of the mineral forms of nitrogen, nitrate nitrogen is dominant although its concentration is lower than the MAC, except for a few samples where it has increased up to twice the MAC. The river is polluted with ammonium and nitrite nitrogen (Khomenko and Emelyanova, 1991).

Table 15.3   Mean total fluxes of particulate nutrients into the Aral Sea from the Amu Darya and Syr Darya rivers for periods between 1911 and 1985

Period	P (10^3 t)	N (10^3 t)	Si (10^3 t)
1911–60	56.0	82.0	292,444
1961–65	61.4	59.3	24,287
1966–70	115.5	112.1	45,585
1971–75	37.7	36.5	14,914
1976–80	14.5	14.3	5,750
1981–85	1.5	1.4	588

Source: Hydrometeorology and Hydrochemistry, 1990

The average seasonal concentration of inorganic dissolved phosphorus in rivers of the basin varies from 0.003 to 0.103 mg l^{-1} P. The highest concentration of phosphorus is in the middle and lower reaches of the river. Observations for 1979–88 show that maximum inorganic phosphorus reached 1 mg l^{-1} P (Khomenko and Emelyanova, 1991).

The Amu Darya has a high content of total suspended matter (TSS). In 1979–88, the suspended matter variations ranged from 1 to 36,000 mg l^{-1}. The maximum values were recorded in September 1987 and 1988 at Dargan Ata and Ilchik (Khomenko and Emelyanova, 1991). Intensive mechanical erosion is a natural characteristic for the Amu Darya upper basin. The development of these processes is promoted by the widespread presence of easily eroded sedimentary rocks, the high gradient slopes of the area and poor vegetation cover. The mean annual TSS transport rate for the Amu Darya tributaries in the upper zone of discharge formation (Zeravshan, Vakhsh, Surkhandarya tributaries) ranges from 100 to 420 t km^{-2} a^{-1}. At the border of the Amu Darya zone of discharge formation and zone of discharge use (near Kerki), the mean annual TSS transport rate reaches 650 t km^{-2} a^{-1} and the mean annual transport 200×10^6 t a^{-1}. This latter value is the highest among the rivers of the former USSR (Babkin *et al.*, 1987).

Information concerning particulate nutrient runoff is limited. At present, the total nitrogen content in Amu Darya suspended matter at the most downstream observation site (Temirbay) is 0.06 per cent, total phosphorus 0.062 per cent, and silicium 24 per cent (Kuznetsov *et al.*, 1978; Kliukanova, 1986). In comparison with the period prior to 1960, when the natural sea regime prevailed, the suspended nutrient concentrations in river waters did not change significantly (Hydrometeorology and Hydrochemistry, 1990). Subsequently, the suspended forms of nutrient runoff to the Aral Sea changed dramatically (Table 15.3). In terms of nutrient runoff, there are three periods of interest, 1961–70, 1971–80 and 1981–85. In 1961–70, despite a significant reduction in water discharge, an increase in suspended nutrient flux occurred. This is explained by an increase in the turbidity of river waters and the corresponding increase in sediment transport to the sea. In 1971–80, while approximately the same suspended matter concentrations occurred in river waters despite a progressive decrease in total water runoff, the suspended nutrient runoff decreased sharply and reached a minimum in 1981–85 (Hydrometeorology and Hydrochemistry, 1990).

15.5.3 Chemical and biochemical oxygen demand

The average seasonal values for chemical oxygen demand (COD) in the rivers of the Amu Darya basin vary from 3 to 21 mg l^{-1} O_2. In the upper reaches of the rivers, COD values do not, as a rule, exceed 10 mg l^{-1} O_2, whereas they increase downstream and in the mouths of the rivers they exceed the MAC. In the Amu Darya, from Termez to Kzyldzhar, the maximum value of COD increases from 1.3 to 3.4 times the MAC (Khomenko and Emelyanova, 1991). High values of COD are found in the lower reaches of the River Zeravshan (3.1 times the MAC), in the mouth of the Surkhandarya (2.3 times the MAC) and other rivers.

The average seasonal values for biochemical oxygen demand (BOD5) vary up to 3.2 mg l^{-1} O_2. The highest mean values are characteristic for the mouth areas of the Rivers Tankhyzdarya (Kattagan sampling station) at 3.2 mg l^{-1} O_2, Kashkadarya (Karatikon station) at 2.49 mg l^{-1} O_2 and the Amu Darya (Kzyldzhar station) at 2.07 mg l^{-1} O_2. Maximum concentrations of BOD5 exceed the MAC (3.0 mg l^{-1} O_2) by 1.1–4.3 times at most sampling stations.

The oxygen regime in rivers of the basin is generally favourable. Occasional cases of oxygen deficit (< 6 mg l^{-1} O_2) have been observed at irregular intervals in a year and are probably related to local anthropogenic impacts (Khomenko and Emelyanova, 1991). There are no cases where the oxygen concentration decreases to values lower than the MAC.

15.5.4 Petroleum products, phenols and pesticides

In 1989–91, the average concentration of total hydrocarbons varied from 0 to 0.2 mg l^{-1}. In most cases (91–98 per cent) they did not exceed the MAC (0.05 mg l^{-1}). Seasonal trends in total hydrocarbon concentrations have not been fully determined although there is a tendency for their increase in the autumn–winter period. Maximum concentrations varied between 1.1 and 5.4 times the MAC. The highest concentration of petroleum products was noted in the Amu Darya at the towns of Termez (3.4 times the MAC) and Kipchak (4.6 times the MAC), in the mouth of the rivers Surkhandarya (3.5 times the MAC) and Khalkadzhara (5.3 times the MAC). Data showed oil pollution is most frequent near towns and ports. According to a regular monitoring network, the average annual concentration of petroleum hydrocarbons varied over a wide range from 0.01 to 0.84 mg l^{-1} between 1979 and 1988. In most cases,

Table 15.4   Average pesticide pollution at two sampling stations on the Amu Darya in 1986–91

River	Month	Concentration (µg l⁻¹)							
		HCH_{total}	γ-HCH	DDE	DDD	DDT	Butiphos	Parathion methyl	Rogor
Amu Darya, Termez	I–III	0.020	0.013	0	0	0	0	0	0
	IV–VI	0.030	0.022	0	0	0.018	0	0	0.183
	VII–IX	0.018	0.011	0	0	0	0	0.008	0.085
	X–XII	0.015	0.012	0	0	0	0	0	0.163
Amu Darya, Nukus	I–III	0.034	0.015	0	0	0.002	0	0	3.22
	IV–VI	0.041	0.022	0	0.020	0	0	0.016	3.85
	VII–IX	0.020	0.021	0	0	0	0	0.051	1.20
	X–XII	0.022	0.017	0	0	0	0	0.001	1.29

DDE   Metabolite of DDT
DDD   Metabolite of DDT
DDT   Dichlorodiphenlytrichloroethane

HCH_{total} Total hexachlorocyclohexane
γ-HCH   1,2,3,4,5,6,-hexachlorocyclohexane
(Lindane)

their total concentration was lower or close to the MAC (Khomenko and Emelyanova, 1991).

The average annual concentrations of phenols varied from 0.001 to 0.040 mg l⁻¹ (MAC = 0.001 mg l⁻¹). In most cases they were lower than 0.020 mg l⁻¹ (Khomenko and Emelyanova, 1991). The highest pollution with phenols in the Amu Darya basin occurred in the mouth of the Surkhandarya and in the lower reaches of the Zeravshan, where concentrations of phenols 32–33 times the MAC were found.

Pesticide pollution is summarised in Table 15.4. The average concentration of HCH in the river is much lower than the MAC for this pollutant (20 µg l⁻¹). However, the very presence of HCH and other pesticides, even in the upper Amu Darya basin where their use is limited, is a matter for concern. It is probable that these pesticides enter the river waters with atmospheric transport by wet and dry deposition. The average value of lindane concentration in the Amu Darya does not exceed 0.025 µg l⁻¹.

The mean rogor concentration ranges from zero up to values that exceed the MAC (1.4 µg l⁻¹). There is a significant increase (25 times on average) in rogor concentrations in the Lower Amu Darya (at Nukus; Zone II) compared with the upper zone of the Amu Darya (at Termez; Zone I) (Table 15.4). Among Amu Darya tributaries, the greatest pollution with rogor is in the mouth of the Surkhandarya where, in 3 out of 44 cases, an excess of MAC of 2.7–5.3 times occurred.

The average seasonal concentrations of DDT (Dichlorodiphenlytrichloroethane) and DDD and DDE (metabolites of DDT) vary from 0.0 to 0.02 µg l⁻¹. Butiphos is not found in river waters. Parathion methyl is present in waters of the Amu Darya in quantities of $0.0–0.051$ μg l^{-1} which is more than 1–2 orders lower than the MAC of 0.3 μg l^{-1} (Table 15.4).

15.5.5 Heavy metals

The distribution of heavy metals along the length of the Amu Darya is rather heterogeneous. Extreme concentrations of heavy metals have occurred only in some years and, as a rule, are only found in certain areas. Thus, Cu concentrations up to 30 μg l^{-1} in filtered water samples were recorded near the towns of Kipchak and Nukus. Local areas with Zn concentrations greater than 0.1 mg l^{-1} and Ni up to 30 μg l^{-1} were found near the towns of Termez and Nukus. Almost at all sites of the river where sampling occurred, the concentration of Cr VI exceeded the MAC (1 μg l^{-1}). In 1979–88, the presence of Pb was found occasionally at concentrations up to 100 times the MAC (Khomenko and Emelyanova, 1991). The MAC was also exceeded for Hg, Cd and Fe (Kulmatov *et al.*, 1983, 1989, 1993; Ismatov *et al.*, 1988). In general, there is a tendency for an increase of heavy metal contamination from the zone of discharge formation (Zone I) to the zone of water use (Zone II) of the Amu Darya.

15.5.6 Groundwater and its connection with surface water

In the upper Amu Darya basin, there is a constant exchange of water between the river and groundwater. Mineralisation and ion composition of the groundwater and surface water are identical. Groundwater mineralisation does not exceed 0.5 g l^{-1} TDS, with mostly HCO_3^--Ca^{2+} as dominant ions. Further downstream, in Zone II where water is used intensively, groundwater is greatly transformed by drainage and industrial and municipal discharges. The concentration of Mg, nitrogen, sulphates and chlorides increases. In the coastal zone, where the river is hydraulically connected with groundwater, groundwater mineralisation corresponds to that of river water and does not exceed 1–1.5 g l^{-1} TDS.

The mineralisation of groundwater in irrigated regions far from the river may exceed 5–10 g l^{-1} TDS (e.g. Karshin steppe). Groundwater from such territories is abstracted and flows into drainage pipes (e.g. Dengizkul), or into canals (Sultandag discharge, Main Bukharo canal, etc.). Mineralisation of these waters in the middle and lower reaches of the Amu Darya varies from 2.6 to 2.7 g l^{-1} TDS.

In the lower reaches of the Amu Darya (including the present and original delta), groundwater with mineralisation of 1.5–3.0 g l^{-1} TDS dominates. Along river beds and canals it is < 1.5 g l^{-1}. The anion content of the groundwater of the middle and lower reaches of the Amu Darya is dominated by sulphates and chlorides, and the cation content by Mg and Na.

The groundwater of the Amu Darya basin is polluted with nitrogen, pesticides, toxic chemicals and heavy metals. Intensive pollution of surface water has led to the qualitative exhaustion of water supply sources. The mineralisation of fresh groundwater used for water supplies in the lower reaches of the Amu Darya reached 1.2–3.0 g l^{-1} TDS, its hardness increased by up to 16–23 meq l^{-1}, and the concentration of nitrate and nitrite nitrogen exceeded the MAC.

15.6 Conclusion

A significant increase in water consumption for irrigation and discharge of highly mineralised drainage waters caused, in recent decades, salinisation of Amu Darya basin waters leading to their transformation from CO_3^{2-}-Ca^{2+} to SO_4^{2-}-Na^+ type and the deterioration of the general environmental situation.

In 1989, Amu Darya discharge to the Aral Sea was only 1.1 km^3. Drinking water supply in the lower reaches of the Amu Darya became critical. Low quality drinking water caused an increase in human morbidity which became one of the most important signs of the ecological crisis in the region.

It should be stressed that water management measures only, cannot improve the environmental situation in the basin. They should be integrated with a set of drastic measures including restructuring the economy and giving up monoculture-orientated agriculture, and development and implementation of economic and legislative mechanisms of water management based on a river basin approach.

The Amu Darya water management system should incorporate water quality protection measures including the following:

- Cessation of untreated wastewater discharge from agricultural, industrial and communal sources and the development of efficient waste and drainage water treatment facilities and closed cycles of industrial water supply;
- Establishment of water protection zones and their afforestation;
- More efficient use of groundwater sources for drinking water supply;
- Development of water desalinisation facilities with low energy consumption and optimal use of local climate conditions (e.g. high sun radiation, use of water from glaciers).

15.7 References

(All references are in Russian unless otherwise stated).

Babkin, V.I., Voskresensky, K.P. and Vuglinsky, V.S. 1987 *USSR Water Resources and Their Utilisation*. Gidrometeoizdat, Leningrad, 302 pp.

Blinov, L.K. 1956 *Aral Sea Hydrochemistry*. Gidrometeoizdat, Leningrad, 256 pp.

Chembarisov, E.I. 1988 *Hydrochemistry of Irrigated Territories (Aral Sea Basin as an Example)*. FAN, Tashkent, 108 pp.

Fedorov, Y.A. 1992 Isotope composition of surface and ground water under the impact of natural and anthropogenic processes. Thesis for the Degree of Doctor of Science, Hydrochemical Institute, Rostov-on-Don, 56 pp.

Fedorov, Y.A., Grinenko, V.A., Ustinov, V.I. and Nikanorov, A.M. 1993 Mass transfer of salt in the Aral Sea region according to isotope characteristics of sulphates. *Academy of Science of Russia Reports,* **338**(2), 246–249.

Glazovsky, N.F. 1990 Aral Sea crisis. *Nature,* **10**, 20.

Grigoriev, A.A. 1985 *Anthropogenic Influence on the Natural Environment According to Satellite Observations.* Nauka, Leningrad, 239 pp.

Hydrometeorology and Hydrochemistry of USSR Seas. Volume VII, Aral Sea 1990 Gidrometeoizdat, Leningrad, 195 pp.

Ismatov, M.S., Kulmatov, R.A. and Kist, A.A. 1988 Research in micro-elements migration trends in natural water of arid zones and the elaboration and application of different radio-analytic methods. *Water Resources,* **4**, 1210–1222.

Izrael, Y.A. and Anokhin, Y.A. 1991 Problems of assessment of ecological and social consequences of natural environment degradation in the Aral region. In: *Natural Environment Monitoring in the Aral Sea Basin.* Gidrometeoizdat, Leningrad, 4–6.

Kes, A.S. 1983 Research on wind erosion and salt and dust transfer processes. *Problems of Desert Reclamation,* **1**, 3–15.

Khomenko, A.N. and Emelyanova, V.P. 1991 Amu Darya and Syr Darya pollution characteristics. In: *Natural Environment Monitoring in the Aral Sea Basin.* Gidrometeoizdat, St Petersburg, 109–115.

Kliukanova, I.A. 1986 Possible changes in some nutrient elements, humus and carbonates entering the Aral Sea through Amu Darya and Syr Darya suspended alluvium. In: *Turkmenistan Hydrometeorology.* Ashkhabad, Ilim, 75–82.

Kulmatov, R.A., Rakhmatov, U. and Kuist, A.A. 1983 On the physico-chemical condition of mercury, cadmium, zinc in the surface waters of the USSR arid zones. *Proceedings of the Academy of Science, USSR,* **272**(5), 1226–1228.

Kulmatov, R.A., Ismatov, E.E. and Volkov, A.A. 1989 Monitoring and analysis of micro-element concentrations in the Amu Darya and Syr Darya in 1983–1985. *Water Resources,* **1**, 178–182.

Kulmatov, R.A., Narsulin, A.V. and Ismatov, M.S. 1993 Research on the temporal and spatial distribution of polluting substances in the Amu Darya. In: *Water Problems of Arid Territories.* FAN, Tashkent, 20–30.

Kuznetsov, N.T., Kluikanova, I.A. and Nikolaeva, R.V. 1978 Some hydrological aspects of the Aral Sea problem. *Water Resources,* **1**, 72–82.

Rubinova, F.E. 1985 Amu Darya runoff disturbance under impact of water melioration in the basin. SANIGMI Publications, Volume 106(187), 113 pp.

Rubinova, F.E. 1987 The influence of water schemes on river runoff and the hydrodynamic regime of the Aral Sea basin. SANIGMI Publications, Volume 125(205), 158 pp.

Shodiemetov, Y. 1993 *Introduction to Socio-Ecology.* Ükituvch, Tashkent, 264 pp.

Tsytsarin, A.G. and Bortnik, V.N. 1991 Contemporary problems of the Aral Sea and perspectives for their solution. In: *Natural Environment Monitoring in the Aral Sea Basin*. Gidrometeoizdat, St Petersburg, 7–28.

Tsytsenko, K.V., Solyanik, N.L. and Malygina, T.V. 1991 The dynamics of anthropogenic and natural runoff losses in the lower reaches of the Amu Darya. In: *Natural Environment Monitoring in the Aral Sea Basin*. Gidrometeoizdat, Leningrad, 45–51.

USSR Surface Water Resources 1971 Central Asia, Amu Darya Basin. Issue 3, Volume 14. Gidrometeoizdat, Leningrad, 471 pp.

Chapter 16[*]

RYBINSK RESERVOIR ON THE VOLGA RIVER

Rybinsk reservoir, the third in a cascade of reservoirs on the Upper Volga, is the largest artificial reservoir built in 1941–47 as a result of dam construction upstream of the town of Rybinsk on the Volga and Sheksna rivers (Figure 16.1 and General Appendix IV). The area of the basin is 150,500 km^2 and is situated between lat 56–62 °N and long 32–40 °E in the northern part of the Central Russian Plain within three administrative regions: Yaroslavl, Vologda and Tver.

16.1 Characteristics of the reservoir and its basin

16.1.1 Basin description

The basin relief is flat and weakly eroded. The area close to the Rybinsk reservoir occupies the deeper part of the Moscow syncline. The lithology of the basin is essentially sedimentary Carboniferous limestone, Permian sand and clay, Triassic clays, and Upper Jurassic and Lower Cretaceous sediments. During the Quaternary ice age, the whole basin had been covered by thick glacial and fluvio-glacial sediments, showing bedrock only in some places, mainly on river banks.

The basin climate is temperate continental. Typically, latitudinal atmospheric circulation dominates and western winds prevail in all seasons. For all regions, strong cyclone activity is characteristic. This is accompanied by the intrusion of Atlantic air masses and by precipitation ranging from 300 to 800 mm a^{-1} (567 mm on average). Precipitation exceeds evaporation by 1.2–1.5 times and, as a result, the basin belongs to the zone of excessive water availability. The coldest month is February (−10.8 to −11.8 °C) and the warmest is July (+16.9 to +17.8 °C).

The soil for most of the water catchment is turf-podsolic, whereas in wetland areas it is gley horizons with different granulometric composition. Different varieties of wetland soil occur mainly in the upper and middle reaches of rivers. The entire water catchment area of the Rybinsk reservoir is situated in a forest zone. Most of the basin and the reservoir itself are in the southern taiga. The extent of forest areas in the basin

---

[*]   *This chapter was prepared by A.A. Bylinkina, V.I. Kozlovskaya and N.N. Osipov*

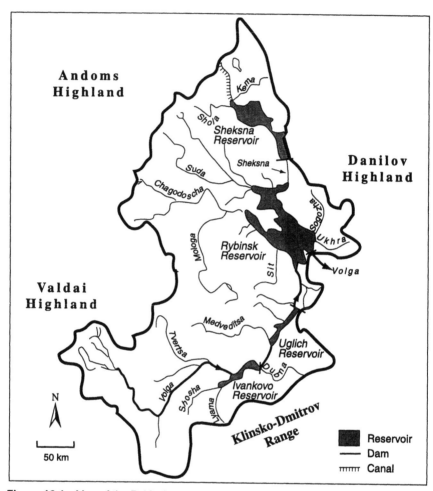

Figure 16.1   Map of the Rybinsk reservoir basin

varies. Forest forms 81 per cent of the basin of the Sheksna river, 54 per cent along the River Mologa, upstream of Vesyegonsk and 40 per cent of the Volga basin. The extent of the wetland areas varies from 2 per cent in the south to 26 per cent in the north.

The location of the basin in a zone of excessive water availability explains the well developed river network. Sixty-four rivers with a length of more than 10 km flow directly into the reservoir. A large part of the basin belongs to the Volga water catchment (40 per cent), the Mologa (19 per cent) and Sheksna (13 per cent) basins. The rest is occupied by small rivers with water catchment areas of less than 2,000 km². The main recharge source is water from snow melt. According to long-term data, flood runoff in spring represents 54 per cent of the annual flow, the

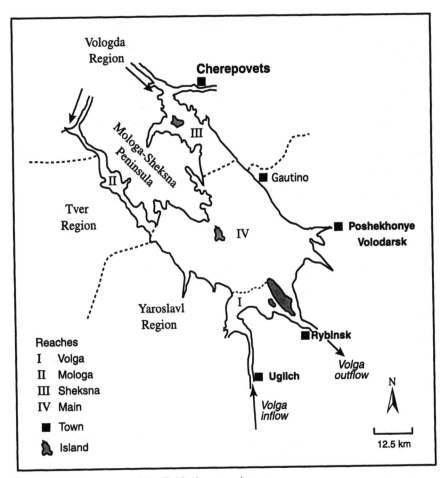

Figure 16.2 Reaches of the Rybinsk reservoir

summer–autumn period 30 per cent and winter 16 per cent. The value of the average annual specific runoff is 6–12 l s^{-1} km^{-2} (Kuzin, 1972).

16.1.2 Reservoir description

The Rybinsk reservoir mainly occupies the Mologa-Sheksna lowland, which also determines its configuration (Figure 16.2). The total length of the shore line is 21,500 km. The main morphometric characteristics of the reservoir are presented in Table 16.1.

Rybinsk is one of the largest European reservoirs, both in terms of area and volume, and is one of the oldest of its type. Unlike other major reservoirs, its water residence time (0.56 years) is still moderate, of the same order of magnitude as that of most smaller reservoirs. This difference is also expressed in the chemical and biological regime of the

Table 16.1 Morphometric characteristics of the Rybinsk reservoir

Surface area	4,550 km^2
Altitude	102 m above sea level
Maximum volume	25.42 km^3
Net volume	16.7 km^3
Average annual volume	18.7 km^3
Water residence time	0.56 years
Average depth	5.6 m
Maximum depth	28.0 m
Maximum length (Uglich dam to Sheksna dam)	250.0 km
Maximum length of the main lake	150.0 km
Maximum width of the main lake	70.0 km

Source: Kuzin, 1972

Table 16.2 Surface area and volume of the Rybinsk reservoir reaches

Reach	Surface area		Volume	
	(km^2)	(%)	(km^3)	(%)
Volga	550.0	12.0	2.62	10.4
Mologa	220.3	4.6	0.74	2.9
Sheksna	696.5	15.3	2.51	9.9
Main	3,077.4	67.7	19.39	76.8

reservoir. The reservoir can be divided into four reaches: Volga, Mologa, Sheksna and the Main reaches (Figure 16.2). The first three are situated along valleys of the corresponding rivers and are long, narrow and deep areas. The Main reach is most lake-like and occupies the central part of the reservoir. The principal features of the four reaches are summarised in Table 16.2.

The reservoir bed is characterised, in its deepest part, by the flooded, narrow, meandering bands of rivers. Medium depths tend to occur in areas which were previously wetlands, with the more shallow parts corresponding to flooded land terraces (depths from 2 to 9 m). Shallow waters (< 2 m) represent about 20 per cent of the total reservoir area. They are found mainly around the Mologa-Sheksna peninsula and in the south-east part of the Main reach. Along the banks of the Main reach are the flooded beds of the Mologa, Sheksna and Volga. In the open part of the reach there is a gradual rise in the relief in the direction of the flooded Mologa-Sheksna watershed, although the minimum depth is about 4 m. The relief at the bottom of the Main reach is complicated by the presence of small rivers, streams and lakes in the valleys of the main rivers prior to flooding and which now cause, in some places, sharp changes in depth.

The most common types of bottom sediments are sands (42 per cent) and grey sandy silt (40 per cent). Other types include peaty silt (8 per cent), silty clay (5 per cent) and organic macrophytic sediments (> 1 per cent) (Zakonnov, 1981). The main origins of the sediments are bank and bed erosion (60–80 per cent) and river inputs (20–40 per cent). This proportion is unusual and is due to the relatively low suspended load of tributaries, including the Volga, and the high wave action caused by wind. It should be noted that, initially, approximately 1,500 km^2 of forests and shrubs and 8,000 km^2 of peat swamps were flooded when filling the reservoir. This later led to the formation of quaking peat bogs over a large area. There has recently been a sharp change in the temporary flooded areas where woody vegetation has practically completely disappeared and the area of quaking peat bogs has been sharply reduced (Tachalov, 1956).

16.1.3 Hydrological regime

The average annual inflow of water into the reservoir is about 37 km^3, of which 90 per cent is surface discharge. The main inputs are the Volga waters, via the Uglich hydropower plant (40 per cent), the Sheksna waters, discharged through the Cherepovets dam (12 per cent), and the unregulated discharge of other tributaries (40 per cent), half of which form the Mologa river. The input from direct atmospheric precipitation is 7.7 per cent, and ground discharge is approximately 1 per cent. The main water output from the reservoir is through the Rybinsk hydropower plant (90 per cent).

The seasonal distribution of the discharge and the water level regime of the reservoir varies over the years. This is caused by natural fluctuations in discharge and by artificial regulation of discharge through the hydropower plants. The most characteristic pattern is for an increase in levels during spring floods (up to the end of May) and then, after a short standing period, levels begins to decrease gradually until the beginning of winter (December) when the water level decreases sharply. The average amplitude of seasonal fluctuations in the water level reaches 3.34 m in some parts of the reservoir. This is connected mainly with changes in water storage and, to some extent, as a result of wind and compensation currents. Depending on the direction and velocity of the wind, the rise in the water level near the windward bank may reach 19 to 30 cm, and during storms 50 to 70 cm (Kuzin, 1972).

The Rybinsk reservoir is strongly influenced by most wind directions. In the Main reach, when the reservoir is open to navigation, the height of waves is 55–65 cm for 50 per cent of the time and 120 cm high for 10 per cent of the time. A maximum wave height of 250 cm has been recorded. Due to the wind and compensation currents, the water mass is mobile with calm days reported in the Main part for less than 2 per cent of the time.

The reservoir freezes in November at temperatures close to 0 °C at all depths. By the end of winter, the ice cover may be about 50–60 cm, reaching 70–80 cm in some years. The break-up of ice starts in mid-April near river mouths and in the last 10 days of April in the Main reach. The reservoir is usually completely ice-free in the first half of May. The spatial variation in water temperature at this time is 8–9 °C. The maximum warming of the water body is from July until the beginning of August, when the average temperature is 19 °C, with a maximum in surface layers of up to 21–27 °C. In large parts of the reservoir, the thermal stratification is weak with temperature differences between the surface and bottom layers usually not exceeding 1–2.5 °C. Only in the deep areas might there be marked stratification (ΔT from 10 °C to 15 °C). In the second half of August, cooling begins leading to renewed homothermic conditions. According to the standard thermal classification, the reservoir is dimictic. Temperature stratification is disturbed by strong wind action, supported by the large surface area and shallow water (a depth of < 6 m over more than half of the reservoir).

16.2   Water use

The Rybinsk reservoir is a source of drinking water supply for the towns of Cherepovets, Uglich and Rybinsk, which have a total population of more than half a million people (Figure 16.2). Cherepovets also consumes significant amounts of water for industrial purposes, with the largest consumers being the enormous metallurgical plant 'Severostal', two plants producing mineral fertilisers and a number of smaller enterprises. The town uses on average 210×10^6 m^3 a^{-1}, including 145×10^6 m^3 for industrial needs and 65×10^6 m^3 for domestic and drinking purposes (Kozlovskaya and Sambursky, 1992). The total water consumption of Cherepovets is, on average, 4 per cent of the annual input of water ($4,900 \times 10^6$ m^3) from the Sheksna reservoir to the Rybinsk.

The fishery is orientated towards catching bream, pike-perch, roach, pike and burbot (Kuzin, 1972). Amateur fishing is intensive and is mainly for perch and roach. After the reservoir was filled, the commercial fish catch gradually increased. It reached its maximum at the end of the 1950s, after which it began to decrease. Such dynamics are well illustrated by an analysis of the bream catch during 1953–86; bream was the main commercial fish catch of the Rybinsk reservoir (Figure 16.3) (Volodin, 1993). The relatively low fish yield was mainly due to unfavourable conditions for the breeding and survival of young fish. A reduction in spawning areas occurred with the disappearance of areas rich in vegetation and epifauna following flooding. The disappearance of these habitats, which were the refuges for yearlings, forced these fish

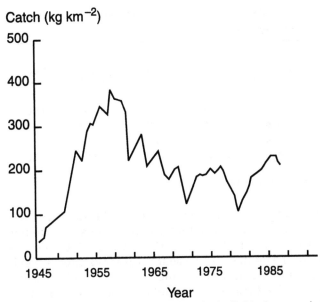

Figure 16.3 Total commercial fish catch of bream in the Rybinsk reservoir (After Volodin, 1993)

into open waters and caused their mass elimination at early stages of development. This trend is now accentuated by increasing pollution caused by agricultural, domestic and industrial discharges. At the same time, the deep areas of the reservoir constitute a food reserve favourable for the survival of fish reaching maturity. This latter fact could allow an increase in the fish yield of sexually mature fish without causing damage to fish stocks provided proper remediation and fish management measures are taken (Volodin, 1993).

The water resources of the reservoir are also used to produce electric power by the Rybinsk hydropower complex. Its capacity is 330 MW; the average power production is 1.1×10^9 kW h^{-1}. The Rybinsk reservoir is part of the Volga-Baltic Sea water system and is used for shipping, via the Sheksna river and by canal to Lake Onega. Annual freight traffic volume through Rybinsk sluices is 10–12×10^6 t (Butorin, 1978) which makes it one of the world's major inland waterways. Use of the Rybinsk reservoir for recreational purposes is not popular mainly because of the small population on the banks of the Main reach. The only developed banks of the reservoir are near the big towns (e.g. Cherepovets, Rybinsk) and adjacent areas, which are accessible by small boats.

16.3 Sources of pollution
The main sources of pollution to the reservoir are the wastewaters of the industrial complexes of Cherepovets. At present, the reservoir's Sheksna

Table 16.3   Composition of effluent from some Cherepovets industries discharging into the Rybinsk reservoir around 1990

Enterprise	Wastewater discharge (10^6 m^3 a^{-1})	NH$_4$-N (t a^{-1})	Petroleum products (t a^{-1})	Suspended matter (t a^{-1})	Fe$_{total}$ (t a^{-1})	BOD$_5$ (t a^{-1})
Severostal	53.2	74.5	13.3	1,600	18.6	431
Amophos	2.6	4.8	2.6	104	1.8	28.1
Azot	8.4	84	3.4	190	9.2	103
Vodocanal	90.2	2,435	63.1	21,650	9.9	6,684

BOD   Biochemical oxygen demand

reach and its tributaries in the vicinity of the town annually receive 159×10^6 m^3 a^{-1} of inadequately treated industrial, municipal and domestic wastewaters through 57 sewers. Data on wastewater discharges for the main pollution sources are presented in Table 16.3. On the whole, the metallurgical and chemical enterprises annually discharge, into the Sheksna reach and its tributaries, about 1.7×10^3 t of suspended substances, 85 t of oil products, 3.5 t of chlorides, 20 t of sulphates, 350 t of nitrogen and other pollutants, including polychlorinated biphenyls (PCBs), polyaromatic hydrocarbons (PAHs), Cu, Ni, Hg, Cr and Pb (Kozlovskaya and Sambursky, 1992). The Volga reach of the Rybinsk reservoir is significantly contaminated by the Tver city wastewaters, in spite of the town's remoteness in the upper reaches of the Ivankovo reservoir.

The main diffuse source of pollution to the Rybinsk is the surface runoff of nutrients. Navigation and dredging also play a significant role in the pollution of the reservoir.

16.4   Factors influencing water quality

As a result of the different water input and morphometric features, all the reservoir reaches have specific features in terms of hydrochemical regime, sedimentation pattern, biological processes and, consequently, water quality.

16.4.1 Major ions

The water of the Rybinsk reservoir is characterised by low mineralisation. The average annual concentration of total dissolved solids (TDS) does not exceed 250 mg l^{-1}, which is characteristic of surface runoff from podsol soils in regions of excessive water availability. The ionic composition of all the reservoir reaches is similar. HCO$_3$$^-$-Ca^{2+} ions represent 65–70 per cent of the sum of ions, Mg^{2+} and SO$_4$$^{2-}$ approximately 23 per cent and Cl$^-$ and Na$^+$+K$^+$ 3 per cent and 5–6 per cent respectively. In the water of the Main reach near the dam, the annual ionic average for

HCO_3^- is 112.0 mg l^{-1}, for SO_4^{2-} 22.6 mg l^{-1}, Cl^- 3.8 mg l^{-1}, Ca^{2+} 32.1 mg l^{-1}, Mg^{2+} 8.0 mg l^{-1}, Na^+ 5.4 mg l^{-1} and K^+ 1.6 mg l^{-1}.

16.4.2 Nutrients

The main source of nutrients to the Rybinsk is river inputs. More than 60 per cent of nitrogen and approximately 40 per cent of phosphorus enter the reservoir annually with river discharge (Zakonnov and Ziminova, 1984). Therefore, along with the hydrological and chemical features of the reservoir in each reach, water quality and nutrient distribution are greatly affected by economic activity in the basin. The other important anthropogenic components are direct wastewater discharge to the reservoir and atmospheric inputs of nitrogen and phosphorus. The impact of watershed inputs through rivers is greater in spring, especially in the upper reservoir sections which are influenced by rivers, while the influence of wastewaters and atmospheric precipitation is greatest during the low water period.

The cultivation of land, coupled with few forests in the Volga basin and the discharge of wastewaters from Tver has led to relatively high concentrations of nitrogen and phosphorus in the Volga reach. According to long-term data from 1966 to 1974, concentrations were 1.55 mg l^{-1} for nitrogen and 0.074 mg l^{-1} for phosphorus, which is approximately 1.5–2 times greater than in other reaches. Many wetlands and forests in the water catchments of the Mologa and Sheksna result in a lower concentration of nutrients in the corresponding reservoir reaches. A comparison of concentrations and fluxes of nutrients in the spring waters of the Volga with the northern rivers in 1965–74 gave a wide-ranging assessment of the impact of land and fertiliser use on the nutrient regime in the Rybinsk reservoir (Drachev et al., 1976; Trifonova and Bylinkina, 1984). Thus, agricultural sources during these years were found to be responsible for 10 per cent of the nitrogen and 15 per cent of the phosphorus of the total annual input. The annual nutrient input from the municipal wastewaters of Cherepovets and Tver cities during the same years was 4,450 t of nitrogen and 234 t of phosphorus, which increased the anthropogenic contribution into the nutrient budget of Rybinsk reservoir by up to 20 per cent for nitrogen and 30 per cent for phosphorus.

In the 1990s, the nutrient discharge from industrial and domestic wastewaters of Cherepovets city increased by about four times. Previously it was comparable with the annual nutrient discharge of the Sheksna river but now it exceeds these inputs. The annual nutrient discharge of wastewaters from Tver is still at the same level as it was in the 1960–70s. The result of the anthropogenic influence in the Sheksna reach is a change in the nutrient regime in recent years, especially during the summer low water periods where the average concentration

Table 16.4   Average concentration of nutrients in the Rybinsk reservoir in summer

	1965–70				1980–82				1989		
	(mg l^{-1})		(µg l^{-1})		(mg l^{-1})		(µg l^{-1})		(mg l^{-1})		(µg l^{-1})
Reach	N_{tot}	N_{min}	P_{tot}	P_{min}	N_{tot}	N_{min}	P_{tot}	P_{min}	N_{tot}	N_{min}	P_{tot}[1]
Volga	1.61	0.36	72	26	1.13	0.43	80	31	1.12	0.39	81
Mologa	1.56	0.16	36	7	0.72	0.11	54	13	1.00	0.29	50
Sheksna	1.51	0.12	52	12	0.78	0.17	64	24	1.71	0.81	85
Main	1.45	0.20	33	4	0.92	0.34	53	14	1.01	0.27	40

N_{tot} Total nitrogen	[1] No P_{min} data available for 1989
N_{min} Mineral nitrogen	
P_{tot} Total phosphorus	Sources: Butorin, 1978; Razgulin, 1985;
P_{min} Mineral phosphorus	Bylinkina, 1993

of inorganic nitrogen and phosphorus has increased greatly (Table 16.4). At the same time, in other reaches, there was a tendency for a reduction in the concentration of total nitrogen associated with organic compounds, and a gradual increase in the concentrations of total phosphorus. In the 1980s, compared with 1970, the average annual concentration of total phosphorus in the reservoir increased (Table 16.4) (Razgulin, 1985). One of the possible reasons for a reduction in the concentration of organic nitrogen might have been a comprehensive land amelioration programme (e.g. drying out wetlands and cutting down vegetation) in the catchment of the reservoir. The reduction in organic nitrogen content in the reservoir was also promoted by the almost complete destruction and disappearance of peat bogs in the catchment.

By contrast, a significant increase in the concentration of inorganic nitrogen and phosphorus may be caused by atmospheric precipitation in summer, when the inflow of river waters decreases significantly. Thus, the input of nutrients from atmospheric sources to the Rybinsk reservoir in the summer of 1988, was estimated at 33 per cent of total nitrogen inputs and 17 per cent of total phosphorus inputs from external sources; during this period, nutrient inputs from the atmosphere were double those from wastewaters (Bylinkina, 1993).

Observations made in the Sheksna reach during summer and autumn 1987–89, showed that high concentrations of nitrogen (up to 7.0 mg l^{-1}) and phosphorus (up to 250 µg l^{-1}) were found mainly near the outlets of wastewaters from Cherepovets. While mixing with the main water mass of the reservoir, concentration decreases rather rapidly and, therefore, in the middle section of the Sheksna reach (Myaksa sampling station) they do not differ from nutrient concentrations in the Main reach. The dispersion rate of high concentrations depends, to a great extent, on the volume of water discharge from the Sheksna reservoir which facilitates their dilution (Koreneva, 1993). However, the inflow of water from the

Table 16.5   Characteristics of phytoplankton production during the autotrophic phase of phytoplankton development in the Rybinsk reservoir in summer, 1989

Reach	Mean ± S.E.			Primary production (t C d^{-1})	
	Biomass (g m^{-3})	Chlorophyll *a* (µg l^{-1})	Photosynthesis (mg O$_2$ l^{-1} h^{-1})	Gross	Net
Volga	2.48 ± 0.79	16.1 ± 2.4	0.34 ± 0.008	647.9	185.9
Mologa	5.39 ± 0.97	21.1 ± 2.2	0.30 ± 0.05	260.6	157.6
Sheksna	6.11 ± 0.56	30.3 ± 4.4	0.48 ± 0.08	1,222.0	452.6
Main	2.68 ± 0.61	16.5 ± 1.9	0.26 ± 0.02	4,909.0	2,104.0

S.E.   Standard error

Sources: Mineeva, 1993; data on algal biomass compiled by Koreneva, 1993

Sheksna reservoir during summer differs greatly from year to year (e.g. in 1987 it was 1.11 km^3 and in 1989 only 0.17 km^3). This factor affects the distribution of pollutants significantly.

16.4.3 Algal biomass and primary production

The results of monitoring at many sampling stations in the summer of 1989 showed the significant impact of massive nutrient discharges by the Cherepovets complex, on the biota of the Sheksna reach. Each of the Rybinsk reservoir reaches had specific characteristics of phytoplankton production (Table 16.5). The Mologa, and in particular the Sheksna, reaches were characterised by higher concentrations of chlorophyll leading to the classification of waters as eutrophic. The main part of the reservoir was mesotrophic (Mineeva, 1993). The intensity of photosynthesis showed a high level of phytoplankton primary production during this period. Maximum values of primary production were also characteristic of the Sheksna reach. During this period primary production dominated over degradation processes (Table 16.5). In the Sheksna reach, net production was 2.5 times greater than in the Volga and Mologa reaches. However, it should be noted that these data relate to the autotrophic stage of phytoplankton development.

The phytoplankton biomass of the Rybinsk reservoir is formed predominantly by three groups of algae, Cyanophyceae, Bacillariophyceae and Chlorophyceae. During 1953–89, 1,017 species of algae were found in the reservoir (Koreneva, 1993). The majority of the algae are species commonly inhabiting freshwater reservoirs and preferring neutral and slightly alkali waters. The algal diversity in the reservoir decreases with an increase in the biomass of the algae. During the summer of 1989, algal biomass was at its maximum in the Sheksna and Mologa reaches (Table 16.5). Regression analysis of long-term phytoplankton data from six standard monitoring stations over 30 years

Table 16.6   Relative concentration of chlorophyll a in wet phytoplankton biomass in the Main and Volga reaches of the Rybinsk reservoir

Year	Chlorophyll a (μg l^{-1})	Biomass (mg l^{-1})	Chlorophyll a/biomass (%)
1969	4.5	1.66	0.27
1970	3.7	1.18	0.31
1971	6.6	3.06	0.22
1972	10.6	3.41	0.31
1973	9.5	1.82	0.51
1974	7.8	2.29	0.34
1975	6.9	1.64	0.42
1976	9.1	2.34	0.39
1977	7.0	2.23	0.31
1978	7.9	1.40	0.56
1979	11.1	2.20	0.50
1980	6.7	–	–
1981	10.0	–	–
1982	8.8	–	–

Data for both reservoir reaches combined   Sources: Koreneva, 1993; data from 1980–82 for chlorophyll a from Pyrina and Mineeva, 1990

showed that there was no real increase in the algal biomass in the Rybinsk reservoir although there was a slight increase in chlorophyll a (Table 16.6) (Koreneva, 1993). The average long-term algal biomass does not exceed the values characteristic of mesotrophic reservoirs. However, the increase in photosynthetic pigments in water and the indication of phases of eutrophication are indirect proof of the slow increase in the trophic state of the reservoir. An increase in the relative concentration of smaller individual algal cells, subject to more intensive grazing by zooplankton, coupled with a reduction in the average size of cells, confirmed the increase in chlorophyll in the phytoplankton biomass of the Rybinsk reservoir.

The results, over 20 years, of systematic observations of the annual primary production of phytoplankton showed significant fluctuations from 31 to 168 g C m^{-2} (Figure 16.4) (Romanenko, 1984; Smirnov et al., 1993). These fluctuations depend on changes in meteorological factors (e.g. sun radiation, temperature, precipitation, wind activity) and on the availability of nutrients. The established 11-year (summer) cycle of climatic parameters is reflected in the inter-annual dynamics of numbers and biomass of phyto- and zooplankton (Smirnov et al., 1993).

16.4.4 Organic matter

The principal input of organic substances to the Rybinsk reservoir is due to river discharge (54 per cent) and phytoplankton production (34 per cent) (Zakonnov and Ziminova, 1984). Decaying plankton

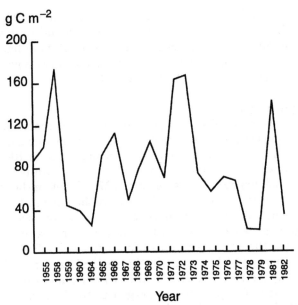

Figure 16.4   Annual phytoplankton production in the Rybinsk reservoir (After Romanenko, 1984)

produce highly labile organic substances easily accessible to bacteria (Romanenko, 1985). They can therefore be mineralised rapidly by biochemical processes. Conversely, river inputs are characterised by more stable coloured organic substances which are humic in nature and are characteristic of surface runoff from podsolic soils (with wetland features) of forest water catchments. The concentration and seasonal distribution of dissolved organic carbon (DOC) and the colour of the reservoir waters in river reaches and the Main reach differ greatly. The highest colour and concentration of organic carbon are characteristic of the waters of the Mologa reach, the seasonal dynamics of which correlate with the volume of discharge of the Mologa river. Maximum values occur in spring (colour up to 120°, DOC 13–16 mg l^{-1}) and minimum values occur in winter and summer low waters (colour 50–80°, DOC 10–12 mg l^{-1}). A similar change in colour and DOC, but not as marked, is characteristic of the Sheksna reach. In the Volga reach, these levels do not vary significantly during the year due to agricultural development in the Volga catchment and the regulation of discharge of the Uglich and Ivankovo reservoirs. The colour of the water in the Main reach has not changed significantly for years. This is a result of the gradual movement of spring waters of the river section and their mixing with winter waters of the Main reach. It is also due to the slow transformation of organic substances under the influence of biochemical processes, the ultraviolet sun radiation and sedimentation processes. Thus, in 1981, colour varied in different parts of the reach between 55° and 65° and in 1982 between 45° and 50° (Bikbulatova, 1984). Earlier it was found that the inter-annual fluctuations of colour depended on the annual water balance. The reduction in coloured organic substances in humid years is greater than in dry years (Fortunatov, 1975; Gapeeva, 1993).

16.4.5 Oxygen regime

Seasonal changes in dissolved oxygen concentrations consist of a reduction in the oxygen concentration from spring to summer on average from 10–13 to 8–10 mg l^{-1}, an increase in the autumn of up to 10–12 mg l^{-1} and a gradual reduction to 5 mg l^{-1} under ice during winter. Minimum concentrations of oxygen occur at the end of March before the start of ice cover break-up in the reservoir and, even in the bottom layers, these concentrations are several milligrams per litre. However, there are a few exceptions in some deeper parts, such as over former river beds where in the bottom layer the oxygen concentration may reach 0.2–0.5 mg l^{-1}. In the summer, there is little change in the vertical oxygen profile due to wind action: oxygen stratification in the reservoir is a rare and short-term phenomenon. The presence of oxygen in the bottom layers is one of the major indicators of biochemical processes occurring at the water–sediment interface. This has a positive influence on water quality.

The impact of water discharge regulation on the oxygen regime during the vegetation growth period results in less (compared with river discharge) oxygen saturation. This is caused by the characteristic domination of oxygen-consuming processes over production processes. However, under the influence of wind and currents, there is an active aeration of water with atmospheric oxygen, compensating for oxygen use in oxidising processes. Oxygen saturation in the main water mass of the reservoir is 75–90 per cent, which is still adequate for the requirements of most biota (Butorin et al., 1984).

16.4.6 Suspended particulates

The average theoretical water residence time is six months and therefore the spring flood discharge can be almost completely accumulated within the Rybinsk reservoir. Most river suspended solids, the major external source of total suspended solids (TSS) to the reservoir, settle in the reservoir. The main sediment sources within the reservoir itself are bank erosion processes and, to a lesser extent, the production of organic matter. The maximum amount of suspended matter is found in river reaches in spring. During high waters, the concentration of TSS in the upper parts of these reaches is 20–40 mg l^{-1} but in the flood plain and lower parts of the reaches it decreases to 4–6 mg l^{-1}. Therefore, most of the river alluvium (in all seasons) is deposited in river reaches and only a small amount of suspended matter enters the central part of the reservoir where concentrations are 3–4 mg l^{-1}. In the western and southern parts of the Main reach, the concentration of suspended matter slightly exceeds these average concentrations. This is caused by the penetration of river waters into these regions and the resuspension of sediment by wind.

At the time of maximum algal production, and by the end of summer, the amount of suspended matter increases due to autochthonous material. In river reaches, this increase can be up to 8–12 mg l^{-1} and in the Main reach it can be up to 4 mg l^{-1}. However, due to the extended wind fetch and wave height, the wind action may cause a resuspension of deposited sediments and as a result, suspended solids in the Main reach may increase to 6–8 mg l^{-1}.

According to the seasonal changes in suspended matter concentrations and water colour in river sections, the minimum Secchi disc transparency is observed in spring at 0.4–0.5 m from the surface. In the vegetation growth period, in all the reaches including the Main reach, transparency is 0.6–1.5 m, and on average 1.1 m. The euphotic layer does not usually exceed 2 m in depth.

16.4.7 Organic micropollutants

Pollution studies of water and sediment in the Rybinsk reservoir since 1990 have determined the most significant organic pollutants (Table 16.7). These include petroleum products, PCBs and PAHs, formaldehyde, anionic detergents and phenols. Petroleum hydrocarbons and resinous components are found in all samples of water and sediment. The frequency of occurrence of 2–3 nuclear and some 4–7 nuclear PAHs and some PCB isomers is greater than 50 per cent in sediment samples.

The distribution of organic contaminants in the reservoir is determined by the location of the main sources of pollution, the dynamics of the water mass and the nature of bottom sediments. The main source of pollution is Cherepovets city, which is why the highest concentrations of PCBs, PAHs and petroleum products are found in water and sediments of the Sheksna reach (Table 16.8), where pollution may be considered as chronic. Intensive navigation is another source of these contaminants (except PCBs) in the other reaches of the reservoir. The Mologa reach is least polluted because industrial enterprises are absent from its banks and navigation there is limited.

The levels of PAHs, PCBs and petroleum products in the reservoir water and sediment have not changed significantly from 1990 to 1993 (Table 16.8). A significant increase in the concentration of PAHs in the Sheksna and Main reaches was caused by the spilling of diesel fuel following a shipping accident in August 1993. This shipwreck led to higher concentrations of petroleum products in the sediments of the Sheksna reach.

The spatial distribution of organic substances in the sediment of the reservoir also depends on the composition and origin of the latter. For example, the highest concentrations of PCBs within the Sheksna

Table 16.7 Principal organic pollutants and frequency of occurrence in water and sediment samples of the Rybinsk reservoir, 1990–93

Substance	Frequency of occurrence (%)	
	Water	Sediment
Petroleum products		
Hydrocarbon	100	100
Resinous acids	100	100
PAHs 2-3 nuclear		
Naphthalene	59	71
Fluorene	53	94
Phenanthrene	58	91
Anthracene	52	97
PAHs 4-7 nuclear		
Fluoranthene	59	98
Pyrene	76	97
Benzo(b)fluoranthene	12	24
Benzo(a)pyrene	9	38
Indenopyrene	18	74
Benzo(g,h,i)perylene	50	82
Chrysene	100	94
Triphenylene	62	88
PCBs		
Chlophen A-40	18	32
Chlophen A-50	20	65
Other compounds		
Formaldehyde	50	–
Detergents	88	–
Methane	90	–
Phenol	16	–

PAHs   Polyaromatic hydrocarbons   PCBs   Polychlorinated biphenyls

reach are found in 'black silt' of anthropogenic origin (i.e. areas of wastewater discharge). The lowest concentrations are found in silty sand (Figure 16.5). Grey silt, grey sandy silt and sand have intermediate concentrations and, moreover their capacity to accumulate PCBs decreases respectively.

Fish accumulate PCBs mainly in the liver. The average total PCB concentration in the liver of bream in the Sheksna reach is 5.4 mg kg^{-1}, while in muscle tissue it is 0.5 mg kg^{-1}. In burbot it is 11.2 and 0.2 mg kg^{-1} respectively, in pike-perch 1.4 and 0.1 mg kg^{-1} and in perch 4.8 and 1.0 mg kg^{-1}. The order of relative bioaccumulation of PCBs in liver is as follows: burbot > bream > perch > pike > pike-perch.

Table 16.8   Average concentrations of PAHs, PCBs and petroleum products in water and sediment of the Rybinsk reservoir

Reach	1990			1993		
	PAHs ($\mu g\ l^{-1}$)	PCBs ($\mu g\ l^{-1}$)	Petrol. prod. ($mg\ l^{-1}$)	PAHs ($\mu g\ l^{-1}$)	PCBs ($\mu g\ l^{-1}$)	Petrol. prod. ($mg\ l^{-1}$)
Water						
Volga	0.06	0	0.04	0.051	0	0.04
Mologa	0.03	0	0.03	–	–	–
Sheksna	0.23	traces	0.11	0.28	traces	0.09
Main	0.01	0	0.06	0.02	0	0.05
Sediment						
Volga	2.7	0	2.5	1.9	0	1.8
Mologa	1.4	0	0.9	1.0	0	0.7
Sheksna	4.3	0.2	1.1	70.0	0.2	3.0
Main	1.4	0	1.4	24.0	0	0.7

PAHs   Polyaromatic hydrocarbons   Petrol. prod.   Petroleum products
PCBs   Polychlorinated biphenyls

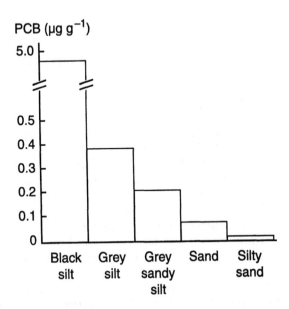

Figure 16.5   Average concentrations of PCBs in different types of sediments in the Sheksna reach

Table 16.9   Concentrations of heavy metals in water of the Rybinsk reservoir

Metal	1988		1989	
	Range ($\mu g\ l^{-1}$)	Average ($\mu g\ l^{-1}$)	Range ($\mu g\ l^{-1}$)	Average ($\mu g\ l^{-1}$)
Cu	0.5–6.2	1.4	0.4–177	2.0
Pb	0.9–3.5	1.4	0.6–87	1.8
Cd	0.0–0.3	0.1	0.0–4.6	0.2
Zn	4.0–800	12.2	6.0–460	18.4
Ni	0.6–95	1.0	0.6–95	1.8

Source: Gapeeva, 1993

Table 16.10 Concentration of heavy metals in sediment of Sheksna reach

Metal	Concentration ($\mu g\ g^{-1}$)	
	1978	1985
Cu	14.5 ± 1.8	33.0 ± 10.0
Ni	19.2 ± 1.6	25.1 ± 4.5
Pb	24.5 ± 3.1	39.0 ± 12.0
Cd	2.2 ± 0.2	1.7 ± 0.4
Cr	16.0 ± 1.5	7.4 ± 1.2
Zn	117.0 ± 13.0	359.0 ± 201.0
Mn	690.0 ± 62.0	760.0 ± 66.0

Source: Gapeeva, 1993

16.4.8 Heavy metals

The concentrations of heavy metals in the water of Rybinsk reservoir vary considerably (Table 16.9). However, in the last decade average concentrations have not changed significantly (Gapeeva, 1993). Maximum concentrations of metals, as a rule, are found in the lower reaches of the reservoir. The greatest concentration of metals in the surface layer of sediments is near Sheksna reach, where there is an overall tendency for an increase in average concentrations over time of Zn, Pb and Cu (Table 16.10). The exceptions are Cd and Cr, the concentrations of which either do not change significantly or decrease.

Editorial note. The absolute concentrations presented in Table 16.9 seem to be very high and reflect general problems connected with the determination of trace amounts of heavy metals in water (see Chapter 8). At the same time, however, relative changes showed a general increase in Cu,

Pb and Zn in sediments, in which heavy metals are easier to analyse (Table 16.10).

16.5   Conclusion and future requirements

At present, in connection with the fast development of different sources of energy, the importance of the Rybinsk reservoir as a source of energy has decreased. The capacity of Rybinsk hydropower station is over 300 MW, which is equal to the capacity of one unit of energy produced by a modern thermal power plant. The use of the reservoir in fisheries is also limited. The annual fish catch does not exceed 4–5 kg ha^{-1}.

The water of Rybinsk reservoir is used for the water supply of the towns of Cherepovets, Uglich and Rybinsk. The Central Industrial region, which is situated nearby and which includes Moscow, is not adequately supplied with drinking and industrial water and its water deficit is constantly increasing. Therefore, the possible use of the reservoir as a potential source of drinking water supply for this region must be considered (Drachev, 1972). The morphometry, hydrological regime and the chemical, physical and biological indicators of water quality of the Rybinsk reservoir meet the demands for industrial and drinking water supply. The use of the reservoir for water supply is not an obstacle to the development of fisheries. The exploitation of the reservoir to cover the deficit in the water balance of the Central Industrial region will significantly change the importance of the reservoir and its relative significance among the other natural resources of the country.

In coming decades the high economic importance of the reservoir will be preserved only if there is no deterioration in water quality. Considering the prospects of development of Rybinsk reservoir as a source of drinking water supply for the Central region, a number of measures are necessary to preserve water quality:

- To prevent further pollution from atmospheric sources, the reservoir should be protected against the construction in the region of large industrial enterprises. At present, because of the prevalence of westward winds, the danger of air pollution from the Cherepovets industrial centre is not significant.
- The long shoreline exposes the reservoir to pollution from mineral fertilisers and pesticides. Therefore special attention should be given to the preservation and expansion of forests and to limiting the increase in agricultural areas in the water catchment of the reservoir. At present agricultural regions occupy relatively small areas (along the Rivers Sit, Sebla, Ukhra and Sogozha).
- To eliminate further pollution of the reservoir with domestic and industrial wastewaters, the location of chemical, petroleum, pulp and paper industries and other similar enterprises on the banks and in close proximity to the reservoir should be banned.

- There must be an improvement in the treatment of domestic and industrial wastewaters from the existing industries and towns, in particular the 'Severostal' metallurgical plant and the waste waters of Cherepovets city.
- With the aim of maintaining high water quality, a sanitary protection zone around drinking water sources should be established.

16.6   References

(All references are in Russian unless otherwise stated).

Bikbulatova, E.M. 1984 Organic substance balance in the Rybinsk reservoir. River water transformation during flow regulation and water transfer. Scientific Report (final). IBVV of Academy of Sciences of the USSR, No. 81015334, Borok.

Butorin, N.V. [Ed.] 1978 *Volga River and its Life*. Nauka, Leningrad, 350 pp.

Butorin, N.V., Bylinkina, A.A., Ziminova, N.A. and Trifonova, N.A. 1984 Abiotic factors of material cycles in the Volga reservoirs. In: *Biological Productivity and Water Quality in the Volga River and its Reservoirs*. Nauka, Moscow, 37–48.

Bylinkina, A.A. 1993 Nitrogen and phosphorus concentrations in the Rybinsk reservoir during its autotrophic stage of functioning. In: *Present State of the Rybinsk Reservoir Ecosystem*. Gidrometeoizdat, St Petersburg, 28–41.

Drachev, S.M. 1972 Sanitary state and water quality of the reservoir as a source of drinking water. In: *Rybinsk Reservoir*. Nauka, Leningrad, 119–128.

Drachev, S.M., Bylinkina, A.A., Trifonova, N.A. and Kudryavtseva, N.A. 1976 The impact of anthropogenic factors on nutrient and salt concentrations of Volga reservoirs. In: *Biological Production Processes in the Volga Basin*. Nauka, Leningrad, 18–24.

Fortunatov, M.A. 1975 Water colour and transparency. In: *The Hydrometeorological Regime of Lakes and Reservoirs of the USSR*. Gidrometeoizdat, Leningrad, 179–186.

Gapeeva, M.V. 1993 Biogeochemical distribution of heavy metals in the Rybinsk reservoir ecosystem. In: *Present State of the Rybinsk Reservoir Ecosystem*. Gidrometeoizdat, St Petersburg, 42–49.

Koreneva, L.G. 1993 Rybinsk reservoir phytoplankton: content, distribution trends, eutrophication consequences. In: *Present State of the Rybinsk Reservoir Ecosystem*. Gidrometeoizdat, St Petersburg, 50–113.

Kozlovskaya, V.I. and Sambursky, V.N. 1992 Realisation of a water protection policy using the example of Cherepovets industrial complex on the Rybinsk reservoir. Russian-German Symposium "A Comparison of the Rhine-Volga Economic Systems", 7–11 September 1992. Borok.

Kuzin, B.S. [Ed.] 1972 *Rybinsk Reservoir and its Life*. Nauka, Leningrad, 365 pp.

Mineeva, N.M. 1993 Primary production in the Rybinsk reservoir in summer. In: *Present State of the Rybinsk Reservoir Ecosystem.* Gidrometeoizdat, St Petersburg, 114–140.

Pyrina, I.L. and Mineeva, N.M. 1990 Phytoplankton pigment concentration in the Rybinsk reservoir water column. In: *Pelagic and Littoral Phytocenoses Productivity and Flora in the Volga Basin Water Reservoirs.* Nauka, Leningrad, 176–188.

Razgulin, S.M. 1985 Nutrient balance of the Rybinsk reservoir and forecast of concentration fluctuations due to the transfer of the northern river's flow. Thesis of Candidate of Geographical Science, Borok, 169 pp.

Romanenko, V.I. 1984 Primary production and photosynthesis processes in the Volga reservoirs cascade. In: *Biological Productivity and Water Quality in the Volga River and its Reservoirs.* Nauka, Moscow, 48–60.

Romanenko, V.I. 1985 Microbiological processes of production and degradation of organic matter in inland water bodies. Nauka, Leningrad, 294 pp.

Smirnov, N.P., Vainovsky, P.A. and Titov, J.E. 1993 Correlation of inter-annual climate fluctuations and water reservoir ecosystem parameters. In: *Present State of the Rybinsk Reservoir Ecosystem.* Gidrometeoizdat, St Petersburg, 20–27.

Tachalov, S.N. 1956 Dynamics of areas of flooded woods and peat bogs around the Rybinsk reservoir. Collected Publications of Rybinsk Hydrometeorology Observatory, Leningrad. Issue 2, 115–122.

Trifonova, N.A. and Bylinkina, A.A. 1984 Nitrogen and phosphorus compounds in bottom sediment of the Upper Volga reservoirs. In: *Water and Sediment Interaction in Lakes and Reservoirs.* Nauka, Leningrad, 146–154.

Volodin, V.M. 1993 Population structure dynamics of *Abramis brama* (L.) in the Rybinsk reservoir. In: *Present State of the Rybinsk Reservoir Ecosystem.* Gidrometeoizdat, St Petersburg, 233–251.

Zakonnov, V.V. 1981 Bottom sediment distribution in the Rybinsk reservoir. Information Bulletin. *Biology of Inland Waters,* **51,** 68–72.

Zakonnov, V.V. and Ziminova, N.A. 1984 Nutrient balances in the Upper Volga reservoirs. In: *Water and Sediment Interaction in Lakes and Reservoirs.* Nauka, Leningrad, 114–122.

Chapter 17[*]

DNIEPER AND ITS CASCADE OF RESERVOIRS

The Dnieper river is third in terms of size, after the Volga and the Danube, on the European continent. Its water catchment area is 509×10^6 km^2 and its length is 2,285 km. The natural, average long-term discharge of the Dnieper is 52 km^3 and the average annual flow is 1,650 m^3 s^{-1} (Voskresensky, 1962). The source of the Dnieper is in the Valdai Highland. On its way to the Black Sea, the Dnieper crosses three natural vegetation zones: forest, forest steppe and steppe. It also flows through the territories of three states: Russia (500 km), Byelorussia (735 km) and the Ukraine (1,050 km). It is the largest river of the Ukraine.

17.1 The Dnieper basin

By its morphological characteristics, the Dnieper is divided into three parts: upper, middle and lower (USSR Surface Water Resources, 1971). The Upper Dnieper includes the part from the source to Kiev (1,375 km). This part receives the main tributaries, the Berezina, Sozh, Pripyat and Desna rivers (Figure 17.1 and General Appendix IV). The Middle Dnieper is the part of the river which runs from Kiev to Zaporozhye and is 570 km long. The largest tributaries of this section of the river are the Ros, Sula, Psel, Vorskla, Orel and Samara. The Lower Dnieper is the part of the river from Zaporozhye to the mouth and is 340 km long. This part includes the lower reaches, i.e. the area from the dam of the Kakhovka hydropower station (HPS) to the mouth (90 km) which, together with the Dnieper estuary, forms the mouth area of the Dnieper.

The relief in the upper part of the Dnieper basin is mainly characterised by plains, and only near the source are there still fragments of ancient moraines. Over the rest of the water catchment area, there are alternating lowlands and highlands and, dividing these, the valleys of the tributaries. The largest plains in the Upper Dnieper basin are the Central-Berezina plain and the Byelorussia low-lying woodland (vast alluvial plain). Significant water availability is characteristic of these regions. The annual amount of rainfall is 560–610 mm. Moreover, there is significant precipitation in the western and north-eastern parts, with up to 200 days of precipitation a year (Levkovsky, 1979). Long, snowy

*   *This chapter was prepared by V.I. Peleshenko and A.I. Denisova*

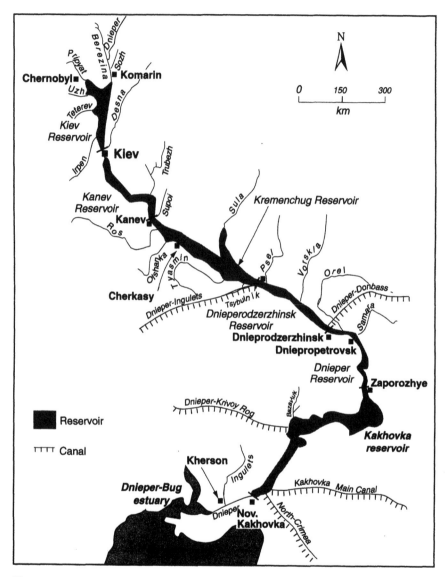

Figure 17.1   Map of the River Dnieper and its reservoirs

and cold winters and moderately warm and humid summers mean that there is more precipitation than evaporation. Thus, within the water catchment basin of the Upper Dnieper, the hydrographic network is well developed with a density of 0.37–0.47 km km^{-2} (Laponogov, 1978). This part of the basin is richly covered with mixed and coniferous forests, which in some places occupy up to 50 per cent of the surface area. Soil cover is

Table 17.1 Characteristics of the main tributaries of the river Dnieper

River	Length (km)	Basin area (10^3 km²)	Annual discharge (km³)	Proportion of total Dnieper discharge (%)
Berezina	613	24.5	4.6	8.7
Sozh	648	42.1	7.0	13.0
Pripyat	761	121.0	14.5	27.0
Teterev	365	15.1	0.8	1.5
Desna	1,130	88.9	11.4	20.0
Sula	363	18.5	1.3	2.5
Psel	717	22.8	1.8	3.5
Vorskla	464	14.7	1.0	1.9
Orel	346	9.8	0.4	0.8
Ingulets	549	13.7	0.3	0.6

represented by podsolic, sandy loam, loam, sandy and peat-swamp soils. The proportion of wetlands is on average 14 per cent of the basin surface, but in the western part (Berezina and Pripyat river basins) it increases by up to 30 per cent.

Over the whole of the Dnieper basin, there are more than 30,000 large and small rivers with a total length of 130,000 km (Denisova *et al.*, 1989). Due to the high humidity and dense hydrographic network in the basin of the Upper Dnieper (occupying 65 per cent of the total area), 80 per cent of the annual discharge is formed in this region, where snow-melt waters dominate in the recharge of rivers. Spring floods deliver up to 36 per cent of the annual discharge volume in years with low flow and up to 60–64 per cent in years with high water flow. Three large tributaries, Berezina, Sozh and Pripyat, the water catchment of which occupies only 37 per cent of the area, form about 49 per cent of the Dnieper discharge. The basin of the River Desna is situated in the forest zone and contribute 20 per cent of discharge (Table 17.1).

There are not many lakes in the Dnieper basin. Small flood plain lakes occur along the valleys of the Upper Dnieper, Pripyat and Desna, but downstream of these, there are mainly reservoirs. Only downstream of the Kakhovka hydropower station are there flood plain and delta lakes.

In the south, climatic conditions change significantly. Air temperatures increase and the annual total precipitation falls from 610 mm to 323 mm, while evaporation increases from 642 mm to 921 mm. Compared with the Upper Dnieper basin, the duration of stable snow cover decreases. The character of the soil cover also changes from north to south. Podsolic soils and chernozem begin to dominate, as do chestnut soils. In some places, saline soils (solonchak) occur.

Table 17.2   Main morphometric characteristics of the Dnieper reservoirs

Reservoir	Distance from dam to Dnieper mouth (km)	Years reservoirs were filled	Length along axis (km)	Average width (km)	Area (km²)
Kiev	836	1965–66	110	8.4	922
Kanev	713	1973–76	123	5.5	564
Kremenchug	564	1960–61	149	15.1	2,250
Dnieprodzerzhinsk	450	1963–64	114	5.1	567
Dnieper	321	1931–34 1947[1]	129	3.2	410
Kakhovka	91	1955–56	230	9.3	2,150

[1]   Secondary filling of reservoir after World War II

In the steppe zone, only small islands of forest vegetation are left and steppe grass occupies the remnants of unploughed land in the basin. Still further south, in the water catchment area of the Black Sea lowland, massive sands occur in the left-bank area. In the valley, flooded forest and bush vegetation has developed.

The chemical composition of natural waters is quite varied. The mineralisation of groundwaters which are situated in the surface sand and clay sediment on the territory of the Byelorussia woodland, does not exceed 300–400 mg l^{-1} total dissolved solids (TDS) and HCO_3^-, Ca^{2+} and Mg^{2+} are prevalent. According to the classification by Alekin (Alekin, 1970) this corresponds to the bicarbonate class of the calcium group of the second type (Chapter 2). Higher mineralisation is characteristic of groundwaters in the middle reaches of the Dnieper, in the forest steppe zone. In Tertiary sediment, over a large area of the Ukraine, the aquifers occur at a depth of 8–120 m. The groundwaters of the Byelorussia woodland belong to the bicarbonate-calcium category of waters. Their TDS values vary from 53 to 676 mg l^{-1}.

17.2   Dnieper reservoir cascade

The creation in 1932 of the first Dnieper reservoir, bearing the same name, marked the start of a construction programme on the Dnieper and the start of its flow regulation. After the Dnieper reservoir, other reservoirs were built: Kakhovka in 1955, Kremenchug in 1960, Dnieprodzerzhinsk in 1963, Kiev in 1965 and Kanev in 1973 (Figure 17.1, Table 17.2.). In the same years, large water-supply and irrigation canals were built in the basin of the Lower Dnieper (Table 17.3).

A considerable volume of water is taken from the cascade of reservoirs to supply water to large cities and industrial centres. At present the Dnieper waters irrigate more than 1×10^6 ha. Almost one third of this

Table 17.2   Continued

Reservoir	Average depth (m)	Volume (km³)	Water residence time (months)	Area of shallow waters (%)	Variation in level (m)
Kiev	4.0	3.73	12–13	40	0.5–1.0
Kanev	3.9	2.62	17–18	24	0.5
Kremenchug	6.0	13.50	2.5–4.0	18	3.0–5.0
Dnieprodzerzhinsk	4.3	2.45	18–20	31	0.5
Dnieper	8.0	3.30	12–14	34	0.5–1.0
Kakhovka	8.5	18.20	2.0–3.0	5	3.0–4.0

Table 17.3   Main canals and other important artificial waterways in the Dnieper basin

Name	Source of intake	Length (km)	Capacity ($m^3 s^{-1}$)	Capacity ($10^6 m^3 a^{-1}$)
Main canals				
Northern Crimea	River Dnieper, Kakhovka reservoir	400.3	300	4,200
Kakhovka main canal	River Dnieper, Kakhovka reservoir	129.7	520	8,200
Dnieper-Donbass	River Dnieper, Dnieprodzerzhinsk reservoir	263.0	120	2,743
Dnieper-Krivoy Rog	River Dnieper, Maryanovsk gulf of Kakhovka reservoir	42.9	41	929
Dnieper-Ingulets	River Dnieper, Tsybulnik gulf of Kremenchug reservoir	40.0	37	1,008
Important artificial waterways				
Dnieper-Donbass-Kharkov	Canal Dnieper-Donbass	142.0	8.6	239
Dnieper-White Church	Dnieper, Kanev reservoir	250.0	2.0	60
Dnieper-Kirovograd	Kremenchug reservoir	116.0	1.7	54
Mezhgornoye reservoir	Northern Crimea canal	78.5	1.5	47.5
Dnieper-West Donbass	River Dnieper	70.0	1.4	48.8

area is situated outside the Dnieper basin, in the Crimea, Donbass and the Black Sea lowland.

The total volume of water consumption in the Dnieper basin is constantly increasing, mainly at the expense of water extraction from the Kakhovka reservoir (Table 17.4). The principal use of this reservoir is for water supply rather than energy production. At present the total volume of water consumed reaches 15–17 km^3 a^{-1}, of which 8 km^3 are abstracted irrevocably (Levkovsky, 1979). The number of major water users in the Dnieper basin is 5,796, which is 46 per cent of the total number of water users in the Ukraine. The amount of water abstracted from the Dnieper and its reservoirs is 16,297 × 10^6 m^3 a^{-1} including 2,218 × 10^6 m^3 a^{-1} for municipal and domestic purposes, 8,595 × 10^6 m^3 a^{-1} for industry, 4,740 × 10^6 m^3 a^{-1} for irrigation and 744 × 10^6 m^3 a^{-1} for agricultural water supply.

17.3   Factors influencing water quality

The hydrochemical regime and water quality of both the Dnieper and the cascade of its reservoirs are determined by processes in the water ecosystem and the catchment area. The location of the reservoir in the cascade, its age, morphology and hydrochemical conditions, also determine overall water quality.

The influence of surface runoff has been recorded only in the Kiev reservoir, which is the uppermost reservoir in the cascade. Further down along the cascade inter-reservoir processes are the most significant, including primary production and degradation, sedimentation, redox and sorption processes. These processes determine the nutrient and organic matter transformations and cycles within the reservoirs.

The position of the reservoir in the cascade and the processes within it determine, to a great extent, the seasonal and inter-annual dynamics of the main components of chemical composition of the water and their distribution in the water body. The lower the flow rate of the reservoir, the more the shoreline is indented and the higher the level of urbanisation and agricultural use of the catchment area, the greater the level of chemical heterogeneity.

17.3.1 Newly created reservoirs

During the filling and the first year of a reservoir's existence, changes in hydrological conditions and the flooding of large areas of soil and vegetation significantly influence water quality. During this period, active hydrogeochemical processes occur at the water–ground interface and there is intensive inflow of all soluble compounds from the ground, underlying soils, rocks and flooded vegetation. These processes contributed 134,000 t of organic carbon, 42,000 t of nitrogen and 2,000 t of phosphorus to the cascade of the Dnieper reservoirs in their first year.

Table 17.4 Water abstraction from reservoirs and wastewater discharge to reservoirs of the Dnieper, late 1980s and early 1990s

	Water abstraction						Wastewater discharge					
	Main body		Tributaries		Total		Main body		Tributaries		Total	
Reservoir	(10⁶ m³ a⁻¹)	(%)	(10⁶ m³ a⁻¹)	(%)	(10⁶ m³ a⁻¹)	(%)	(10⁶ m³ a⁻¹)	(%)	(10⁶ m³ a⁻¹)	(%)	(10⁶ m³ a⁻¹)	(%)
Kiev	15.66	0.1	118.5	6.9	134.21	0.8	13.12	0.2	89.29	11.4	102.41	1.2
Kanev	2,098.38	13.4	555.16	32.4	2,653.54	15.3	2,453.08	32.0	84.80	10.8	2,537.88	30.0
Kremenchug	576.38	3.7	106.64	6.2	683.02	3.9	279.52	3.6	77.26	9.8	356.78	4.2
Dnieprodzerzhinsk	771.19	4.9	25.86	1.5	797.05	4.6	15.14	0.2	118.16	15.0	133.30	1.6
Dnieper	2,911.52	18.6	162.95	9.6	3,074.47	17.7	1,998.21	26.1	22.61	28.4	2,220.82	26.3
Kakhovka	9,181.71	58.7	–	–	9,181.71	52.9	2,886.27	37.7	6.74	0.8	2,893.01	34.3
Lower Dnieper and Ingulets	92.22	0.6	742.36	43.4	834.58	4.8	17.9	0.2	187.21	23.8	205.11	2.4
Total	15,647.06	100.0	1,711.52	100.0	17,358.58	100.0	7,663.24	100.0	786.07	100.0	8,413.81	100.0

For nitrogen and phosphorus these quantities were equal to the annual inflow from the water catchment area (Denisova, 1979).

17.3.2 Anthropogenic factors

The Dnieper and its reservoirs, situated within densely populated regions with intensive agriculture and industry, are under the increasing influence of anthropogenic factors. This influence begins with the regulation of discharge because this is the most active human interference with natural processes and leads to significant changes in the character of a natural water body. Due to the inflow of contaminants with industrial, municipal and domestic wastewaters and the discharge from agricultural lands and livestock facilities into the reservoirs, there is an accumulation of nutrients, organic substances and heavy metals. Eutrophication is increasing, leading to algal blooms and a deterioration in the dissolved gas regime (primarily a decrease in oxygen concentrations), with a subsequent significant influence on the self-purification processes of the water body. Under these conditions, the redox potential of the water decreases, especially in bottom layers where anaerobic zones may appear. The breakdown of organic substances under anaerobic conditions is much slower and results in the final production of hydrogen sulphide, ammonium and other reducing compounds, which accumulate in the reservoir and further degrade its condition. Under anaerobic conditions, the nutrient flux from sediments into the water is enhanced.

Many contaminants, such as petroleum products and phenols, are only processed by micro-organisms under aerobic conditions. When water exchange is reduced and oxygen levels decline, the processes of degradation and oxidation of such organic substances slow down. The number of anaerobic bacteria increases significantly, by 10 or 100 fold, with the degradation of organic substances. Sanitary indicators of water quality also deteriorate as a result of the accumulation of decay products.

Industrial, municipal and domestic wastewaters

With increasing urbanisation and industrialisation, the inflow to natural waters of nutrients and organic substances with industrial, municipal and domestic wastewaters increases. In municipal sewage, the concentration of organic nitrogen varies from 30 to 90 mg l^{-1} and the concentration of phosphorus from 1.7 to 2.6 mg l^{-1}. Fifty per cent of the phosphorus comes from organic compounds. Higher nutrient levels are present in the wastewaters of different industrial enterprises. The chemical industry produces 30–80 mg l^{-1} of nitrogen, nitrogen-fertiliser industries produce 100–140 mg l^{-1} and the coke production industry 800–1,000 mg l^{-1} (Drachev, 1964).

In the Dnieper basin, 85 of the most significant polluters discharge approximately $8,500 \times 10^6$ m^3 a^{-1} of wastewaters treated to different levels (Table 17.4) leading to a high level of water pollution in these reservoirs. The greatest amount of wastewater is discharged into the Dnieper and Kakhovka reservoirs (and tributaries), at 26.3 per cent and 34.3 per cent respectively. In other reservoirs, the input of wastewater volume ranges from 1.2–4.2 per cent of the total volume discharged to the Dnieper basin. In the area of the Kiev reservoir, there are no industrial enterprises discharging wastewater directly to the reservoir. The industries are situated on the Upper Dnieper and on tributaries feeding into the reservoir.

The present practice of biological treatment of wastewaters does not eliminate nutrients. In secondary settling tanks, the nitrogen concentration of the water remains at 22–27 mg l^{-1} (of which 14–17 mg l^{-1} is inorganic nitrogen) and the total phosphorus concentration remains at 1.3–3.7 mg l^{-1}. Where wastewaters are discharged into the Dnieper via the Bortnik canal (Kanev reservoir), the treated wastewaters contain 3.4–8.8 mg l^{-1} of inorganic nitrogen, up to 12.3 mg l^{-1} of organic nitrogen and 0.1–0.4 mg l^{-1} of phosphorus. This is an increase of 10–15 times the original background concentration. In some years the amount of inorganic nitrogen and phosphorus entering the Dnieper reservoir cascade with wastewaters is almost equal to the inflow of these elements with surface runoff.

Agricultural discharge
The amount of mineral fertilisers used on agricultural fields situated in the Dnieper water catchment area within the Ukraine increases from year to year. Generally, 20–40 per cent of nitrogen from nitrogen fertilisers applied in the catchment enters the river. Less of the phosphorus from phosphorus fertilisers is discharged to the river; the total amount does not exceed 1 kg ha^{-1} (Demchenko *et al.*, 1969; Burov, 1971). In this assessment, the discharge of nitrogen and potassium has been taken as 30 per cent of the total amount of fertilisers used.

The amount of nutrients which can leach from the Dnieper catchment area into the reservoir cascade is 202,000 t of nitrogen, 13,000 t of phosphorus and 88,000 t of potassium. The calculated increase in total nitrogen, total phosphorus (organic and mineral fertilisers) and potassium concentrations are, on average, 2.66 mg l^{-1} of nitrogen, 0.18 mg l^{-1} of phosphorus and 1.1 mg l^{-1} of potassium. These values highlight the significant role of agricultural discharge in the total balance of organic and mineral compounds in the Dnieper reservoirs.

17.3.3 Sedimentation processes
Decreasing water velocity in the reservoir leads to the sedimentation of allochthonous and autochthonous substances and their accumulation in sediments (e.g. heavy metals with clays, iron and manganese hydroxides). Suspended substances (inorganic and organic) are able to adsorb most chemical compounds and this leads to the extraction of a number of

dissolved substances from the water mass, and their accumulation in bottom sediment. Thus, bottom sediment and especially silts, the volume of which increases from year to year in the reservoir, become one of the important factors influencing water quality.

17.3.4 Bottom sediment

Sediments have a double role in the formation of the hydrochemical regime. On the one hand, as accumulators of many substances (heavy metals, some inorganic and organic substances) sediments promote the self-purification of the water body. The role of sediments in self-purification depends on numerous factors including their level of dispersion, organic matter content, the presence of iron and manganese hydroxides and on the hydrochemical regime of the water–sediment interface. On the other hand, sediments act as sinks for almost all chemical compounds and, in certain circumstances, they act as sources of secondary contamination in reservoirs. For example, sediments with a high silt fraction act as sources of nutrient pollution. The intensity of the adsorption–desorption processes depends on many factors, mainly the gas regime at the water–sediment interface, flow rate, temperature, pH and Eh, the composition and thickness of the bottom sediment and the chemical composition of the water.

Organic compounds in sediment undergo biochemical decay with the formation of inorganic substances and simpler organic compounds. Some organic substances are transformed into insoluble and poorly assimilated forms and become buried. In an aerobic environment, the breakdown of organic substances is usually complete, i.e. leading to the formation of carbon dioxide, ammonium, phosphates, etc. In an anaerobic environment the breakdown may cease at an intermediate stage. This leads to the accumulation in the sediment and subsequent transfer into the water, of amino acids, mono- and polysaccharides, organic acids and other compounds. The most favourable conditions for breakdown processes in aerobic waters occur at a temperature of 20–26 °C.

A principal role of sediment is its absorption of organic forms of nitrogen and the emission of ammonium ions into the water. The greatest release of NH_4^+ occurs under anaerobic conditions. The pH value has no significant influence on the formation and desorption of ammonium ions. However, an increase in temperature of 10 °C increases by two fold the rate of emission of NH_4^+. If the concentration of nitrates in water is greater than 1 mg l^{-1}, the desorption of phosphates under anaerobic conditions slows down significantly or stops altogether. The presence of oxidised manganese compounds also slows down the desorption of phosphates from sediment (Denisova et al., 1987, 1989).

Under weakly reducing conditions in spring and autumn, 19–27 mg m^{-2} d^{-1} of ammonium nitrogen, 2–3 mg m^{-2} d^{-1} of iron and 0.6 mg m^{-2} d^{-1} of inorganic phosphorus are emitted from silty sediments.

In summer, when reducing conditions become more pronounced, the sediment layer produces approximately 80 mg m^{-2} d^{-1} of ammonium nitrogen, 15 mg m^{-2} d^{-1} of iron and 9 mg m^{-2} d^{-1} of mineral phosphorus. Thus, desorption from sediment is significantly higher in summer. Taking into account water volume and the chemical composition of the sediment, an increase in the average annual concentration of phosphorus, nitrogen and iron in the water of the Dnieper reservoir, as a result of their release from the sediment, is in the range 0.014–0.057 mg l^{-1} for phosphorus, 0.12–0.48 mg l^{-1} for nitrogen and 0.025–0.106 mg l^{-1} for iron. In a number of cases, large sinks of nutrients and organic substances in the sediment have led to secondary pollution and the deterioration of water quality.

The role of sediments in the self-purification process of reservoirs is clearly demonstrated when considering heavy metals. The self-purification process is determined by the level of metal dispersion and the amount of silt present. The binding of metals with sediment tends to be rather stable and their reverse transformation into the water body is possible only under certain conditions. Such specific behaviour of metals in sediments determines the stability of metal concentrations in the reservoir water. However, as soon as water flow is regulated there is an accumulation of heavy metals in the sediments which then become a potential source of secondary pollution, and could influence water quality and the survival of aquatic biota. Continuous eutrophication, which changes the hydrochemical regime of a water body, may be the principal reason for the remobilisation of metals from the sediment.

17.3.5 Phytoplankton

In reservoirs where abiotic conditions are favourable, intensive algal growth occurs in the very first year of reservoir operation. The algae, in turn, become a significant source of organic and nutrient substances. Cyanophyceae (e.g. *Microcystis, Aphanizomenon, Anabaena*), at low or average biomass (up to 10–20 g m^{-3}), have a significant, positive role as photosynthesising organisms. However, in reservoirs these organisms attain such a high biomass (up to 330–500 g m^{-3}) that the products of their growth and decomposition become sources of water pollution. Where algal blooms occur and decaying Cyanophyceae accumulate, the biomass can reach 5–10 kg m^{-3}, the concentration of organic nitrogen increases by up to 30–150 times, ammonium nitrogen and inorganic phosphorus increase by up to 10–15 times and chemical oxygen demand (COD) increases tens of times. The concentration of phenols also increases as do the concentrations of cyanides and other toxic substances. During the breakdown of phytoplankton, significant amounts of certain metals enter the water and the total number of bacteria increases 25–100 times and the number of anaerobic bacteria increases by up to 400 times (Denisova, 1979).

In the Dnieper reservoirs, algal blooms are most intensive in the lake-like low flows of the Kremenchug and Kakhovka reservoirs (Primachenko, 1981; Sirenko *et al.*, 1989) where the biological consequences of pollution are well demonstrated. In reservoirs with greater flow rates, algal blooms do not reach such intensity. However, in the 'inter-cascade' position between the Dnieper and Dnieprodzerzhinsk reservoirs, intensive blooms develop due to the discharge of nutrients and organic substances and the discharge of algae from reservoirs upstream (Primachenko, 1981).

An increase in intensive blue-green algal blooms is favoured by a number of hydrological and hydrochemical factors, namely low velocities (< 0.2 m s^{-1}), high transparency of water (low content of suspended matter), and high concentrations of labile forms of dissolved organic substances (oxygen demand > 10 mg l^{-1} O$_2$). Other factors which favour blooms include decreased oxygen saturation and decreased redox potential, sedimentation of dissolved iron and silicon (which promotes the development of diatoms), and conditions encouraging the inflow of nutrients from the catchment area (e.g. wastewaters) and from bottom sediment.

17.3.6 Atmospheric precipitation

The average composition of inputs due to precipitation is 4.5 mg l^{-1} organic carbon, 0.4 mg l^{-1} organic nitrogen, 1.25 mg l^{-1} ammonia nitrogen, 0.6 mg l^{-1} nitrate nitrogen, and 0.106 mg l^{-1} phosphate phosphorus. Taking into account the available data, the amount of nutrients and organic substances entering the water of the Dnieper reservoirs and originating from atmospheric precipitation varies from 11,000 to 23,000 t for organic carbon, from 1,000 to 2,000 t for organic nitrogen, from 3,000 to 6,000 t for ammonium nitrogen, from 1,500 to 2,000 t for nitrate nitrogen and from 300 to 500 t for phosphorus. Considering the annual volume of water, these inputs will increase the concentration of these substances by 0.255–0.516 mg l^{-1} for organic carbon, by 0.022–0.046 mg l^{-1} for organic nitrogen, by 0.107–0.143 mg l^{-1} for ammonia nitrogen, by 0.034–0.069 mg l^{-1} for nitrate nitrogen, and by 0.006–0.012 mg l^{-1} for phosphate phosphorus.

17.4 Water quality assessment

Detailed information of the hydrochemical regime and water quality in the Dnieper and its reservoirs, based on systematic observations of the river before and after the creation of the reservoirs and subsequent discharge regulation, is given in numerous articles and monographs (Denisova and Maistrenko, 1962; Maistrenko, 1965; Almazov *et al.*, 1967; Peleshenko, 1975; Denisova, 1979; Nakhshina, 1983; Denisova *et al.*, 1987, 1989; Sirenko *et al.*, 1989). These publications describe the formation of the hydrochemical regime in different periods of the reservoirs' existence, including changes occurring over years and seasons and changes in water volume and depth. This information enabled the main trends in the formation of hydrochemical regimes of the reservoirs in the cascade to be defined. This, in turn, could be helpful in understanding changes in the hydrochemical regime of other large rivers when creating reservoir cascades.

17.4.1 Major ions and mineralisation of water
The creation of reservoirs leads to changes in water discharge of rivers in different seasons. In connection with this, there are changes in the dynamics of mineralisation and in the concentration of major ions. As a result of poorly-mineralised flood water accumulating in the reservoirs and its mixture with an inflow of more highly mineralised river water, there was a decrease in the annual amplitude of mineralisation and the concentrations of major ions, especially in reservoir areas close to dams. The greatest amplitude of mineralisation occurs in the Kiev reservoir.

In 1985–90, the mineralisation of water in the Dnieper reservoir cascade varied from 101 to 715 mg l^{-1} TDS (Table 17.5). An insignificant increase in mineralisation was found near discharges of industrial waste-waters. On entering a reservoir, tributaries with higher mineralisation have a slight impact in bays but barely influence the main reaches of the reservoirs because of the low volume of their discharge (0.04–0.8 km^3).

The dominating cation in the Dnieper water and its reservoirs is Ca^{2+} (20–75 mg l^{-1}), and the dominating anion is HCO$_3^-$ (42–255 mg l^{-1}). The concentrations of other ions is lower: Mg^{2+} 2–32 mg l^{-1}, Na$^+$+K$^+$ 1–59 mg l^{-1}, SO$_4^{2-}$ 7–94 mg l^{-1}, Cl$^-$ 1–60 mg l^{-1}, and CO$_3^{2-}$ 0–52 mg l^{-1}. The average ion discharge of the Lower Dnieper (near the town of Berislav) for 1956–68 was 11.56 × 10^6 t, of which 1.89 × 10^6 t were Ca^{2+}, 0.44 × 10^6 t were Mg^{2+}, 0.63 × 10^6 t were Na$^+$+K$^+$, 6.53 × 10^6 t were HCO$_3^-$+CO$_3^{2-}$, 1.34 × 10^6 t were SO$_4^{2-}$ and 0.73 × 10^6 t were Cl$^-$.

At present, the water in all the Dnieper reservoirs meets the maximum allowable concentration (MAC) requirements for ion composition, mineralisation and hardness (CMEA, 1982; Water Supply Sources, 1987; Sanitary Conditions, 1988) and may be used for different purposes.

17.4.2 Dissolved gases and pH
The concentration of dissolved oxygen in the water of the Dnieper reservoirs during the sampling period varied over a wide range from 2 to 23.5 mg l^{-1} (14–178 per cent saturation). In the area from the Kakhovka HPS dam to Kherson town it varied from 4.5 to 20.8 mg l^{-1}. In winter the upper reservoirs (Kiev and the upper part of Kanev) experience oxygen stress. This is due to the inflow of the rivers Pripyat and Desna, the waters of which have a very low oxygen content and a higher CO$_2$ content in the winter season. Wastewaters from Kiev also have a negative impact. The water at this time does not always meet the requirement of the MAC for water bodies used for water supply. This is true mainly of the Kiev reservoir where, during winter, anoxic conditions occur.

Table 17.5 Measured ranges of selected water quality variables in the Dnieper and its reservoirs, 1985–90

Variable	Kiev	Kanev	Kremenchug	Dnieprodzerzhinsk	Dnieper	Kakhovka	1 km upstream of Kherson town
pH	6.00–8.20	7.00–8.00	6.30–8.85	6.85–8.50	7.40–8.50	7.25–9.86	7.60–8.65
CO_2 (mg l^{-1})	0.8–23.8	1.76–17.6	0.20–20.2	0–10.2	0–39.6	0–26.4	1.8–6.60
O_2 (mg l^{-1})	2.0–16.2	4.0–18.4	3.47–23.5	6.4–15.8	5.0–22.7	3.6–19.6	5.51–20.8
TDS (mg l^{-1})	167–550	101–442	205–553	225–589	211–715	227–561	290–569
Total hardness (meq l^{-1})	1.94–5.48	4.26–4.80	2.04–5.80	2.29–5.51	2.51–5.57	2.5–5.51	3.02–4.46
NO_2^- (mg l^{-1} N)	0–0.106	0–0.131	0–0.195	0.006–0.199	0.002–0.197	0–0.478	0.03–0.146
NH_4^+ (mg l^{-1} N)	0–1.25	0–1.50	0.040–3.30	0.006–1.36	0.02–1.37	0.03–2.23	0.1–0.78
NO_3^- (mg l^{-1} N)	0–0.32	0.01–0.33	0.01–0.90	0–0.86	0.15–3.81	0.02–3.06	0.01–1.22
Dissolved P_{inorg} (mg l^{-1} P)	0.001–0.201	0–0.240	0.005–0.220	0.010–0.950	0.010–0.140	0–0.280	0–0.213
Total P (mg l^{-1} P)	0.018–0.631	0.04–1.11	0.008–1.97	0.010–2.60	0.068–2.70	0–0.642	0–0.280
Si (mg l^{-1})	0.10–4.5	0.9–4.7	0–14.8	0.80–6.7	1.0–14.20	0.60–8.4	1.0–10.2
Fe_{total} (mg l^{-1})	0–1.62	0–0.74	0–1.15	0–1.70	0–0.28	0–0.30	0–0.36
COD (mg l^{-1} O_2)	0.9–93.0	8.6–168.0	7.4–191.0	2.0–73.0	12.0–80.0	0.2–159.0	4.0–96.0
BOD_5 (mg l^{-1} O_2)	0–4.84	0.42–4.18	0.28–11.4	0.70–8.70	1.04–11.4	0.70–9.76	0.34–10.1
Petroleum products (mg l^{-1})	0–0.17	0–0.23	0–0.76	0–0.05	0.01–0.49	0–1.39	0.05–0.41
Phenols (mg l^{-1})	0.001–0.04	0–0.016	0–0.100	0–0.020	0.001–0.011	0–0.025	0–0.016
Detergents (mg l^{-1})	0–0.22	0.01–0.22	0–0.29	0–0.61	0–0.42	0–0.15	0–0.13
Cu (µg l^{-1})	0–27.0	0–36.0	0–29.0	1.0–31.0	0–10.5	0–64.1	0–25.0
Zn (µg l^{-1})	6.0–46.0	0–100.0	0–30.0	2.0–22.0	–	0–63.0	0–130.0
Cr^{6+} (µg l^{-1})	30.0–70.0	0–14.0	0–10.0	0–12.0	0.5–14.0	0–29.0	0.5–9.3
Colour (°)	18–110	18–80	12–80	14–88	14–67	12–75	7–51
TSS (mg l^{-1})	1.5–50.0	0.8–25.8	0.15–79.1	1.0–117.9	2.5–27.0	0–117.9	–

TDS Total dissolved solids   COD Chemical oxygen demand

TSS Total suspended solids   BOD Biochemical oxygen demand

In the low-flow Kremenchug and Kakhovka reservoirs, a large oxygen deficit occurs in summer in the bottom layers at the centre of the reservoir and near dams. Water at this time does not meet the MAC requirements for water bodies used for water supply. At high temperatures, and in calm weather, the mass development of phytoplankton subsequently leads to anoxic conditions.

An unfavourable gas regime occurs in the Dnieper reservoirs in low-water years in stagnant areas when there is a combination of long-lasting calm weather, high water temperature and the mass growth and accumulation of Cyanophyceae.

In 1985–90, the concentration of CO_2 in the water was 0–39.6 mg l^{-1}, depending on the season and on the biochemical and biological processes occurring in the water and sediment. The CO_2 concentration also changed over 24-hour and seasonal cycles. Maximum concentrations were recorded during ice cover periods.

Water pH varies over a wide range according to changes in CO_2 (Table 17.5). Seasonal changes in pH are caused by changes in the carbonate balance. Minimum values occur in winter and maximum values in summer. Photosynthetic processes, under which dissolved CO_2 in water is assimilated, lead to the decrease and sometimes to the complete disappearance of CO_2 in surface layers. The pH then rises to 8.8–9.9 and CO_3^{2-} appears in the water, at concentrations varying from 0 to 50 mg l^{-1}.

17.4.3 Nutrients

The concentration of nutrients in the Kiev reservoir, the most upstream reservoir in the cascade, depends mainly on the concentrations of nutrients in tributaries feeding the reservoir. In reservoirs downstream, nutrient concentrations depend on their inflow from reservoirs upstream and the transformation processes within each reservoir. The maximum concentration of most nutrients occurs in winter especially before floods when the mineralisation products of organic substances in the water and sediment accumulate.

Inorganic compounds of nitrogen (NH_4^+, NO_2^-, NO_3^-) enter the reservoir mainly with surface runoff. They are easily assimilated by the phytoplankton and bacteria populations and are changed into protein nitrogen. During the vegetation growth period, as a result of the intensive development of phytoplankton, all forms of nitrogen are usually present in small quantities. In autumn, due to the processes of mineralisation of organic substances and the weakening of biological processes, their concentration increases again. By the end of the winter, nitrogen concentrations reach maximum values.

The dominant form of nitrogen is ammonium. Its concentration in 1985–90 (Table 17.5) in the Kiev, Kanev, Dnieprodzerzhinsk and Dnieper reservoirs exceeded the MAC by 2–3 times (CMEA, 1982). In the Kremenchug and Kakhovka reservoirs, the concentrations of NH_4-N exceeded the MAC for water bodies used for water supply by 2–3 times and for fisheries by 5–6 times. The concentration of nitrate nitrogen in 1985–90 in the water of the Dnieper reservoirs did not exceed the MAC. There was a decrease in nitrate nitrogen concentrations from 1986 to 1990. The worst conditions were recorded for nitrite concentrations, which increased from the upper Kiev to the lower Kakhovka reservoir by five times. Values for nitrite exceeded the MAC for reservoirs used for drinking water by 5 to 25 times (Sanitary Conditions, 1988) and occasionally by up to 243 times (CMEA, 1982). The trends in the annual average concentrations of all forms of nitrogen are presented in Figures 17.2A,B,C.

Depending on the season and the reservoir, the concentration of dissolved inorganic phosphorus for the period of study ranged from 0.001 up to 0.95 mg l^{-1} . The concentration of total phosphorus ranged between 0.63 to 2.7 mg l^{-1}. In the Dnieprodzerzhinsk, Dnieper and Kakhovka reservoirs, the concentration of inorganic phosphorus exceeded the MAC by 1.5–5 times. The concentration of total phosphorus in all the reservoirs exceeded the MAC values. The trends in the average annual concentrations of total and inorganic phosphorus are presented in Figures 17.2D,E,F.

The creation of a cascade of reservoirs on the Dnieper which promoted, due to internal processes, sedimentation of iron and its accumulation in the sediment has led to the decrease in iron content in water for all the Dnieper reservoirs. The greatest concentrations of iron are in the waters of the Kiev reservoir, while further down the cascade they decrease as a result of sedimentation processes (Figures 17.2D,E,F).

In 1985–90, concentrations of silica in the three upper reservoirs were lower than the MAC values. However, in the other reservoirs, in some periods, the silica concentration exceeded permissible levels.

The discharge of nutrients by the Dnieper river on average for 1956–69 was 22,300 t for ammonia nitrogen, 600 t for nitrite nitrogen, 13,900 t for nitrate nitrogen, 4,100 t for mineral phosphorus (dissolved + suspended), 5,500 t for iron and 149,000 t for silica.

17.4.4 Organic matter
The concentration of organic matter as defined by the COD in the Dnieper water and its reservoirs was 0.2–191.0 mg l^{-1} O$_2$ (Table 17.5, Figure 17.3). The greatest concentration was in the low-flow Kremenchug and Kakhovka reservoirs. At all monitored times, the concentration of organic matter in the Dnieper reservoirs exceeded the MAC by 2–5 times. The concentration of labile organic substances defined by the biochemical oxygen demand (BOD$_5$) did not meet the MAC criteria and exceeded the standards by 2–5 times (Table 17.5, Figure 17.3).

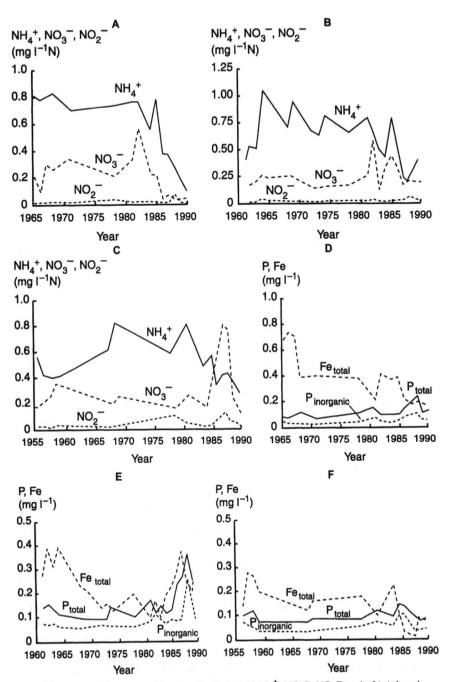

Figure 17.2 Average annual concentrations of NH_4^+, NO_2^-, NO_3^- and of total and inorganic phosphorus and total iron in: **A.** and **D.** The Kiev reservoir, from its creation in 1965 up to 1990; **B.** and **E.** The Kremenchug reservoir, from its creation in 1960 up to 1990; **C.** and **F.** The Kakhovka reservoir, from its creation in 1955 up to 1990

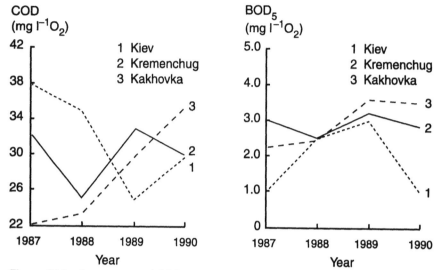

Figure 17.3 Average annual COD and BOD5 values in the Kiev, Kremenchug and Kakhovka reservoirs, 1986–90

17.4.5 Organic pollutants

The concentration of petroleum products in the Dnieper reservoirs varied from 0 to 1.39 mg l^{-1}, increasing from the upper Kiev to the lower Kakhovka reservoir (Table 17.5, Figure 17.4). Maximum allowable concentrations for drinking water supply were not exceeded in the Kiev, Kanev and Dnieprodzerzhinsk reservoirs, but MAC values were exceeded slightly in reservoirs used for fisheries. In the other reservoirs, at times, the concentration of petroleum products exceeded the standards. Especially high concentrations were characteristic of the Kakhovka reservoir.

The main source of phenols to the Dnieper reservoirs are industrial wastewaters and the products of the decomposition of phytoplankton, macrophytes and terrestrial vegetation. The concentration of phenols in 1985–90 exceeded the MAC by tens of times (CMEA, 1982). In the Kiev, Kremenchug and Kakhovka reservoirs, the concentration of phenols exceeded the MAC. In the Kiev and Kanev reservoirs, phenol concentrations had decreased by 1990 whereas in the other reservoirs they had increased (Figure 17.4).

In 1986–90, the concentration of detergents in all the reservoirs of the Dnieper cascade, and in the River Dnieper itself upstream of Kherson town, was lower than the MAC, although values were exceeded slightly in the Dnieprodzerzhinsk reservoir. In the Kiev reservoir, there was a tendency for a decrease in detergent concentrations in the years up to 1990, whereas in the other reservoirs they had increased (Figure 17.5).

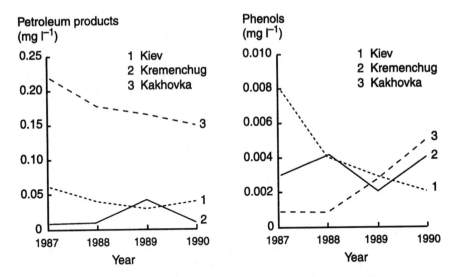

Figure 17.4 Average annual concentrations of petroleum products and phenols in the Kiev, Kremenchug and Kakhovka reservoirs, 1986–90

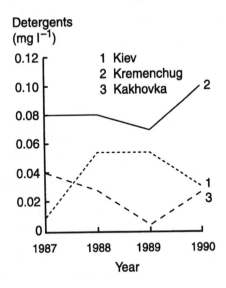

Figure 17.5 Average annual concentrations of detergents in the Kiev, Kremenchug and Kakhovka reservoirs, 1986–90

Table 17.6   Concentrations of dissolved forms of heavy metals in the Dnieper reservoirs, 1985–90

Reservoir	Fe[1](μg l^{-1}) Range	Average	Mn[1](μg l^{-1}) Range	Average	Zn (μg l^{-1}) Range	Average
Kiev	25.5–96.0	42.3	1.0–9.0	3.5	6.0–34.0	15.4
Kanev	27.0–83.0	61.0	3.0–36.0	16.0	13.0–33.0	20.0
Kremenchug	27.0–85.0	52.0	3.0–30.0	13.0	14.0–42.0	30.0
Dnieprodzerzhinsk	24.0–90.0	51.0	3.0–30.0	13.0	13.0–37.0	28.0
Dnieper	27.0–73.0	56.0	6.0–34.0	20.0	17.0–36.0	28.0
Kakhovka	22.0–42.0	37.0	5.0–23.0	11.0	24.0–44.0	36.0

[1]   Excluding concentrations of Fe and Mn in winter

17.4.6 Heavy metals

Fluctuations and the average concentration of dissolved forms of heavy metals in the Dnieper reservoirs are presented in Table 17.6 (for problems associated with determining trace metal concentrations in water, see Chapter 8). Analysis of the data showed that, for reservoirs of the Dnieper cascade, elements may be ranked in order of decreasing concentrations as follows: Fe > Zn > Mn > Cu > Pb > Ni > Co > Cd.

In order to characterise metal speciation, the ratio of the concentrations of metals migrating in readily-dissolved form to their total concentration in water (expressed in per cent), was considered. Table 17.7 gives averaged data for the migration of dissolved and suspended forms of Fe, Mn, Cu and Zn. Whether in dissolved or suspended form, these heavy metals can be divided into three groups. The first group includes Fe and Mn, i.e. elements with changing valency. The proportion of dissolved forms of Fe in winter under long-lasting ice cover and low Eh values reaches 50 per cent of the total content. For Mn, this value increases up to 93 per cent. During the rest of the year, these elements are found mainly in suspension. The proportion of dissolved forms of Fe and Mn to the total concentration at this time does not exceed 18 per cent and 11 per cent respectively. The second group of metals includes Zn, Cu, Pb, Ni, Co. The proportion of dissolved forms of these metals is 46.0 per cent, 53.1 per cent, 59.7 per cent, 60.9 per cent and 53.0 per cent respectively of their total concentration. For this group, the amount of suspended matter transported by river flow is of great importance with regard to the change in speciation correlation. The third group includes Cd. This element is different from Fe, Mn, Zn, Cu, Pb, Ni, because it has an obvious dissolved migration form. The average concentration of dissolved Cd in the reservoirs sampled was 0.3 μg l^{-1}. The proportion of the dissolved form of Cd (in relation to the total) reaches 88 per cent.

Table 17.6 Continued

Reservoir	Ni (µg l⁻¹)		Pb (µg l⁻¹)		Cu (µg l⁻¹)	
	Range	Average	Range	Average	Range	Average
Kiev	1.0–16.0	10.0	2.0–9.0	6.0	2.0–11.0	6.0
Kanev	6.0–19.0	11.0	2.0–19.0	8.0	3.0–10.0	7.0
Kremenchug	7.0–16.0	11.0	3.0–17.0	11.0	5.0–9.0	7.0
Dnieprodzerzhinsk	6.0–14.0	11.0	3.0–16.0	10.0	4.0–10.0	8.0
Dnieper	8.0–14.0	11.0	7.0–13.0	10.0	3.0–8.0	7.0
Kakhovka	10.0–14.0	12.0	3.0–9.0	7.0	8.0–10.0	9.0

Table 17.7 Concentrations of dissolved and suspended forms of heavy metals in the Dnieper reservoirs in summer

Reservoir	Fe				Mn			
	Diss. (µg l⁻¹)	Susp. (µg l⁻¹)	Total (µg l⁻¹)	Diss.[1] (%)	Diss. (µg l⁻¹)	Susp. (µg l⁻¹)	Total (µg l⁻¹)	Diss.[1] (%)
Kanev	37.4	197.0	234.4	16.0	8.1	43.0	51.1	15.9
Kremenchug	33.6	128.8	162.4	20.7	3.7	37.6	41.3	9.0
Dnieprodzerzhinsk	31.0	143.5	174.5	17.8	3.2	24.0	27.2	11.8
Kakhovka	25.7	45.0	70.7	36.4	3.3	27.0	30.3	10.9

Reservoir	Zn				Cu			
	Diss. (µg l⁻¹)	Susp. (µg l⁻¹)	Total (µg l⁻¹)	Diss.[1] (%)	Diss. (µg l⁻¹)	Susp. (µg l⁻¹)	Total (µg l⁻¹)	Diss.[1] (%)
Kanev	37.5	18.3	55.8	67.2	15.4	4.3	19.7	78.2
Kremenchug	28.6	18.7	47.3	60.5	17.7	4.6	22.3	79.4
Dnieprodzerzhinsk	32.2	8.1	40.3	79.9	14.7	4.2	18.9	77.8
Kakhovka	24.3	3.9	28.2	86.2	13.5	2.2	15.7	86.0

Diss. Dissolved form
Susp. Suspended form

[1] Proportion of the dissolved form, expressed as a percentage of the total

In the regulated reservoirs of the Dnieper, there are a number of factors influencing the specific behaviour of heavy metals, and the change in their speciation. The most important of these are a decrease in flow velocity and rate of water exchange with the intensification of sedimentation processes; the decrease in the specific surface runoff; the intensification of intra-reservoir processes; the increase in biological productivity; the change in gas regime; and the shift in pH values towards more alkaline (especially in the vegetation growth period).

Because of their shallow waters (Table 17.2) the Dnieper reservoirs are characterised not only by complete vertical mixing, but also by periodic resuspension of sediment in shallow places. When the wind influence is strong, this leads to an increase in the share of suspended forms of heavy metals in the water.

17.4.7 Colour and transparency

The colour index of water in all the reservoirs of the Dnieper cascade in 1985–90 did not meet standards. The greatest colour index was registered in the Kiev reservoir, where there is a significant influence from humus substances discharged by the Pripyat. The content of humic substances reduces as it moves down the cascade, thereby leading to a decrease in the colour index.

Suspended matter in water, its composition, concentration and dynamics is an important factor in the formation of natural water quality. It significantly affects many processes which take place in the reservoirs including the activity of the aquatic biota. Total suspended solids concentrations in the Dnieper reservoirs vary over a wide range (Table 17.5).

The lowest water transparency in the Kiev reservoir, depending on the colour index and the quantity of suspended substances, is in the upper part of the reservoir in spring. This is when the reservoir receives a large amount of suspended substances and where its colour index is at its highest. In other reservoirs, the lowest transparency is in winter and during the mass accumulation of Cyanophyceae and the formation of algal blooms.

It should be noted in conclusion, that the water of the Dnieper reservoirs, in terms of mineralisation, chloride concentrations, total iron, and pH value (with some exceptions) meets MACs and may be used for different economic needs, including irrigation.

17.5   Conclusion

Using the available information on the present regime and the water quality in the reservoirs, together with the results of long-term observations (including the study of both the river and reservoirs at different stages in their development), water protection measures to preserve and improve water quality can be formulated. These measures include:

- Creation of littoral water protection zones which would decrease the discharge of nutrients from the water catchment area;
- Reduction of the negative influence of bottom sediment on water quality by aeration of bottom water layers and the periodic removal of silty sediment from the reservoirs;

- Complete treatment of industrial, municipal and domestic discharges to remove heavy metals, nutrients and organic substances;
- Technological changes in applying mineral fertilisers on fields to reduce their inflow into the water bodies;
- Consolidation of shore zones to curb erosion processes; and
- Appropriate biological and physical management of shallow waters.

Specific demands should be made concerning the composition of water when used as a source of municipal and drinking water supply. These demands should be strict and should comply with certain conditions related to the technical possibilities of eliminating pollution in conjunction with effective methods of processing, purifying and disinfecting water.

17.6   References

(All references are in Russian unless otherwise stated).

Alekin, O.A. 1970 *Fundamentals of Hydrochemistry*. Gidrometeoizdat, Leningrad, 444 pp.

Almazov, A.M., Denisova, A.I., Maistrenko, Y.G. and Hakhshina, E.P. 1967 *Hydrochemistry of the Dnieper River, Its Reservoirs and Tributaries*. Naukova Dumka, Kiev, 316 pp.

Burov, V.S. 1971 Mineral fertiliser discharge from rural lands by surface runoff. *Transactions of the State Hydrological Institute*, **198**, 176–196.

CMEA 1982 *Unified Water Quality Criteria*. Conference of the Council for Mutual Economic Assistance countries, Moscow, 69 pp.

Demchenko, A.S., Brazhnikova, L.V., Tarasov, M.N. and Demidov, A.D. 1969 Studies of nitrogen fertiliser transformations in irrigated soil areas. *Hydrochemical Materials*, **52**, 49–54.

Denisova, A.I. 1979 *The Hydrochemistry of the Dnieper River Reservoir Regime and Methods for its Forecast*. Naukova Dumka, Kiev, 290 pp.

Denisova, A.I. and Maistrenko, Y.G. 1962 *Kakhovka Reservoir Hydrochemistry*. Academy of Science, Ukraine, Kiev, 200 pp.

Denisova, A.I., Nakhshina, E.P., Novikov, B.I. and Ryabov, A.K. 1987 *Bottom Sediments of Reservoirs and Their Impact on Water Quality*. Naukova Dumka, Kiev, 164 pp.

Denisova, A.I., Timchenko, V.M., Nakhshina, E.P., Novikov, B.I., Ryabov, A.K. and Bass, Y.I. 1989 *Hydrology and Hydrochemistry of the Dnieper River and Its Reservoirs*. Naukova Dumka, Kiev, 216 pp.

Drachev, S.M. 1964 *Prevention of River, Lake and Reservoir Pollution by Industrial and Municipal Waste Water*. Nauka, Moscow, 272 pp.

Laponogov, A.N. 1978 *The Dnieper River Yesterday, Today and Tomorrow*. Gidrometeoizdat, Leningrad, 72 pp.

Levkovsky, S.S. 1979 *Ukraine Water Resources*. Vyssh Shkola, Kiev, 200 pp.

Maistrenko, Y.G. 1965 *Organic Matter in Water and Bottom Sediments of Ukrainian Rivers and Reservoirs (The Dnieper and Danube Basins)*. Naukova Dumka, Kiev, 240 pp.

Nakhshina, E.P. 1983 *Micro-elements in the Dnieper Reservoirs*. Naukova Dumka, Kiev, 158 pp.

Peleshenko, V.I. 1975 *Assessment of the Inter-relationships of Chemical Composition of Different Natural Water Types*. Visha Skola, Kiev, 292 pp.

Primachenko, A.D. 1981 *Phytoplankton and Primary Production of the Dnieper and Its Reservoirs*. Naukova Dumka, Kiev, 278 pp.

Sanitary Conditions and Norms of Surface Waters Protection from Pollution 1988 San Pin No. 4830-88. USSR Ministry of Health Protection, Moscow, 70 pp.

Sirenko, L.A., Korelyakova, I.L. and Mikhailenko, L.E. 1989 *Vegetation and Bacterial Populations of the Dnieper River and its Reservoirs*. Naukova Dumka, Kiev, 231 pp.

USSR Surface Water Resources 1971 Ukraine and Moldova, Middle and Lower Dnieper. Issue 2, Volume 6. Gidrometeoizdat, Leningrad, 654 pp.

Voskresensky, K.P. 1962 *Average Annual Runoff and Its Fluctuations in USSR Rivers*. Gidrometeoizdat, Leningrad, 546 pp.

Water Supply Sources 1987 Hygienic and Technical Requirements and Selection Guidelines. GOST 2761-84. State Standard Committee, Moscow, 13 pp.

Chapter 18[*]

LAKE BAIKAL

The first studies of the Lake Baikal ecosystem date back to the late 19th century. Since the 1920–30s studies of the limnological features became more regular and extensive, and since the 1960s the studies have acquired a new approach, i.e. an integrated assessment of anthropogenic impacts. Kozhov (1947, 1962a,b, 1970, 1972) was the first to characterise the general scheme of the Lake Baikal ecosystem structure and function, its description and evolution. He also formulated the requirements to protect the unique Lake Baikal ecosystem (Kozhov, 1966, 1971). The Limnological Institute of the Siberian Branch of the Russian Academy of Sciences, the Research Institute of Biology of the Irkutsk State University and the Russian Federal Hydrometeorological and Environmental Service (ROSGYDROMET) play a key role in the implementation of integrated monitoring of Lake Baikal.

18.1 Lake Baikal and its region

Located in the central part of Asia, the Lake Baikal basin area is 557,000 km^2. The lake basin lies within the borders of two countries, Russia and Mongolia at lat 46°20'N and lat 56°40'N, long 96°50'E and long 114°10'E (Figure 18.1 and General Appendix IV). Lake Baikal is situated in a moderately high mountainous region with most of the basin area over 1,000 m above sea level. A considerable part of the water catchment area, 448,000 km^2 is occupied by the basin of the lake's major tributary, the Selenga river. The lower section of the Selenga water catchment area (148,000 km^2 and 409 km of the river length out of the total length of 1,024 km) is situated within the Russian border (Figure 18.1).

The Lake Baikal region is characterised by high seismic activity with up to 2,000 earthquakes a year. Earthquakes up to 8 points on the Richter scale are possible within a 100 km zone east of Lake Baikal, and up to 6–7 points beyond this zone. The lake is situated in the Baikal rift system between the Siberian platform and the Baikal fold region. A feature of the tectonic origin of the lake (rift valley) is the presence of three steep depressions similar in size; the maximum depth of the

---

*   This chapter was prepared by O.M. Kozhova and A.A. Matveev

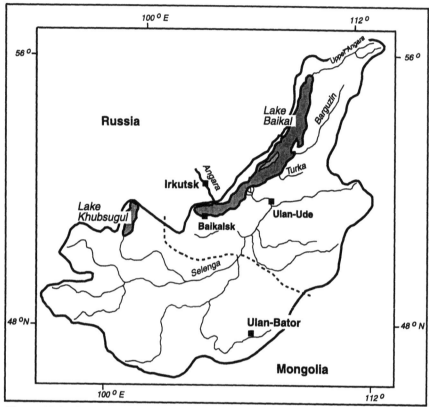

Figure 18.1   The Lake Baikal basin

southern depression is 1,424 m, the middle depression is 1,620 m and the northern depression is 890 m.

The age of Lake Baikal, within its present boundaries, is approximately 1 million years old; the rift opening began about 30 million years ago. The Lake Baikal rift is still being formed, manifested by the recent formation of the Proval bay (in the Selenga river delta) as a result of an earthquake in 1861 and a 20 m submergence of the bottom in the Lake Baikal middle region in 1959.

The main morphometric features of Lake Baikal include the elongated shape in the SW–NE direction (Figure 18.2). The length of the lake is 636 km; its average width is about 50 km and the maximum width is 79 km. The average depth is 707 m and maximum depth is 1,620 m. The lake surface area is 31,500 km^2 (including bays and shore areas comprising approximately 2,000 km^2) and the lake volume is 23,000 km^3. The lake surface is 455 m above sea level whereas the lake bottom and the bed rock underlying a few kilometres of lacustrine sediments are well below sea level. Baikal has the largest volume of all

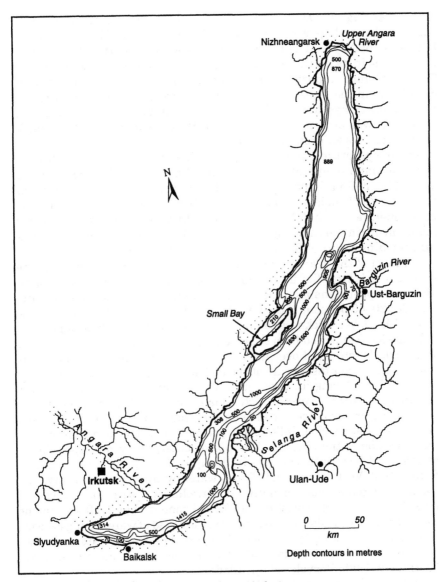

Figure 18.2 Lake Baikal and some morphometric features

freshwater lakes in the world (only the saline Caspian Sea has a greater volume) and it is also the deepest and has the greatest biodiversity. However, it is only the eighth largest of all lakes (Caspian Sea included) in terms of area.

Lake Baikal has ice cover for 4–6 months of the year. Stable snow cover forms on the surrounding banks at the end of October and the lake starts freezing in October with ice build-up continuing until mid-January. Ice and snow cover thaw begins at the end of March and the lake is completely ice-free in mid-May in its southern part and in the first half of June in its northern part. Average ice thickness is between 30 and 70 cm, reaching a maximum of 90–130 cm in the north of the lake and 60–100 cm in the southern part. However, ice cover is not extensive at river mouths, bays, areas of thermal groundwater outlets and in the area of treated wastewater discharge of the Baikal paper mill at Baikalsk (where open water areas can occur).

The temperature regime is most variable in the lake's upper layers. Seasonal water temperature changes are almost not noticeable at depths below 200–300 m. Maximum water temperature fluctuations occur in the upper 25–30 m layer. Here the water temperature often exceeds 6 °C in summer, reaching 15 °C in some places. In winter, the under-ice water temperature in the same layer is 0.3–2 °C. Below 300 m, water temperature remains stable at 3.4–3.6 °C over the entire year.

For most of the year, approximately 10 months, the water mass of the pelagic zone is stratified into an epilimnion, a metalimnion and a hypolimnion. An upper dynamic zone, where increased water exchange occurs, includes 30–40 per cent of the lake depth. There is also a bottom dynamic zone with a layer thickness comprising 10 per cent of the lake depth.

In the ice-free period from the end of June, thermal stratification occur which can be observed over the entire lake in the middle of July. This temperature layer disappears at the end of November, and the beginning of December. During ice cover, an inverse temperature stratification is observed in the water mass which remains until the beginning of June.

Lake water exchange depends on the system of horizontal currents, vertical exchange, river inflows and precipitation. The mean theoretical water residence time in Baikal is approximately 300 years. The scheme of dominant horizontal currents in the upper 50 m layer is presented in Figure 18.3. Vertical renewal of deep water by surface water is diverse. Vertical circulations, usually downwellings, with a speed of 0.01–0.1 cm s^{-1} are typical of the littoral zone within 2–10 km. In the central part of the lake, on the contrary, the speed of water rising exceeds that of water sinking by 10 times. The period of deeper water substitution by surface waters is evaluated at not less than 10–20 years (Votintsev, 1985; Verbolov et al., 1986).

Data on the distribution of temperature, chemical and suspended substances and plankton indicate their considerable spatial and temporal heterogeneity. Large homogeneous hydrophysical, hydrochemical and hydrobiological fields are noted in the upper dynamic layer of the lake.

The Lake Baikal basin is mostly characterised by a continental climate, with a cold winter with little snow cover and a hot dry summer. However, high mountains with ridges up to 2,500 m encircle the lake and

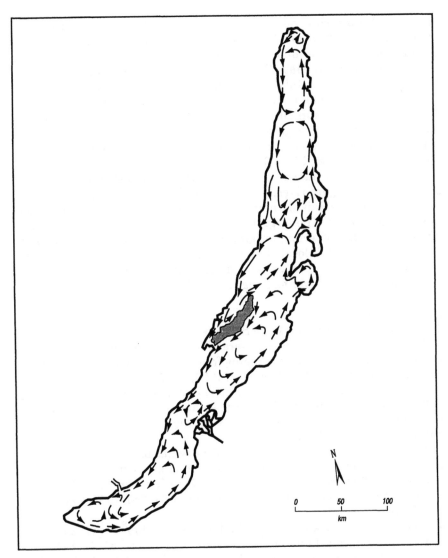

Figure 18.3   Dominant water currents in Lake Baikal

influence the formation of a specific climate pattern near the lake. The annual amplitudes of air temperature at islands and along the shoreline do not exceed 30–35 °C. Mean monthly air temperatures are –31 to –16°C in January and 13 to 20 °C in June. Mean long-term annual precipitation in different parts of the lake range from 170 to 1,400 mm, and are usually below 300 mm in the Trans-Baikal region. The annual precipitation at the wind-swept slopes of the surrounding Khamar-

Daban and Barguzin ridges and in the tributary valleys reaches 1,000 mm, in the inter-montane valleys 200–300 mm, and in the lake's central part about 200 mm. Approximately 86 per cent of the annual precipitation falls in May–November, the maximum precipitation of 17–29 per cent falls in July and the minimum of 0.8–3 per cent falls in February. Approximately 200–250 mm of precipitation falls in the Selenga river valley annually, and 200–400 mm in the basin of its tributaries. The annual precipitation in the basin of other large Baikal tributaries is 300–350 mm in the Upper Angara basin, 200–400 mm in the Barguzin basin and 500–600 mm in the Turka basin.

A total of 336 permanent tributaries flow into Baikal which has only one outlet, the Angara river, a tributary of the Yenisey (Figure 18.2).

Apart from Lake Baikal, there are numerous smaller freshwater lakes in the Baikal basin. The largest amongst these freshwater lakes is the Khubsugul, a lake of tectonic origin, with a surface area of 2,760 km^2 and a volume of 381 km^3, located at 1,650 m above the sea level in northwestern Mongolia. It is connected with the Selenga river via the outflowing Egin-Gol river forming the Khubsugul-Baikal system.

18.2   Water balance

The first water balance for Lake Baikal was compiled by Vereschagin in 1936. In the following years several authors made numerous water balance estimates for periods ranging from 3–4 years to 70 years starting from 1889, 1901 and 1904 (Afanasev, 1960, 1967). The latest calculation was performed by the State Hydrological Institute for the period 1962 to 1988 inclusive. The average long-term balance of chemicals presented in this chapter was calculated on the basis of the average, long-term water balance in the period 1980–88.

In 1980–88 the basic components of the water balance varied within the following ranges (years with extreme values are shown in brackets):

- Surface inflow from 44.16 km^3 a^{-1} (1987) to 79.55 km^3 a^{-1} (1988);
- Atmospheric precipitation from 10.45 km^3 a^{-1} (1987) to 15.49 km^3 a^{-1} (1988);
- Lake outflow from 40.98 km^3 a^{-1} (1982) to 77.89 km^3 a^{-1} (1985);
- Evaporation from 11.04 km^3 a^{-1} (1988) to 16.37 km^3 a^{-1} (1984).

The groundwater discharge to the lake is estimated at 1.64 km^3 a^{-1}.

The 1980–88 annual data correlate with the range of average long-term components of the water balance throughout the observation period of 1901–88. The average long-term water balance of Lake Baikal in 1980–88 is summarised in Table 18.1.

The largest tributaries, in terms of water inflow, are the Selenga river with an average long-term water flow (1980–88) of 28.9 km^3 a^{-1} of which one third is formed in Mongolia; the Upper Angara river with a water flow of 8.63 km^3 a^{-1}; the Barguzin river with a water flow of

Table 18.1   Average long-term water balance of Lake Baikal, 1980–88

Input	(km³ a⁻¹)	Output	(km³ a⁻¹)
Direct runoff and river inputs	61.79	Outflow by the Angara river	57.45
Atmospheric precipitation	12.55	Evaporation	13.45
Groundwater discharge	1.64	Accumulation in lake and errors in balance estimations	5.08
Total	**75.98**	**Total**	**75.98**

$4.31 \text{ km}^3 \text{ a}^{-1}$; and the Turka river with a water flow of $1.82 \text{ km}^3 \text{ a}^{-1}$. The water flow of the other Lake Baikal tributaries which have been studied varies from 0.02 to $1.44 \text{ km}^3 \text{ a}^{-1}$.

According to the distribution of the annual water flow, the rivers of the Lake Baikal basin are classified as the East-Siberian type. Approximately 80 per cent of the total water flow is formed in April–September and flow is minimal from January till March (only 5 per cent of the annual flow). River floods usually take place in April–May during the snow melt period and occasionally in July–September.

18.3   Hydrochemistry

18.3.1   Baikal tributaries
The chemical composition of water in most Lake Baikal tributaries is dominated by bicarbonate (up to 75–90 per cent eq) and calcium (60–75 per cent eq), while total dissolved solids (TDS) usually vary from 50–150 mg l⁻¹. River water contains mostly 8–13 mg l⁻¹ of dissolved oxygen and this concentration may decrease to 6–7 mg l⁻¹ during floods. The pH of the tributaries usually varies within 6.8–7.8. Characteristics of the chemical composition of the principal Baikal tributaries are presented in Table 18.2.

Anthropogenic changes in the chemical composition of river water have been recorded over the last 30 years. An increase in SO_4^{2-}, Cl⁻ and organic compounds was recorded in some rivers, including the major tributaries. The tributaries are constantly contaminated by petroleum hydrocarbons and phenols, and detergents are also often detected. Average long-term concentrations of these compounds from 1979 to 1988 are presented in Table 18.3. DDT and its metabolites, α- and γ-HCH were detected in 4–32 per cent of samples collected in different years. Maximum concentrations did not exceed 4–6 ng l⁻¹ and were recorded mostly in the rivers of southern Lake Baikal and the Selenga river. The main source of lake contamination is chemical runoff whereas the Selenga river is the principal pollution source amongst the tributaries.

Table 18.2 Average chemical composition of the major tributaries of Lake Baikal

River	Period of observation[1]	Concentration (mg l^{-1})							TDS	Si	Fe$_{total}$	Organic matter
		HCO$_3^-$	SO$_4^{2-}$	Cl$^-$	NO$_3^-$	Ca^{2+}	Mg^{2+}	Na$^+$ + K$^+$				
Selenga	1950–62	88.67	6.50	1.23	0.43	20.69	4.55	5.17	127	3.5	0.89	10.62
	1979–88	79.28	13.63	2.28	0.25	18.96	4.46	8.20	127	5.1	0.66	13.0
Upper Angara	1950–62	57.34	3.70	0.27	0.20	14.90	2.06	2.65	81	1.95	0.15	5.04
	1979–88	42.00	8.10	1.16	0.25	11.94	2.30	2.60	68	2.4	0.46	9.3
Barguzin	1950–62	89.22	7.90	1.06	0.39	24.05	2.74	5.52	131	2.8	0.30	8.15
	1979–88	68.93	9.50	1.63	0.30	18.12	3.60	4.30	106	2.7	0.48	10.2
Turka	1950–62	36.06	3.14	0.62	0.54	8.55	1.40	3.50	54	3.6	0.23	6.48
	1979–88	27.15	5.83	1.85	0.05	12.40	3.33	4.22	55	6.5	0.47	10.7

TDS   Total dissolved solids

[1]   1950–62: Votintsev et al., 1965; 1979–88: Data of the USSR Hydrometeorological Committee

Table 18.3   Average concentrations of polluting substances in Lake Baikal's major tributaries

	BOD$_5$ (mg l^{-1})	Petroleum products (mg l^{-1})	Resins and asphaltenes (mg l^{-1})	Phenols (µg l^{-1})	Detergents (µg l^{-1})
Upper Angara	1.85	0.14	0.04	5	17
Barguzin	2.31	0.12	0.03	4	20
Turka	1.74	0.14	0.02	4	18
Selenga	1.44	0.13	0.03	3	24
Other small tributaries	1.05–3.00	0.03–0.16	0.02–0.04	0.31–6	8–30

BOD   Biochemical oxygen demand

Table 18.4   Average long-term concentrations and fluxes for selected variables in Angara River water, 1980–88

Substance	Average concentration (mg l^{-1})	Average annual outflow (10^3 t a^{-1})
HCO$_3^-$	62.2	3,573
SO$_4^{2-}$	5.7	327
Cl$^-$	0.8	46
Ca^{2+}	15.4	885
Mg^{2+}	3.4	195
Na$^+$+ K$^+$	2.7	155
TDS	90.2	5,181
Organic compounds	5.1	292
N$_{min}$	0.024	2.9
P$_{min}$	0.009	0.52
Suspended matter	0.6	34.5

TDS   Total dissolved solids   P$_{min}$ Mineral phosphorus
N$_{min}$ Mineral nitrogen

18.3.2 The outflowing Angara river

The mean long-term concentrations and outputs of chemicals from the lake by the Angara river are presented in Table 18.4 for monitoring data for the period 1980–88.

18.3.3 Groundwater

The hydrochemistry of the groundwater inflow to the lake remains to be studied. The mineralisation of groundwater (20–60 mg l^{-1} total dissolved solids (TDS)) may be lower than that of the rivers and the lake water. The maximum salt concentration in groundwater, up to 400 mg l^{-1} has

Table 18.5   Average annual concentrations of components of precipitation, 1980–88

Component	Range (mg l^{-1})	Component	Range (mg l^{-1})
pH	5.6–7.4[1]	Cl$^-$	0.5–10.6
Dust	3.5–1,030	Ca^{2+}	0.4–9.3
Organic matter	3.7–22	Mg^{2+}	0.2–9.2
TDS	6.4–145	Na$^+$	0.2–35
HCO$_3^-$	0.7–47	K$^+$	0.1–5.4
SO$_4^{2-}$	0.8–67	N$_{min}$	0.2–8.2

1   pH scale
TDS   Total dissolved solids

N$_{min}$ Mineral nitrogen

been recorded at a few local sites due to carbonated sediments (Khaustov *et al.*, 1990). The present groundwater input of TDS is estimated at approximately 100,000 t a^{-1}.

18.3.4 Atmospheric precipitation

Regular monitoring of the input of atmospheric deposition in the Lake Baikal region has been performed since 1971 through monthly sampling of wet and dry atmospheric deposition at eight sites located around the lake. Snow samples are collected in winter at different sites along the shoreline and the lake for chemical analyses. The 1980–88 mean annual precipitation components are shown in Table 18.5. Concentrations in random and monthly collected samples may differ occasionally from the values in Table 18.5 by 10-fold (in either direction). Atmospheric dry deposition varies from 10 per cent to 76 per cent of the total fallout, and the Lake Baikal average is close to 30 per cent.

18.3.5 Lake Baikal water

The chemical composition of the basic water mass has not changed significantly over the last 30–40 years. This is not surprising considering the long water residence time. Chemical composition of the lake water, as in the tributaries, is dominated by HCO$_3^-$ and Ca^{2+} ions.

A high concentration of dissolved oxygen is characteristic of the open lake, with 11.7–11.9 mg l^{-1} in the upper 50 m layer and 9.5–10 mg l^{-1}, on average, in deeper layers (over 1,000 m). Free CO$_2$ concentration changes from 1.5–1.7 mg l^{-1} in the surface waters and up to 4.3–4.6 mg l^{-1} at the bottom. Values for pH vary from 7.4 to 8.5, being around 7.6 on average. Silica concentration increases with depth from 1.1 mg l^{-1} up to 2.5 mg l^{-1}. In the photic zone, the concentration of nitrate nitrogen (NO$_3$-N) is approximately 0.05 mg l^{-1}, phosphate phosphorus (PO$_4$-P) is 0.02 mg l^{-1} and total iron is 0.03 mg l^{-1} (Votintsev, 1961).

Table 18.6   Average chemical composition of Lake Baikal water

Layer	Period of observation	Concentration (mg l^{-1})						
		HCO$_3^-$	SO$_4^{2-}$	Cl$^-$	Ca^{2+}	Mg^{2+}	Na$^+$+K$^+$	TDS
Mean for water column[1]	1961	66.5	5.2	0.6	15.2	3.1	5.8	96.4
Mean for water column[2]	1980–88	62.2	5.7	0.8	15.4	3.4	2.7	90.2
Surface layer[3]	1981–84	67.1	4.9	0.6	15.7	3.1	4.5	96.1
Mean for 0–200 m layer[2]	1980–88	59.6–65.5	5.4–6.4	0.7–1.1	14–15	3.1–3.6	1.0–4.2	–

TDS   Total dissolved solids
[1]   Votintsev, 1961
[2]   Tarasova and Mescheryakova, 1992
[3]   Unpublished data of the Irkutsk Regional Board of Hydrometeorology

Table 18.7   Average concentrations (± S.E.) of nitrogen, phosphorus, chlorophyll *a*, silica and organic carbon in southern Lake Baikal

Period of observation	NO$_3$-N (µg l^{-1})	N$_{org}$ (µg l^{-1})	PO$_4$-P (µg l^{-1})	P$_{org}$ (µg l^{-1})	Chlorophyll *a* (µg l^{-1})	Si (mg l^{-1})	C$_{org}$ (mg l^{-1})
1967–69	92 ± 5	81 ± 7	11 ± 1	4 ± 1	0.19 ± 0.15	1.34 ± 0.08	0.87 ± 0.5
1982–84	81 ± 10	63 ± 30	–	–	0.19 ± 0.15	–	1.04 ± 0.18
1986–89	84 ± 10	103 ± 50	10 ± 2	9 ± 3	0.16 ± 0.11	1.41 ± 0.14	1.04 ± 0.13

S.E.   Standard error   Source: Tarasova and Mescheryakova, 1992

Characteristics of the basic chemical components of the lake water measured in different years are presented in Table 18.6.

Chemical analysis of the lake water confirms an increase in sulphates which correlates with a long-term trend of increasing concentrations in the surface runoff. The concentration of nutrients, subject to internal hydrobiological and biochemical processes, has strong spatial and temporal variability. Recent data on the annual concentration of nitrogen, phosphorus and silica in southern Lake Baikal are presented in Table 18.7 (Tarasova and Mescheryakova, 1992). The 1980–88 average data of the ROSGYDROMET monitoring network on the whole, coincide with the values listed in Table 18.7. However, it should be noted that the average inorganic nitrogen concentration estimated by ROSGYDROMET for the whole lake was < 55 µg l^{-1}, with values ranging from 13–119 µg l^{-1}.

18.4   Hydrobiology

Numerous factors influence Lake Baikal's hydrobiology. The large volume of lake water (23,000 km^3) is responsible for the very slow water turnover (c. 300 years). Water temperatures are relatively low: 10–15 °C in summer in the surface layer and 3.5–4 °C in the deeper main water mass throughout the year. Water transparency is very high and allows photosynthesis down to 50–70 m depth. The stable chemical composition and the low salt content promote an exceptionally stable biocoenoses. The lake population comprises over 2,000 species. The majority, 60 per cent of fauna and 28 per cent of flora, are endemic species. On the whole, the endemic fauna in the open lake comprise 84 per cent of the total species.

18.4.1 Phytoplankton

Lake Baikal not only has a typical seasonal vegetation growth pattern, with a specific under-ice 'bloom' of diatoms and dinoflagellates, but it also has inter-annual successions. Algal growth is not observed under ice-cover on an annual basis. In some years, the diatoms *Aulacoseira baicalensis, A. islandica helvetica* and *Stephanodiscus binderanus* produce a high level of biomass, up to 5–6 g m^{-3}. This high algal biomass in the ice-cover period has an important impact on the lake ecosystem which can last several years. In such years, the lake water acquires a eutrophic nature. Other years are marked by intensive under-ice development of the diatoms *Cyclotella baicalensis, C. minuta, Synedra acus radians,* of the dinoflagellates *Peridinium aciculiferum, P. baicalensis,* and particularly carapace-free dinoflagellates *Gymnodinium baicalensis, G. coeruleum,* and in some years the golden Chrysophycae *Dinobryon cylindricum* and green algae e.g. *Monoraphidium irregulare, Koliella longiseta.* The greatest biomass (up to 100 g m^{-3} and above) in the top surface layer is produced by the carapace-free dinoflagellates.

In Lake Baikal, there are years when the maximum algal biomass has occurred under-ice in the spring (estimated by chlorophyll *a* concentrations). These were defined by Izmestyeva (1979) as 'harvest years'. In shallow waters, the localised outbursts of under-ice development of spring algae are more frequent than in the open parts of the lake. Shallow water areas are therefore important areas for the preservation of algae during inter-vegetation growth periods.

The 1960–70s were marked by a decrease in the proportion of *Aulacoseira baicalensis* in the total phytoplankton and by an increase in the amount of smaller algae present. One of the major smaller species is *Synechocystis limnetica.* The reasons for these annual variations remain unclear.

The algal assemblage during the ice-free period is represented by a broader species variety than in the ice-cover period. The ratio of species number varies from year to year in different parts of the lake due to thermal and wind features and the impacts of cyclonic currents. The summer algal development is usually followed by an autumn biomass peak characterised by the endemic diatom, *Cyclotella minuta*. During the ice-free period, the greatest variety of phytoplankton occurs between the upper pelagic zone, the littoral zone and the main lake body. The algal assemblage found in the littoral zone is generally similar to the summer phytoplankton of many lakes and reservoirs. In cold years, the open pelagic zone is dominated both by endemic Lake Baikal species and also by common lake species which have a low temperature optimum, whereas in warm years the pelagic zone is dominated by species commonly inhabiting the littoral zone. A major factor in the spatial variability of phytoplankton is the thermal regime against a background of annual climatic variations. Beyond the shallow water zone, the more thermophillic algae are disseminated by currents and shifts in the water mass. In addition to variations in the lake warming pattern, particularly in the north-south direction, the hydrodynamic mobility of the lake brings the phytoplankton of the littoral, the pelagic and the open lake into contact creating the heterogeneous distribution of the phytoplankton. Natural algal aggregations are also important in the heterogeneity of algal distribution.

Despite the variety in phytoplankton distribution, no stable or consistent differences have been found between different parts of the open lake. Since 1960 and up to the 1990s, the dominant phytoplankton species in late summer and early autumn were *Cyclotella minuta*, *Nitzschia acicularis*, *Chroomonas acuta*, species of the genus *Cryptomonas*, several species of the genus *Stephanodiscus*, *Monoraphidium*, *Koliella longiseta*, *Elakatothrix lacustris*, and Chrysophycae including *Chrysococcus*. The phytoplankton composition is more varied in the central part of the lake which is morphometrically more differentiated with a higher variability in abiotic conditions.

Seasonal and annual dynamics of phytoplankton have been detected in the biomass and also in the chlorophyll *a* concentration. The under-ice concentration of chlorophyll *a* in 'harvest years' reaches 6 mg m^{-3} (absolute maximum 16.5 mg m^{-3}) but is 10-fold less in non 'harvest years' (maximum 0.56 mg m^{-3}). Maximum concentrations of chlorophyll *a* in ice-free periods are generally 1–2 mg m^{-3} and always < 3 mg m^{-3}, often < 0.5 mg m^{-3}. Minimum concentrations often coincide with the homothermic period following the break up of ice on the lake.

The spatial distribution of chlorophyll *a* in the main body of Lake Baikal during late summer and early autumn is relatively even. The average concentration throughout the water column in the upper photic layer is 0.56 ± 0.03 mg m^{-3}. The littoral zone has a higher chlorophyll *a* concentration, with values up to 9.2 mg m^{-3} (Izmestyeva, 1983; Izmestyeva and Kozhova, 1985, 1988; Izmestyeva *et al.*, 1990).

Photosynthetic pigments are at a maximum in the upper 50 m water layer, but not directly near the surface. The maximum concentration is found at a depth of 0.25 Zs, where Zs is the Secchi disc water transparency value. This shows a photo-inhibition of the photosynthesis process in the uppermost subsurface layers. This has been proved experimentally by special observations using luminescent techniques (Venediktov *et al.*, 1989).

Primary daily production estimated by radiocarbon methods (^{14}C) under ice in good 'harvest years' may reach 54.4 mg C m^{-3} d^{-1} while in non 'harvest years' it is < 2.9 mg C m^{-3} d^{-1} and in open water periods it is 95.3 and 22.9 mg C m^{-3} d^{-1} respectively. In 1977–80, annual production varied from 24 to 124 g C m^{-3} a^{-1} (Kozhova and Pautova, personal communication). Despite the high production and high algal biomass typical of good 'harvest years', daily primary production during an ice-free period is higher than that in the ice-cover period. Annual variations in both primary production and photosynthetic pigments in the summer/autumn period are typical of the whole lake.

Since 1946 and up until the present time, changes have been observed in the dynamics of the total number of algae in Lake Baikal. Mean annual phytoplankton numbers in 'harvest years' in the 1950s were below 70,000 cells per litre; in the 1960s they were 150,000 cells per litre and in the 1970s they were below 400,000 cells per litre (excluding picoplankton). The amount of picoplankton, mostly represented by *Synechocystis limnetica* (cell diameter of 1.5–3 µm), reaches tens of millions of cells per litre. Variations in the total numbers of algae correspond with marked changes in species ratio.

Long-term annual variations in phytoplankton biomass show two fluctuation cycles, one of about 20 years and another of 3–4 years for *Aulacoseira baicalensis*, *A. islandica helvetica* and *Stephanodiscus binderanus*. No increasing or decreasing trends in the biomass have been found. In the 1970s, some new phytoplankton species were found in the lake, namely *Chroomonas acuta* (*Rhodomonas pusilla*) and species of the genus *Cryptomonas*, Chrysophycae including *Chrysococcus*, green algae *Koliella longiseta* and *Elakatothrix lacustris*.

18.4.2 Zooplankton

The most important in the pelagic zooplankton is the famous endemic phytophagous copepod *Epischura baicalensis* which comprise up to 80–90 per cent of the zooplankton biomass. Another, although less common, zooplankton species is the carnivorous palaearctic copepod *Cyclops kolensis*. Other *Cyclops* species living in the shallow water zone are rarely found in the open pelagic. Typical of the pelagic zone is the endemic amphipod *Macrohectopus branickii*. Of the Cladocera, *Daphnia longispina* and *Bosmina longirostris* are common in the summer–autumn period. Rotifera, including endemic species, are also typical of the Lake Baikal zooplankton. Rotifer numbers are occasionally higher than that of *E. baicalensis*, yet the Rotifera occupy a lower ranking in terms of significance, not exceeding 20–30 per cent of the total biomass. The species found all the year round are *Keratella quadrata*, *Filinia terminalis*, *Kellicottia longispina* and *Keratella cochlearis*. Species found primarily in spring include *Synchaeta pachypoda*, *Notholca grandis*, and *N. intermedia*. Species found both in summer and autumn include *Synchaeta*, *Asplanchna priodonta*, *Collotheca mutabilis* and *Conochilus unicornis*.

In Baikal, the pelagic zone does not have a rich zooplankton population. *Epischura baicalensis* has two abundance peaks corresponding to two seasons of maximum reproduction whereas *Cyclops kolensis* does not reproduce extensively every year in the open lake. Long-term observations of zooplankton communities have found periods with high and low biomass production, as well as periodic variations in the species composition. The annual abundance of some groups of *E. baicalensis* had been decreasing for a long time, but since the 1970s it has been increasing and, significantly, the seasonal dynamics of its population numbers are gradually becoming smoother. Different trends have been observed in the annual dynamics of *C. kolensis*. Since 1965, a long-term decrease has been observed that coincided with a decrease in numbers of the diatoms *Aulacoseira baicalensis*. In the late 1980s and early 1990s, an increase in *Cyclops* number occurred. A trend showing an increase in the minimum number of *Epischura* copepods coincided with a trend of increasing numbers of algae *Cyclotella minuta* and species of the genus *Synedra acus*. Kozhov (1947, 1962a,b) was a pioneer in estimating Lake Baikal zooplankton and observed that in abundant zooplankton years, the zooplankton production can increase to 1.6×10^6 t of wet biomass.

18.4.3 Benthos
The highest concentration of living organic matter and biological life is concentrated in the upper 250 m of the lake, where active mixing of pelagic waters occurs. Typical of the pelagic zone is a relatively poor species composition compared with the benthic zone. The overwhelming majority of Lake Baikal endemic species are found in the bottom or bentho-pelagic layers. Lake Baikal has two benthic communities of European-Siberian flora and fauna as well as relicts of the Tertiary period. Representatives of the former benthic community mostly occupy the littoral zone from the shore down to 20 m. The zone comprises approximately 1.5 per cent of the total bottom surface and approximately 14 per cent of the littoral surface.

Phytobenthos
The benthic flora differs in the southern and northern parts of Lake Baikal which is possibly due to the former isolation of these depressions. The dominant communities and total algal biomass also vary along the littoral zone. Up until the present time, 76 species of algae have been discovered along the open shoreline of Lake Baikal. The most common

are the green algae, of which 35 species are endemic. Several algal communities can be distinguished in the lake based on distribution pattern, species composition and phytomass. Along the eastern shoreline, *Draparnaldioides* communities, common at 8–10 m, are dominant. *Stratonostok verrucosum*, with *Draparnaldioides* and *Calothrix* are common in the northern part of the western coast, and *Tetraspora cylindrica* var. *bullosa* with *Cladophora glomerata* and *C. kursanovii* are common in the southern part.

Most benthic algae are concentrated at a depth of 2.5–15 m. Sixty-five different species alone have been discovered at these depths. The average phytomass along the eastern coast is approximately 70 g m^{-2} (dry weight organic matter). Along the western coast, the average phytomass is 280 g m^{-2}, with a maximum up to 700 g m^{-2}. In bays, gulfs and river deltas, the average phytomass is 160 g m^{-2}, occasionally reaching up to 3,000 g m^{-2}. In thin clumps the average phytomass is lower, approximately 100 g m^{-2}, and in more dense clumps it reaches 330 g m^{-2}.

At depths from 0–1.5 m and from 15–70 m (occasionally up to 100 m), 28 and 38 algal species respectively have been found of which 10 and 20 species respectively are endemic. At a depth of 0–1.5 m, phytomass is usually < 10 g m^{-2}. *Ulothrix zonata* is common throughout Lake Baikal at this depth. The algal biomass decreases sharply in the littoral layers below 15 m. At a depth of 15 m down to 50–70 m numerous diatoms are found. In this zone of restricted light penetration, the total phytomass rarely exceeds 100 g m^{-2} and often ranges between 0.5–4 g m^{-2} but occasionally goes up to 30–40 g m^{-2}. Here, at stony sites, the phytomass is highly variable with an average of 14 g m^{-2} and occasionally as little as 1 g m^{-2}. The biomass is higher in clumps at 29 g m^{-2} and in bays in the northern part of the lake it reaches 23 g m^{-2}.

The largest quantities of phytobenthos are observed from May to August and the maximum phytomass occurs in June–July at a depth of 8–15 m occasionally reaching 1–3 kg m^{-2}, but the average is 200–500 g m^{-2} (Izhboldina, 1990). Votintsev *et al.* (1975) found that the phytobenthos produces 26,000 t a^{-1} of organic carbon.

Zoobenthos

The bulk of macroinvertebrates in Lake Baikal are found within the bottom layer, down to 250 m depth. Kozhov (1962a,b, 1970), Bekman (1971), Cherepanov (1978), Kozhova *et al.* (personal communication) and Popovskaya (personal communication) estimated the average benthic biomass (excluding sponges) in parts of Lake Baikal at different depths and the results are given in Table 18.8. When sponges are included, the biomass at 1–20 m depth may be twice as high. Biomass and species combinations may vary significantly in relation to depth and also in relation to the bottom substrate, e.g. stony sites, sandy or silt sediments.

Table 18.8   Zoobenthic biomass in Lake Baikal

Depth (m)	Biomass (g m^{-2} dry weight)			
	Central Baikal	Southern Baikal	Selenga river delta	Small Bay gulfs
0–20	21–53	22–33	24	18–35
20–100	9–23	14–21	25	32–33
100–250	9–12	12–15	20	20–22
> 250	1–3	2–5	–	–

Biomass variations along the littoral zone (excluding sponges) may be 2.5–69 g m^{-2}, and along the sublittoral zone it may be 3.6–35 g m^{-2}. The pelagic zone typically demonstrates relative species scarcity compared with the benthic zone.

Approximately 800 zoobenthic species are found in Lake Baikal. The dominant species in the macrobenthos are Gammaridae, Mollusca, Oligochaeta as well as Chironomidae, Trichoptera and Turbellaria. Important species in the meiobenthos are Cyclopoida, Harpacticidae, Ostracoda and Nematoda.

Compared with the bioproductivity in the pelagic zone, productivity of the benthic zone is lower and ranges from 9,200 t C a^{-1} to 12,000 t C a^{-1}, of which approximately 70 per cent is due to the littoral area. The total zoobenthic biomass is 300,000 t (as wet weight). A productivity/biomass (P/B) ratio at 0–50 m depth is 0.8, and in the deeper zones down to the maximum depth it is 0.35. Despite a low P/B ratio, the productivity of the benthic invertebrate population, down to 50 m depth is similar to that of eutrophic reservoirs. In the remaining deeper part of Lake Baikal, zoobenthos production corresponds to that of oligotrophic lakes (Kozhov, 1947; Votintsev, 1961; Cherepanov, 1978).

18.4.4 Bacterioplankton
Micro-organisms in the water column of Lake Baikal and in bottom sediments are important for self-purification processes and ecosystem sustainability. Long-term studies of the Lake Baikal microbiological features have revealed that the seasonal cycle and quantitative variations correspond, up to the present time, to baseline values in most parts of the lake (Kozhova, 1982, 1983, 1986). Calculations of pelagic microbiological productivity in 1978–87 have shown that in the 0–10 m depth layer productivity is 2 µg C l^{-1} d^{-1} or 7 g C m^{-2} a^{-1}; in the 10–50 m depth layer productivity is 1.04 µg C l^{-1} d^{-1} (16 g C m^{-2} a^{-1}); and from 50 m down to the bottom it is 0.52 µg C l^{-1} d^{-1} (121 g C m^{-2} a^{-1}). Total bacterial production within the water column is 144 g C m^{-2} a^{-1}. The P/B ratio is 32 (Maksimova and Maksimov, 1989).

18.4.5 Fish

Lake Baikal is inhabited by 52 fish species, half of which are of European-Siberian origin. Seventeen species are fished including the omul (Baikal cisco, *Coregonus autumnalis migratorius* Georgi), other whitefish (*Coregonus* sp.), grayling (*Thymallus*) and taimen (*Hucho*). The greatest species variety is observed among the Gobiidae which comprise 25 species of which 4 inhabit the pelagic zone. Twenty-seven fish species are endemic to Baikal. Bullheads (Cottidae) are the most numerous. In addition to the Baikal omul, other endemics include Baikal sculpin-oilfishes (Comephoridae), sturgeon (*Acipenser baeri stenorhynchus* var. *baicalensis* Nik.) and Arctic charr (*Salvelinus alpinus erythrinus*).

Long-term data show fish catches have been decreasing in Lake Baikal. For example, in the 1830s, the average fish catch was 8,700–10,000 t a^{-1} of omul and at the end of the 1960s it had decreased to 1,000–1,800 t a^{-1}. A ban on catching omul was imposed in 1969 and a limit (1,000 t of total fish catch) was placed on the other fisheries. It is expected that by the year 2000 the potential for the omul fishery will be 2,000–2,500 t a^{-1}.

18.4.6 Mammals

The endemic Baikal seal (*Pusa sibirica* Gm.) population is estimated at 60,000–80,000 species. In the 1970s, seals were estimated at 68,000. They are a valuable species, with the annual catch estimated at 2,500–5,000 t a^{-1}.

18.5 Lake pollution sources

The majority of polluting substances enter the lake by means of three sources, the Selenga river, as a result of direct wastewater discharge by the Baikalsk combined pulp and paper mill (BPPM) and by atmospheric pollution, mainly by pollutants from the BPPM. Also of significant influence is pollution from timber rafting, shipping, surface runoff from urbanised sites on or near the lake shore, including port areas, and part of the Baikal-Amur railroad outlet to the lake.

The Selenga river

The Selenga basin is the most developed part of the Baikal region. Pollution of the Selenga river begins in Mongolia. In Russia, the large Ulan-Ude industrial district is located only 150 km from Baikal and the Selenga Pulp and Cardboard Plant is situated even closer, at 50 km from Baikal. The close location of industrial zones to the lake creates a great threat to that part of the lake and to the Selenga river and delta. Some of these regions are important spawning sites for Baikal fish. Mongolia's contribution to river water pollution is significant and includes petroleum products, phenols and persistent organic substances.

Approximately 60 per cent of all inorganic substances, 40 per cent of organic substances (including 40 per cent of certain groups such as hydrocarbons, phenols and detergents) and over 40 per cent of suspended matter enter the lake with Selenga waters. The nutrient contribution varies from 20 per cent for total phosphorus up to 60 per cent for silica.

In spite of the large river delta (total area approximately 600 km^2), the contaminated Selenga waters can be traced from the delta over a large area up to several hundred square kilometres in extent.

The Baikalsk combined pulp and paper mill
The Baikalsk pulp and paper mill is situated on the southern shore of the lake (Baikalsk town) and is a major source of contaminant inputs to Lake Baikal. This plant started operating in 1965 and was established particularly to manufacture highly resistant cellulose. Treated waste-water discharged from the plant and dust and gas emissions to the atmosphere, pollute southern parts of Lake Baikal and an area of approximately 1,500 km^2 around the town of Baikalsk. Routine control of chemical wastewater composition, atmospheric precipitation and snow cover contamination has been performed since 1967.

Over 28 years, 2 km^3 of treated wastewater have been discharged into Lake Baikal. These contained 1.2×10^6 t of inorganic substances including approximately 600,000 t of SO_4^{2-}, 160,000 t of Cl^-, 350,000 t of Na^+, 150,000 t of persistent organic compounds, 50 t of phenols, 300 t of toxic methylsulphur compounds and 17,000 t of suspended matter. Organic substances discharged annually into the lake include 4,800 t of lignin, of which 70 per cent is deposited on the lake bottom and is subject to slow breakdown and burial. Annual discharges also contain 400 t of total nitrogen and 250 t of total phosphorus.

Annual atmospheric emissions from the BPPM comprise approximately 18,000 t of dust, 6,800 t of SO_2, 900 t of sulphur compounds including hydrogen sulphide, methylmercaptane, dimethylsulphide and sulphuric acid, 1,400 t of NO_x and 100 t of hydrocarbons.

Timber rafting
Timber rafting on some rivers and on Lake Baikal has occurred for over 50 years. The annual input of contaminants resulting from timber rafting is estimated at 7,800 t of organic substances, 300 t of inorganic nitrogen, 600 t of organic nitrogen, 130 t of inorganic phosphorus and 160 t of organic phosphorus.

Shipping
The lake, over its 630 km length, is a major transport route during the ice-free period. The average annual cargo shipping turnover in the 1980s

Table 18.9   Chemical balance of Lake Baikal

Balance	Total mineral compounds	SO_4^{2-}	Cl^-	N_{min}	P_{min}
Inputs (10^3 t a^{-1})					
Total input by rivers[1]	6,128 (100)	672 (100)	104 (100)	4.80 (100)	0.26 (100)
Input from Selenga[1]	3,663 (60)	394 (59)	66 (63)	1.22 (25)	0.13 (50)
Input from Upper Angara[1]	584 (9.5)	70 (10)	10 (9.6)	0.50 (10.4)	0.023 (8.8)
Input from Barguzin[1]	489 (8)	43 (6.4)	7.5 (7.2)	0.26 (5.4)	0.003 (1.2)
Input from Turka[1]	91 (1.5)	14 (2)	2.5 (2.4)	0.13 (2.7)	0.003 (1.2)
Atmospheric precipitation and dust	273	80	19	2.5 (3.1)[2]	0.06 (0.3)[2]
Primary production					
Point sources of pollution	45	26	10.7	3.8	0.11
Total	**6,446**	**778**	**134**	**11.1**	**0.43**
Outputs (10^3 t a^{-1})					
Output by the Angara	5,181	327	46	2.9	0.52
Accumulation and breakdown in the lake	1,265	451	88	8.2	−0.09 (0.15)[3]
Total	**6,446**	**778**	**134**	**11.1**	**0.43**

N_{min} Mineral nitrogen
P_{min} Mineral phosphorus

[1] Figures in brackets show percentage of runoff by all tributaries

was estimated at 200,000 t. The contribution of shipping to Baikal pollution from petroleum products is estimated between 700 and 1,000 t a^{-1}.

Anthropogenic surface runoff
The annual surface runoff from surrounding land, including urban areas, contains over 8,000 t of total inorganic substances including 3,300 t of Cl^- and 1,600 t of SO_4^{2-}. It also includes 3,100 t of inorganic nitrogen, 1,000 t of total phosphorus, 2,000 t of organic matter, 1,000 t of petroleum products, 100 t of detergents and 1,400 t of suspended matter. Approximately 60 per cent of contaminated surface discharge is attributed to the southern part and 20 per cent to the northern part of Lake Baikal.

18.6   Chemical balance
The chemical balance of Lake Baikal is based on the assessment of basic components, i.e. chemicals transported via tributaries, atmospheric dry and wet deposition, resulting in direct contamination of the lake water and the runoff of the Angara river. Chemical inputs from the Baikal tributaries and, separately, from the largest tributaries (Selenga, Upper Angara, Barguzin and Turka) are presented in Table 18.9.

The primary production in the lake involves 286,000 t a^{-1} of inorganic nitrogen and 62,000 t a^{-1} of inorganic phosphorus which depend on a

Table 18.9   Continued

Balance	P_{tot}	Organic matter	Petroleum products	Phenols	Suspended matter
Inputs (10³ t a⁻¹)					
Total input by rivers[1]	1.68 (100)	694 (100)	9.0 (100)	0.20 (100)	1,634 (100)
Input from Selenga[1]	0.60 (36)	376 (54)	4.4 (49)	0.08 (40)	1,173 (72)
Input from Upper Angara[1]	0.177 (10.5)	80 (11.5)	1.6 (18)	0.04 (20)	163 (10)
Input from Barguzin[1]	0.057 (3.4)	47 (6.8)	0.6 (6.7)	0.02 (10)	67 (4.1)
Input from Turka[1]	0.015 (0.9)	19 (2.7)	0.3 (3.3)	0.01 (5)	28 (1.7)
Atmospheric precipitation and dust	0.17 (0.7)[2]	159	0.1	0.05	1,012
Primary production		7,902			
Point sources of pollution	1.54	111	1.7	0.01	2
Total	**3.39**	**8,866**	**10.8**	**0.26**	**2,648**
Outputs (10³ t a⁻¹)					
Output by the Angara	0.86	292	1.7	0.03	34.5
Accumulation and breakdown in the lake	2.53	8,574	9.1	0.23	2,613
Total	**3.39**	**8,866**	**10.8**	**0.26**	**2,648**

[2]   Figures in brackets show data obtained by the   [3]   See text for explanation
Limnological Institute

turnover rate for these nutrients of 10 to 12 times per year. Up to 75 per cent of the annual input of organic matter in the water column, and 50–60 per cent in bottom sediments, is subject to immediate breakdown. Residual organic matter is subject to breakdown within 5–10 years.

The data on nutrients and organic matter inputs for the purposes of the chemical balance require further assessment in view of their importance for determining inter-basin processes and the trophic status of the lake. Of the listed chemical constituents in Table 18.9, the inorganic phosphorus input via atmospheric precipitation into the balance estimate needs to be defined more precisely. Because some uncertainty exists in this case, a range of inorganic phosphorus balance from −0.09 to 0.15×10^3 t a⁻¹ has been presented. The relatively high (in comparison with other major ions) rates of SO_4^{2-} and $Na^+ + K^+$ accumulation have also been noted. These substances slowly increase in the lake in relation to the increasing trend of SO_4^{2-} and Na^+ concentrations in river runoff.

The data available for Lake Baikal also enable the evaluation of accumulation rates of some of the chemical components in the lake.

18.7   Review of anthropogenic impacts on the lake

Until the mid-1950s, anthropogenic impacts on the Lake Baikal ecosystem were considered mostly in relation to fish abundance and fish catches. When the Irkutsk hydropower plant was constructed in 1956, regulation of the Angara river resulted in a permanent lake level elevation of 1 m. This elevation has resulted in an additional 10 per cent abrasion in the shore zone compared with the pre-regulation period. Marked changes have also occurred in areas of soft sediment and shoreline, in the water movement between bays and shores, and in plankton and benthos. Other anthropogenic factors which were difficult to forecast included shoreline abrasion and erosion processes in the lake catchment area resulting in a higher sediment load for incoming rivers.

In 1965, the Baikalsk pulp and paper mill commenced operation on the southern coast of Lake Baikal. In the 1960–70s the Baikal-Amur railway was constructed in the northern part of the lake basin which resulted in increased transportation and a growing population along the shore zone. Fertiliser application and agricultural and municipal waste discharges increased in various rivers basins, especially the Selenga river.

18.7.1 The Selenga river and delta

According to integrated monitoring data, the major sources of pollution in Lake Baikal include discharge of the Selenga river, direct pollution of the lake (particularly by the BPPM), the inflow of chemicals with other tributaries and atmospheric fallout. However, the largest impact on the lake ecosystem is due to the Selenga river (Table 18.9).

The large Selenga river delta (with numerous branches) and the Proval bay which serves as a natural settling pond for sediments discharged by the delta's northern branches create favourable conditions for pollutant sedimentation in the delta zone. Due to this, Selenga river contamination levels exceed, at present, those of shallow delta waters. Comprehensive hydrochemical, geochemical and microbiological monitoring from 1989 to 1991 estimated the extent of the shallow zone affected by pollutant inputs from the Selenga river to be 700–750 km^2.

The most hazardous chemicals transported by the Selenga river are polycyclic aromatic hydrocarbons (PAHs), organochlorine pesticides and heavy metals. A survey along the 400 km lower reaches of the river and 1,500 km^2 of the outer delta in the lake detected PAHs in 56 per cent of water samples and 16 per cent of shallow water bottom sediments. The PAH concentrations ranged from 1 to 9 ng l^{-1} in water and were approximately 72 ng g^{-1} in bottom sediments. Maximum allowable concentrations were exceeded in 12 per cent of samples. The pesticide concentration in samples of shallow water and sediments did not exceed 7 ng l^{-1} and 0.2 ng g^{-1} respectively. Organochlorine pesticides were detected in 60 per cent of water samples and 80 per cent of sediment samples.

Heavy metal concentrations in the water and sediment of the river and shallow water area have certain distinctive features. The Hg, Cd, Pb, Zn, Cu, Mn, Fe and Cr concentrations in water in the outer delta are 1.5 times lower than those in the river water. However, Hg and Pb concentrations in the outer delta sediments are slightly higher (1.1–1.3 times) than those in the Selenga river sediments, whereas for the other metals the concentrations are similar in delta and river sediments. It can be concluded that the delta purification capacity is due to metal sorption by suspended matter with subsequent sedimentation in the shallow water. Metal concentration in the shallow water sediments are: Hg 0.02–0.06 µg g^{-1}; Cd 0.01–0.03 µg g^{-1}; Pb 1.1–4.8 µg g^{-1}; Zn 3–38 µg g^{-1}; Cu 2.4–7.8 µg g^{-1}; Mn 45–307 µg g^{-1}; Cr 6–19 µg g^{-1} and Fe 2.6–10 mg g^{-1} (see editorial note at the end of the chapter).

The total number of heterotrophic bacteria in water and bottom sediments at the Selenga buffer zone is 3–50 times higher than the maximum baseline value defined in lake water, and 2–5 times higher in the lake sediments. Mean long-term numbers of heterotrophic bacteria have been estimated. At a 4 km distance off the delta, bacterial numbers are 300 cells per ml, at 4–7 km numbers are 160 cells per ml and at 7–12 km they are approximately 100 cells per ml, while the maximum baseline value is 30 cells per ml. Mean average values for bottom sediments at the same distances were 25×10^3 cells per gram, 20×10^3 cells per gram and 19×10^3 cells per gram wet sediment respectively, while the maximum baseline value is 15×10^3 cells per gram of wet sediment.

18.7.2 The Baikalsk combined pulp and paper mill

The BPPM is the second largest source of lake pollution. Wastewater discharges into the lake were described in detail in section 18.5. Only the impact of BPPM discharges on the lake ecosystem are described here.

Littoral plant communities near the BPPM have changed little since the start of its operation except for maximum phytomass values (> 500 g m^{-2}) which have decreased. This was due to the decrease in Tetraspora cylindrica var. bullosa and Draparnaldioides communities.

The composition and qualitative features of invertebrate communities changed markedly and exceeded the limits of natural fluctuations. This was due to drastic changes in the bottom sediment composition. Several zones have been identified which differ to various extents from the baseline and which reflect the sphere of influence of the BPPM. The first zone is relatively restricted (c. 0.3 km^2) and is characterised by a crucial restructuring of biocoenoses leading to a modified composition of the major taxonomic units and a sharp reduction in biomass. This site is where most suspended matter discharged with the wastewater is accumulated. In the early years after the operation of the BPPM, the macrozoobenthos biomass (at 5–20 m depth) decreased 10-fold compared with the reference period prior to the plant operation. At depths of 20–50 m and 50–100 m it decreased by three and five times respectively. At 20–100 m depth, the community was originally dominated by molluscs and oligochaetes but was transformed to a community dominated by Gammaridae. The invertebrate communities that survived in the zone underwent significant changes in species composition with many species disappearing; for example, only a single gammarid species *Echiuropus rhodophtalmus microphtalmus* (= *Carinogammarus rhodophtalmus*) remained.

The second zone, estimated at a few square kilometres, is a transition zone where pollutants constantly disperse due to hydrodynamic factors and where there is complete or partial restoration of the original biocoenoses. Communities in this zone exist either in a stable or unstable state depending on prevailing conditions. Occasionally, there is a higher zoobenthos biomass in this zone compared with control areas under similar conditions because of the increased numbers of some pollution-resistant species, including endemic molluscs. However, these mollusc populations have a significant proportion of abnormalities and chromosome aberrations that testify to the genetic dangers of the wastewater discharges from the Baikalsk pulp and paper mill.

The third zone (a few square kilometres in extent) is where intensive self-purification occurs and where the impact of pollutants is neutralised by biotic communities. Self-purification processes do not have an impact on the structure of the invertebrate communities, although the intensified organic matter turnover due to microbial activity results in an increase in zoobenthos biomass. In different years, the location of this zone has varied along the shoreline up to a maximum distance of 5 km from the BPPM wastewater discharge site.

Data from microbiological, phyto- and zooplankton observations have shown that the polluted water plume near Baikalsk has varied between 1 and 24 km^2 in extent and was from 9 to 15 km^2 on average, depending on the year. The average number of heterotrophic bacteria was 130 cells per ml and the maximum number was 430–1,600 cells per ml. The average number of heterotrophs in the sediment at the polluted site was 13,000 cells per gram wet sediment and the maximum was 20,000–80,000 cells per gram.

Monitoring of the Baikalsk area bottom water layer and sediments is regularly and effectively performed at sites of excessive anthropogenic impact. According to data from the Limnological Institute, dissolved oxygen concentrations in the bottom layer in the southern part of Lake Baikal in the 1950–60s never dropped below 10 mg l^{-1} and the O_2 saturation never went below 70–80 per cent. The particulate organic carbon (C_{org}) concentration in the upper 2 cm layer of sediments varied within 0–3.5 per cent, with an average value of 1.29 per cent in sand and 2.29 per cent in silt. Particulate organic nitrogen (N_{org}) did not exceed 0.2 per cent and sulphide sulphur did not exceed 0.005 per cent.

Since 1969, the geochemical survey of bottom sediments near the BPPM has shown a 5–12 km^2 zone with oxygen concentrations below 9 mg l^{-1}, which was registered in ≥ 50 per cent of oxygen determinations.

By the end of the 1980s, the zone with depleted oxygen levels in the bottom layers expanded eastwards. In the 1990s, the average C_{org} concentration in bottom sediments remains at 1.55 per cent while the regional baseline is 1.2 per cent. The N_{org} concentration is ubiquitous at approximately 0.2 per cent. Also of significance is the fact that a bottom area contaminated with sulphur compounds has been formed. By the end of 1994, the lake bottom area, where sulphide sulphur average concentrations were > 0.005 per cent, reached 15 km^2 with a maximum concentration up to 1.4 per cent. The wastewater discharged by the BPPM contains suspended particles of lignin that, upon deposition on the bottom, create a sediment site with a higher content of both persistent and easily-hydrolysed hydrocarbons. By the late 1980s, this pollution site was estimated at about 20 km^2. Since 1981, bottom sediments near the BPPM have also been analysed for the highly carcinogenic chemical, benzo(a)pyrene. The site polluted by this compound is estimated at 6–18 km^2.

The wastewater impact on the microflora of the water and the bottom sediments is not well-defined; the impact can either stimulate or inhibit the development of microflora. In the first case the number of microorganism groups may increase by 10–20 times the baseline level. A similar situation can be observed for phytoplankton and zooplankton which can be attributed to the eutrophic impact of wastewaters on aquatic biota (Kozhova, 1983, 1986).

The long-term impact of the wastewaters on the pelagic community has caused no qualitative changes in composition of phytoplankton and zooplankton in the control area. Of note, however, is the decreasing trend in the average biomass of zooplankton from 205 mg m^{-3} in 1976–80 to 181 mg m^{-3} in 1981–85 to 114 mg m^{-3} in 1986–89. A relatively high level of mortality of Lake Baikal plankton has also been noted. The qualitative changes in plankton in the impact zone are about twice those of the baseline site. All age groups of *Epischura* have been damaged in this zone.

The zoobenthos at the 5 km^2 control site on the lake bottom in the wastewater discharge area show a marked inhibition in development. Prior to the pulp and paper mill operation, the zoobenthos biomass averaged 14.4 g m^{-2}. In 1976–80, the biomass decreased to 13.8 g m^{-2} and in 1986–89 it decreased to 7.4 g m^{-2}. Live molluscs have almost completely disappeared at the impacted site at a depth of 0–27 m, whereas an estimated 85 per cent of live molluscs occur at 27–70 m depth and up to 95 per cent below 70 m. However, below 70 m the population of molluscs is low. The diversity of the Gammaridae has been reduced noticeably in the wastewater discharge area from 32 species originally to 10 species in the late 1980s. Chironomidae numbers have decreased and their larvae are almost extinct. At present, the zoobenthos is dominated by oligochaetes, comprising 55–65 per cent of the total zoobenthos biomass.

The zone of polluted bottom sediments is gradually spreading along the bottom towards a greater depth (over 400 m) and in an eastward direction. It is characterised by high C_{org} and sulphide sulphur concentrations in the bottom sediments. The presence of the stable pollution zones in the wastewater discharge area can be confirmed by tracer experiments and by evaluating the zone's borders using sulphur isotope composition (Vetrov and Dekin, 1977; Fedorov et al., 1992; Nikanorov et al., 1992).

On the basis of the key indicators discussed above, it may be concluded that Lake Baikal near the BPPM has markedly polluted surface water, has a partly modified littoral zone and a little modified pelagic zone.

Up until the present time, Lake Baikal has maintained oligotrophic features. However, in the 1960s discrepancies arose in defining the trophic status of the lake due to a high average production/biomass ratio (P/B) of 304, which is more typical of mesotrophic rather than oligotrophic lakes. In addition, some researchers noted changes in phytoplankton community structure, the large-scale emergence of species previously unknown in the lake open water, and more than a twofold increase in total biomass. The changes in community structure were observed in the Selenga shallow waters, in the Proval and Barguzin bays and in the middle of Lake Baikal (Popovskaya, personal communication). However, the anthropogenic origin of the ongoing changes of phytoplankton have not been proved (Kozhova, 1986). According to Votintsev (1992), there is no present threat of Lake Baikal becoming eutrophic. However, the phosphorus input to Lake Baikal, the total load for which is close to 2,500 t a^{-1}, is the most serious potential threat to the lake. Aschepkova and Kozhova (1984) estimated a eutrophication threshold for Baikal of 1,300 t a^{-1} and that the phosphorus load therefore exceeds the critical load by twofold. Other estimates of critical phosphorus loads have been proposed, particularly based on phosphorus runoff through the Selenga river (estimated at 1,300 t a^{-1}) and resulting in a total load of 3,300 t a^{-1}.

It is difficult to forecast the outcome of a continuous increase in sulphate and chloride concentrations in the tributaries and lake water. Their relative concentration in Lake Baikal is, so far, below 6 per cent of the total dissolved mineral substances. However, their concentrations are increasing continuously.

A danger of extinction or impoverishment of biota has arisen from the input of chemicals at concentrations exceeding baseline values for anthropogenic chemicals. The compounds concerned are primarily highly toxic sulphur-containing organic compounds contained in the pulp and paper mill wastewater, phenols, petroleum products, detergents, heavy metals transported to the lake with surface runoff, and chloride-containing organic compounds transported by river discharge and from the atmosphere. According to ROSGYDROMET data, heavy metal concentrations in Baikal tributaries are 0.07–0.35 µg l^{-1} for Hg, 0.2–1.1 µg l^{-1} for Cd and 0.4–4.4 µg l^{-1} for Cu (Izrael et al., 1985). The concentrations of the same metals are 1.5–7 times higher in the BPPM wastewater. The Na^+ concentration in the BPPM wastewater is 40 times higher than in the lake.

Although organochlorine pesticides have rarely been detected in river water (or when detected they were at concentrations of 1–2 ng l^{-1}), they are actively accumulated in living biota. They have been detected in the tissue of Baikal seals and in zooplankton. Polychlorinated biphenyls (PCBs) also accumulated in Baikal seal tissue and concentration in organs and tissues were estimated at 0.02–5.6 mg kg^{-1} wet mass (Bobovnikova et al., 1985).

The evaluation of the omul (Coregonus autumnalis migratorius) stock in Lake Baikal is relatively difficult. Despite scientifically-regulated fishing, overall catches of this valuable fish have decreased. From an initially high number of fish caught, the total fish catch dropped 4–5 times. In 1936–63, the average fish catch was estimated at 9,500 t, including 5,400 t of omul. Subsequently the total fish catch decreased to 2,000 t. In 1969, a catch limit of 1,000 t was imposed. More recently, the total fish catch was permitted at 2,000–2,500 t. There are several reasons for the decrease in fish catch. The principal reasons include poaching (according to some estimates, this amounts to 50 per cent of the fish catch), uncontrolled catches, the decline in the number of gobies (Gobiidae) which are the basic food of the omul, and pollution of spawning areas in streams. Over the last 50 years, the average body weight of the omul has also decreased from 470 g to 350 g and the fish roe numbers have decreased from 39×10^9 to 19×10^9. Gonad mass, spawning capacity and fertility have also decreased. All of these factors led, in the early 1980s, to a decrease in omul spawning in the major tributaries of Lake Baikal.

18.8 Biological invasions

Apart from chemical pollution, Lake Baikal is threatened by biological invasions which are very difficult to forecast (Kozhova and Beim, 1993). A major biological change in the Lake Baikal ecosystem was manifested for the first time in the 1970s with the emergence of Elodea canadensis (Canadian pondweed) in the lake. In the 1980s it appeared in the Selenga shallow waters, in 1981 in the Small Bay and along the southern Lake Baikal shore and in 1983 in the Barguzin and Chivyrkui bays and the open waters. At present, E. canadensis appears in northern Lake Baikal near Nizhneangarsk and on the western shore at the Angara river outflow. The E. canadensis introduction is a typical example of an unpredicted consequence of eutrophication and a disaster on a large scale.

Other examples of biological introductions are the appearance of predatory fish like the Amur sleeper, Perccottus glehni Dyb., in the Selenga river and the peled, Coregonus peled (Gm.), in the lake. The latter may compete in the future with the Baikal omul. Tens of previously unknown parasitic animals (Myxosporidia, tapeworms) and microflora have been discovered in Lake Baikal in the last decade. In 1988, a spontaneous viral disease killed approximately 10,000 Baikal seals, which was about 12 per cent of the total seal population of the lake.

18.9   Conclusions

Long-term monitoring of Lake Baikal shows that changes in its ecosystem are generally of local character. On the whole, the lake maintains characteristic features of an oligotrophic water body. At the same time, however, increased loads of autochthonous and allochthonous organic matter and high production to biomass ratios are characteristic of a mesotrophic status. Over several decades, chemical substances with concentrations above characteristic background levels or of anthropogenic origin have been discharged to Lake Baikal. Introduction of new plants and fish species, intensive erosion processes in the water catchment area and abrasion of the banks are amongst the factors which may ultimately have an environmental impact on the lake ecosystem, although this is difficult to predict.

As a result of rising lake levels, shore abrasion has increased by 10 per cent and changes have occurred in areas of soft soils and along banks. This has led to changes in the water circulation in bays and in the plankton and benthic communities. In addition, changes in omul migratory behaviour have also been noted. The major factors influencing the Lake Baikal ecosystem are summarised in Table 18.10. The factors endangering the unique lake ecosystem include poorly regulated anthropogenic activities in the lake basin, environmental protection measures which are not effectively employed or enforced, and a decrease in government concern for scientific research and environmental issues. Not a single governmental resolution on Lake Baikal has been completely fulfilled since 1969.

Note from the editors

Whilst this chapter was being reviewed, the results of a Russian-American field study on the metal balance in Baikal, performed with the help of the Irkutsk Limnological Institute, was published by Falkner *et al.* (1997). The water column and some Baikal tributaries were sampled with the application of ultra-clean techniques. Average lake concentrations for dissolved elements were extremely low (Ni 0.12 µg l^{-1}, Cu 0.22 µg l^{-1}, Al 0.1 µg l^{-1}, V 0.4 µg l^{-1}, Cr 0.07 µg l^{-1}, Ba 9.8 µg l^{-1}, Cd 0.001 µg l^{-1} and Hg 0.0008 µg l^{-1}). Discharge weighted dissolved concentrations for tributaries were 0.2 µg l^{-1} for Cr, 1.1 µg l^{-1} for U, 0.4 µg l^{-1} for Ni and 0.9 µg l^{-1} for Cu. Only Cd is found at a higher levels in Baikal surface waters (0.005 µg l^{-1}) but this may be due to sample contamination. From these results, the anthropogenic impacts on Lake Baikal waters in terms of heavy metals seem to be very low.

Table 18.10 Summary of factors influencing Lake Baikal

Factors	Time of first appearance and duration of influence	Characteristic impact and area of influence
Lake level changes	1956, c. 40 years	Changes in ecosystem, littoral and shore areas
Abrasion and erosion of bank	Intensified after 1956	Bank zone ecosystem
BPPM wastewater discharge	1965, 30 years	Water and sediment pollution of southern Baikal, 30–250 km^2 (according to various data)
Atmospheric pollution by BPPM emissions	1965, 30 years	Pollution of lake and adjacent banks, c. 500 km^2
Pollution of lower Selenga river (150 km) from Ulan-Ude to the lake	c. 100 years	Impact on fish spawning and reproduction, water and sediment pollution, 500 km^2 of delta and 500–1,500 km^2 surrounding the delta
Baikal-Amur railway	20–30 years	Disturbance of 100–150 km^2 of the north-west bank of northern Baikal
Input of polluting substances with river runoff	30–50 years	Hydrochemical modifications, impact on aquatic biota affecting all the lake
Fertiliser, pesticide and herbicide application in the lake basin	c. 50 years	Threat of eutrophication intensification and toxic effects on aquatic biota
Unregulated, increased input of organic substances and nutrients with surface runoff	c. 100 years	Danger of an increase in trophic status of the lake
Poorly regulated fisheries and fish-breeding	Over 60 years	Fish population and biomass changes, introduction of exotic species
Shipping	c. 50 years	Lake pollution by oil products

BPPM Baikalsk pulp and paper mill

18.10 References

(All references are in Russian unless otherwise stated).

Afanasev, A.N. 1960 The Lake Baikal water balance. *Transactions of the Baikal Limnological Station of Academy of Sciences of the USSR*, **17**, 155–241.

Afanasev, A.N. 1967 *Hydrometeorological Fluctuations in the USSR and in the Baikal Basin in Particular*. Nauka, Moscow, 232 pp.

Aschepkova, L.Y. and Kozhova, O.M. 1984 Prognosis of Baikal Phytoplankton Dynamics. In: *Ecological Systems Prognosis Methods*. Nauka, Novosibirsk, 90–110.

Bekman, M.Y. 1971 Benthos quantitative characteristics. In: *Limnology of Delta-Connected Areas of Baikal*, Volume 12(32), LIN Publications, Nauka, Leningrad, 114–126.

Bobovnikova, C.I., Virchenko, E.P. and Dibtseva, A.V. 1985 Sea mammals as indicators of pollution by organochlorine pesticides and polychlorobiphenyls. In: *Improvement of the Regional Monitoring System of Lake Baikal*. Gidrometeoizdat, Leningrad, 49–54.

Cherepanov, V.V. 1978 Ecological structure and productivity of the benthic population. The Baikal Problems. LIN Publications, Volume 16 (36), Nauka, Novosibirsk, 293 pp.

Falkner, K.V., Church, M., Measures, C.I., Le Baron, G., Thouron, D., Jeandel, C., Stordal, M.C., Gill, G.A., Mortlock, R., Frælich, P., Chan, L.-H. 1997 Minor and trace element chemistry of Lake Baikal, its tributaries and surrounding hot springs. *Limnology and Oceanography*, **42** (2), 329–345.

Fedorov, Y.A., Grinenko, V.A., Krouse, R. and Nikanorov, A.M. 1992 Use of hydrochemical and isotopic criteria for evaluation of the influence of anthropogenic sulphur on surface waters. In: *Proceedings of the International Symposium on Isotope Techniques in Water Resources Development, 1991*. International Atomic Energy Agency, Vienna, 477–494. (In English).

Izhboldina, L.A. 1990 Macro-phytobenthos of the open littoral of the lake. In: *The Baikal Underwater Landscapes*. Nauka, Novosibirsk, 142–165.

Izmestyeva, L.R. 1979 Some trends in chlorophyll *a* distribution in Angara reservoirs. In: *Ecological Problems of the Baikal Region*. Abstract of the Republic Conference, Part 1, 10–13 September 1979, Irkutsk, Irkutsk, 73–75.

Izmestyeva, L.R. 1983 Chlorophyll *a* content in the water reservoirs of the Baikal region. Candidate of Medical Science Thesis, Kiev, 24 pp.

Izmestyeva, L.R. and Kozhova, O.M. 1985 Trends in temporal and spatial dynamics of chlorophyll *a* and its assimilation in South Baikal in the region of Baikalsk town. VINITI Dep., N2462-B-87, Baikalsk, 39 pp.

Izmestyeva, L.R. and Kozhova, O.M. 1988 Phytoplankton structure and successions. In: *Ecosystems Condition Prognoses*. Nauka, Novosibirsk, 97–129.

Izmestyeva, L.R., Kozhova, O.M. and Usenko, N.B. 1990 Photosynthetic pigments as components of the Baikal monitoring programme. *Water Resources*, **3**, 89–95.

Izrael, Y.A., Anokhin, U.A. and Ostromogilskiy, A.K. 1985 Selected results of environmental monitoring in the Lake Baikal region. In: *Regional Monitoring of Lake Baikal*. Gidrometeoizdat, Leningrad, 4–22.

Khaustov, A.P., Fedorov, V.N., Zimina, T.Y. and Ilyicheva, E.A. 1990 Method for assessment of groundwater discharge to large lakes (using Lake Baikal as an example). *Water Resources*, **3**, 33–43.

Kozhov, M.M. 1947 *Fauna of Lake Baikal*. Oblgiz, Irkutsk, 303 pp.

Kozhov, M.M. 1962a *Lake Baikal and Its Life*. Academy of Sciences of the USSR Publication, Moscow, 295 pp.

Kozhov, M.M. 1962b *The Baikal Lake Biology*. Academy of Sciences of the USSR Publication, Moscow, 315 pp.

Kozhov, M.M. 1966 Possible pollution investigations of Lake Baikal by the pulp industry waste water discharge. In: *Sanitary and Technical Hydrobiology*. Nauka, Moscow, 44–49.

Kozhov, M.M. 1970 The southern Baikal benthos. *Trans. Biol. Geogr. Sci. Research Institute of the Irkutsk State University*, **237**(1), 3–12.

Kozhov, M.M. 1971 The present state of the Baikal flora and fauna in the area of waste water discharge of the Baikal pulp and paper mill. In: *Hydrobiological Regime of East Siberian Water Reservoirs Research*, Irkutsk, 3–9.

Kozhov, M.M. 1972 *Essays on Lake Baikal*. East Sib. Publication, Irkutsk, 254 pp.

Kozhova, O.M. 1982 Major criteria of aquatic ecosystems microbiological analyses. The Baikal Branch of the Ecological Toxicology Institute, VNP Obumprom Minlesbumprom USSR Funds, Baikalsk, 22 pp.

Kozhova, O.M. 1983 Problems of normal and pathological states of ecosystems in terms of aquatic microbiology. The Baikal Branch of the Ecological Toxicology Institute, VNP Obumprom Minlesbumprom USSR Funds, Baikalsk, 35 pp.

Kozhova, O.M. 1986 Study of plankton variability in lake water. V Congress VGBO: Abstract of Rep., Tolyatti, 16–19 September 1986, Kuybyshev, 254–256.

Kozhova, O.M. and Beim, A.M. 1993 *Ecological Monitoring of Lake Baikal*. Ecology, Moscow, 350 pp.

Maksimova, E.A. and Maksimov, V.N. 1989 *Microbiology of the Baikal Waters*. Irkutsk University Publication, 167 pp.

Nikanorov, A.M., Fedorov, Y.A., Grinenko, V.A. and Krouse, R. 1992 Primary data on the isotopic content of sulphate sulphur in Lake Baikal water. *R. A. Sci. Reports*, **325**(4), 844–847.

Tarasova, E.N. and Mescheryakova, A.I. 1992 *Present State of Lake Baikal Hydrochemical Regime*. VO Nauka, Novosibirsk, 141 pp.

Venediktov, P.S., Izmestyeva, L.R., Matorin, A.N. and Vavilin, A.N. 1989 Luminescent study of Baikal phytoplankton physiological state. In: *Freshwater Phytoplankton Study*. Nauka, Novosibirsk, 77–78.

Verbolov, V.I., Pokatilova, T.N. and Shimaraev, M.N. 1986 *Formation and Dynamics of the Baikal Water*. Nauka, Novosibirsk, 117 pp.

Vetrov, V.A. and Dekin, S.A. 1977 Investigations of admixture transport using radioactive indicators. In: *The Baikal Currents*. Nauka, Novosibirsk, 133–143.

Votintsev, K.K. 1961 Lake Baikal Hydrochemistry. *The Baikal Limnological Station Transactions*, Volume XX, 310 pp.

Votintsev, K.K. 1985 Substance cycles in Lake Baikal as major factor in formation of water quality. Substance and energy cycles in water bodies. The VI All-Union Limnological Conference, Abstracts, 4–6 September, 1985, Listvenichnoye-on-Baikal, Irkutsk, 22 pp.

Votintsev, K.K. 1992 Does eutrophication threaten Baikal? *Transactions of the USSR Academy of Science, Biology Series*, **4**, 616–627.

Votintsev, K.K., Glazunov, I.V. and Tolmacheva, A.A. 1965 *The Baikal Basin Hydrochemistry*. Nauka, Moscow, 492 pp.

Votintsev, K.K., Mescheryakova, A.I. and Popovskaya, G.I. 1975 *Organic Matter Cycle in Lake Baikal*. Nauka, Novosibirsk, 188 pp.

Chapter 19*

LAKE LADOGA

Lake Ladoga is the largest lake in Europe. It is one of the most northern of the world's largest lakes and it is situated between lat 59°54'N and lat 61°47'N (Figure 19.1 and General Appendix IV). The state of the lake ecosystem is the result of complex processes in the lake and its basin, determined by both natural and anthropogenic factors. Natural conditions have resulted in a high quality of water within the lake. However, since the early 1960s, signs of anthropogenic eutrophication have appeared.

19.1 Physico-geographical characteristics

The basin of Lake Ladoga is situated in the temperate, taiga zone, between lat 63–56°N and long 30–37°E. In the temperate zone, cyclone activity is the most important factor for atmospheric circulation processes. Contrasting air masses of different origins meet in this zone giving rise to unstable climatic conditions. Overall, air masses from the Atlantic Ocean prevail, although Arctic air influences are also frequent.

The surface area of Lake Ladoga is 17,680 km^2 which puts it tenth in terms of world lake size (Caspian Sea included). The maximum depth is 228 m. Depths exceeding 100 m are characteristic of the northern part, where sharp changes in depth occur over small distances. Towards the southern shore, the lake is less deep and the bottom is more even. The average depth of the lake is 51 m and the volume is 908 km^3.

The Lake Ladoga drainage basin covers 258,000 km^2 and comprises four secondary basins, Lake Ladoga's own basin (28,400 km^2), Onega-Svir basin (83,200 km^2), Ilmen-Volkhov basin (80,200 km^2) and the Saima-Vuoksa basin located partly in Finland (66,700 km^2) (Figure 19.1). Any changes over this vast territory inevitably influence Lake Ladoga.

Annual water inputs, originating from rivers and streams, are 71.13 km^3 and direct precipitation onto the lake surface is 11.6 km^3 (656 mm a^{-1}). The main river inflow comes from the Rivers Svir (34 per cent), Vuoksa (27 per cent) and Volkhov (23 per cent) (Kirillova, 1984). Groundwater inflow is insignificant and is therefore not taken into account. Annual outputs include evaporation from the lake surface (7.2 km^3 or 405 mm) and the lake outlet, the River Neva (75.7 km^3),

---

*   *This chapter was prepared by V.G. Drabkova and V.A. Rumyantsev*

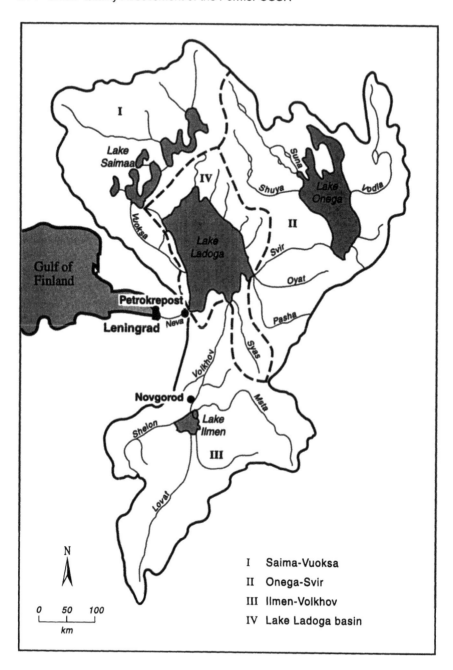

Figure 19.1   Lake Ladoga basin

which flows through St Petersburg (formerly called Leningrad) to the Baltic Sea. The water renewal rate is 0.08 times per year, i.e. an average water residence time of 12.5 years, which is a relatively short time for such an extensive water body.

As Lake Ladoga is situated at a more northern latitude than many other great lakes, the total annual sun radiation in the basin is not high, i.e. 300.6–334.1 kJ cm^{-2}. This radiation level determines to a large extent the thermal regime of the lake; the lake freezes each winter. The deep waters of the lake are characterised by constant, low temperatures around 4 °C, and which extend to the whole lake volume in winter. Lake Ladoga is characterised by lateral thermal heterogeneity. In spring, littoral areas warm first, forming a thermal bar at their border with the cold central part of the lake. This thermal bar gradually gets mixed further away from the shore. The difference in temperature on either side of the thermal bar may reach 20 °C. By mid-July, the whole surface of the lake has warmed up and in August its temperature reaches 16 °C (Tikhomirov, 1982) and the lake develops classical vertical stratification. However, the complex morphology of the lake bottom leads to uneven warming of the water mass.

The lake may be classified into four zones which have characteristic features: i) the littoral zone (depth <15 m, total area 3,700 km^2); ii) the sub-littoral zone (15–52 m, 5,300 km^2); iii) the profundal zone (52–89 m, 5,800 km^2); and iv) the ultra-profundal zone (> 89 m, 3,000 km^2) (Gusakov and Terzhevik, 1992) (Figure 19.2).

Waters of Lake Ladoga have low total dissolved solids (TDS) (average 62 mg l^{-1}) with bicarbonate and calcium as the major ions. The average transparency of the waters is relatively low (2.5 m), mostly due to naturally high dissolved organic carbon.

19.2   Water use

The Lake Ladoga basin is characterised by a high level of economic development, particularly for industrial production which is higher than the Russian average. This, in itself, makes the protection of this unique lake difficult. Moreover, the quality of drinking water for the 5 million inhabitants of St Petersburg (which is located on the mouth of the River Neva, the lake's outlet), depends on the environmental state of Lake Ladoga. Although, the natural qualities of Lake Ladoga and its fishery resources need to be preserved, the water resources of the lake are also intensively exploited for industry and agriculture. The lake receives wastewaters from 594 industrial and 680 agricultural enterprises. The proportion of processing industries and primary processing of raw materials in the lake basin is significant. Such industries have the greatest impact on natural resources, because they are characterised by high levels of waste and subsequent pollution. In the St Petersburg (Leningrad) region, heat and power, timber, pulp and paper, chemical and machinery industries are well developed (Table 19.1). In Karelia, timber

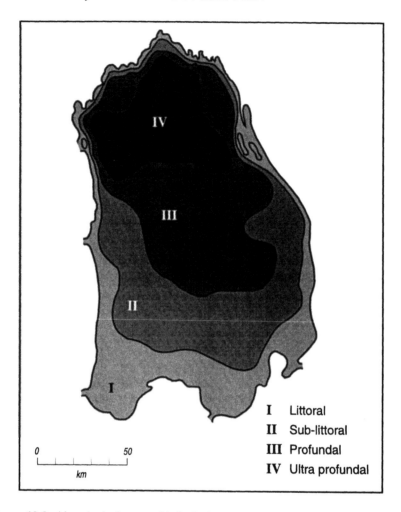

Figure 19.2   Limnological zones of Lake Ladoga

and wood-processing industries are particularly numerous and account for 42 per cent of industrial production.

The 13 largest enterprises of the timber and the pulp and paper industries are situated in the lake's basin, discharging more than 300 toxic compounds. The input of toxic organochlorine compounds alone into the lake is 2.5 t a^{-1} (Voropaeva and Rumyantsev, 1991). In addition to industry, agriculture is also rather intensive in the Lake Ladoga basin where it is mainly orientated towards cattle breeding in association with land melioration, peat extraction and peat use. The total volume of polluted wastewater flowing into Lake Ladoga is 400×10^6 m^3 a^{-1}, of which 228.4×10^6 m^3 is from the St Petersburg region.

Table 19.1   Industrial activities in the Leningrad region,1980 and 1985

Industry	Proportion of total industry (%)	
	1980	1985
Electric power	11.9	12.2
Fuel	18.8	16.3
Chemical, petrochemical	7.4	8.7
Ferrous, non-ferrous metallurgy	6.3	5.4
Machinery, metal-working	15.1	17.0
Timber, wood-processing, pulp and paper	13.2	14.7
Industrial construction materials	6.4	5.8
Light industry	7.9	6.9
Food industry	7.6	6.9
Other	5.4	6.1
Total	**100.0**	**100.0**

Source: Economy of Leningrad, 1986

In addition, Ladoga is important for navigation between the Volga and the Baltic Sea. The main traffic is between the Neva (the Lake Ladoga outlet) and the Svir (the Lake Onega outlet). The regions around navigable routes are highly polluted with oil products.

19.3   Pollution sources
Significant anthropogenic impact on the lake's ecosystem has led to both anthropogenic eutrophication and to pollution with toxic compounds. The basin of Lake Ladoga was populated in the late Stone Age and therefore anthropogenic influence on the lake has been ongoing for around 5,000 years (Dolukhanov and Timofeev, 1972). Analysis of the upper sediment (0–30 cm) of central areas of Lake Ladoga, which characterises an accumulation period of approximately 750 years, has shown that there were no significant changes in the chlorophyll content and, therefore, no changes in the saprobic level of the lake sediment (i.e. an oligosaprobic status occurred throughout the core). In littoral regions, the sediment layer below the top 5 cm contained 13–17 per cent xenosaprobic and oligosaprobic species (i.e. inhabitants of clear water), whereas in the upper 5 cm layer the proportion was only 1–4 per cent. These results demonstrate that the process of eutrophication, under anthropogenic influence, has occurred at different rates in deep water and littoral regions of the lake, and is more clearly seen in the latter (Davidova and Trifonova, 1982).

The chemical composition of water in Lake Ladoga results mostly from river discharges, which account for 87 per cent of the total water input. Pollutants such as phenols, petroleum products, detergents and

Table 19.2   Annual average discharge of substances from the main rivers flowing into Lake Ladoga, 1991 and 1992

Substance	1991	1992
Suspended matter (10^3 t a^{-1})	366.0	335.1
NH$_4$-N (t a^{-1})	626.0	806.4
NO$_2$-N (t a^{-1})	438.2	438.2
NO$_3$-N (t a^{-1})	25,168.4	13,749.6
Inorganic phosphorus (t a^{-1})	1,330.6	2,279.5
Phenols (t a^{-1})	132.3	242.3
Detergents (t a^{-1})	985.9	958.0
Petroleum products (t a^{-1})	3,488.6	3,707.1
Copper (t a^{-1})	323.0	244.9
Lead (t a^{-1})	281.4	302.8
Manganese (t a^{-1})	2,812.4	1,681.9
Cadmium (t a^{-1})	32.6	30.0

Average water discharge is 71 km^3 Source: Ecological Situation, 1993

heavy metals originate mainly from these sources (Table 19.2). Substances causing eutrophication and pollution originate both from point and diffuse sources. Industrial wastewater and domestic sewage, as well as sewage from livestock farms, are generally collected and controlled discharges (point sources). The major impact of agriculture on water bodies results from water runoff (diffuse sources), including runoff from livestock farms, uncollected sewage and losses of fertiliser during storage and transportation.

To evaluate the level of anthropogenic eutrophication of the lake, it is necessary to define sources and rates of nutrient inflow (primarily phosphorus) into the lake ecosystem (Table 19.3) (see also Chapter 6). Cattle breeding is the main source of phosphorus to the lake. Losses of mineral fertiliser are also significant. The leaching of phosphorus from agricultural lands and precipitation constitutes only a minor part of the total inflow of phosphorus to the lake.

In addition to the inflow of nutrients, which cause anthropogenic eutrophication, another serious problem is toxic pollution. The latter may lead to unfavourable socio-economic and ecological consequences, which impose significant restrictions on economic activity in the region. The most dangerous classes of hazardous compounds are organochlorine compounds, polycyclic aromatic hydrocarbons (PAHs), heavy metals (Hg, Cd, Pb) and radionuclides. In Lake Ladoga, organochlorine compounds pose the greatest danger (Ikonnikov, 1990). They originate mostly from the use of more than 80 types of pesticide in the lake's basin, including strong organochlorine compounds such as γ-HCH (lindane)

Table 19.3 Phosphorus inputs into Lake Ladoga, 1991–92

Source	Load (t a^{-1})
Point sources	
Municipal sewage	536–958
Industrial wastewater	310–1,070
Livestock farms	2,742–2,759
Total inflow	3,588–4,787
Diffuse sources	
Livestock farms	2,389
Mineral fertiliser losses	411
Leaching from fields	120
Leaching from meadows	98
Untreated municipal sewage waters	222
Total inflow	3,240
Natural sources	
Precipitation	890
Leaching from forests	226
Total inflow	1,116
Total	**7,944–9,147**

Source: Russian-Finnish Joint Commission on Environmental Protection, 1993

and endosulfan. Average pesticide loads are comparatively high in the basin at 4.3 kg ha^{-1} (1.6 kg ha^{-1} for the most toxic compounds) and in some regions these loads reach 20.3 kg ha^{-1} (5.4 kg ha^{-1} for the most toxic compounds). In spite of the large numbers of pesticides used in agriculture, the monitoring of their concentration in the lake water is highly inadequate. Only α- and γ-HCH concentrations are measured regularly. Lindane is widely used in large quantities in the basin of Lake Ladoga and leads to an excess of allowable concentrations of α- and γ-HCH in the entire lake (Figure 19.3).

There are regions in Lake Ladoga with high levels of metals in the water, although prevailing concentrations are 1–3 times lower than maximum allowable concentrations (MACs). Nevertheless, levels exceeding the MACs at some sampling stations, both in the lake and in its tributaries, cause serious concern (Table 19.4).

19.4   Structural changes in the lake ecosystem

Significant anthropogenic impacts on Lake Ladoga have resulted in serious damage to the lake's natural ecosystem. In the 1970s, significant anthropogenic eutrophication started due to the massive inputs of phosphorus. In the last decade, the problem of toxic pollution has also became urgent.

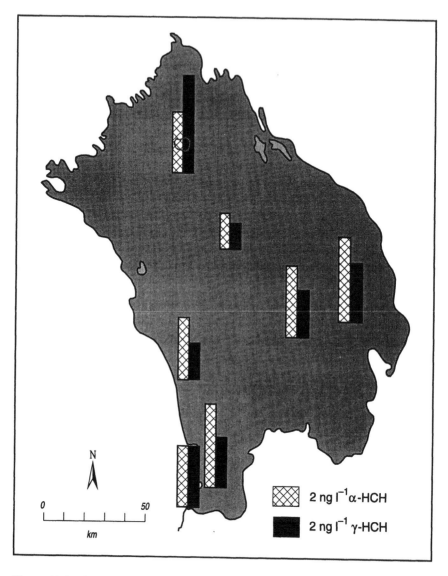

Figure 19.3   Concentration of α- and γ-isomer HCH in the water of Lake Ladoga, 1985

19.4.1 Anthropogenic eutrophication and its effect on biota

Phosphorus levels
The period 1976–83 is characterised by significant changes and restructuring in the lake's ecosystem resulting from anthropogenic eutrophication due to increasing nutrient inputs. From 1959 to 1962, the average phosphorus load per unit lake area was 0.14 g m^{-2} a^{-1}, whereas

Table 19.4 Exceedances of maximum allowable concentrations of some heavy metals and organochlorine compounds in Lake Ladoga, its tributaries and in the River Neva

Area	Percentage of samples exceeding MAC (%)							
	Cu	Pb	Mn	Cd	Hg	α-HCH	γ-HCH	DDT
Lake Ladoga								
Surface	25.6	0	17.1	1.8	1.3	97.7	95.8	4.7
Middle	27.0	0	15.0	2.8	1.5	75.0	87.5	25.0
Bottom	26.4	0	9.1	2.1	0	90.0	90.0	33.0
Rivers								
Vuoksa	44.4	0	30.7	0.7	–	99.1	100.0	29.3
Svir	46.8	0	20.1	0	–	100.0	100.0	0
Pasha	62.3	0	31.4	0	–	100.0	83.3	41.2
Syas	44.1	0	30.5	0	–	94.2	92.0	18.1
Volkhov	47.6	0	37.2	6.0	0	97.5	88.4	31.2
Neva								
Upstream	51.8	3.0	40.1	3.1	0	97.1	91.5	24.0
Leningrad (St Petersburg)	55.3	0	41.8	4.2	0	97.5	87.4	26.1

MAC   Maximum allowable concentration
HCH   Hexachlorocyclohexane
DDT   Dichlorodiphenyltrichloroethane

Source: Data of the north-western Hydrometeorological District, 1986–1987. See Ikonnikov, 1990

in 1976–79 it reached 0.39 g m^{-2} a^{-1}. As a result, annual total phosphorus increased from 10 mg l^{-1} in 1959–62 to 27 mg l^{-1} in 1976–79 (Raspletina and Gusakov, 1982). This sharp increase in phosphorus input to Lake Ladoga in the early 1960s resulted from major changes in the raw material used by the Volkhov aluminium plant. Apatite-nepheline ore, rich in phosphorus, was used and phosphorus-rich wastewaters were subsequently discharged into the River Volkhov. The phosphorus inputs into the lake exceeded the permissible level and sometimes even the critical level (Figure 19.4). Since 1983–84, phosphorus levels have stabilised in the lake, but at a lower level than in the 1970s. The concentration of total phosphorus in the lake has decreased to 21–22 µg l^{-1}.

As a result of the overall increase in nutrient loads, there have been significant structural and physiological changes in the biological communities of the lake.

Phytoplankton

During the period of intensive anthropogenic eutrophication of the lake (1976–83) some species characteristic of eutrophic conditions appeared in addition to phytoplankton species typical of oligotrophic conditions.

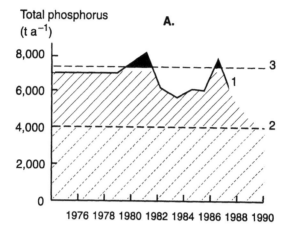

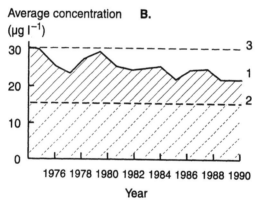

Phosphorus levels: 1 Actual 2 Permissible 3 Critical

Figure 19.4 A. Input of total phosphorus to Lake Ladoga; **B.** Average phosphorus concentration in lake water (After Petrova *et al.*, 1992b)

This was most clearly observed in the summer, where blue-green algae prevailed and diatoms were almost completely absent. A new phase was noted after 1984 when there was a noticeable displacement of algal species characteristic of eutrophic lakes by diatoms characteristic of oligotrophic conditions. By 1989, the summer composition of dominant phytoplankton species became practically identical to those of the oligo-trophic period of 1956–62 (Petrova *et al.*, 1992a).

The period of intensive anthropogenic eutrophication of the lake was characterised by maximum concentrations of chlorophyll *a* (average annual concentrations reached 2.6–2.8 µg l^{-1} in 1976–83 compared with 1.6–2.0 µg l^{-1} in 1984–89). During this period, the greatest seasonal variations in chlorophyll concentrations occurred, with the maximum

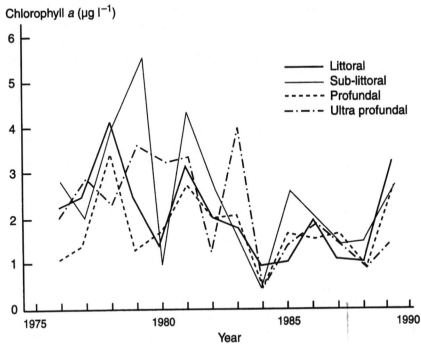

Figure 19.5 Average annual concentration of chlorophyll a in plankton in the littoral, sub-littoral, profundal and ultra-profundal zones (After Petrova *et al.*, 1992a)

concentration of chlorophyll *a* exceeding the average concentration by 15–17 times, whereas after 1984 it was exceeded by only 7–9 times. At the initial stages of anthropogenic eutrophication, higher chlorophyll *a* values were recorded in littoral and especially sub-littoral zones, but when the plankton of the oligotrophic stage became dominant, the differences diminished (Figure 19.5).

Bacteria
The quantitative characteristics of bacteria in different parts of the lake vary, from the lowest in mesotrophic lake sections (i.e. in profundal and ultra-profundal zones) to higher values which are characteristic of eutrophic water bodies (in southern parts of the lake and regions affected by pulp and paper industries and agriculture) (Kapustina, 1992). The bacterioplankton population has been stable in recent years. The average number of bacteria in the lake epilimnion increased during the period 1977–82 from 0.4 to 0.9–1.0 × 10^6 cells per ml. Since 1982, levels have been stable at 0.7–0.86 × 10^6 cells per ml. In the hypolimnion, bacterial numbers kept increasing up until 1985. Despite this, in recent years, there has been a tendency for increasing rates of carbon dioxide fixation in the deep waters of the lake.

In addition to the high total bacteria counts, large quantities of pathogenic bacteria have been recorded in polluted littoral areas near the towns of Priozersk, Pitkyaranta and Petrokrepost, especially in sediments. Thus, the concentrations of faecal coliforms and streptoccoci reach tens of thousands per kilogram wet sediment (Selyuzhitsky et al., 1990).

Invertebrates
The composition of zooplankton and benthic communities has changed little over the last 20–30 years. Quantitative values of zooplankton vary in different parts of the lake and range from values characteristic of mesotrophic lakes to high values typical of eutrophic water bodies. The average zooplankton biomass in a water column (0–50 m) in August 1983 was 6.1–19.4 g m^{-2}, while in August 1948 these values ranged from 3.4 to 10.1 g m^{-2} (Smirnova, 1987). Maximum values were registered in 1978 at 13.8–37.6 g m^{-2}.

The benthic fauna in the ultra-profundal and profundal zones has either changed very little or not at all in the last 10 years. The average number of macroinvertebrates in the profundal zone in 1975–79 was 507 species per m^2 with a biomass of 1.2 g m^{-2}. In 1986, these values were 240–780 species per m^2 and 0.5–2.2 g m^{-2}. The ultra-profundal zone is less productive and, up until the present time, the level of development of benthic fauna in this zone was similar to that of ultra-oligotrophic water bodies. In littoral and sub-littoral zones, an increase in number and biomass of benthos has been noted. In littoral zones of southern bays of Ladoga in 1975–86, the number of macroinvertebrates increased from 620–1,360 species per m^2 to 1,276–4,136 species per m^2 and biomass increased from 2.4–5.3 g m^{-2} to 1.4–8.2 g m^{-2}. In the sub-littoral zone, the numbers and the biomass of the benthos reached maximum values. At sampling stations where large accumulations of Amphipoda have occurred, the number of organisms present reached 6,000 species per m^2 and the benthic biomass was over 20 g m^{-2}. In 1975–79, these values were 1,520 species per m^2 and 4.7 g m^{-2} respectively (Slepukhina, 1992).

Fish
Intensive economic activity in the Lake Ladoga basin has had a negative impact on fisheries. In the last decade, stocks of whitefish (*Coregonus lavaretus baeri* Kessler), salmon (*Salmo salar* L. m. *sebago* Girard) and trout (*Salmo trutta* L. m. *lacustris*) have been greatly reduced. In the 1930s, salmon catches reached 160 t a^{-1} and trout catches reached 40 t a^{-1}. At present, Ladoga trout are not marketable and salmon catches are only 1–5 t a^{-1}. In the 1970s, the whitefish catch in the lake reached 600 t a^{-1} whereas in 1986–90, the catch was only 300 t a^{-1} or less (Kudersky, 1991). Changes in catches have also occurred for the most marketable fish in Lake Ladoga, the pike-perch (*Strizostedion lucioperca* (L.)). From 1971, catches of pike-perch increased reaching over 1,100–1,200 t a^{-1} in

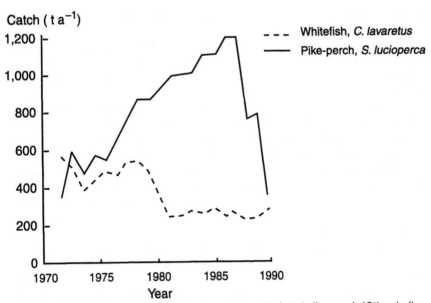

Figure 19.6 Catches of whitefish (*Coregonus lavaretus*) and pike-perch (*Stizostedion lucioperca*) in Lake Ladoga, 1971–90 (After Kudersky, 1991)

1984–87. However, after 1987, pike-perch catches decreased significantly. In 1990, the catch was only 388 t, 3.1 times less than in 1987 (Figure 19.6). At present, overall fish catches in the lake reach 5,000–6,000 t a^{-1}.

19.4.2 Response of the lake to toxic compounds

In parallel with eutrophication, there are various pollutants which lead to an increase in the saprobic level and toxicity of Lake Ladoga. This is most evident in the bays of the lake which are exposed the most to pollution. Water and sediments, taken from different regions of the lake, are characterised by significant concentrations of heavy metals, the presence of high molecular organic furane and terpene derivatives and polychlorinated biphenyls (PCBs). The presence of PCBs, up to five times the MAC, shows that areas near Priozersk, Pitkyaranta and Petrokrepost can be considered as zones of high health risk (Selyuzhitsky *et al.*, 1990).

All these factors affect the biological communities of the lake. For example, zooplankton data from the 1940s to the 1980s show that the species composition has not changed significantly. However, there have been significant structural changes with the small-size fraction (Rotifera) becoming dominant particularly near the Petrokrepost and Volkhov bays. 'Dead zones' and vast polysaprobic areas have also been formed on the lake bed. Under the influence of toxic pollutants, the aquatic biota, especially chironomid larvae, show serious morphological changes. Pathological signs in the zooplankton community near the town of Pitkyaranta have also been recorded (Andronikova, 1991).

In connection with the pollution levels of Lake Ladoga, many fish species have marked toxicosis. In the Volkhov bay, 70–80 per cent of whitefish, pike-perch, bream (*Abramis brama* (L.)), roach (*Rutilus rutilus* (L.)) and ruffe (*Gymnocephalus cernua* (L.)) have severe toxicosis. In this bay, fish tissue (in 20–60 per cent of fish sampled) had an odour of petroleum products. In the Svir bay, 50–60 per cent of fish have toxicosis and clinical evidence of toxicosis was observed for 30–60 per cent of fish from the mouth of the River Vidlits (Arshanitsa, 1988). These fish showed irreversible pathological changes in most vital organs including oedema and even disintegration of cardiac muscle, brain hyperaemia, pathological changes in liver and kidneys and tumours in different organs. High mortality rates and interruptions in the development of juvenile fish also occur.

Water quality deterioration also affects human health. Epidemiological research has shown a high level of sickness and death rates in some regions of Lake Ladoga due to diseases directly related with water such as gastrointestinal and genitourinary diseases. These are mostly noted in regions where drinking water is contaminated by pulp and paper mill effluents. During 25 years of monitoring in these regions, the proportion of the above mentioned diseases was constantly increasing and was higher than in a control population in the town of Vyborg, outside the Ladoga basin. For example, gastric cancer was 1.5 times higher in the Ladoga region than in the control region. There was a significant increase in gastrointestinal diseases from 1960 to 1985. In areas of pulp and paper mills, the frequency of these diseases increased sevenfold whereas the increase in the control was only twofold during this period.

19.5   Management of Lake Ladoga

Due to its unique ecosystem, Lake Ladoga is of national heritage value. It has vast resources of freshwater and it is also of strategic importance as a source of drinking water for St Petersburg. The critical ecological situation is due to the existing management practice and to the policy of economic development in this region. The situation can be stabilised and improved only by implementing major changes in regional development, restructuring water consumption patterns and setting new aims and priorities (Voropaeva and Rumyantsev, 1991).

At present, the problem of restoring the lake ecosystem is acute, especially in certain bays and littoral zones. This demands urgent measures such as the significant reduction in toxic waste discharges directly to the lake or through groundwaters. Measures should be taken to reduce the influence of atmospheric pollution, to set higher standards for protecting forests and marshes and to set legislative and economic criteria to control water usage.

Measures which have been taken in the Lake Ladoga basin have already produced positive results. In 1980–83, measures were taken to reduce phosphorus concentrations in wastewaters of the Volkhov aluminium plant, and these led to a 2,000–3,000 t a^{-1} decrease in phosphorus input to the lake. After closing down the Priozersk pulp and paper mill in 1986, the benthic fauna rapidly recovered in a region previously marked as a 'dead zone' and the concentration of low molecular organic compounds in water and sediment was also reduced. However, the concentration of high molecular pollutants, especially PCB, is still higher than the MAC. These examples show that Lake Ladoga can be saved, although its restoration and recovery may be slow. If the appropriate actions are not implemented, further pollution would exceed the stable limits of current pollution loads and the consequences for the lake ecosystem would be irreversible.

19.6   References

(All references are in Russian unless otherwise stated).

Andronikova, I.N. 1991 Structural-functional indices of zooplankton in the assessment of toxic pollution levels in lake ecosystems. In: *Methodology of Ecological Standardisation.* VNIIVO, Kharkov, 111–113.

Arshanitsa, N.M. 1988 Ichthyo-toxicological research in the Lake Ladoga basin. GOSNIORH Transactions, **285**, 12–23.

Davidova, N.N. and Trifonova, I.S. 1982 Changes in diatom structures and chlorophyll *a* concentrations in sediments of different areas of the lake. In: *Anthropogenic Eutrophication of Lake Ladoga.* Nauka, Leningrad, 202–206.

Dolukhanov, P.M. and Timofeev, V.I. 1972 Chronology of the neolithic. In: *Problems of Absolute Dates in Archaeology.* Nauka, Moscow, 28–75.

Ecological Situation in the Leningrad Region in 1992 (Analytical Review) 1993 St Petersburg, 270 pp.

Economy of Leningrad and the Leningrad Region in the XI five-year period 1986 Leningrad, 230 pp.

Gusakov, B.L. and Terzhevik, A.J. 1992 Limnological zones and their features in lake processes. In: *Lake Ladoga: State of Ecosystem Criteria.* Nauka, Leningrad, 21–26.

Ikonnikov, V.V. 1990 The problem of Lake Ladoga toxic pollution. In: *Ways for Improving the Use of Natural Resources in Basins of Large Lakes.* Nauka, Leningrad, 34–53.

Kapustina, L.L. 1992 Trends in spatial and temporal distribution and functional characteristics of bacterioplankton. In: *Lake Ladoga: State of Ecosystem Criteria.* Nauka, Leningrad, 146–179.

Kirillova, V.A. 1984 Lake Ladoga: hydrological regime. In: *Natural Resources of the USSR's Large Lakes and Their Possible Changes*. Nauka, Leningrad, 15–19.

Kudersky, L.A. 1991 Fish in danger: some consequences of economic activity on inland water bodies. GOSNIORH Transactions, **310**, 45–64.

Petrova, N.A., Antonov, S.E. and Protopopova, E.V. 1992a Structural and functional phytoplankton characteristics. In: *Lake Ladoga: State of Ecosystem Criteria*. Nauka, Leningrad, 119–146.

Petrova, N.A., Raspletina, G.F., Tregubova, T.M., Kapustina, L.L., Iofina, I.V., Kulish, T.P. and Judin, E.A. 1992b Major stages in lake ecosystem changes under the influence of anthropogenic eutrophication. In: *Lake Ladoga: State of Ecosystem Criteria*. Nauka, Leningrad, 240–251.

Raspletina, G.F. and Gusakov, B.L. 1982 Application of direct and indirect methods for calculation of nutrient load and concentrations in Lake Ladoga. In: *Anthropogenic Eutrophication of Ladoga Lake*. Nauka, Leningrad, 222–242.

Russian-Finnish Joint Commission on Environmental Protection 1993 Working group for the protection of aquatic ecosystems against pollution. Russian Report, 78 pp.

Selyuzhitsky, G.V., Vorobeva, L.V., Ermolaeva-Makovskaya, A.P. and Chernova, G.I. 1990 Hygiene and toxicological assessment of water and bottom sediments in the Lake Ladoga basin. In: *Ecological Situation in the North-west Part of Lake Ladoga*. Abstracts, Leningrad, 20–21.

Slepukhina, T.D. 1992 Macroinvertebrate trends in different lake zones. In: *Lake Ladoga: State of Ecosystem Criteria*. Nauka, Leningrad, 214–218.

Smirnova, T.S. 1987 Zooplankton. In: *Contemporary State of Lake Ladoga*. Nauka, Leningrad, 119–126.

Tikhomirov, A.I. 1982 *Thermal Characteristics of Large Lakes*. Nauka, Leningrad, 232 pp.

Voropaeva, G.M. and Rumyantsev, V.A. 1991 *Concept of Improving the Use of the Natural Resources in the Lake Ladoga Basin*. INOS, Leningrad, 27 pp.

Chapter 20[*]

WATER RESOURCES OF THE MOSCOW REGION

The Moscow Region (47,000 km^2) is divided into three main zones. The Moscow agglomeration itself, including the city of Moscow and its suburbs, is 13,300 km^2 and is inhabited by approximately 12 million people. Sixty-six per cent of the population lives in the city, within the Moscow ring motorway (MRM), which occupies 7 per cent of the agglomeration area, i.e. 878 km^2. The city is surrounded by a forest and park protection belt (FPPB) which is 1,700 km^2. The remaining part of the region beyond the FPPB is the outer belt of the suburban zone (OBSZ) (Figure 20.1 and General Appendix IV).

The proportion of developed land within the Moscow agglomeration, decreases from its core to its periphery and ranges from 60 to 20 per cent respectively. The proportion of land in the three Moscow zones used for municipal and rural economic activity is similar (60 per cent for the city, 70 per cent for FPPB, and 60 per cent for OBSZ). The proportion of agricultural lands in comparison with urbanised land increases towards the outer borders of the agglomeration (Table 20.1). Surface water covers 5 per cent of the total area of the agglomeration.

The main types of landscapes within the Moscow agglomeration are urban, agricultural and forest. These make up the main topography of the Moskva and Klyazma river basins. The basins of the Moskva river's smaller tributaries (after the confluence of the River Istra), the northern slope of the Pakhra river basin and the basin of the River Klyazma downstream of the Uchinsk reservoir, are mostly industrially developed.

Central Moscow, within the 'Garden Ring' road (an area with an approximate diameter of 6 km), is a highly urbanised area. This city area is characterised by maximum housing density and roads and pavements are practically all paved with asphalt. In connection with this, the impervious surface area is high, reaching up to 80 per cent in some places (Chernogaeva, 1976; Lvovich and Chernogaeva, 1977).

20.1 Hydrological features of the region

During urbanisation, a new anthropogenic environment is created which significantly influences hydrological processes (see Chapter 3). This influence is determined by specific features of the natural landscape and

*   *This chapter was prepared by G.M. Chernogaeva*

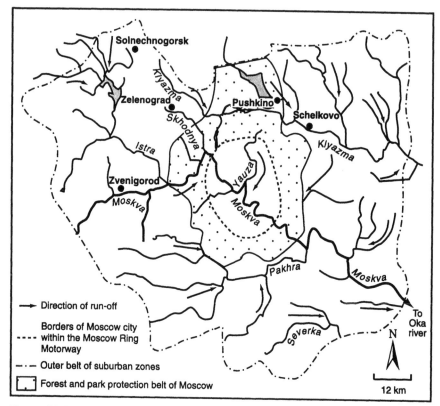

Figure 20.1   Scheme of the Moscow agglomeration (After Hydrolithosphere Models, 1991)

Table 20.1   Land use in the Moscow agglomeration

	Moscow		FPPB		OBSZ	
Land type	(km²)	(%)	(km²)	(%)	(km²)	(%)
Built-up land[1]	531	60	550	32	2,166	20
Open areas[2]	347	40	1,152	68	8,664	80
Total	878	100	1,702	100	10,830	100

FPPB   Forest and park protection belt
OBSZ   Outer belt of suburban zone

[1]   Developed land, industrial land and transport areas
[2]   Forests, parks, agricultural land and water

Table 20.2   Water balance of rural and urban landscapes

Landscape	Proportion of total precipitation (%) lost to:			
	Evaporation	Surface runoff	Ground runoff	River discharge[1]
Forest (Moscow region)	80	8	12	20
Fields (Moscow region)	70	22	8	30
City of Moscow (in general)	57	36	7	43
Central Moscow (within the 'Garden Ring')	27	68	5	73

[1] River discharge is the sum of surface and ground runoff.

Source: Chernogaeva, 1982

the climatic zone in which the city is situated. The Moscow Region is situated in a forest zone of temperate continental climate. Between the city and surrounding regions, there are considerable deviations in climate indicators. For example, the average annual temperature is 1.5 °C higher in the city and the average annual wind speed is twice as low in the city than outside it (Gerburg-Geybovich, 1988).

Characteristics of the water balance structure formed under varying conditions in urban and rural landscapes of the Moscow Region are presented in Table 20.2. These values were monitored at the Moscow suburban water balance station near the town of Odintsovo. Other gauging stations are shown in Figure 20.2. Evidently, significant changes occur in the formation of surface water runoff. The increase in impervious areas has led to a considerable increase in surface water runoff which, on average, is approximately 250×10^6 m^3 a^{-1} for the Moscow city area. This is twice the surface runoff volume of natural water catchment areas in the Moscow Region (Chernogaeva, 1983).

Local surface water resources in the Moscow Region are formed by the river discharge of the basins of the Oka and Upper Volga (Figure 20.3). In the Moscow Region (47,000 km^2), 7.31 km^3 of river discharge are formed annually (Table 20.3). The inflow through the water diversions (or transfers) network from adjacent regions is 10.7 km^3 a^{-1}. Thus, the average long-term natural resources of river discharge in the region are approximately 18 km^3 a^{-1}, which is 1.5–2 times less than in adjacent northern regions such as Yaroslavl, Kostroma and Ivanovo. In these regions, water resources per capita are 10,000–20,000 m^3 a^{-1}. Provision of the inhabitants of Moscow Region with water formed locally is considerably lower and is 500 m^3 a^{-1} per capita (Water Resources of the USSR, 1987). This has necessitated the transfer of large volumes of river water from adjacent regions and the creation of water reservoir systems which totally regulate the natural river flow of the Moskva river (Figure 20.2).

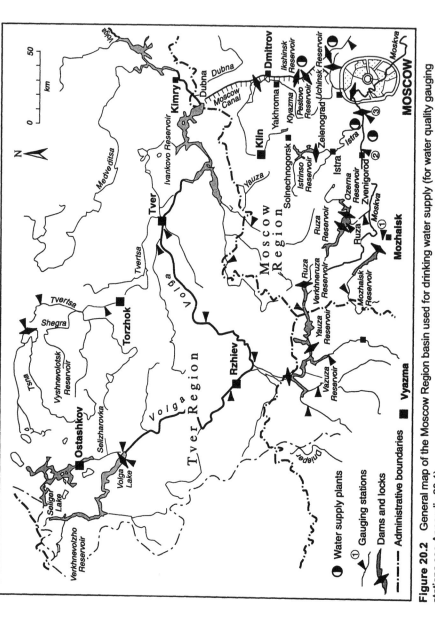

Figure 20.2 General map of the Moscow Region basin used for drinking water supply (for water quality gauging stations see Appendix 20.1)

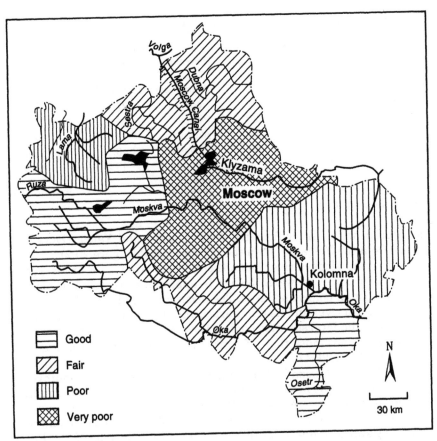

Figure 20.3 State of water resources in the Moscow Region. For a definition of criteria, see Table 20.6 (After Assessment of Surface Water Quality and Resources, 1989)

Table 20.3 Natural water balance of the Moscow region

| | Precipitation | Total river discharge | | Evaporation |
		Surface	Ground	
Input/output (km³ a⁻¹)	33.8	4.61	2.7	26.5
Input/output (mm a⁻¹)	720	98.1	57.4	564

Source: Subbotin and Dygalo, 1982

As a result, the discharge of the Moskva river has increased considerably. At present, the guaranteed minimum Moskva river discharge is 29 m³ s⁻¹ compared with a natural level of 9 m³ s⁻¹ in dry years. The river hydrological regime has also changed. Spring discharge has decreased by 1.5 times and winter discharge has increased more than three times in

Table 20.4 Water reservoirs in the Moscow region

Reservoir	Surface area (km²)	Total volume (10⁶ m³)	Reservoir	Surface area (km²)	Total volume (10⁶ m³)
Upper part of the Moskva river basin			*Water reservoirs of the Moscow Canal*		
Mozhaisk	31.0	240	Ivankovo	327.0	1,120
Ruza	32.7	220	Pestovo	4.9	54
Ozerna	23.1	144	Akulovo	19.3	14.5
Istra	33.6	183	Pyalovo	6.3	18.0
Rublyevo	3.1	–	Klyazma	15.2	86.5
			Khimki	3.5	29

Source: Samoilova, 1983

connection with winter flooding of the Moskva and Yauza rivers with water from the Volga river via the Moscow canal (Figure 20.2).

Within the Moskva river basin there are no large lakes or reservoirs. The total number of water bodies is approximately 1,115, consisting mainly of small ponds and lakes. The reservoirs regulating the discharge of the Moskva river and its tributaries, and the reservoirs of the Moscow Canal, are the largest water bodies. However, the area of most of these reservoirs is not more than 35 km², the exception being the Ivankovo reservoir (Table 20.4).

In 1978, the Vazuza hydrotechnical system was built to improve the water supply to the Moscow agglomeration. It is a complex construction, including two water reservoirs, the Vazuza and Yauza, with a useable volume of 550×10^6 m³, two canals with a total length of 23 km and three pumping stations (Reservoirs of the Moskva River System, 1985) (Figure 20.2). Thus, as mentioned above, the Moskva river within the city and its lower flow is completely regulated. The level of water in the city is controlled by two constantly active dams; the speed of water flow is approximately 0.1 m s⁻¹.

Water from the River Volga is diverted to the Moskva river through the Moscow Canal. This amounts to approximately 60 per cent of the total discharge of the Moskva river. Twenty per cent belongs to the discharge of the river itself and almost as much water is added by tributaries and surface runoff from the water catchment area. Downstream of the city, the average long-term discharge is approximately 112 m³ s⁻¹ (Kozlova *et al.*, 1982).

The volume of water discharge downstream of the city is determined mainly by the discharge of water through the Pererva hydrosystem and from the treatment installations of Kyryanovo and Lyublino aeration stations. In connection with this, the discharge is characterised by a high level of regulation, both in the long-term and in seasonal distribution. From 1985 to 1988, the average annual discharge was 110 m³ s⁻¹ in

Table 20.5   Monthly distribution of the Moskva River discharge, downstream of the city, 1985–88

	I	II	III	IV	V	VI	VII	VIII	IX	X	XI	XII
Discharge (m³ s⁻¹)	104	100	109	198	121	113	107	102	105	101	103	102
Discharge (%)	7.6	7.3	8.0	14.5	8.9	8.3	7.8	7.5	7.7	7.4	7.5	7.5

1985, 125 m^3 s^{-1} in 1986, 107 m^3 s^{-1} in 1987 and 114 m^3 s^{-1} in 1988. Monthly discharge is almost evenly distributed, except in April during the main flood period (Table 20.5). Under natural conditions, discharge in the region is distributed unevenly, with spring floods which usually last 2–3 months accounting for 60–70 per cent of the annual discharge in larger rivers and for 80–85 per cent in small rivers.

20.2   Water use in the region

At present, Moscow uses two independent sources of water supply: the basins of the Volga and Moskva rivers and groundwater. The proportion of groundwater in total water consumption is not large, being approximately 3 per cent. However, in the Moscow Region as a whole, groundwater mainly of Carboniferous rocks is used as a source of domestic and drinking water supply in most towns. According to State water statistics, the total use of groundwater in the region is approximately 3.8 × 10⁶ m^3 d^{-1}, of that 3.6 × 10⁶ m^3 d^{-1} is used by towns and villages, 0.25 × 10⁶ m^3 d^{-1} is used in mines and open pits and approximately 0.2 × 10⁶ m^3 d^{-1} is used in rural areas. The greatest groundwater use is in eastern, north-eastern and southern regions.

Intensive exploitation of groundwater aquifers has led to the disruption of natural interconnections of surface and groundwater, which in turn has caused land subsidence (about tens of metres deep) and the reduction of river discharges. In some cases, this reduction is also related to the constant outflow of water from river beds into groundwater aquifers (Assessment of Surface Water Resources and Quality, 1989).

The Moscow Region has been divided into four zones according to different water resources and the changes in aquifer recharge (Figure 20.3). The division is based on four categories reflecting the state of the water basins (Table 20.6).

The Moscow Region is characterised by significant use of surface and groundwater resources, urbanisation, hydrotechnical construction and population growth. In the whole region, there is a decrease in the total annual discharge by 5–25 per cent. For some rivers (e.g. Moskva, Ruza, Yauza) water discharge increases significantly due to the transfer of waters from adjacent regions (Ustyuzhanin, 1983). In the most urbanised zones, groundwater discharge has decreased 3–4 times in comparison with natural background levels (basin of the River Severka)

Table 20.6   Status of water resources in the Moscow region

Criteria	Good	Fair	Poor	Very poor
% of groundwater abstraction for use	< 20	20–30	30–70	> 70
% change in natural groundwater recharge	< 20	20–40	40–70	> 70 (upper aquifers) 100 (lower aquifers)
% decrease in minimal river discharges as a result of groundwater exploitation	< 15	15–20	50–100	–

and, in some cases, losses in river discharge have occurred. The intensity of losses increases nearer the areas of intensive groundwater exploitation (e.g. towns of Moscow, Podolsk, Domodedovo).

Changes in recharge and discharge of groundwater in the Moscow agglomeration are related to factors such as groundwater use and infiltration losses in the water supply network. Minimal water infiltration losses (up to 40–50 mm a^{-1}) are characteristic in areas of modern construction (Moscow Water Supply, 1983). Maximum losses are characteristic of industrial zones and areas of water distribution installations, sometimes reaching up to 500 mm a^{-1}. In forest and park zones and in regions of older construction and low population density, losses almost correspond to natural infiltration values.

The structure of water consumption in the Moscow Region and city is shown in Figure 20.4. Moscow city has the highest water consumption ($3,189 \times 10^6$ m^3 a^{-1}) and it occupies 20 per cent of the region. Approximately 60 per cent of water consumption is accounted for by municipal and drinking purposes. Within the Moscow Region (excluding Moscow city), about half of the water consumption is for industrial purposes and only 27 per cent is used for municipal and drinking purposes. In terms of industrial water consumption, the industrial potential of the city is almost equal to the industrial potential of the rest of the Moscow Region; this is reflected in the quality of surface and groundwater.

20.3   Quality of water resources

20.3.1 Surface waters

There are over 3,000 water-consuming enterprises in the Moscow Region. These are mainly municipal enterprises, and energy, aviation, chemical, automobile, machinery, textile and food industries. The wastewaters of all these are one of the main sources of surface water pollution (Zayets *et al.*, 1990). Another significant source of pollution by nutrients and pesticides is agriculture. The main types of agriculture in the region

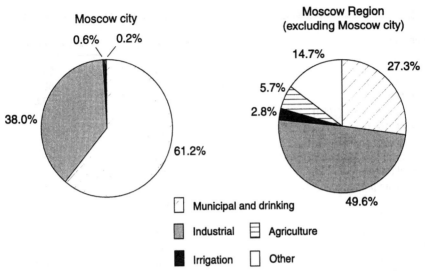

Figure 20.4   Water consumption in Moscow city and Region in 1993 (based on State statistical data)

are dairy livestock breeding, pig husbandry, and vegetable and potato growing. The untimely use of fertilisers and pesticides on agricultural lands, poor storage conditions and the location of practically all livestock farms near water are reasons for the high levels of pollution of surface water bodies and upper aquifers in the region (Yegorenkov, 1990).

Over half of all water consumers of the Moscow Region discharge their wastewaters into the Moskva river (towns of Moscow, Lytkarino, Zhukovsky, Lyubertsy, Ramenskoye, Voskresensk, Kolomna, etc.), into the Pakhra river (towns of Podolsk, Vidnoye, Domodedovo, etc.) and into the Klyazma river (towns of Schelkovo, Pavlovsky Posad, Noginsk, Electrostal, Losino-Petrovsky, Orekhovo-Zuyevo, etc.) and into their tributaries. The highest proportion of wastewaters of industrial and agricultural enterprises of the south enter the River Oka (towns of Serpukhov, Kashira, Stupino, Ozyery, Kolomna, etc.).

Monitoring data show that, for practically all water bodies of the region, standards are exceeded for most variables. Processes of self-purification in the Rivers Moskva (downstream of the city to its mouth), Pakhra (from the town of Podolsk to the river mouth), Klyazma (from the town of Schelkovo to Orekhovo-Zuyevo) are practically suppressed by the anthropogenic loads from wastewater discharges and by the inflow of significant amounts of pollutants (Chernogaeva, 1992; Chernogaeva *et al.*, 1994) (see also Appendix 20 Table I).

The recycled water supply within Moscow city is twice the volume of annual water intake from all water sources. However, in spite of the rather high and constantly increasing volume of recycled water, the

Table 20.7   Pollutants in wastewaters from Moscow city and the Moscow region, 1993

Pollutant	Moscow city	Moscow region
Chloride (10^3 t)	220	88
Sulphate (10^3 t)	138	88
NO_3-N (10^3 t)	23	3.5
Total phosphorus (10^3 t)	5	4
NO_2-N (10^3 t)	2	0.2
Suspended matter (10^3 t)	23	15
Organic substances (BOD_5) (10^3 t)	23	10
NH_4-N (10^3 t)	18	5
Petroleum products (10^3 t)	2	0.5
Iron (t)	538	250
Copper (t)	415	14
Detergents (t)	399	181
Zinc (t)	180	22
Phenols (t)	25	0.7

BOD Biochemical oxygen demand   Source: Based on State statistics

volume of discharged wastewaters is large. Approximately 70 per cent of the wastewaters of Moscow city are polluted, untreated or inadequately treated (Surface Water Quality Yearbook, 1990). Approximately 1.73 km^3 a^{-1} of wastewaters are discharged into the water bodies of Moscow city, which is comparable with the Moskva river discharge within the city. There is a wide spectrum of contaminants in water. Characteristics of contaminants discharged into water bodies in the city and Moscow Region from point sources of wastewaters are shown in Table 20.7.

A major source of pollution of the Moskva river and its tributaries is surface runoff from the city formed by melting snow and rain water. The total surface runoff in Moscow city is 0.25 km^3 a^{-1} (Chernogaeva, 1983; Chernogaeva and Mikhlin, 1987a). The volume of wastewaters discharged into the Moskva river and its tributaries is 1.73 km^3 a^{-1}, i.e. seven times more than the surface runoff value. As a rule, the proportion of impervious surfaces increases from suburban districts to central districts. Therefore, urban surface runoff values in central districts increase (Figure 20.5).

Examples of the average distribution of some pollutants in the Moscow Region are shown in Figures 20.6 and 20.7. The maximum specific pollutant load is, as a rule, in central districts of the city, which is to a large extent correlated with the distribution of pollution sources (Figure 20.8). Besides these central districts, two zones occur in the north and south-east of the city with higher than average pollutant loads. The northern zone is the main part of the water catchment of the Yauza river and the south-eastern zone is close to the Moskva river. As a

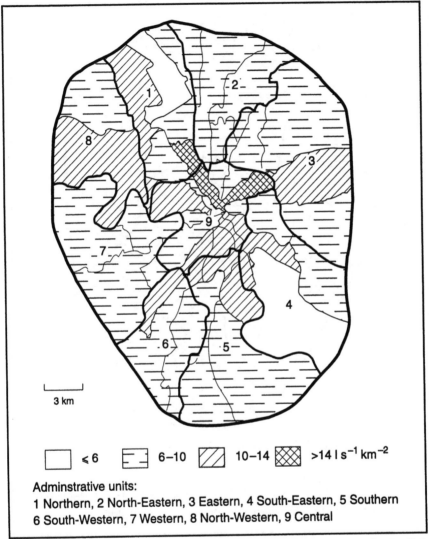

Figure 20.5   Schematic distribution of average annual urban surface runoff according to nine administrative units of Moscow city (After Chernogaeva and Mikhlin, 1987a)

result, in spite of the distribution of surface runoff as a source of pollution, the Moskva river pollution level increases greatly in the city centre due to the inflow of a mass of contaminants from central districts and with the confluence of the Yauza.

Most of the surface runoff from the city is not treated and directly enters water bodies. A study of surface runoff composition showed that it was greatly polluted with a number of substances (Petrov *et al.*, 1982). In

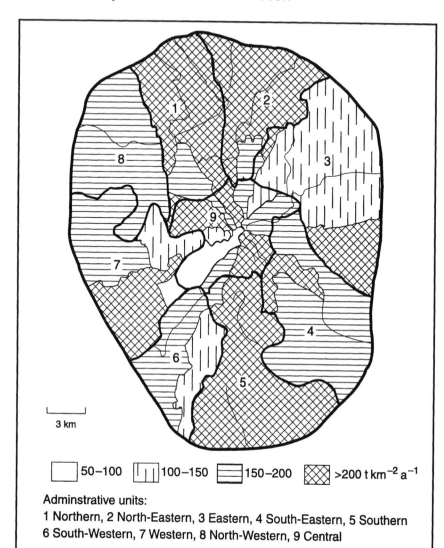

Figure 20.6 Schematic distribution of urban chloride output according to nine administrative units of Moscow city (After Chernogaeva and Mikhlin, 1987b)

summer, concentrations of petroleum products in surface runoff in city districts of modern municipal construction are 12 mg l^{-1}, in industrial zones 18 mg l^{-1} and in motorway runoff 20 mg l^{-1}. Corresponding figures for BOD_{20} are 40 mg l^{-1} O_2, 80 mg l^{-1} O_2 and 90 mg l^{-1} O_2 respectively. Assessments show that, in Moscow city, 3,840 t of petroleum products, 452,080 t of suspended substances, 173,280 t of chlorides and 18,460 t of organic substances (by BOD_{20}) are discharged into water bodies with

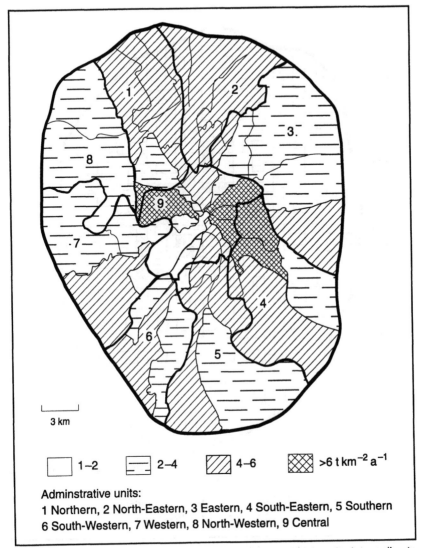

Figure 20.7 Schematic distribution of urban petroleum products output according to nine administrative units of Moscow city (After Chernogaeva and Mikhlin, 1987b)

surface runoff. As a result, city water bodies receive 1–8 times more petroleum products via surface runoff of than from the wastewaters of enterprises. For suspended substances almost 24 times more arises from surface runoff. The majority of the polluting substances occur in surface runoff in the winter–spring period. For petroleum products this proportion is 63 per cent, for suspended substances it is 75 per cent, for organic substances it is 64 per cent, and for chlorides it is 95 per cent. This occurs

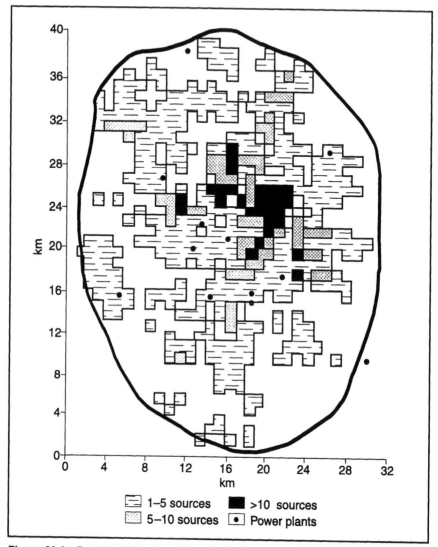

Figure 20.8 Density distribution of the principal industrial pollution sources per km^2 of Moscow city (After Bekker *et al.*, 1987)

at a time when the self-purifying capacity of the water has decreased significantly because of low water temperatures. In addition to the above mentioned substances, surface runoff has a significant concentration of heavy metals, ammonium nitrogen and other harmful substances.

The Moskva river is classified as a polluted water body, whereas downstream of the town it is classified as very polluted (Alexinskaya *et al.*, 1982; Kozlova, 1983). The level of river pollution increases along the

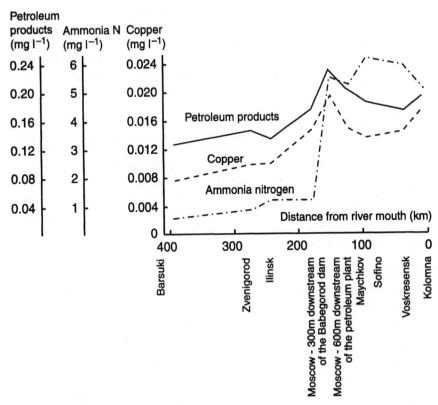

Figure 20.9   Changes in concentrations of pollutants along the course of the Moskva river (Distances are measured from the Moskva-Oka confluence)

course up to its confluence with the Oka river. There is a sharp increase in pollutant concentrations especially around the borders of the city of Moscow (Figure 20.9).

20.3.2 Groundwaters

Changes in the hydrodynamic structure of water flow in the Moscow artesian basin led to the deterioration in groundwater quality in Quaternary aquifers. This was the result of the inflow of pollutants from the aeration zone from deep aquifers with highly mineralised waters and also as a result of the infiltration of polluted surface waters (Zeegofer and Likhacheva, 1984; Yazvin, 1984, 1989; Bogoslovsky *et al.*, 1991). However, the main water abstraction in the region is from Carboniferous sedimentary rocks. For Carboniferous aquifers a straightforward hydrochemical zonation is characteristic, under which mineralisation of groundwater regularly increases with depth and their composition changes from bicarbonate calcium and magnesium to sulphate and sodium chloride (see also Chapter 7). The zone of fresh groundwater in the region, with mineralisation less than 1,000 mg l^{-1} on average, is at 100–150 m, increasing in central parts of the Moscow Region to 200–250 m. Groundwater in the upper Carboniferous aquifer under natural conditions, unaffected by anthropogenic activity, has a mineralisation of 300–400 mg l^{-1} and as a rule meets the requirements for drinking water quality.

At the same time, when aquifers deepen, changes in chemical composition occur related to the increase of gypsum and other specific minerals in water-containing rocks. This leads to the enrichment of groundwater with fluorine and sometimes with strontium, the content of which in some regions may exceed standards. The high concentration of fluorine, at 2 to 3 mg l^{-1}, is characteristic of middle and deep Carboniferous aquifers. Surface layers are characterised by higher iron concentrations (up to several mg l^{-1}), as a result of the discharge into these aquifers of iron-rich groundwater of Meso-Cenozoic sediments (Yazvin, 1984).

20.4 Problems related to drinking water supply

Water in the Moscow Region has distinctive features which affect overall water quality and quantity. These include:

- Limited water resources, which have led to the import of additional sources of water from supplies situated at a distance of 150–200 km from Moscow;
- A complex system of water use under regulated discharge conditions, where water bodies are simultaneously used for economic, drinking, recreational and domestic purposes, navigation, electric power, agricultural irrigation and as receptors of increasing quantities of inadequately treated waste waters;
- Stable pollution levels of water bodies, including sources of drinking water supply;
- Population growth and an increasing demand for recreational use of water bodies.

The water supply into the Moscow mains is by four pumping stations, of which Rublyevo and western stations use the Moskva river source and eastern and northern stations use the Volga river source. The length of the water pipe network was 8,041 km in 1989. The total distribution from the four stations is 2.43 km^3 a^{-1} (Figure 20.2).

An assessment of the state of surface water sources has shown progressive deterioration in their water quality (Sinelnikov, 1982). Over the last decade, there has been a noticeable deterioration in water quality based on microbiological indicators at the Moskva river source. This is also true for tributaries of the Moskva river such as Lusyanka, Ozerna, Teschenka and Istra, where bacterial pollution has increased by 10–100 times. The main reasons for this are agricultural enterprises situated near the water source (Novikov, 1987; Matveev, 1990). In the basin of the Moskva river water source, there are 56 agricultural enterprises, of which seven are poultry farms. On the whole, there are 240 farms with 86,000 cows and 50,000 calves, 33,000 pigs and 9 million hens. In addition, there are 250 wastewater treatment plants which discharge 150,000 $m^3 d^{-1}$ of wastewaters.

In the basin of the Volga water source, there are 19 agricultural enterprises, of which three are poultry farms. There are 75 farms with 22,000 cows, 10,000 calves, 2.5 million hens and 30,000 pigs. The amount of wastewaters discharged from 42 treatment plants is 24,000 $m^3 d^{-1}$. On the whole, 80 per cent of wastewaters from agricultural enterprises are not adequately treated because of overload and severe exploitation of the treatment facilities. The proportion of farms with adequate manure storage facilities is 35–40 per cent. Inadequate treatment of wastewaters from towns, villages and military units has a negative impact on water quality. Of 170,000 $m^3 d^{-1}$ of wastewaters discharged, 150,000 $m^3 d^{-1}$ are inadequately treated. Treatment plants in the towns of Mozhaisk, Istra and Dedovsk are overloaded by almost three times.

High levels of pollution of water sources used for economic and drinking water supply and the limited technical capacity of water supply facilities inevitably affects the quality of drinking water. According to laboratory data of Mosvodocanal, water quality at water abstraction sites of Moscow water supply stations does not meet the requirements for sources of drinking water supply (State Report, 1992).

As was noted above, Carboniferous aquifers are widely exploited for water supply purposes. At some water supply sites, the abstraction of groundwater exceeds the resources set for exploitation which leads to the unacceptable decrease in groundwater level, exhaustion of resources, pollution of groundwater and land subsidence causing damage to buildings and construction (Figure 20.10).

The most serious problem is the change in groundwater quality. At a number of water supply points in the Moscow Region, higher values of total dissolved solids and general water hardness occur, there is an increase in sulphate concentrations and petroleum products, phenols, some heavy metals and nitrogen compounds occur in the water. In most cases, pollution is sporadic. However, at water supply points situated directly in areas of towns and industrial and agricultural enterprises, there is constant pollution of groundwater. Nevertheless, pollution of groundwaters has a local character which cannot serve as a basis for the general reduction in groundwater use in the region.

Water consumption in Moscow and the Moscow Region, principally for drinking purposes, is constantly increasing. From the 1930s, efforts to curb the development of industry and population growth first in Moscow and then in the region as a whole have failed. Improving the industrial structure of the Moscow Region to decrease high water-consuming production, by introducing clean or low-waste technologies and water

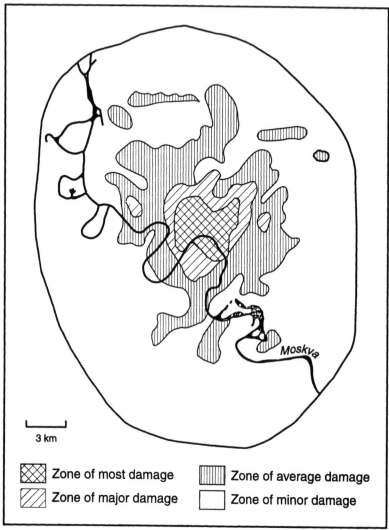

Figure 20.10   Damage to buildings and construction due to land subsidence in
Moscow city (After Hydrolithosphere Models, 1991)

recycling schemes has led to some reduction in industrial water
consumption (Table 20.8). However, for an overall reduction in water
consumption, it is necessary to create a unique joint water management
system for Moscow and the Moscow Region.

A reduction in water consumption by industry in 1970 in comparison
with 1960 can be explained by the introduction in 1964 of a limited
supply of water to enterprises consuming 100 m^3 d^{-1} of drinking water
(Table 20.8). A reduction in tap water consumption also occurred due to

Table 20.8   Major sources of water consumption in Moscow, 1940–89

Year	Water consumption (%)		
	Domestic	Municipal	Industrial
1940	35.3	18.1	46.6
1960	39.3	21.3	39.4
1970	54.8	15.7	29.5
1980	55.1	17.5	27.4
1985	56.8	18.2	25.0
1987	59.6	16.4	24.0
1988	59.3	17.2	23.5
1989	61.2	17.0	21.8

Source: Based on State statistics

the introduction in industry of water-recycling systems, and the construction of a separate system for industrial water supplies and the elimination of water loss during industrial construction and in housing.

The joint water management system will be created gradually, but the use of new sources for the Moscow Region may be necessary in the near future. At present, there are different projects for creating new sources of water supplies, including water supply from the Oka to southern districts of the region and to the south of Moscow. As future sources of water supply for the Moscow agglomeration in the north-east, north and north-west, water bodies such as the Uglich and especially the Rybinsk reservoirs are being considered. However, in addition to significant economic expense, the realisation of these projects is difficult because of the pollution of future water supply sources and the unstable water balance of the Oka.

The most rational option is for an increase in the groundwater proportion of total water use. Existing exploitable groundwater resources in the Moscow Region are adequate to satisfy the demand both in the Moscow Region and partially for the economic and drinking needs of Moscow city.

20.5   References

(All references are in Russian unless otherwise stated).

Alexinskaya, L.N., Vilensky, V.D. and Saet, J.E. 1982 Pollution zones in the basin of the Moskva river according to chemical elements. Proceedings of the II Moscow City Conference on Environmental Protection. Gidrometeoizdat, Moscow, 33–34.

Assessment of Surface Water Resources and Quality (Using the Moscow Region as an Example) 1989 Moscow University, Moscow, 196 pp.

Bekker, A.A., Reznichenko, T.I., Pavlov, A.A., Volkov, L.O. and Smekalova, E.A. 1987 Inventory of the Moscow city air pollution sources. In:

Moscow City and the Region's Environmental Protection. CVGMO, Moscow, 9–13.

Bogoslovsky, V.A., Orlov, M.S. and Zhigalin, A.D. 1991 Integrated geophysical and hydrogeological supply of treatment installations in connection with ground water pollution studies. Engineering Geology. Academy of the USSR, No. 4. Moscow, 69–73.

Chernogaeva, G.M. 1976 Water resources. In: *Moscow, the Capital of the Soviet Union.* Progress Publishers, Moscow, 49–54.

Chernogaeva, G.M. 1982 The influence of urbanisation on surface runoff quality in cities. In: *Geographical Aspects of Investigation and Use of Water Resources in the USSR.* Geograph. Soc., Moscow Branch, 11–17.

Chernogaeva, G.M. 1983 Assessment of pollution carried by surface runoff from the territory of Moscow. Proceedings of the Conference on Hydrological Investigations and the Water Economy in the Moskva River Basin. Geograph. Soc., Moscow Branch, 93–95.

Chernogaeva, G.M. 1992 Sources and causes of river pollution in the Moscow Region. Proceedings of the International Symposium on Ecoinformatic Problems. Russian Academy of Science, Zvenigorod, 169–172.

Chernogaeva, G.M. and Mikhlin, A.A. 1987a Assessment of surface runoff in Moscow. CVGMO, Moscow, 60–69.

Chernogaeva, G.M. and Mikhlin, A.A. 1987b Assessment of economic losses due to river water pollution in Moscow city. Proceedings of the XXIX All-Union Hydrochemical Conference 'State and prospects of the development of methodological principles of chemical and biological surface and groundwater monitoring'. Rostov-on-Don, 83–87.

Chernogaeva, G.M., Gorelova, L.P. and Shemyakina, N.K. 1994 Water quality in small rivers of the Moscow Region. In: *Small Rivers of Russia.* Geograph. Soc., Moscow Branch, 49–54.

Gerburg-Geybovich, A.M. 1988 Mesoscale temperature and wind zones in Moscow and its suburbs. *Transactions of the USSR Hydrometeorological Centre,* **233**, 107–110.

Hydrolithosphere Models of City Agglomerations (Using the Moscow Agglomeration as an Example) 1991 Nauka, Moscow, 47–68.

Kozlova, N.M. 1983 Trends in water chemical composition in lower reaches of the Moscow river and prospects for improving water quality. Proceedings of the Conference on Hydrological Research and Water Economy in the Moskva River Basin. Geograph. Soc., Moscow Branch, 27–30.

Kozlova, N.M., Khramova, E.E. and Babich, K.A. 1982 State and perspectives of the Moskva river water quality research within the borders of Moscow city. Proceedings of the II Moscow City Conference on Environmental Protection. Gidrometeoizdat, Moscow, 31–33.

Lvovich, M.I. and Chernogaeva, G.M. 1977 The water balance of Moscow. In: *Effect of Urbanisation on the Hydrological Regime and Water Quality.* AIHS Publication No. 23, 48–52.

Matveev, N.P. 1990 Small Rivers of the Moscow Region and Their Rational Use. In: *Moscow Regional Landscapes and Their Rational Utilisation.* Geograph. Soc., Moscow Branch, 90–118.

Moscow Water Supply 1983 Moskovskii Rabochii, Moscow, 140 pp.

Novikov, J.V. 1987 *For the City of Hygiene and Health.* Moskovskii Rabochii, Moscow, 109–160.

Petrov, N.A., Krisanov, O.P. and Artemiev, U.A. 1982 Pollution investigation and methods of surface runoff treatment in the city of Moscow. Proceedings of the II Moscow City Conference on Environmental Protection. Gidrometeoizdat, Moscow, 39–41.

Reservoirs of the Moskva River System 1985 In: *Integrated Reservoir Research.* Issue VI. Moscow, 266 pp.

Samoilova, Y.K. 1983 Hydrological observations in the Moskva river basin and Moscow canal. Proceedings of the Conference on Hydrological Research and Water Economy in the Moskva River Basin. Geograph. Soc., Moscow Branch, 6–11.

Sinelnikov, V.E. 1982 Analysis of the factors associated with eutrophication in the Moscow water reservoirs. Proceedings of the II Moscow City Conference on Environmental Protection. Gidrometeoizdat, Moscow, 42–44.

State Report 1992 *Moscow City, State of the Environment in 1992.* Moscow, 57–93.

Subbotin, A.I. and Dygalo, V.S. 1982 Long-term characteristics of the Moscow Region hydrometeorological regime. The Moscow water-balance station observations. CVGMO, Moscow, 320 pp.

Surface Water Quality Yearbook and the Effectiveness of Water Protection Actions (1990–1993) 1990 MosCGNS, Moscow, 99 pp.

Ustuzshanin, B.S. 1983 Influence of water management on the Moskva river discharge. Proceedings of the Confluence on Hydrological Research and Water Economy in the Moskva River Basin. Geograph. Soc., Moscow Branch, 62–64.

Water Resources of the USSR and Their Use 1987 Gidrometeoizdat, Leningrad, 302 pp.

Yazvin, L.S. 1984 'Contemporary state of groundwater use' and 'Protection of ground waters from exhaustion'. In: *Hydrogeological Principles of Groundwater Protection.* UNESCO/UNEP. CIP, 113–150.

Yazvin, L.S. 1989 Systematic principles of groundwater regime and resources management. In: *Geoecological Research in the USSR.* VSEGINGEO, Moscow, 59–68.

Yegorenkov, L.I. 1990 Moscow Region agricultural landscapes; research, rational use and protection problems. In: *Landscapes of the Moscow Region, Their Use and Protection.* Geograph. Soc., Moscow Branch, 52–68.

Zayets, E.S., Makhrova, A.G. and Pertsik, E.N. 1990 Moscow Region: basic problems of development. In: *Geographic Aspects of the Influence of*

Scientific and Technological Progresses on the Development of Productive Forces in the Moscow Agglomeration. Geograph. Soc., Moscow Branch, 8–17.

Zeegofer, Y.O. and Likhacheva, E.A. 1984 On the prognostic models of geological environment. Basin approach on the example of the Moscow agglomeration. In: *Natural and Environmental Trends in Moscow City and Moscow Region and Their Use in the National Economy.* Geograph. Soc., Moscow Branch, 41–61.

Appendix 20

Table I   Characteristics of Moskva river pollution, 1989–90

Appendix 20 Table I Characteristics of Moskva river pollution, 1989–90

Monitoring site details	Water quality variable	Station number	Concentration (mg l⁻¹)	
			Average	Maximum
Water body, location: Moskva river, Barsuki village Monitoring site: Category IV[1] Number of stations: 1 station Location of stations: Station 1: Hydrological station	Dissolved oxygen	1	10.5	7.78
	BOD$_5$	1	3.78	5.03
	COD	1	20.6	34.0
	Suspended matter	1	41.3	88.5
	Phenols	1	0.005	0.010
	Petroleum products	1	0.36	0.69
	NH$_4$-N	1	0.7	0.9
	NO$_2$-N	1	0.013	0.022
	NO$_3$-N	1	0.73	1.91
	Phosphorus	1	0.022	0.062
	Copper	1	0.010	0.016
Water body, location: Moskva river, Zvenigorod town Monitoring site: Category III[1] Number of stations: 2 stations Location of stations: Station 1: 0.3 km upstream of Zvenigorod and 0.2 km upstream of the River Storozhev inflow Station 2: 1.4 km below Zvenigorod and 1.4 km below industrial discharge	Dissolved oxygen	1	10.0	6.80
		2	9.60	6.89
	BOD$_5$	1	3.39	6.90
		2	3.73	5.41
	COD	1	17.4	28.1
		2	20.7	28.1
	Suspended matter	1	41.2	97.5
		2	50.4	106.5
	Phenols	1	0.004	0.008
		2	0.006	0.009
	Petroleum products	1	0.32	0.59
		2	0.32	0.66
	NH$_4$-N	1	0.6	1.0
		2	0.7	1.1
	NO$_2$-N	1	0.019	0.039
		2	0.051	0.162
	NO$_3$-N	1	0.48	1.14
		2	0.45	1.06
	Phosphorus	1	0.035	0.080
		2	0.034	0.081
	Copper	1	0.005	0.015
		2	0.004	0.011

Continued

Appendix 20 Table 1 Continued

Monitoring site details	Water quality variable	Station number	Concentration (mg l⁻¹) Average	Maximum
Water body, location: Moskva river, Moscow city	Dissolved oxygen	1	9.60	6.29
Monitoring site: Category I[1]		2	9.40	5.19
Number of stations: 3 stations		3	7.41	3.25
Location of stations: Station 1: 19.0 km upstream of Moscow city and 0.5 km upstream of water intake	BOD_5	1	3.26	9.63
		2	5.30	10.2
		3	6.47	9.85
Station 2: 0.1 km upstream of Ilinsk town and 0.3 km downstream of Babegorod dam, within Moscow city	COD	1	10.7	65.2
		2	26.3	60.8
		3	31.1	45.4
Station 3: 0.6 km downstream of oil factory and 0.01 km upstream of MRM highway bridge, within Moscow city	Suspended matter	1	39.7	103.0
		2	38.2	89.5
		3	48.2	104.0
	Phenols	1	0.006	0.018
		2	0.007	0.024
		3	0.003	0.021
	Petroleum products	1	0.28	0.80
		2	0.27	0.93
		3	0.37	1.23
	NH_4-N	1	0.8	1.9
		2	0.9	4.1
		3	4.7	7.6
	NO_2-N	1	0.162	0.324
		2	0.090	0.312
		3	0.204	0.406
	NO_3-N	1	0.68	4.58
		2	0.86	5.86
		3	1.82	4.69
	Phosphorus	1	0.051	0.129
		2	0.063	0.163
		3	0.232	0.823
	Copper	1	0.006	0.019
		2	0.009	0.027
		3	0.011	0.027
	Formaldehyde	1	0.113	0.164
		2	0.108	0.181
		3	0.110	0.150

Continued

Appendix 20 Table I Continued

Monitoring site details

Water body, location: Moskva river, New Maychkov village
Monitoring site: Category III[1]
Number of stations: 2 stations
Location of stations: Station 1: 0.1 km upstream of New Maychkov village and
1.5 km upstream of River Pakhra
Station 2: 0.1 km downstream of the inflow of the River Pekhorka

Water quality variable	Station number	Concentration (mg l^{-1})	
		Average	Maximum
Dissolved oxygen	1	7.96	6.10
	2	8.15	5.52
BOD$_5$	1	5.90	8.20
	2	5.31	9.10
COD	1	30.1	47.0
	2	34.2	55.4
Suspended matter	1	44.7	73.0
	2	50.8	105.0
Phenols	1	0.009	0.018
	2	0.010	0.012
Petroleum products	1	0.34	1.00
	2	0.35	0.93
NH$_4$-N	1	4.6	6.2
	2	5.4	10.6
NO$_2$-N	1	0.165	0.287
	2	0.215	0.404
NO$_3$-N	1	1.30	3.48
	2	1.41	2.73
Phosphorus	1	0.314	0.548
	2	0.452	0.729
Copper	1	0.012	0.016
	2	0.010	0.020
Formaldehyde	1	0.103	0.125
	2	0.110	0.167

Continued

Appendix 20 Table I Continued

Monitoring site details	Water quality variable	Station number	Concentration (mg l⁻¹) Average	Maximum
Water body, location: Moskva river, Sofino village	Dissolved oxygen	1	8.46	5.90
Monitoring site: Category III[1]		2	8.88	7.49
Number of stations: 2 stations	BOD₅	1	4.37	8.14
Location of stations: Station 1: 0.2 km upstream of Sofino village and 3.0 km		2	3.83	7.07
upstream of the inflow of the River Gzshelka	COD	1	31.8	38.3
Station 2: 1.0 km downstream of Sofino village and 0.3 km		2	30.5	41.9
downstream of the inflow of the River Gzshelka	Suspended matter	1	43.6	117.0
		2	30.9	68.0
	Phenols	1	0.010	0.014
		2	0.011	0.022
	Petroleum products	1	0.37	0.80
		2	0.34	0.76
	NH₄-N	1	5.5	7.4
		2	5.5	8.2
	NO₂-N	1	0.185	0.382
		2	0.202	0.397
	NO₃-N	1	1.58	2.52
		2	1.51	2.33
	Phosphorus	1	0.513	0.795
		2	0.475	0.719
	Copper	1	0.008	0.015
		2	0.010	0.016
	Formaldehyde	1	0.112	0.140
		2	0.129	0.132

Continued

Appendix 20 Table I Continued

Monitoring site details	Water quality variable	Station number	Concentration (mg l^{-1})	
			Average	Maximum
Water body, location: Moskva river, Voskresensk town	Dissolved oxygen	1	9.50	6.32
Monitoring site: Category II[1]		2	9.24	6.19
Number of stations: 2 stations	BOD$_5$	1	4.82	7.94
Location of stations: Station 1: 0.2 km upstream of Voskresensk city and		2	4.92	8.89
0.5 km downstream of highway bridge	COD	1	32.4	46.1
Station 2: 1.0 km downstream of Voskresensk city and		2	32.8	48.1
0.5 km downstream of cement factory discharge	Suspended matter	1	47.2	86.0
		2	49.6	77.5
	Phenols	1	0.010	0.021
		2	0.010	0.016
	Petroleum products	1	0.42	1.02
		2	0.40	0.73
	NH$_4$-N	1	2.7	7.0
		2	3.7	7.0
	NO$_2$-N	1	0.167	0.290
		2	0.246	0.835
	NO$_3$-N	1	1.40	2.44
		2	2.05	7.90
	Phosphorus	1	0.404	0.706
		2	0.456	0.696
	Copper	1	0.008	0.020
		2	0.008	0.017

Continued

Appendix 20 Table I Continued

Monitoring site details			

Water body, location: Moskva river, Kolomna city
Monitoring site: Category III[1]
Number of stations: 1 station
Location of stations: Station 1: 0.1 km upstream of the mouth of the River Moskva

Water quality variable	Station number	Concentration (mg l⁻¹) Average	Concentration (mg l⁻¹) Maximum
Dissolved oxygen	1	9.01	8.13
BOD$_5$	1	5.68	8.53
COD	1	31.7	45.6
Suspended matter	1	53.0	160.0
Phenols	1	0.009	0.019
Petroleum products	1	0.38	0.64
NH$_4$-N	1	4.8	7.4
NO$_2$-N	1	0.241	0.373
NO$_3$-N	1	2.09	4.38
Phosphorus	1	0.337	0.666
Copper	1	0.011	0.019
Formaldehyde	1	0.125	0.225

Source: Surface Water Quality Yearbook, 1990

[1] See Chapter 4 for classification of monitoring sites

BOD   Biochemical oxygen demand
COD   Chemical oxygen demand

GENERAL APPENDICES

Appendix I   List of pesticides monitored in freshwater bodies of the former USSR

Name used in the assessment	Other commercial or common names	Type	Chemical formula
Atrazine	–	Herbicide	$C_8H_{14}ClN_5$
Butiphos	Folex; S,S,S,tributyl-phosphorotrithioate	Defoliant	$C_{12}H_{27}OPS_3$
Carbophos; malathion	Maldison; mercaptothion; phosthion; sumitox	Insecticide, acaricide	$C_{10}H_{19}O_6PS_2$
2,4-D	–	Herbicide	$C_8H_6Cl_2O_3$
DDT	Chlorphenothane; dicophane; gesarol; guesarol; neocid	Insecticide	$C_{14}H_9Cl_5$
Dihydro-heptachlor	β-dihydro-heptachlor; dichlor	Insecticide	$C_{10}H_7Cl_7$
Dimethoate; Rogor	Perfection; phosphamide; phosthione-MM; roxione	Insecticide, acaricide	$C_5H_{12}NO_3PS_2$
Diuron	Dichlorfenidim; karmex	Herbicide	$C_9H_{10}Cl_2N_2O$
Fluometuron	Cotoran; cottonex	Herbicide	$C_{10}H_{11}F_3N_2O$
Granozan	Cerezan	Fungicide, bactericide	C_2H_5ClHg
HCB; hexachlorobenzene	–	Fungicide, bactericide	C_6Cl_6
HCH; hexachlorocyclohexane	BHC; benzene hexachloride; cotol; dolmix; hexachloran; hexaklor; hexatox	Insecticide	$C_6H_6Cl_6$, mixture of isomers
γ-HCH	Agrisert; gammafex; lindatox; gammexane; lindagame, lindane	Insecticide	$C_6H_6Cl_6$, γ-isomer
Molinate	Ordram; yapane	Herbicide	$C_9H_{17}NOS$
Parathion-methyl; metaphos	Bladan-M; folidol-M; methacide; methyl-parathon; nitrox-80; vofatox	Insecticide, acaricide	$C_8H_{10}NO_5PS$
Phosalone	Benzophos; benzophosphate; rubitox; zolone	Insecticide, acaricide	$C_{12}H_{15}ClNO_4PS_2$
Prometryne	Caparol; gesagard; prometryn	Herbicide	$C_{10}H_{19}N_5S$
Propanil	DCPA; DPA; propanide; rogue; stam F-34; surcopur	Herbicide	$C_9H_9Cl_2NO$
Ramrod	Propachlor	Herbicide	$C_{11}H_{14}ClNO$
Simazine	Aquazine; CAT; gesatop; princepe; weedex	Herbicide	$C_7H_{12}ClN_5$
TCA	Trichloroacetate	Herbicide	$C_2Cl_3NaO_2$
Thiobencarb; saturn	Benthiocarb; bolero	Herbicide	$C_{12}H_{16}ClNOS$
Triazine	Anilazine; direz; dyrene; kemate	Fungicide	$C_9H_5Cl_3N_4$
Trichlorfon (trichlorophon)	Chlorophos; DEP; dipterex; dylox; metriphonate; neguvon; reciphon; tugon	Insecticide	$C_4H_2Cl_3O_4P$
Treflan; trifluralin	Elancolan; nitran K; olitref	Herbicide	$C_{13}H_{16}F_3N_3O_4$
Zineb	Aspore; dithane Z-78; lonacol; parzate; tiezene	Fungicide, bactericide	$(C_4H_6N_2S_4Zn)_x$

Sources: Melnikov, N.N., Novozhilov, K.V., Belan, S.R., Pylova, T.N. 1985 *Reference Book on Pesticides*. Chimiya, Moscow, 352 pp
Worthing, C.R. [ed.] 1987 *The Pesticide Manual. A World Compendium*. 8th Edition, British Crop Protection Council

Appendix II Sample pretreatment and storage requirements for chemical variables measured by the USSR water quality monitoring network

Variable	Type of bottle[1]	Conservation method	Maximum storage time
Suspended matter	G, P	Cooling to 4°C	7 days
pH	G, P	No conservation, determination at the sampling site	2 hours
Redox potential (Eh)	G	No conservation, determination at the sampling site	1 hour
Transparency	G, P	No conservation	24 hours
Colour	G, P	Cooling to 4°C	48 hours
Conductivity	G, P	Cooling to 4°C	24 hours
Hardness	G, P	No conservation	6 months
Potassium	P	No conservation	6 months
Sodium	P	No conservation	6 months
Carbonate, hydrocarbonate	G, P	Cooling to 4°C	24 hours
Carbon dioxide	G, P	In situ determination	–
Sulphate	P	No conservation	6 months
Chloride	P	No conservation	6 months
Dissolved oxygen	G	$MnCl_2$ and basic KI solution	6 hours
Heavy metals (excluding Hg)	P	HNO_3 (in some cases HCl)	6 months
Mercury	G	$KMnO_4$ or $K_2Cr_2O_7$ solutions in concentrated H_2SO_4	28 days
Boron	P	No conservation	6 months
Silica	P	Cooling to 4°C	3 days
Sulphide	G, P	Chemical stabilisation or conservation with $Cd(CH_3COO)_2$	3 days
Fluoride	P	No conservation	7 days
Nitrate	G, P	Cooling to 4°C, chloroform	3 days
Nitrite	G, P	Cooling to 4°C	24 hours
Ammonia	G, P	Cooling to 4°C, H_2SO_4	24 hours
Phosphate, total phosphorus	G	Cooling to 4°C, chloroform	3 days
BOD	G, P	Cooling to 4°C	4 hours
COD	G, P	Cooling to 4°C or H_2SO_4 (pH > 2)	24 hours / 7 days
Petroleum products	G	Liquid extraction within 4 hours or adding of solvent with shaking (2–3) min; keeping in dark place	3 months, 15 days
Organochlorine pesticides	G	Cooling to 4°C	7 days
Organophosphorus pesticides	G	Cooling to 4°C, HCl	24 hours
Triazine pesticides	G	Cooling to 4°C	24 hours
2,4-D type pesticides	G	Cooling to 4°C	6 days
Thiocarbamate pesticides	G	Cooling to 4°C	3 days
Phenols (photometric method)	G	Cooling to 4°C, NaOH	3 days
Phenols (gas chromatography)	G	Na_2S, $CuSO_4$, H_2SO_4	14 days
Anionic surfactants	G	Chloroform, cooling to 4°C	7 days
Non-ionic surfactants	G	Adding of solvent	10 days
Xantogenates	G	No conservation	3 hours
Methanol	G, P	No conservation	4 hours
		H_2SO_4, cooling to 4°C	10 day
Formaldehyde	G	H_2SO_4	10 days
Chlorinated hydrocarbons	G	Cooling to 4°C	7 days

[1] G Glass bottles; P Polyethylene bottles

Appendix III Analytical methods for the determination of major chemical variables used by the USSR water quality monitoring network

Variables	Determination method	Main analytical procedures	Minimum determined concentration C_{min}	Relative error at concentration (%)		Source
				$10 \times C_{min}$	$20 \times C_{min}$	
pH	Colourimetry	Comparison with scale		0.1 pH		1, 2
	Potentiometry	Measurement of a glass electrode potential		0.1 pH		1, 2
Conductivity	Conductometry	Measurement of electric conductivity	Depend on characteristics of the instrument			1, 2
Colour	Colourimetry	Visual (cobalt-dichromate scale) or spectrophotometric measurements	5 ± 2 units	2		
Dissolved oxygen	Titrimetry	Oxygen fixation by $Mn(OH)_2$; deposit dissolving; interaction with KI; titration of I_2 by $Na_2S_2O_3$	1 ± 0.1 mg l^{-1}	4		1*, 2*
Chloride	Titrimetry	Titration by $AgNO_3$ with K_2CrO_4 as indicator. Titration by $Hg(NO_3)_2$ with diphenylcarbazone as indicator	10 ± 1 mg l^{-1}	4	4	1, 2
Sulphate	Titrimetry	Ion-exchange removal of cations; titration by $Pb(NO_3)_2$ in water-ethanol mixture with dithizone as indicator.	2 ± 0.4 mg l^{-1}	17	17	1, 2
	Titrimetry	Ion-exchange removal of cations; titration by $BaCl_2$ in water-ethanol mixture with anthranile K as indicator.	50 ± 8 mg l^{-1}	8	7	
	Turbidimetry (315 nm)	Turbidity measurement after adding $BaCl_2$ in ethanol-glycerine mixture	50 ± 7 mg l^{-1}	8	7	1*, 2
Hydrocarbonate	Titrimetry	Adding of excess borax solution; removal of CO_2; titration of remaining borax with HCl	2 ± 0.5 mg l^{-1}	17	17	1*, 2
Calcium	Potentiometric titration	Titration by HCl with glass electrode	10 ± 3 mg l^{-1}	8	6	
	Titrimetry	Titration by EDTA in basic solution with murexide as indicator	10 ± 3 mg l^{-1}	5	4	1, 2
Hardness	Titrimetry	Titration by EDTA in ammonia solution with eriochrome black T as indicator	1 ± 0.3 mg l^{-1}	6	5	1, 2
			0.5 ± 0.04 mg eq l^{-1}	3	3	1, 2

Continued

Appendix III Continued

Variables	Determination method	Main analytical procedures	Minimum determined concentration C_{min}	Relative error at concentration (%)		Source
				$10 \times C_{min}$	$20 \times C_{min}$	
Sodium	Flame photometry	Direct sample injection into flame	1 ± 0.05 mg l^{-1}	4	4	1*, 2*
Potassium	Flame photometry	Direct sample injection into flame	1 ± 0.1 mg l^{-1}	6	6	1*, 2*
Ammonia and ammonium ions	Photometry (440 nm)	Formation of coloured solution with the Nessler reagent in basic media.	0.3 ± 0.05 mg l^{-1}	4	2	1, 2
	Photometry (630 nm)	Formation of indophenol blue in reaction with phenol and hypochlorite at pH 10	0.02 ± 0.01 mg l^{-1}	20	10	1, 2
Nitrite	Photometry (520 nm)	Reaction with Griss reagent	10 ± 5 µg l^{-1}	17	15	1*, 2*
Nitrate	Photometry (520 nm)	Nitrate reduction to nitrite in a cadmium-copper column, determination of nitrite with Griss reagent	10 ± 6 µg l^{-1}	28	26	1*, 2*
Orthophosphate	Photometry (882 nm)	Formation of antimony-phospho-molibdate complex with its further reduction by ascorbic acid	10 ± 5 µg l^{-1}	6	4	1, 2
Polyphosphate	Photometry (882 nm)	Hydrolysis to orthophosphate in acidic media and determination of orthophosphate	10 ± 7 µg l^{-1}	8	5	1*, 2*
Total phosphorus	Photometry (882 nm)	Oxidation with potassium persulphate in acidic media to orthophosphate and determination of orthophosphate	40 ± 20 µg l^{-1}	15	14	1*, 2*
Silica	Photometry (400 nm)	Reaction with ammonium molybdate in acidic media with formation of silica-molybdate hetero-polyacid.	0.5 ± 0.1 mg l^{-1}	10	9	1*, 2*
	Photometry (815 nm)	Reaction with ammonium molybdate in acidic media with formation of silica-molybdate hetero-polyacid; reduction of hetero-polyacid with sodium sulphite-methanol mixture	0.1 ± 0.05 mg l^{-1}	10	7	1*, 2*

Continued

Appendix III Continued

Variables	Determination method	Main analytical procedures	Minimum determined concentration C_{min}	Relative error at concentration (%)		Source
				$10 \times C_{min}$	$20 \times C_{min}$	
Cyanide, rodanide (thiocyanide)	Photometry (530 nm)	Formation of CNBr in reaction with bromine water; interaction with benzidine in presence of pyrydine. In case of rodanide determination, cyanide should be distilled	0.05 ± 0.03 mg l^{-1}	20	15	
Fluoride	Photometry (610 nm)	Reduction of lanthanum (or cerium)-alizarine complex colour in presence of fluoride.	0.02 ± 0.01 mg l^{-1}	10	8	1, 2
	Ionometry	Potential measurement of fluoride-selective electrode in Brois buffer solution	0.3 ± 0.04 mg l^{-1}	10	10	1*, 2*
Borate	Photometry (415 nm)[1]	Reaction with azometyne-N	0.1 ± 0.05 mg l^{-1}	8	8	1*, 2*
Hydrogen sulphide and sulphide ion	Photometry (667 nm)	Reaction with N,N-dimethyl-p-phenylendiamine and FeCl$_3$ in strong acidic media	0.05 ± 0.01 mg l^{-1}	6	5	1, 2
	Photometry (656 nm)[2]	Additional chloroform extraction in presence of lauryl-sulphate	2 ± 1 µg l^{-1}	12	9	
Iron	Photometry (510 nm)	Reduction of Fe(III) to Fe(II) by hydroxylamine; reaction with 1,10-phenantroline	50 ± 10 µg l^{-1}	6	4	1, 2
Copper	Photometry (430 nm)	Boiling with ammonium persulphate in acidic media; chloroform extraction of diethyldithiocarbamate complex.	3 ± 2 µg l^{-1}	30	20	
	Photometry (455 nm)[3]	Boiling with hydrogen peroxide in acidic media; chloroform extraction of diquinolyldisulphide complex	1 ± 0.3 µg l^{-1}	11	10	
Zinc	Photometry (535 nm)	CCl$_4$ extraction of ditizone complex after extraction of copper complex	2 ± 1 µg l^{-1}	20	20	
Manganese	Photometry (535 nm)	Sample evaporation with nitric acid; oxidation by ammonium persulphate	50 ± 30 µg l^{-1}	25	16	1*, 2*

Continued

Variables	Determination method	Main analytical procedures	Minimum determined concentration C_{min}	Relative error at concentration (%)		Source
				$10 \times C_{min}$	$20 \times C_{min}$	
Cobalt	Photometry (540 nm)	Boiling with ammonium persulphate, toluene extraction of β-nitrozo-αnaphtole	2 ± 1 µg l⁻¹	25	16	
Nickel	Photometry (445 nm)	Boiling with ammonium persulphate; chloroform extraction of dimethylglioxyme (1) complex; acid re-extraction; reaction with (1) in presence of bromine	5 ± 2 µg l⁻¹	10	8	1*, 2*
Lead	Photometry (520 nm)	CCl₄ extraction of ditizone complex after extraction of copper complex	2 ± 1 µg l⁻¹	20	20	
Molybdenum	Photometry (490 nm)[4]	Sample evaporation with nitric and sulphuric acids; sedimentation of iron; reduction of molybdenum by thiourea; reaction with rodanide	0.1 ± 0.05 mg l⁻¹	10	10	1*, 2*
Cadmium	Photometry (540 nm)	Boiling with ammonium persulphate; CCl₄ extraction of ditizone complex; acid re-extraction; neutralisation; extraction with ditizone	1 ± 0.5 µg l⁻¹	20	20	1*, 2*
Fe, Cu, Mn, Co, Ni, Al, Mo, Cd, Pb, Sn, Ti, V, Bi	Emission spectrography	Boiling with ammonium persulphate; chloroform extraction of diethyldithio-carbamate and 8-hydroxyquinoline; extract evaporation; mixing with graphite matrix; spectrographic determination	2 ± 1 µg l⁻¹ (Fe 3 ± 2 µg l⁻¹) (Cu 1 ± 0.5 µg l⁻¹) (Sn 4 ± 2 µg l⁻¹)	30	20	1*, 2*
Fe, Cu, Mn, Ni, Cd, Pb, Zn	Atomic absorption spectrometry[5]	Sample concentration with evaporation or extraction of complexes; flame atomisation	$0.1 - 10$ µg l⁻¹	10–20		
Chromium	Photometry (540 nm)	Cr(III) oxidation to Cr(VI) by KMnO₄; reduction of permanganate excess; determination with diphenylcarbazide; in case of Cr concentration < 20 µg l⁻¹ – preliminary extraction with iso-amyl alcohol	Direct 20 ± 3 µg l⁻¹; With extraction 1 ± 0.2 µg l⁻¹	11 / 5	10 / 5	1*, 2*

Continued

Appendix III Continued

Variables	Determination method	Main analytical procedures	Minimum determined concentration C_{min}	Relative error at concentration (%)		Source
				$10 \times C_{min}$	$20 \times C_{min}$	
Mercury	Photometry (490 nm)	Boiling with KMnO$_4$; reduction of permanganate excess; chloroform extraction of ditizone complex.	1 ± 0.5 µg l^{-1}	20	10	1*, 2*
	Atomic absorption spectrometry	Reduction by SnCl$_2$ in strong acid media; aeration from solution in closed system; measurement of vapour absorption	0.1 ± 0.02 µg l^{-1}	20	15	1, 2
Arsenic	Photometry (540 nm)	Reduction by SnCl$_2$ in strong acid media; aeration in presence of iodine in closed system; trapping by chloroform solution of ephedrine with silver diethyldithiocarbamate	10 ± 5 µg l^{-1}	20	20	1*, 2*
Biochemical oxygen demand	Titrimetry	Determination of dissolved oxygen before and after 5 days of incubation at 20 °C	1 ± 0.4 mg l^{-1}	10		1*, 2*
Chemical oxygen demand	Titrimetry	Oxidation of organic compounds in boiling conditions by K$_2$Cr$_2$O$_7$ in sulphuric acid; titration of dichromate excess by Mohr salt	4 ± 2 mg l^{-1}	10	8	1*, 2*
Petroleum hydrocarbons	IR-photometry (3.42 µm)	Extraction with CCl$_4$; column or thin-layer chromatographic separation of matrix; photometric measurement of CCl$_4$ solution.	0.04 ± 0.02 mg l^{-1}	21	20	1*
	UV-photometry or luminescence, λ_{ex}-390 nm, λ_t-480 nm	Thin layer chromatographic separation in the system hexane–CCl$_4$-acetic acid; eluation; measurement	0.05 ± 0.03 mg l^{-1}	25	24	
Resins and asphaltenes	Luminescence, λ_{ex}-480 nm, λ_t-520 nm	Analogous to determination of petroleum hydrocarbons	10 ± 5 µg l^{-1}	17	15	

Continued

Variables	Determination method	Main analytical procedures	Minimum determined concentration C_{min}	Relative error at concentration (%)			Source
				$10 \times C_{min}$	$20 \times C_{min}$		
Volatile phenols	Photometry (460 nm)	Distillation of phenols in acidic media; reaction with 4-aminoantipyrine and hexacyanoferrate (III) or amidopyrine and ammonium persulphate; chlorofom extraction.	2 ± 1 µg l^{-1}	11	8		1*
	Gas chromatography[2]	Butylacetate extraction from acidic media; base re-extraction; acylation by acetic anhydide; hexane extraction of ethers; evaporation; chromatographic determination	0.5 ± 0.3 µg l^{-1}	15–20	10–15		
Anionic surfactants	Photometry (650 nm)	Chloroform extraction of associate with methylene blue	15 ± 7 µg l^{-1}	20	15		1*
Non-ionic surfactants	Photometry (690 nm)	Extraction by chloroform-buthanol mixture; deposition by molybdo-phosphoric acid in the extract; deposit separation; dissolving in base; reaction of molybdenum with pyrocatechol violet in presence of dodecylpyridine	20 ± 10 µg l^{-1}	15	15		
Methanol	Photometry (590 nm)	Distillation; oxidation to formaldehyde by KMnO4; interaction with chromotropic acid in sulphuric acid media. Interference of formaldehyde is eliminated by silver nitrate in basic media	0.1 ± 0.04 mg l^{-1}	15	13		
Formaldehyde	Photometry (540 nm)	Interaction with phenyl-hydrazine in basic media in presence of potassium hexacyanoferrate (III); extraction by iso-propanol	40 ± 20 µg l^{-1}	15	10		
Xantogenates	Photometry (301 nm)	Toluene extraction of nickel complex; ammonia re-extraction	15 ± 10 µg l^{-1}	15	10		
Furfurol	Photometry (520 nm)	Distillation; interaction with aniline in acetic media	0.15 ± 0.05 mg l^{-1}	10	8		

Continued

Appendix III Continued

Variables	Determination method	Main analytical procedures	Minimum determined concentration C_{min}	Relative error at concentration (%)		Source
				$10 \times C_{min}$	$20 \times C_{min}$	
Organochlorine pesticides	Gas chromatography	Hexane extraction; extract treatment with concentrated sulphuric acid; evaporation				1*, 2*
HCB; HCH			2 ± 1 ng l^{-1}	23	20	
DDT			20 ± 11 ng l^{-1}	25	25	
DDD			10 ± 4 ng l^{-1}	23	22	
DDE			5 ± 2 ng l^{-1}	15	11	
Organo-phosphorus pesticides	Gas chromatography	Hexane and chloroform extraction; extract evaporation; termo-ionic or termo-aerozole detection				
Parathion-methyl			0.20 ± 0.12 µg l^{-1}	11	9	
Carbophos			0.40 ± 0.14 µg l^{-1}	9	8	
Phosalon			0.50 ± 0.07 µg l^{-1}	10	9	
Dimethoate			2.0 ± 0.6 µg l^{-1}	10	9	
2,4-D	Gas chromatography	Diethyl-ether extraction; derivatisation by boron trifluoride in methanol (1) or by ethanol (2); hexane extraction	$1 - 50$ 20 ng l^{-1}	27	20	
			$2 - 200$ 70 ng l^{-1}	8	6	

1 Since 1988
2 Since 1991
3 Since 1989
4 Method was used only in Armenia
5 C_{min} for certain elements and determination error depends on type of analytical instrument used

Sources:
1 Standard Methods for the Examination of Water and Waste Water 1989 17th edition, American Public Health Association, Washington D.C., 1268 pp
2 GEMS/Water Operational Guide 1987 WHO, Geneva

Methods have the same principles but can differ in some procedures

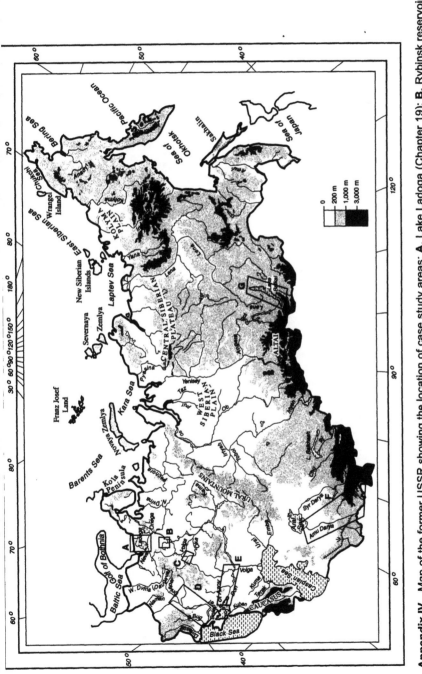

Appendix IV Map of the former USSR showing the location of case study areas: **A.** Lake Ladoga (Chapter 19); **B.** Rybinsk reservoir on the Volga river (Chapter 16); **C.** Moscow region (Chapter 20); **D.** Dnieper reservoir cascade (Chapter 17); **E.** Lower Don Basin (Chapter 14; **F.** Amu Darya river (Chapter 15); and **G.** Lake Baikal (Chapter 18)

Index